# Springer Complexity

Springer Complexity is a publication program, cutting across all traditional disciplines of sciences as well as engineering, economics, medicine, psychology and computer sciences, which is aimed at researchers, students and practitioners working in the field of complex systems. Complex Systems are systems that comprise many interacting parts with the ability to generate a new quality of macroscopic collective behavior through self-organization, e.g., the spontaneous formation of temporal, spatial or functional structures. This recognition, that the collective behavior of the whole system cannot be simply inferred from the understanding of the behavior of the individual components, has led to various new concepts and sophisticated tools of complexity. The main concepts and tools – with sometimes overlapping contents and methodologies – are the theories of self-organization, complex systems, synergetics, dynamical systems, turbulence, catastrophes, instabilities, nonlinearity, stochastic processes, chaos, neural networks, cellular automata, adaptive systems, and genetic algorithms.

The topics treated within Springer Complexity are as diverse as lasers or fluids in physics, machine cutting phenomena of workpieces or electric circuits with feedback in engineering, growth of crystals or pattern formation in chemistry, morphogenesis in biology, brain function in neurology, behavior of stock exchange rates in economics, or the formation of public opinion in sociology. All these seemingly quite different kinds of structure formation have a number of important features and underlying structures in common. These deep structural similarities can be exploited to transfer analytical methods and understanding from one field to another. The Springer Complexity program therefore seeks to foster cross-fertilization between the disciplines and a dialogue between theoreticians and experimentalists for a deeper understanding of the general structure and behavior of complex systems.

The program consists of individual books, books series such as "Springer Series in Synergetics", "Institute of Nonlinear Science", "Physics of Neural Networks", and "Understanding Complex Systems", as well as various journals.

# Springer Series in Synergetics

## SSSyn – An Interdisciplinary Series on Complex Systems

The success of the Springer Series in Synergetics has been made possible by the contributions of outstanding authors who presented their quite often pioneering results to the science community well beyond the borders of a special discipline. Indeed, interdisciplinarity is one of the main features of this series. But interdisciplinarity is not enough: The main goal is the search for common features of self-organizing systems in a great variety of seemingly quite different systems, or, still more precisely speaking, the search for general principles underlying the spontaneous formation of spatial, temporal or functional structures. The topics treated may be as diverse as lasers and fluids in physics, pattern formation in chemistry, morphogenesis in biology, brain functions in neurology or self-organization in a city. As is witnessed by several volumes, great attention is being paid to the pivotal interplay between deterministic and stochastic processes, as well as to the dialogue between theoreticians and experimentalists. All this has contributed to a remarkable cross-fertilization between disciplines and to a deeper understanding of complex systems. The timeliness and potential of such an approach are also mirrored – among other indicators – by numerous interdisciplinary workshops and conferences all over the world.

Vadim S. Anishchenko  Vladimir Astakhov
Alexander Neiman  Tatjana Vadivasova
Lutz Schimansky-Geier

# Nonlinear Dynamics of Chaotic and Stochastic Systems

## Tutorial and Modern Developments

**Second Edition**

With 222 Figures

Springer

Professor Dr. Vadim S. Anishchenko
Professor Dr. Vladimir Astakhov
Professor Dr. Tatjana Vadivasova

Saratov State University
Faculty of Physics
Astrakhanskaya ul. 83
410026 Saratov, Russia
E-mail: wadim@chaos.ssu.runnet.ru
          astakhov@chaos.ssu.runnet.ru
          tanya@chaos.ssu.runnet.ru

Professor Dr. Lutz Schimansky-Geier

Humboldt-Universität Berlin
Institut Physik
LS Stochastische Prozesse
Newtonstr. 15
12489 Berlin, Germany
E-mail: alsg@physik.hu-berlin.de

Professor Dr. Alexander Neiman

University of Missouri at St. Louis
Center for Neurodynamics
8001 Natural Bridge Road
St. Louis, MO 63121, USA
E-mail: neiman@neurodyn.umsl.edu

ISSN 0172-7389
ISBN-10  3-642-42467-8 Springer Berlin Heidelberg New York
ISBN-13  978-3-642-42467-0 Springer Berlin Heidelberg New York

Springer is a part of Springer Science+Business Media
springer.com
© Springer-Verlag Berlin Heidelberg 2007
Softcover re-print of the Hardcover 2nd edition 2007

Typesetting: by the author and techbooks using a Springer LaTeX macro package
Cover design: Erich Kirchner, Heidelberg

Printed on acid-free paper       SPIN: 11833901       54/techbooks       5 4 3 2 1 0

To our teachers and friends:

Werner Ebeling, Yuri L. Klimontovich
and Frank Moss

# Preface to the Second Edition

We present an improved and enlarged version of our book *Nonlinear Dynamics of Chaotic and Stochastic Systems* published by Springer in 2002. Basically, the new edition of the book corresponds to its first version. While preparing this edition we made some clarifications in several sections and also corrected the misprints noticed in some formulas. Besides, three new sections have been added to Chapter 2. They are *"Statistical Properties of Dynamical Chaos,"* *"Effects of Synchronization in Extended Self-Sustained Oscillatory Systems,"* and *"Synchronization in Living Systems."* The sections indicated reflect the most interesting results obtained by the authors after publication of the first edition.

We hope that the new edition of the book will be of great interest for a wide section of readers who are already specialists or those who are beginning research in the fields of nonlinear oscillation and wave theory, dynamical chaos, synchronization, and stochastic process theory.

Saratov, Berlin, and St. Louis
November 2006

*V.S. Anishchenko*
*A.B. Neiman*
*T.E. Vadiavasova*
*V.V. Astakhov*
*L. Schimansky-Geier*

# Preface to the First Edition

This book is devoted to the classical background and to contemporary results on nonlinear dynamics of deterministic and stochastic systems. Considerable attention is given to the effects of noise on various regimes of dynamic systems with noise-induced order.

On the one hand, there exists a rich literature of excellent books on nonlinear dynamics and chaos; on the other hand, there are many marvelous monographs and textbooks on the statistical physics of far-from-equilibrium and stochastic processes. This book is an attempt to combine the approach of nonlinear dynamics based on the deterministic evolution equations with the approach of statistical physics based on stochastic or kinetic equations. One of our main aims is to show the important role of noise in the organization and properties of dynamic regimes of nonlinear dissipative systems.

We cover a limited region in the interesting and still expanding field of nonlinear dynamics. Nowadays the variety of topics with regard to deterministic and stochastic dynamic systems is extremely large. Three main criteria were followed in writing the book and to give a reasonable and closed presentation: (i) the dynamic model should be minimal, that is, most transparent in the physical and mathematical sense, (ii) the model should be the simplest which nevertheless clearly demonstrates most important features of the phenomenon under consideration, and (iii) most attention is paid to models and phenomena on which the authors have gained great experience in recent years.

The book consists of three chapters. The first chapter serves as a brief introduction, giving the fundamental background of the theory of nonlinear deterministic and stochastic systems and a classical theory of the synchronization of periodic oscillations. All basic definitions and notions necessary for studying the subsequent chapters without referring to special literature are presented.

The second chapter is devoted to deterministic chaos. We discuss various scenarios of chaos onset, including the problem of the destruction of two- and three-frequency quasiperiodic motion. Different aspects of synchronization and chaos control as well as the methods of reconstruction of attractors and dynamic systems from experimental time series are also discussed.

The third chapter is concerned with stochastic systems whose dynamics essentially depend on the influence of noise. Several nonlinear phenomena are discussed: stochastic resonance in dynamic systems subjected to harmonic and complex signals and noise, stochastic synchronization and stochastic ratchets, which are the noise-induced ordered and directed transport of Brownian particles moving in bistable and periodic potentials. Special attention is given to the role of noise in excitable dynamics.

The book is directed to a large circle of possible readers in the natural sciences. The first chapter will be helpful for undergraduate and graduate students in physics, chemistry, biology and economics, as well as for lecturers of these fields interested in modern problems of nonlinear dynamics. Specialists of nonlinear dynamics may use this part as an extended dictionary. The second and the third chapters of the book are addressed to specialists in the field of mathematical modeling of the complex dynamics of nonlinear systems in the presence of noise.

We tried to write this book in such a manner that each of the three chapters can be understood in most parts independently of the others. Particularly, each chapter has its own list of references. This choice is based on the desire to be helpful to the reader. Undoubtedly, the lists of references are incomplete, since there exists an enormously large number of publications which are devoted to the topics considered in this book.

This book is a result of the long-term collaboration of the Nonlinear Dynamics Laboratory at Saratov State University, the group of Applied Stochastic Processes of Humboldt University at Berlin, and the Center for Neurodynamics at the University of Missouri at St. Louis. We want to express our deep gratitude to W. Ebeling, Yu.L. Klimontovich and F. Moss for their support, scientific exchange and constant interest. We acknowledge fruitful discussions with C. van den Broeck, P. Hänggi, J. Kurths, A. Longtin, A. Pikovski and Yu.M. Romanovski. The book has benefited a lot from our coauthors of the original literature. We wish to thank A. Balanov, R. Bartussek, V. Bucholtz, I. Dikalmin, J.A. Freund, J. García-Ojalvo, M. Hasler, N. Janson, T. Kapitaniak, I. Khovanov, M. Kostur, P.S. Landa, B. Lindner, P. McClintock, E. Mosekilde, A. Pavlov, T. Pöschel, D. Postnov, P. Reimann, R. Rozenfeld, P. Ruszczynsky, A. Shabunin, B. Shulgin, U. Siewert, A. Silchenko, O. Sosnovtseva, A. Zaikin and C. Zülicke for regular and fruitful discussions, criticism and valuable remarks which give us deeper insight into the problems we study.

We acknowledge the Series Editor H. Haken for fruitful comments on the manuscript. We thank P. Talkner, F. Marchesoni, M. Santos and coworkers and students from the group of Humboldt-University at Berlin for helpful remarks and comments during proofreading.

We are especially grateful to Ms. Galina Strelkova for her great work in preparing the manuscript and for translating several parts of this book into English, and to A. Klimshin for technical assistance.

# Contents

V.S. Anishchenko, T.E. Vadivasova and V.V. Astakhov acknowledge support from the US Civilian and Development Foundation under Grant No. REC-006, and the Russian Foundation for Basic Research; V.S. Anishchenko acknowledges support from the Alexander von Humboldt Foundation. A.B. Neiman was supported by the Fetzer Institute and by the US Office of Naval Research, Physics Division. L. Schimansky-Geier acknowledges support from the Deutsche Forschungsgemeinschaft (Sfb 555 and GK 268).

Saratov, Berlin and St. Louis                            *V.S. Anishchenko*
December 2001                                                  *A.B. Neiman*
                                                          *T.E. Vadivasova*
                                                            *V.V. Astakhov*
                                                     *L. Schimansky-Geier*

# 1. Tutorial

## 1.1 Dynamical Systems

### 1.1.1 Introduction

The knowledge of nonlinear dynamics is based on the notion of a *dynamical system* (DS). A DS may be thought of as an object of any nature, whose state evolves in time according to some dynamical law, i.e., as a result of the action of a *deterministic* evolution operator. Thus, the notion of DS is the result of a certain amount of idealization when random factors inevitably present in any real system are neglected.

The theory of DS is a wide and independent field of scientific research. The present section addresses only those parts, which are used in the subsequent chapters of this book. The main attention is paid to a linear analysis of the stability of solutions of ordinary differential equations. We also describe local and nonlocal bifurcations of typical limit sets and present a classification of attractors of DS.

The structure of chaotic attractors defines the properties of regimes of deterministic chaos in DS. It is known that the classical knowledge of dynamical chaos is based on the properties of robust hyperbolic (strange) attractors. Besides hyperbolic attractors, we also consider in more detail the peculiarities of nonhyperbolic attractors (quasiattractors). This sort of chaotic attractor reflects to a great extent the properties of deterministic chaos in real systems and serves as the mathematical image of experimentally observed chaos.

### 1.1.2 The Dynamical System and Its Mathematical Model

A DS has an associated *mathematical model*. The latter is considered to be defined if the *system state* as a set of some quantities or functions is determined and an *evolution operator* is specified which gives a correspondence between the initial state of the system and a unique state at each subsequent time moment. The evolution operator may be represented, for example, as a set of differential, integral and integro-differential equations, of discrete maps, or in the form of matrices, graphs, etc. The form of the mathematical model of the DS under study depends on which method of description is chosen.

Depending on the approximation degree and on the problem to be studied, the same real system can be associated with principally different mathematical models, e.g., a pendulum with and without friction. Moreover, from a qualitative viewpoint, we can often introduce into consideration a DS, e.g., the cardio-vascular system of a living organism, but it is not always possible to define its mathematical model.

DS are classified based on the form of state definition, on the properties and the method of description of the evolution operator. The set of some quantities $x_j$, $j = 1, 2, \ldots, N$, or functions $x_j(r)$, $r \in \mathbf{R}^M$ determines the state of a system. Here, $x_j$ are referred to as *dynamical variables*, which are directly related to the quantitative characteristics observed and measured in real systems (current, voltage, velocity, temperature, concentration, population size, etc.). The set of all possible states of the system is called its *phase space*. If $x_j$ are variables and not functions and their number $N$ is finite, the system phase space $\mathbf{R}^N$ has a finite dimension. Systems with finite-dimensional phase space are referred to as those with *lumped parameters*, because their parameters are not functions of spatial coordinates. Such systems are described by ordinary differential equations or return maps.

However, there is a wide class of systems with infinite-dimensional phase space. If the dynamical variables $x_j$ of a system are functions of some variables $r_k$, $k = 1, 2, \ldots, M$, the system phase space is infinite-dimensional. As a rule, $r_k$ represent spatial coordinates, and thus the system parameters depend on a point in space. Such systems are called *distributed parameter* or simply *distributed* systems. They are often represented by partial differential equations or integral equations. One more example of systems with infinite-dimensional phase space is a system whose evolution operator includes a time delay, $T_d$. In this case the system state is also defined by the set of functions $x_j(t)$, $t \in [0, T_d]$.

Several classes of DS can be distinguished depending on the properties of the evolution operator. If the evolution operator obeys the property of superposition, i.e., it is linear and the corresponding system is *linear*; otherwise the system is *nonlinear*. If the system state and the evolution operator are specified for any time moment, we deal with a *time-continuous* system. If the system state is defined only at separate (discrete) time moments, we have a system with *discrete time* (*map* or *cascade*). For cascades, the evolution operator is usually defined by the *first return function*, or *return map*. If the evolution operator depends implicitly on time, the corresponding system is *autonomous*, i.e., it contains no additive or multiplicative external forces depending explicitly on time; otherwise we deal with a *nonautonomous* system. Two kinds of DS are distinguished, namely, *conservative* and *nonconservative*. For a conservative system, the volume in phase space is conserved during time evolves. For a nonconservative system, the volume is usually contracted. The contraction of phase volume in mechanical systems corresponds to lost of energy as result of dissipation. A growth of phase volume implies a supply of energy to the system which can be named negative dissipation. Therefore, DS in which the energy or phase volume varies are called *dissipative systems*.

Among a wide class of DS, a special place is occupied by systems which can demonstrate oscillations, i.e., processes showing full or partial repetition. Oscillatory systems, as well as DS in general, are divided into *linear* and *nonlinear*, *lumped* and *distributed*, *conservative* and *dissipative*, and *autonomous* and *nonautonomous*. A special class includes the so-called *self-sustained systems*.

Nonlinear dissipative systems in which nondecaying oscillations can appear and be sustained without any external force are called *self-sustained*, and oscillations themselves in such systems are called *self-sustained oscillations*. The energy lost as dissipation in a self-sustained system is compensated from an external source. A peculiarity of self-sustained oscillations is that their characteristics (amplitude, frequency, waveform, etc.) do not depend on the properties of a power source and hold under variation, at least small, of initial conditions [1].

**Phase Portraits of Dynamical Systems.** A method for analyzing oscillations of DS by means of their graphical representation in phase space was introduced to the theory of oscillations by L.I. Mandelstam and A.A. Andronov [1]. Since then, this method has become the customary tool for studying various oscillatory phenomena. When oscillations of complex form, i.e., dynamical chaos, were discovered, this method increased in importance. The analysis of phase portraits of complex oscillatory processes allows one to judge the topological structure of a chaotic limit set and to make sometimes valid guesses and assumptions which appear to be valuable when performing further investigations [2].

Let the DS under study be described by ordinary differential equations

$$\dot{x}_j = f_j(x_1, x_2, \ldots, x_N), \tag{1.1}$$

where $j = 1, 2, \ldots, N$, or in vector form

$$\dot{\boldsymbol{x}} = \boldsymbol{F}(\boldsymbol{x}). \tag{1.2}$$

$\boldsymbol{x}$ represents a vector with components $x_j$, the index $j$ runs over $j = 1, 2, \ldots, N$, and $\boldsymbol{F}(\boldsymbol{x})$ is a vector-function with components $f_j(\boldsymbol{x})$. The set of $N$ dynamical variables $x_j$ or the $N$-dimensional vector $\boldsymbol{x}$ determines the system state which can be viewed as a point in state space $\mathbf{R}^N$. This point is called a *representative* or *phase* point, and the state space $\mathbf{R}^N$ is called the *phase space* of DS. The motion of a phase point corresponds to the time evolution of a state of the system. The trajectory of a phase point, starting from some initial point $\boldsymbol{x}_0 = \boldsymbol{x}(t_0)$ and followed as $t \to \pm\infty$, represents a *phase trajectory*. A similar notion of integral curves is sometimes used. These curves are described by equations $\mathrm{d}x_j/\mathrm{d}x_k = \Phi(x_1, x_2, \ldots, x_N)$, where $x_k$ is one of the dynamical variables. An integral curve and a phase trajectory often coincide, but the integral curve may consist of several phase trajectories if it passes through a singular point. The right-hand side of (1.2) defines the velocity vector field $\boldsymbol{F}(\boldsymbol{x})$ of a phase point in the system phase space. Points in phase space for which $f_j(\boldsymbol{x}) = 0$, $j = 1, 2, \ldots, N$, remain unchanged with time. They are called *fixed points, singular points* or *equilibrium points* of the DS. A set of characteristic phase trajectories in the phase space represents the *phase portrait* of the DS.

Besides the phase space dimension $N$, the *number of degrees of freedom* $n = N/2$ is often introduced. This tradition came from mechanics, where a system is considered as a set of mass points, each being described by a second-order equation of motion. $n$ generalized coordinates and $n$ generalized impulses are introduced so that the total number of dynamical variables $N = 2n$ appears to be even and the number of independent generalized coordinates $n$ (the number of freedom degrees) an integer. For an arbitrary DS (1.1) the number of degrees of freedom will be, in general, a multiple of 0.5.

Consider the harmonic oscillator

$$\ddot{x} + \omega_0^2 x = 0 . \tag{1.3}$$

Its phase portrait is shown in Fig. 1.1a and represents a family of concentric ellipses (in the case $\omega_0 = 1$, circles) in the plane $x_1 = x$, $x_2 = \dot{x}$, centered at the origin of coordinates:

$$\frac{\omega_0^2 x_1^2}{2} + \frac{x_2^2}{2} = H(x_1, x_2) = \text{const.} \tag{1.4}$$

Each value of the total energy $H(x_1, x_2)$ corresponds to its own ellipse. At the origin we have the equilibrium state called a *center*. When dissipation is added to the linear oscillator, phase trajectories starting from any point in the phase plane approach equilibrium at the origin in the limit as $t \to \infty$. The phase trajectories look like spirals twisting towards the origin (Fig. 1.1b) if dissipation is low and the solutions of the damped harmonic oscillator correspond to decaying oscillations. In this case the equilibrium state is a *stable focus*. With an increasing damping coefficient, the solutions become aperiodic and correspond to the phase portrait shown in Fig. 1.1c with the equilibrium called a *stable node*.

By using a potential function $U(x)$, it is easy to construct qualitatively the phase portrait for a nonlinear conservative oscillator which is governed by

$$\ddot{x} + \frac{\partial U(x)}{\partial x} = 0 .$$

An example of such a construction is given in Fig. 1.2. Minima of the potential function conform to the center-type equilibrium states. In a potential well about each center, a family of closed curves is arranged which correspond to different values of the integral of energy $H(x, \dot{x})$. In the nearest neighborhood

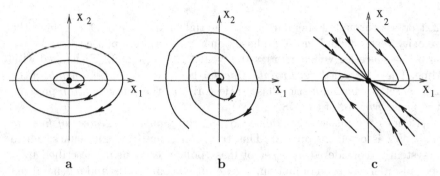

**Fig. 1.1.** Phase portraits of linear oscillators: (**a**) without dissipation, (**b**) with low dissipation, and (**c**) with high dissipation

of the center these curves have an ellipse-like shape but they are deformed when moving away from the center. Maxima of $U(x)$ correspond to equilibria called *saddles*. Such equilibrium states are unstable. Phase trajectories tending to the saddle $Q$ (Fig. 1.2) as $t \to \pm\infty$ belong to singular integral curves called *separatrices of saddle $Q$*. A pair of trajectories approaching the saddle forwardly in time forms its *stable manifold* $W_Q^s$, and a pair of trajectories tending to the saddle backwardly in time its *unstable manifold* $W_Q^u$. Separa-

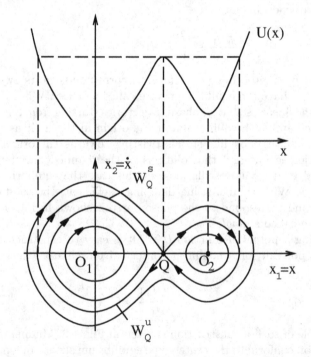

**Fig. 1.2.** Qualitative construction of the phase portrait of a nonlinear conservative oscillator using the potential function $U(x)$

trices divide the phase space into regions with principally different behavior of phase trajectories. They can also close, forming separatrix loops (Fig. 1.2).

Phase portraits of nonautonomous systems have some peculiarities. One of them is that phase trajectories are not located within a certain bounded region of the phase space, since the time variable $t$ varies from $-\infty$ to $+\infty$. A periodically driven nonautonomous system can be reduced to an autonomous one by introducing the phase of the external forcing $\Psi = \Omega t$ and by adding the equation $\dot{\Psi} = \Omega$. However, if $\Psi$ is taken unwrapped such a new variable yields nothing and the phase trajectories remain unbounded as before. For phase trajectories to be bounded, one needs to introduce a cylinderical phase space (in general, multi-dimensional) taking into account that $\Psi \in [0, 2\pi]$. Figure 1.3 represents phase trajectories of a nonautonomous system lying on a cylinder (for visualization purposes only one dynamical variable, $x$, is shown).

Consider the Van der Pol oscillator

$$\ddot{x} - (\alpha - x^2)\dot{x} + \omega_0^2 x = 0. \tag{1.5}$$

For $\alpha > 0$ and as $t \to \infty$, the self-sustained system (1.5) demonstrates periodic oscillations independent of initial conditions. These oscillations follow a closed isolated curve called the *Andronov–Poincaré limit cycle* [1]. All phase trajectories of (1.5), starting from different points of the phase plane, approach the limit cycle as $t \to \infty$. The only exception is the equilibrium at the origin. For small $\alpha$ the limit cycle resembles an ellipse in shape and the equilibrium at the origin is an *unstable focus*. The corresponding phase portrait and time dependence $x(t)$ are shown in Fig. 1.4a. With increasing $\alpha$ the limit cycle is distorted and the character of equilibrium is changed. For $\alpha > 2\omega_0$, the unstable focus is transformed into an unstable node and the duration of transient process significantly decreases (Fig. 1.4b).

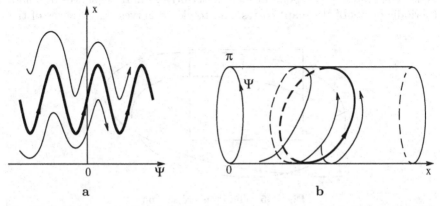

**Fig. 1.3.** Phase portraits of the nonautonomous system $\dot{x} = f(x) + B\sin\Omega t$ (**a**) for $\Psi = \Omega t$ defined in the interval $(-\infty, +\infty)$ and (**b**) for $\Psi \in [0, 2\pi]$

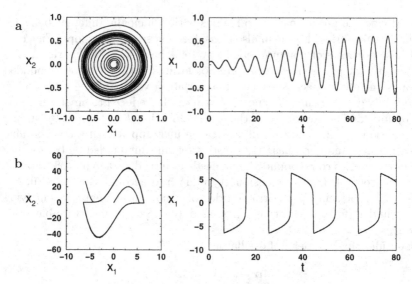

**Fig. 1.4.** Phase portraits and waveforms in the Van der Pol oscillator (1.5) with $\omega_0 = 1$ and (**a**) for $\alpha = 0.1$ and (**b**) for $\alpha = 10$. $x_1 = x, x_2 = \dot{x}$

Phase portraits of three-dimensional systems are not so illustrative. In this case it is reasonable to introduce a plane or a surface of section such that all trajectories intersect it at a nonzero angle. On the secant plane we obtain a set of points corresponding to different phase trajectories of the original system and which can give us an idea of the system phase portrait. The points usually considered are those which appear when the system trajectory intersects the secant plane in one direction, as shown in Fig. 1.5. The evolution operator uniquely (but not one-to-one) determines a map of the secant plane to itself, called the *first return function* or the *Poincaré map* [3]. The Poincaré map reduces the dimension of the sets being considered by one and thus makes the system phase portrait more descriptive. Finite sequences of points (periodic orbits of the map) correspond to closed curves (limit cycles of the

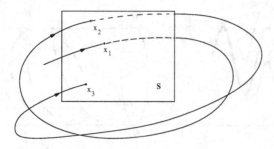

**Fig. 1.5.** The Poincaré section

initial system) and infinite sequences of points correspond to non-periodic trajectories.

### 1.1.3 Stability – Linear Approach

The property of *stability* or *instability* is one of the important characteristics of motions realized in a system. There are several notions of stability, namely, stability according to Lyapunov, asymptotic stability, orbital stability and stability according to Poisson [2, 4, 5]. The motion $\boldsymbol{x}^*(t)$ is said to be *stable according to Lyapunov* if for any arbitrary small $\varepsilon > 0$ there is such $\delta(\varepsilon)$ that for any motion $\boldsymbol{x}(t)$, for which $\|\boldsymbol{x}(t_0) - \boldsymbol{x}^*(t_0)\| < \delta$, the inequality $\|\boldsymbol{x}(t) - \boldsymbol{x}^*(t)\| < \varepsilon$ is satisfied for all $t > t_0$. The sign $\|o\|$ denotes the vector norm. Thus, a small initial perturbation does not grow for a motion that is stable according to Lyapunov. If additionally the small perturbation $\delta$ vanishes as time goes on, i.e., $\|\boldsymbol{x}(t) - \boldsymbol{x}^*(t)\| \to 0$ as $t \to \infty$, the motion possesses a stronger stability property, namely, *asymptotic stability*. Any asymptotically stable motion is stable according to Lyapunov but, in the general case, the opposite is not true.

The definition of *orbital stability* is somewhat different from the definition of stability according to Lyapunov. In the last case the distance between phase points of the studied and the perturbed motion is considered at the same time moments. The orbital stability takes into account the minimal distance between the phase point of the perturbed trajectory at a given time moment $t$ and orbit $\Gamma^*$ corresponding to the motion under study. The motion that is stable according to Lyapunov is always orbitally stable. In the general case, the opposite statement is not valid.

The weakest requirement is the *stability* of motion $\boldsymbol{x}^*(t)$ *according to Poisson*. Its definition assumes that the phase trajectory does not leave a bounded region of the phase space as $t \to \infty$. Spending an infinitely long period of time inside this region, the phase trajectory inevitably returns to an arbitrarily small neighborhood of the initial point. Return times may correspond to the period or quasiperiod of a regular motion or represent a random sequence in the regime of dynamical chaos.

The stability properties of phase trajectories belonging to limit sets, e.g., attractors, are of special importance for understanding the system dynamics. The change of stability character of one or another limit set can lead to a considerable modification of the system phase portrait.

**Limit Sets of a Dynamical System.** Let the state of a system be specified by vector $\boldsymbol{x}_0$ at the time moment $t_0$, and by vector $\boldsymbol{x}(t) = T_{\Delta t}\boldsymbol{x}_0$ at the instant $t$, where $T_{\Delta t}$ is the evolution operator on the interval $\Delta t = t - t_0$. Assume that there are two sets, $V$ and $L \in V$, where $V$ is a set of all points $x_0$ in the phase space for which $\boldsymbol{x}(t) \in L$ as $t \to +\infty$ or $t \to -\infty$. We call $L$ the *limit set* of DS.

Consider the possible types of limit sets of a dissipative DS. If all points of set $V$ belong to $L$ in the limit as $t \to +\infty$, then limit set $L$ is *attracting*, i.e., an *attractor*. Consequently, $V$ is the basin of attraction of attractor $L$. If points of $V$ belong to $L$ in the limit as $t \to -\infty$, set $L$ is *repelling*, i.e., a *repeller*. If set $V$ consists of two subsets, $V = W^s \cup W^u$, and points belonging to $W^s$ approach $L$ forward in time, while points belonging to $W^u$ tend to $L$ backward in time, then $L$ is called a *saddle limit set* or simply a *saddle*. The sets $W^s$ and $W^u$ are the stable and unstable manifolds of the saddle, respectively. Under time inversion, $t \to -t$, regular attractors of the system become repellers, repellers are transformed into attractors, and the roles of stable and unstable manifolds of saddles are interchanged [6].

The simplest limit set of a DS is an equilibrium. There are equilibria which can be attractors (stable focus and stable node), repellers (unstable focus and unstable node), and saddles (simple saddle or a saddle-focus realized in phase space with the dimension $N \geq 3$). A center-type point is neither attractor nor repeller nor saddle because there is no set of points tending to the center forward or backward in time. This is a particular case of the limit set for which $V = L$.

A limit set in the form of a closed curve, a *limit cycle*, can also be an attractor, a repeller or a saddle. The toroidal limit sets, corresponding to quasiperiodic oscillations, and the chaotic limit sets are similarly subclassified.

**Linear Analysis of Stability – Basic Concepts.** Stability according to Lyapunov and asymptotic stability are determined by the time evolution of a small perturbation of a trajectory, namely, whether the perturbation decreases, grows or remains bounded with time. The introduction of perturbation allows one to linearize the evolution operator in the vicinity of the trajectory being studied and to perform a linear analysis of its stability.

Let us have an autonomous DS in the form

$$\dot{x} = F(x, \alpha), \tag{1.6}$$

where $x \in \mathbf{R}^N$, and $\alpha \in \mathbf{R}^M$ is the parameter vector. We are interested in analyzing the stability of a solution $x^0(t)$. Introduce a small perturbation $y = x(t) - x^0(t)$, for which we may write

$$\dot{y} = F(x^0 + y) - F(x^0). \tag{1.7}$$

Expanding $F(x^0 + y)$ into a series in the neighborhood of $x^0$ and taking into account the fact that the perturbation is small, we arrive at the following linearized equation with respect to $y$:

$$\dot{y} = \hat{A}(t)y, \tag{1.8}$$

where $\hat{A}$ is a matrix with elements

$$a_{j,k} = \left. \frac{\partial f_j}{\partial x_k} \right|_{x^0}, \qquad j, k = 1, 2, \ldots, N, \tag{1.9}$$

called the *linearization matrix* of the system in the neighborhood of solution $x^0(t)$, and $f_j$ are the components of function $F$. Matrix $\hat{A}$ is characterized by eigenvectors $e_i$ and eigenvalues $\rho_i$:

$$\hat{A}e_i = \rho_i e_i, \quad i = 1, 2, \ldots, N. \tag{1.10}$$

The eigenvalues $\rho_i$ are roots of the characteristic equation

$$\mathrm{Det}\left[\hat{A} - \rho\hat{E}\right] = 0, \tag{1.11}$$

where $\hat{E}$ is a unit matrix. The initial perturbation specified at the time moment $t^*$ and along the $i$th eigenvector evolves as follows:

$$y^i(t) = y^i(t^*)\exp{(t - t^*)\rho_i}. \tag{1.12}$$

The growth of or decrease in the magnitude of perturbation $\|y^i(t)\|$ is determined by the sign of the real part of $\rho_i$. In general, $\hat{A}$ is a matrix with time-varying elements, and, consequently, its eigenvalues and eigenvectors also change. Therefore, (1.12) is fulfilled only in the limit $t - t^* \to 0$, i.e., locally, in the neighborhood of $x^0(t^*)$. As $t^*$ is varied, the index of the exponent $\rho_i$ takes a different value. Hence, in general, it is plausible that a small perturbation $y(t) = \sum_{i=1}^{N} y^i(t)$ grows exponentially at some points of the trajectory under study $x^0(t)$, whereas it decreases at others.

Assume that at the time moment $t_0$ we have an infinitesimal initial perturbation $y^i(t_0)$ of the trajectory being studied along the $i$th eigenvector of matrix $\hat{A}$ and a perturbation $y^i(t)$ at an arbitrary time moment $t$. The stability of the trajectory along the eigenvector $e_i(t)$ is characterized by the *Lyapunov characteristic exponent* $\lambda_i$,

$$\lambda_i = \overline{\lim_{t \to \infty}} \frac{1}{t - t_0} \ln\|y^i(t)\|, \tag{1.13}$$

where the bar means the upper limit. If the trajectory $x^0(t)$ belongs to $N$-dimensional phase space, the linearization matrix has the $N \times N$ dimension and thus $N$ eigenvectors. In this case the trajectory stability is defined by a set of $N$ Lyapunov exponents. The set of $N$ numbers arranged in decreasing order, $\lambda_1 \geq \lambda_2 \geq \ldots \geq \lambda_N$, forms the so-called *Lyapunov characteristic exponents spectrum (LCE spectrum)* of phase trajectory $x^0(t)$.

Let us elucidate how Lyapunov exponents are related to the eigenvalues of the linearization matrix $\rho_i(t)$. Consider (1.12) at the initial time moment $t^* = t_0$ assuming the interval $\Delta t = t - t_0$ to be small. Now we move to the point $x(t_0 + \Delta t)$ and take the following as the initial perturbation:

$$y^i(t_0 + \Delta t) = y^i(t_0)\exp{\rho_i(t_0)\Delta t}.$$

We suppose that since $\Delta t$ is small, the direction of eigenvectors $e_i$ does not change over this time interval, and vector $y^i(t_0 + \Delta t)$ may be viewed to be directed along the $i$th eigenvector. The initial perturbation $\boldsymbol{y}^i(t_0)$ is taken to be so small that it can also be considered to be small at subsequent time moments. Moving each time along the curve $\boldsymbol{x}^0(t)$ with small step $\Delta t$, we can obtain an approximate expression describing the evolution of the small perturbation along the $i$th eigenvector:

$$\boldsymbol{y}^i(t) \approx \boldsymbol{y}^i(t_0) \exp \sum_k \rho_i(t_k)\Delta t\,. \tag{1.14}$$

Passing to the limits $\|\boldsymbol{y}^i(t_0)\| \to 0$ and $\Delta t \to 0$, we deduce the rigorous equality

$$\boldsymbol{y}^i(t) = \boldsymbol{y}^i(t_0) \exp \int_{t_0}^t \rho_i(t')\mathrm{d}t'\,. \tag{1.15}$$

Substituting (1.15) into (1.13) we derive

$$\lambda_i = \varlimsup_{t\to\infty} \frac{1}{t-t_0} \int_{t_0}^t \mathrm{Re}\,\rho_i(t')\mathrm{d}t'\,. \tag{1.16}$$

Thus, the $i$th Lyapunov exponent $\lambda_i$ can be thought of as the value, averaged over the trajectory under study, of the real part of the eigenvalue $\rho_i$ of linearization matrix $\hat{A}(t)$. It shows what happens with an appropriate component of the initial perturbation, on average, along the trajectory. The divergence of the flow and, consequently, the phase volume evolution are determined by the sum of Lyapunov exponents. It can be shown that

$$\sum_{i=1}^N \lambda_i = \varlimsup_{t\to\infty} \frac{1}{t-t_0} \int_{t_0}^t \mathrm{div}\,\boldsymbol{F}(t')\mathrm{d}t'\,. \tag{1.17}$$

If trajectory $\boldsymbol{x}^0(t)$ is stable according to Lyapunov, then an arbitrary initial perturbation $\boldsymbol{y}(t_0)$ does not grow, on average, along the trajectory. In this case it is necessary and sufficient for the LCE spectrum not to contain positive exponents. If an arbitrary bounded trajectory $\boldsymbol{x}^0(t)$ of the autonomous system (1.6) is not an equilibrium or a saddle separatrix, then at least one of the Lyapunov exponents is always equal to zero [5, 7]. This is explained by the fact that the small perturbation remains unchanged along the direction tangent to the trajectory. A phase volume element must be contracted for phase trajectories on the attractor. Consequently, the system has a negative divergence and the sum of Lyapunov exponents satisfies the following inequality:

$$\sum_{i=1}^N \lambda_i < 0\,. \tag{1.18}$$

**Stability of Equilibrium States.** If the particular solution $x^0(t)$ of system (1.6) is an equilibrium, i.e., $F(x^0, \alpha) = 0$, the linearization matrix $\hat{A}$ is considered at only one point of phase space, and, consequently, it is a matrix with constant elements $a_{i,j}$. The eigenvectors and the eigenvalues of matrix $\hat{A}$ are constant in time, and the Lyapunov exponents coincide with the real parts of the eigenvalues, $\lambda_i = \text{Re}\rho_i$. The LCE spectrum signature indicates whether the equilibrium is stable or not. To analyze the behavior of phase trajectories in a local neighborhood of the equilibrium one also needs to know the imaginary parts of the linearization matrix eigenvalues. In a phase plane, $N = 2$, the equilibrium is characterized by two eigenvalues of matrix $\hat{A}$, namely, $\rho_1$ and $\rho_2$. The following cases can be realized in the phase plane: (i) $\rho_1$ and $\rho_2$ are real negative numbers; in this case the equilibrium is a stable node. (ii) $\rho_1$ and $\rho_2$ are real positive numbers; the equilibrium is an unstable node. (iii) $\rho_1$ and $\rho_2$ are real numbers but with different signs; the equilibrium in this case is a saddle. (iv) $\rho_1$ and $\rho_2$ are complex conjugates with $\text{Re}\rho_{1,2} < 0$; the equilibrium is a stable focus. (v) $\rho_1$ and $\rho_2$ are complex conjugates with $\text{Re}\rho_{1,2} > 0$, the equilibrium is an unstable focus. (vi) $\rho_1$ and $\rho_2$ are pure imaginary, $\rho_{1,2} = \pm i\omega$; the equilibrium in this case is a center. Figure 1.6 shows a diagram of equilibria which are realized in the plane depending on the determinant and the trace of matrix $\hat{A}$ (respectively, $\text{Det}\hat{A} = \rho_1\rho_2$ and $\text{Sp}\hat{A} = \rho_1 + \rho_2$).

Besides the aforementioned states, in phase space with dimension $N \geq 3$, other kinds of equilibria are possible, e.g., an equilibrium state that is unstable according to Lyapunov, which is called a saddle-focus. Figure 1.7 shows

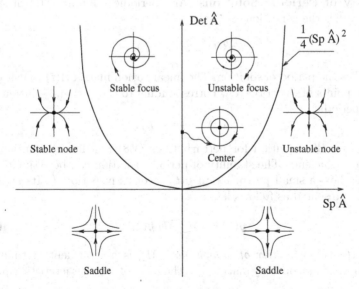

**Fig. 1.6.** Diagram of equilibria in the plane (phase portraits are shown in transformed coordinates [1])

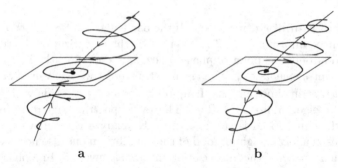

**Fig. 1.7.** Saddle-foci in three-dimensional phase space: (**a**) $\rho_1$ is real and negative, $\rho_{2,3}$ are complex conjugate with $\mathrm{Re}\rho_{2,3} > 0$; (**b**) $\rho_1$ is real and positive, $\rho_{2,3}$ are complex conjugate with $\mathrm{Re}\rho_{2,3} < 0$

two possible types of a saddle-focus in three-dimensional phase space, which are distinguished by the dimensions of their stable and unstable manifolds.

To identify which type of limit set the equilibrium is, it is enough to know the Lyapunov exponents. The equilibrium is considered to be an attractor if it is asymptotically stable in all directions and its LCE spectrum consists of negative exponents only (stable node and focus). If the equilibrium is unstable in all directions, it is a repeller (unstable node and focus). If the LCE spectrum includes both positive and negative exponents, the equilibrium is of saddle type (simple saddle or a saddle-focus). In addition, the exponents $\lambda_i \geq 0$ ($\lambda_i \leq 0$) determine the dimension of the unstable (stable) manifold.

**Stability of Periodic Solutions.** Any periodic solution $x^0(t)$ of system (1.6) satisfies the condition

$$x^0(t) = x^0(t+T),\qquad(1.19)$$

where $T$ is the period of solution. The linearization matrix $\hat{A}(t)$ that is calculated at points of the trajectory corresponding to the periodic solution $x^0(t)$ is also periodic:

$$\hat{A}(t) = \hat{A}(t+T).\qquad(1.20)$$

In this case the equation for perturbations (1.8) is a linear equation with periodic coefficients. The stability of periodic solution can be estimated if it is known how a small perturbation $y(t_0)$ evolves in period $T$. Its evolution can be represented as follows [8]:

$$y(t_0+T) = \hat{M}_T y(t_0),\qquad(1.21)$$

where $\hat{M}_T$ is the *matrix of monodromy*. $\hat{M}_T$ is independent of time. The eigenvalues of monodromy matrix, i.e., the roots of the characteristic equation

$$\mathrm{Det}[\hat{M}_T - \mu\hat{E}] = 0,\qquad(1.22)$$

are called *multipliers* of periodic solution $x^0(t)$ and define its stability. Indeed, the action of the monodromy operator (1.21) is as follows: The initial perturbation of a periodic solution, considered in projections onto the matrix $\hat{A}(t_0)$ eigenvectors, is multiplied by an appropriate multiplier $\mu_i$ in period $T$. Thus, a necessary and sufficient requirement for the periodic solution $x^0(t)$ to be stable according to Lyapunov is that its multipliers $|\mu_i| \leq 1, i = 1, 2, \ldots, N$. At least one of the multipliers is equal to $+1$. The multipliers being the eigenvalues of the monodromy matrix obey the following relations:

$$\sum_{i=1}^{N} \mu_i = \mathrm{Sp}\hat{M}_T, \qquad \prod_{i=1}^{N} \mu_i = \mathrm{Det}\hat{M}_T. \qquad (1.23)$$

They are related to the Lyapunov exponents of periodic solution as follows:

$$\lambda_i = \frac{1}{T} \ln |\mu_i|. \qquad (1.24)$$

One of the LCE spectrum exponents of a limit cycle is always zero and corresponds to a unit multiplier. The limit cycle is an attractor if all the other exponents are negative. If the LCE spectrum includes different sign exponents, the limit cycle is a saddle. The dimension of its unstable manifold is equal to the number of non-negative exponents in the LCE spectrum, and the dimension of its stable manifold is equal to the number of exponents for which $\lambda_i \leq 0$. If all $\lambda_i > 0$, then the limit cycle is a repeller.

**Stability of Quasiperiodic Solutions.** Let a particular solution $x^0(t)$ of system (1.6) correspond to quasiperiodic oscillations with $k$ independent frequencies $\omega_j$, $j = 1, 2, \ldots, k$. Then the following is valid:

$$x^0(t) = x^0\big(\varphi_1(t), \varphi_2(t), \ldots, \varphi_k(t)\big)$$
$$= x^0\big(\varphi_1(t) + 2\pi m, \varphi_2(t) + 2\pi m, \ldots, \varphi_k(t) + 2\pi m\big), \qquad (1.25)$$

where $m$ is an arbitrary integer number and $\varphi_j(t) = \omega_j t$, $j = 1, 2, \ldots, k$. The stability of the quasiperiodic solution is characterized by the LCE spectrum. The linearization matrix $\hat{A}(t)$ is quasiperiodic, and the Lyapunov exponents are strictly defined only in the limit as $t \to \infty$. The periodicity of the solution with respect to each of the arguments $\varphi_j$ in the case of *ergodic quasiperiodic oscillations* results in the LCE spectrum consisting of $k$ zero exponents. If all other exponents are negative, the toroidal $k$-dimensional hypersurface (we shall use the term "$k$-dimensional torus" for simplicity) on which the quasiperiodic trajectory being studied lies is an attractor. When all other exponents are positive, the $k$-dimensional torus is a repeller. The torus is said to be a saddle[1] if the LCE spectrum of quasiperiodic trajectories on the torus has, besides zero exponents, both positive and negative ones.

---

[1] This situation should be distinguished from the case of chaos on a $k$-dimensional torus, which is observed for $k \geq 3$.

**Stability of Chaotic Solutions.** A chaotic trajectory, whether it belongs to a chaotic attractor, or a repeller or a saddle, is always unstable at least in one direction. The LCE spectrum of a chaotic solution always has at least one positive Lyapunov exponent. The instability of phase trajectories and the attracting character of the limit set to which they belong do not contradict one another. Phase trajectories starting from close initial points in the basin of attraction tend to the attractor but they are separated on it. Hence, chaotic trajectories are unstable according to Lyapunov but stable according to Poisson.

Such behavior of trajectories is typical for attractors with a complicated geometrical structure and, therefore, they are called *strange attractors*. A strange attractor can be modeled by a limit set arising in the so-called *horseshoe map (Smale's map)* [9]: a unit square is contracted in one dimension and stretched in another one but the area in this case is decreased. The strip obtained in doing so is bent in the form of a horseshoe and folded again in the initial square as shown in Fig. 1.8. This procedure is repeated many times. In the limit, a set with non-zero area is formed which is not a collection of a countable set of points or lines and has a *Cantor structure* in its cross-section.

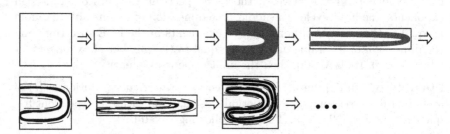

**Fig. 1.8.** Appearance of a strange attractor in Smale's horseshoe map

**Stability of Phase Trajectories in Discrete Time Systems.** Let a discrete time system be described by the return map

$$x(n+1) = P(x(n), \alpha), \qquad (1.26)$$

where $x \in \mathbf{R}^N$ is the state vector, $n$ is a discrete time variable, $P(x)$ is a vector function with components $P_j, j = 1, 2, \ldots, N$, and $\alpha \in \mathbf{R}^M$ is the system parameter vector. Analyze the stability of an arbitrary solution $x^0(n)$. Introducing a small perturbation $y(n) = x(n) - x^0(n)$ and linearizing the map in the vicinity of solution $x^0(n)$, we deduce the linear equation for perturbation:

$$y(n+1) = \hat{M}(n)y(n), \qquad (1.27)$$

where $\hat{M}(n)$ is the linearization matrix with elements

$$m_{j,k} = \left. \frac{\partial P_j(\boldsymbol{x}, \boldsymbol{\alpha})}{\partial x_k} \right|_{\boldsymbol{x} \in \boldsymbol{x}^0(n)} . \tag{1.28}$$

It follows from (1.27) that the initial perturbation evolves according to the law

$$\boldsymbol{y}(n+1) = \hat{M}(n)\hat{M}(n-1)\ldots\hat{M}(1)\boldsymbol{y}(1) . \tag{1.29}$$

By analogy with differential systems we introduce into consideration the Lyapunov exponents of solution $\boldsymbol{x}^0(n)$:

$$\lambda_i = \overline{\lim_{n \to \infty}} \frac{1}{n} \ln \|\boldsymbol{y}^i(n)\| . \tag{1.30}$$

Taking into account the fact that

$$\hat{M}(n)\boldsymbol{y}^i(n) = \mu_i(n)\boldsymbol{y}^i(n) , \tag{1.31}$$

where $\mu_i$ is the eigenvalue of matrix $\hat{M}(n)$, corresponding to the $i$th eigenvector, from (1.29) and (1.30) we derive

$$\lambda_i = \overline{\lim_{n \to \infty}} \frac{1}{n} \sum_{k=1}^{n} \ln |\mu_i(k)| . \tag{1.32}$$

The stability of fixed points and limit cycles of the map is characterized by multipliers. The sequence of states $\boldsymbol{x}_1^0, \boldsymbol{x}_2^0, \ldots, \boldsymbol{x}_l^0$ is called a *period-l-cycle* of the map, or simply *l-cycle*, if the following condition is fulfilled:

$$\boldsymbol{x}_1^0 = \boldsymbol{P}(\boldsymbol{x}_l^0) . \tag{1.33}$$

If $l = 1$, i.e.,

$$\boldsymbol{x}^0 = \boldsymbol{P}(\boldsymbol{x}^0) , \tag{1.34}$$

the state $\boldsymbol{x}^0$ is a *fixed point* or *period-1-cycle*. The linearization matrix $\hat{M}$ along periodic solution $\boldsymbol{x}^0(n)$ is periodic, i.e., $\hat{M}(n+l) = \hat{M}(n)$. The perturbation component $\boldsymbol{y}^i(1)$ is transformed in the period $l$ as follows:

$$\boldsymbol{y}^i(l+1) = \hat{M}(l)\hat{M}(l-1)\ldots\hat{M}(1)\boldsymbol{y}^i(1) = \hat{M}_l\boldsymbol{y}^i(1) . \tag{1.35}$$

The matrix $\hat{M}_l$ does not depend on the initial point and is an analogue of the monodromy matrix in a differential system. The eigenvalues $\mu_i^l$ of matrix $\hat{M}_l$ are called *multipliers of the l-cycle* of the map. They characterize how projections of perturbation vector onto the eigenvectors of linearization matrix $\hat{M}$ change in the period $l$. The multipliers $\mu_i^l$ are connected with the Lyapunov exponents by the relation

$$\lambda_i = \ln \frac{1}{l} |\mu_i^l| . \tag{1.36}$$

The $l$-cycle of the map is asymptotically stable if its multipliers $|\mu_i^l| < 1$, $i = 1, 2, \ldots, N$. Thus, the LCE spectrum involves negative numbers only.

If the map under study with the phase space dimension $(N - 1)$ is the Poincaré map in a section of some $N$-dimensional continuous-time system, then it possesses the following property: The matrix $\hat{M}_l$ eigenvalues $\mu_i^l$, $i = 1, 2, \ldots, (N - 1)$, of the $l$-cycle, complemented by the unit multiplier $\mu_N^l = 1$, are strictly equal to the eigenvalues of the monodromy matrix of the corresponding limit cycle in this continuous-time system. On this basis, the stability of periodic oscillations in differential systems can be quantitatively described by the multipliers of a relevant cycle in the Poincaré map.

### 1.1.4 Bifurcations of Dynamical Systems, Catastrophes

Mathematical modeling of most of practical problems in nonlinear dynamics is most often accomplished by using differential equations which depend on a number of parameters. Variation of a system parameter may cause a qualitative change of the system phase portrait, called *bifurcation* [2, 6, 7, 10]. By the qualitative change of phase portrait we mean its structural rebuilding, which breaks the *topological equivalence*. The parameter value at which the bifurcation takes place is called the *bifurcation value* or the *bifurcation point*. Besides the phase space, a DS is also characterized by its parameter space. The particular set of parameter values $\alpha_1, \alpha_2, \ldots, \alpha_M$ is associated with a radius-vector $\alpha$ in this space. In a multi-dimensional parameter space of the system, bifurcation values may correspond to certain sets representing points, lines, surfaces, etc. Bifurcation is characterized by conditions which impose certain requirements on system parameters. The number of such conditions is called the *codimension* of bifurcation. For example, codimension 1 means that there is only one bifurcation condition.

Local and *nonlocal bifurcations* of DS can be distinguished. *Local bifurcations* are associated with the local neighborhood of a trajectory on a limit set. They reflect the change of stability of individual trajectories as well as of the entire limit set or the disappearance of the limit set through merging with another one. All the above-listed phenomena can be found in the framework of linear analysis of stability. For example, when one of the Lyapunov exponents of a trajectory on a limit set changes its sign, this testifies to a local bifurcation of the limit set. *Nonlocal bifurcations* are related to the behavior of manifolds of saddle limit sets, particularly, formation of separatrix loops, homoclinic and heteroclinic curves, and tangency between an attractor and separatrix curves or surfaces. Such effects cannot be detected by using the linear approach only. In this situation one must also take into consideration the nonlinear properties of the system under study.

Bifurcations can happen with any limit sets, but of considerable interest are bifurcations of attractors, because they cause experimentally observed regimes to change. Attractor bifurcations are usually classified into *internal* (*soft*) bifurcations and *crises* (*hard*) bifurcations [6, 11]. *Internal bifurcations* result in topological changes in attracting limit sets but do not affect their basins of attraction. Attractor *crises* are accompanied by a qualitative modification of the basin boundaries.

The concept of *robustness* (*structural stability*) of DS is closely related to bifurcations. DS are called *robust* or *structurally stable* if small smooth perturbations of the evolution operator lead to topologically equivalent solutions. A DS bifurcation can be thought of as a transition of the system from one structurally stable state to another one through a nonrobust (structurally unstable) state at the bifurcation point.

The analysis of DS bifurcations when varying system parameters enables one to construct a *bifurcation diagram* of the system. The *bifurcation diagram* is a set of points, lines and surfaces in parameter space which corresponds to certain bifurcations of the system limit sets. If several limit sets are realized in the same parameter range, the bifurcation diagram appears to be multi-sheeted. The co-existence of a large (even infinite) number of limit sets is typical for systems with complex dynamics. In this case the bifurcation points may appear to be dense everywhere in the parameter space. Under such conditions, the construction of a complete bifurcation diagram of the system becomes impossible and only its separate leaves and parts can be considered. Besides bifurcation diagrams in parameter space, so-called *phase-parametric diagrams* are often used for descriptive representation. Such a diagram is constructed by plotting parameter values as abscissas and dynamical variables or related quantities as ordinates.

Abrupt changes of system state variables, caused by smooth perturbations of the evolution operator, in particular, by a slight parameter variation, are called *catastrophes*. Thus, crises and catastrophes are very close and probably identical notions. The theory of catastrophes [12, 13], which includes the ideas of Whitney's theory of singularities [14], was developed by the topologist Thom [13]. He showed that there is a small number of elementary catastrophes whereby the system behavior can be locally described. A significant contribution to the development of this theory was made by Arnold [6, 15].

Consider the basic local bifurcations of equilibria.

**Saddle-Node Codimension One Bifurcation.** The codimension one bifurcation can be described by using only one control parameter, $\alpha$. Assume that for $\alpha < \alpha^*$ the system has two equilibria, a stable node $Q$ and a saddle $S$, shown in Fig. 1.9a. At $\alpha = \alpha^*$ the node and the saddle merge to form a nonrobust equilibrium state called a *saddle-node* (Fig. 1.9b). The latter disappears for $\alpha > \alpha^*$ (Fig. 1.9c). Since the attractor (node) disappears in the bifurcation, the basin boundaries have to be qualitatively modified. Consequently, this bifurcation is referred to as a crisis. The simplest model system describing this bifurcation can be presented by the first-order equation

$$\dot{x} = \alpha - x^2 . \tag{1.37}$$

$x_{1,2}^0 = \pm\sqrt{\alpha}$ are the equilibrium coordinates and $\rho_{1,2} = \mp 2\sqrt{\alpha}$ are the eigenvalues of the linearization operator at relevant points, i.e., $x_1^0$ is a stable

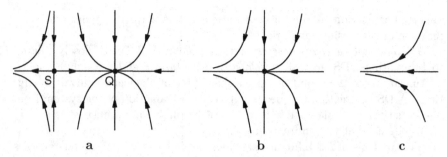

**Fig.1.9a–c.** Qualitative illustration of saddle-node codimension one bifurcation

equilibrium and $x_2^0$ is an unstable equilibrium. At $\alpha = 0$, $x_1^0 = x_2^0 = 0$, and the eigenvalues at this point are equal to zero. The bifurcation has codimension one since it is distinguished by only one bifurcation condition $\rho(\alpha) = 0$.

**Codimension Two Bifurcation – a Triple Equilibrium Point.** This bifurcation consists in merging three equilibria, nodes $Q_1$ and $Q_2$, and a saddle $Q_0$ located between them and in forming a stable node at the point $Q_0$. This is illustrated in Fig. 1.10. The codimension two bifurcation is described by two control parameters. The model system for this bifurcation is defined by

$$\dot{x} = \alpha_1 + \alpha_2 x + x^3 \,. \tag{1.38}$$

The analysis of equilibria shows that for $\alpha_2 > 0$ and for any $\alpha_1$ the system possesses the unique equilibrium state $Q_0$ with the eigenvalue $\rho_{Q_0} < 0$, i.e., the equilibrium is asymptotically stable. For $\alpha_2 < 0$ there exists a parameter $\alpha_1$ range (the shaded region in the bifurcation diagram shown in Fig. 1.11a) where the system has three equilibria, $Q_0, Q_1$ and $Q_2$. One of them, $Q_0$, is unstable since $\rho_{Q_0} > 0$, and the other two, $Q_1$ and $Q_2$, are stable as $\rho_{Q_{1,2}} \leq 0$. The bistability region in the bifurcation diagram of Fig. 1.11a is bounded by curves $l_1$ and $l_2$, which correspond to saddle-node bifurcations of nodes $Q_{1,2}$

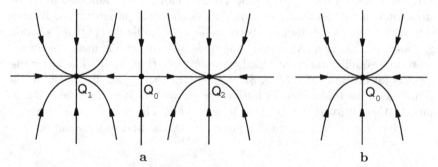

**Fig. 1.10.** Illustration of bifurcation "triple equilibrium". (**a**) Two stable nodes and a saddle before the bifurcation, and (**b**) one stable node after the bifurcation

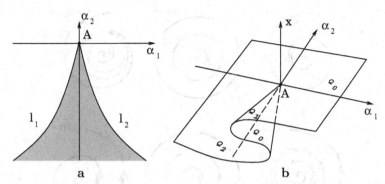

**Fig. 1.11.** Illustration of triple equilibrium bifurcation. (**a**) Bifurcation diagram and (**b**) phase-parametric diagram

with saddle $Q_0$. The curves $l_1$ and $l_2$ converge to the point $A$ $(\alpha_1 = \alpha_2 = 0)$ that is of the cusp type. In this point two bifurcation conditions are fulfilled simultaneously: $\rho_{Q_1}(\alpha_1, \alpha_2) = 0$ and $\rho_{Q_2}(\alpha_1, \alpha_2) = 0$. Hence, this bifurcation is called the triple equilibrium and has codimension two. The structure shown in Fig. 1.11b appears in the phase-parametric space of system (1.38). In the bistability region the upper and the lower leaves correspond to the stable equilibria and the central one to the unstable equilibrium.

**Andronov–Hopf Bifurcation.** In DS with the dimension $N \geq 2$ the situation is possible when a pair of complex-conjugate eigenvalues of the equilibrium of "stable focus" type intersects the imaginary axis. This implies that the bifurcation condition $\mathrm{Re}\rho_{1,2} = 0$ is satisfied. Additionally, $\mathrm{Im}\rho_{1,2} \neq 0$. This case corresponds to *Andronov–Hopf bifurcation* [1, 16] or the *limit cycle birth (death) bifurcation*. This bifurcation was first explored by A.A. Andronov for the case $N = 2$ and then generalized by E. Hopf to systems with arbitrary dimension $N$. There are two different kinds of Andronov–Hopf bifurcation, namely, *supercritical*, or *soft* bifurcation, and *subcritical*, or *hard* bifurcation. Supercritical bifurcation is internal; subcritical bifurcation is an attractor crisis. Andronov–Hopf bifurcation is defined by only one bifurcation condition and has thus codimension one.

*Supercritical Andronov–Hopf bifurcation* is illustrated in Fig. 1.12a–c and is as follows. For $\alpha < \alpha^*$ there exists a stable focus $F$, which at the bifurcation point $\alpha = \alpha^*$ has a pair of pure imaginary eigenvalues $\rho_{1,2} = \pm i\omega_0$. For $\alpha > \alpha^*$ the focus $F$ becomes unstable ($\mathrm{Re}\rho_{1,2} > 0$), and a stable limit cycle $C_0$ is born in the near vicinity.

*Subcritical Andronov–Hopf bifurcation* takes place when at $\alpha = \alpha^*$ the unstable (in the general case for $N > 2$, saddle) limit cycle $C_0$ is "striking" in the focus $F$, being stable for $\alpha < \alpha^*$. As a result, the cycle no longer exists, and the focus becomes unstable (Fig. 1.12d–f).

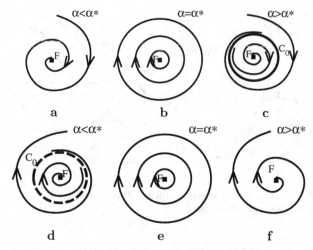

**Fig. 1.12.** (a–c) Supercritical and (d–f) subcritical Andronov–Hopf bifurcation

The model system for Andronov–Hopf bifurcation has the following form:

$$\dot{a} = (\alpha + i\omega_0)a + L_1 a|a|^2, \quad \omega_0 \neq 0, \; L_1 \neq 0, \tag{1.39}$$

where $a$ is the instantaneous complex amplitude and $L_1$ is called the *first Lyapunov quantity* of equilibrium. If $L_1 < 0$, the bifurcation is supercritical. If $L_1 > 0$, the bifurcation is subcritical.[2] For the real instantaneous amplitude and phase of oscillations, we derive the following from (1.39):

$$\dot{A} = \alpha A + L_1 A^3, \quad \dot{\Phi} = \omega_0, \tag{1.40}$$

where $A = |a|$ and $\Phi = \mathrm{Arg}(a)$. From the equation for stationary amplitudes $\alpha A + L_1 A^3 = 0$, we obtain amplitude values for the equilibrium ($A_F = 0$) and for the limit cycle ($A_0 = \sqrt{-\alpha/L_1}$). The limit cycle exists provided that $-\alpha/L_1 > 0$. The quantity $\omega_0$ defines its period $T = 2\pi/\omega_0$. The analysis of the linearized equation for amplitude perturbation enables us to find the eigenvalues for the solutions $A = A_F$ and $A = A_0$: $\rho_{F,0} = \alpha + 3L_1 A_{F,0}^2$. If $L_1 < 0$, the cycle is thus seen to exist and be stable for $\alpha > 0$, whereas the focus is stable for $\alpha < 0$ and unstable for $\alpha > 0$. When $L_1 > 0$, both the unstable cycle and the stable focus exist for $\alpha < 0$, while only the unstable focus exists for $\alpha > 0$.

**Bifurcations of Limit Cycles.** Consider local codimension one bifurcations of nondegenerate limit cycles which have only one multiplier equal to 1. We eliminate the unit multiplier from our consideration and arrange the

---

[2] The character of bifurcation in a particular (degenerate) case $L_1 = 0$ has to be additionally analyzed with higher powers of $a$ taken into consideration.

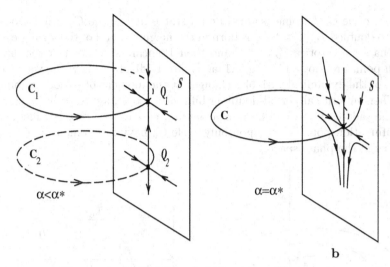

**Fig. 1.13a,b.** Saddle-node bifurcation of limit cycles

remaining multipliers in decreasing order in absolute value. In this case limit-cycle bifurcations are related to one or two (complex conjugate) first multipliers of this sequence, $\mu_{1,2}$. Since codimension one bifurcation assumes only one bifurcation condition and the change of cycle stability is determined by the equality $|\mu_1| = 1$, we have three different kinds of limit-cycle bifurcations occurring at the bifurcation parameter value $\alpha = \alpha^*$: $\mu_1 = +1$, $\mu_1 = -1$, and $\mu_{1,2} = \exp(\pm i\varphi)$. To analyze the limit-cycle bifurcations, it is more convenient to use a Poincaré surface of section technique. Fixed points of the corresponding Poincaré map are characterized by the same multipliers and the transition to the Poincaré section makes the analysis more instructive [2].

**Saddle-Node Bifurcation.** Multiplier $\mu_1$ of a stable cycle becomes $+1$ when the parameter $\alpha$ achieves its bifurcation value $\alpha = \alpha^*$. We illustrate this bifurcation using three-dimensional phase space, $N = 3$. For $\alpha < \alpha^*$ there exist two cycles, a stable cycle $C_1$ and a saddle cycle $C_2$ (Fig. 1.13a). They are associated with stable $Q_1$ and unstable $Q_2$ fixed points in the Poincaré section. The condition $\mu_1 = 1$ determines the bifurcation which is similar to the saddle-node bifurcation of equilibrium, considered above. At the bifurcation point $\alpha = \alpha^*$ cycles $C_1$ and $C_2$ merge forming a nonrobust closed trajectory $C$ of the saddle-node type (see Fig. 1.13b). Both cycles disappear for $\alpha > \alpha^*$. When varying parameter $\alpha$ in the reverse order, a pair of cycles $C_1$ and $C_2$ is born from the phase trajectory concentration.

**Period-Doubling Bifurcation.** Multiplier $\mu_1(\alpha^*)$ becomes $-1$ at the critical point $\alpha = \alpha^*$ provided that $d\mu/d\alpha|_{\alpha^*} \neq 0$. The latter expression is the bifurcation condition for *period-doubling bifurcation*. This bifurcation can be supercritical (internal) or subcritical (crisis). Supercritical bifurcation is as follows: Let us for $\alpha < \alpha^*$ have a stable limit cycle $C_0$ with period $T_0$. When

$\alpha > \alpha^*$, cycle $C_0$ becomes saddle, and a stable limit cycle $C$ with period $T$ close to doubled $T_0$ ($T \approx 2T_0$) is born in the neighborhood of the former cycle $C_0$. Phase trajectories $C_0$ and $C$ and their Poincaré sections near the bifurcation point are plotted in Fig. 1.14a. Figure 1.14b shows how the waveform of one of the dynamical variables changes at the moment of period doubling.

When subcritical period-doubling bifurcation occurs, the stable cycle $C_0$ and the saddle cycle $C$ with doubled period, existing for $\alpha < \alpha^*$, merge at the bifurcation point, whereupon only cycle $C_0$ having become a saddle one remains in the phase space.

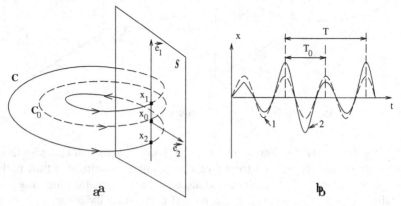

**Fig. 1.14.** Supercritical period-doubling bifurcation. (a) Cycles $C_0$ and $C$ and their Poincaré sections and (b) waveforms before (*curve 1*) and after (*curve 2*) the bifurcation

**Two-Dimensional Torus Birth (Death) Bifurcation (Neimark Bifurcation).** This bifurcation is realized when a pair of complex-conjugate multipliers of a limit cycle go out to the unit circle. At the bifurcation point $\alpha = \alpha^*$, the following relation takes place: $\mu_{1,2}(\alpha^*) = \exp(\pm i\varphi)$, where $\varphi \in [0, 2\pi]$, and $\varphi(\alpha^*) \neq 0, \pi, 2\pi/3$ (so-called *strong resonances* are excluded). This bifurcation can also be supercritical (internal) and subcritical (crisis). Different situations are realized depending on which kind of bifurcation occurs. In the case of supercritical bifurcation a stable two-dimensional (2-D) torus $T^2$ is born from a stable limit cycle $C_0$, the latter becoming unstable thereafter (in general, it is of the saddle-focus type). Subcritical bifurcation takes place when an unstable (in general, saddle) torus $T^2$ is "sticking" into the stable cycle $C_0$ at the moment of its stability loss. The torus birth from a limit cycle is pictured in Fig. 1.15a. A small perturbation $\boldsymbol{y}$ of cycle $C_0$ near the bifurcation point $\alpha = \alpha^*$ is rotated through angle $\varphi$ in one revolution of trajectory $C_0$. At the same time, its magnitude remains unchanged since $|\mu_{1,2}(\alpha^*)| = 1$. Thus, a representative point in the Poincaré section moves along a circle $L$ called the *invariant circle* of the map. The quantity $\theta(\alpha) = \varphi/2\pi$ is called the *winding number* on torus $T^2$ (or on the corresponding invariant circle).

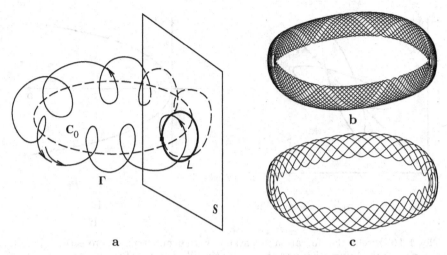

**Fig. 1.15.** Torus birth bifurcation from limit cycle $C_0$: (**a**) the trajectory $\Gamma$ on the torus in the neighborhood of unstable cycle $C_0$, (**b**) ergodic torus, and (**c**) resonance on torus

If the winding number $\theta(\alpha^*)$ takes an irrational value, any trajectory $C$ on the torus does not close on itself and the torus born is *ergodic* (Fig. 1.15b). If $\theta(\alpha^*) = p/q$, where $p$ and $q$ are any positive integers, the resonance phenomenon of the order $p/q$ is said to take place on the torus. An example of resonance on a torus is shown in Fig. 1.15c.

**Symmetry-Breaking Bifurcation.** The bifurcations of limit cycles, described by the conditions $\mu_1(\alpha^*) = \pm 1$ and $\mu_{1,2}(\alpha^*) = \exp(\pm i\varphi)$, may lead to the situation when a limit cycle loses its symmetry. Such bifurcations are typical, for example, for systems consisting of two or more identical partial subsystems. The symmetry property of a limit set is related to the existence of some invariant manifold $U$ in the system phase space. For instance, examine the following two coupled identical subsystems:

$$\dot{x} = F(x, \alpha) + \gamma g(y, x),$$
$$\dot{y} = F(y, \alpha) + \gamma g(x, y), \tag{1.41}$$

where $x, y \in \mathbf{R}^N$ are the state vectors of the subsystems and $\alpha$ is the parameter vector. Function $g$ is responsible for the coupling between the subsystems and $g(x, x) = 0$. In this case the subspace $x = y$ is referred to as an invariant symmetric manifold. If multipliers of limit cycles in $U$ are connected with eigenvalues not lying in $U$ and appear to be bifurcational, then the symmetric attractor undergoes a bifurcation leading to the birth of a nonsymmetric attractor, i.e., one not lying in $U$. The bifurcation is said to result in the loss of attractor symmetry. The symmetry-breaking bifurcations determined by the conditions $\mu_1(\alpha^*) = -1$ and $\mu_{1,2}(\alpha^*) = \exp(\pm i\varphi)$ are very similar to

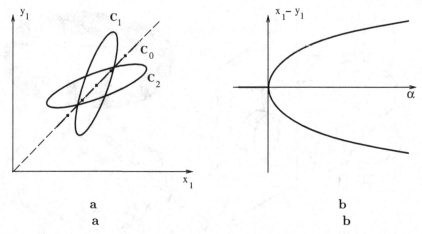

**Fig. 1.16.** Pitchfork bifurcation in a system with symmetry: (**a**) projection of limit cycles after the bifurcation and (**b**) qualitative illustration of the phase-parametric diagram

the analogous bifurcations in systems without symmetry. Bifurcation defined by the condition $\mu_1(\alpha^*) = +1$ represents a special case. As a result of this bifurcation, the symmetric cycle $C_0 \in U$ still exists but becomes a saddle one. It gives birth to two stable cycles with the same period. They do not lie in $U$ but are mutually symmetric. Such a bifurcation is known under the name *pitchfork bifurcation*. Figure 1.16a shows phase portraits of cycles after the pitchfork bifurcation. A phase-parametric diagram of this bifurcation is sketched in Fig. 1.16b, where the ordinate is the difference of appropriate coordinates $(x_1^s - y_1^s)$ in some section of the cycles.

We have considered the local bifurcations of equilibria and of limit cycles. More complicated sets (tori, chaotic attractors) also undergo different bifurcations. However, their studies are more often based on experimental results, and the theory of bifurcations of quasiperiodic and chaotic attractors remains to be developed.

**Nonlocal Bifurcations. Homoclinic Trajectories and Structures.** Nonlocal bifurcations are associated with the form of stable and unstable manifolds of saddle limit sets in phase space. They do not cause the saddle limit sets themselves to topologically change but may significantly affect the system dynamics. Consider the basic nonlocal bifurcations [2, 10].

**The Separatrix Loop of a Saddle Equilibrium State.** This bifurcation can be realized in its simplest form even in a phase plane. Let us have a saddle equilibrium $Q$ whose stable $W_Q^s$ and unstable $W_Q^u$ manifolds come close together with increasing parameter $\alpha$ and touch one another at $\alpha = \alpha^*$. At the tangency point a bifurcation takes place, leading to the formation of a singular double-asymptotic phase trajectory $\Gamma_0$, called the *separatrix loop of the saddle point* (Fig. 1.17a). The fulfillment of this tangency condition

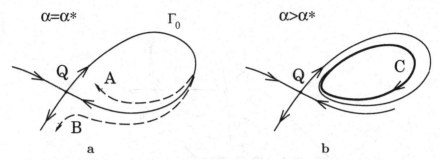

**Fig. 1.17.** Bifurcation of separatrix loop formation (**a**) at the moment of bifurcation and (**b**) after the bifurcation

is conformed by the codimension one bifurcational manifold in parameter space. The separatrix loop in a dissipative system is a nonrobust structure and collapses at $\alpha \neq \alpha^*$. What will happen thereafter depends on which way the separatrices proceed after their splitting, i.e., loop breaking, and on the *saddle quantity* $\sigma_Q$ of equilibrium at the bifurcation point. The saddle quantity is specified as $\sigma_Q(\alpha) = \rho_1(\alpha) + \rho_2(\alpha)$, where $\rho_{1,2}$ are the eigenvalues of the linearization matrix at point $Q$. If $\sigma_Q(\alpha^*) < 0$, then when the loop is broken down in direction $A$ as shown in Fig. 1.17a, the only stable cycle $C$ arises from the loop (Fig. 1.17b). No cycle is born when the loop is destroyed in direction $B$. If $\sigma_Q(\alpha^*) > 0$, the loop $\Gamma_0$ is called *unstable*, and as $\Gamma_0$ is destroyed, an unstable limit cycle can arise from it.

The bifurcation of separatrix loop formation, considered in reverse order, may be treated as a crisis of limit cycle $C$ when it touches the saddle $Q$. At the moment of tangency the loop $\Gamma_0$ is created. When approaching the bifurcation point, the cycle period tends to infinity, and the cycle multipliers approach zero [17].

A more complicated case of nonlocal bifurcation of similar type is possible in phase space with dimension $N \geq 3$. We confine ourselves to $N = 3$. Let $Q$ be a saddle-focus with one-dimensional unstable and two-dimensional stable manifolds and with the so-called *first saddle quantity* $\sigma_1(\alpha) = \text{Re}\rho_{1,2}(\alpha) + \rho_3(\alpha)$, where $\rho_{1,2} = \text{Re}\rho_{1,2} \pm i\text{Im}\rho_{1,2}$ and $\rho_3$ are the eigenvalues of the linearization matrix at point $Q$. Let a *saddle-focus separatrix loop* $\Gamma_0$, shown in Fig. 1.18, be formed at $\alpha = \alpha^*$ and $\sigma_1(\alpha^*) \neq 0$. Under the assumptions made, the theorem by L.P. Shilnikov [18] is valid and states the following:

- $\sigma_1(\alpha^*) < 0$ (the loop is not dangerous). If the loop destruction corresponds to case $A$ as shown in Fig. 1.18, a stable cycle $\Gamma$ arises from it. Nothing happens when the loop is broken down in direction $B$.
- $\sigma_1(\alpha^*) > 0$ (the loop is dangerous). A complicated structure of phase trajectories emerges in the vicinity of loop $\Gamma_0$ when it exists and then is destroyed in any direction. This structure consists of a countable set of periodic attractors, repellers, and saddles as well as of a subset of chaotic

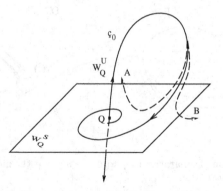

**Fig. 1.18.** Saddle-focus separatrix loop and possible ways to destroy it

trajectories, a so-called *nontrivial hyperbolic subset.* Such a structure is connected with the presence of a set of Smale's horseshoe-type maps in the vicinity of the loop in the Poincaré section.

If the system to be studied has a saddle-focus with one-dimensional stable and two-dimensional unstable manifolds, Shilnikov's theorem can be applied through replacement of $t$ by $-t$.

**The Saddle-Node Separatrix Loop.** This bifurcation is also possible for $N = 2$. Assume that for $\alpha < \alpha^*$ there exist two equilibria, namely, a saddle $Q_1$ and a stable node $Q_2$. In addition, a separatrix loop is formed by the closure of unstable separatrices of the saddle running to the stable node, as shown in Fig. 1.19a. At the bifurcation point $\alpha = \alpha^*$, a saddle-node bifurcation of equilibria occurs, and a nonrobust equilibrium of the saddle-node type arises. Here, the saddle-node has a double-asymptotic homoclinic trajectory $\Gamma_0$, i.e., the separatrix loop (Fig. 1.19b). When $\alpha > \alpha^*$, the saddle-node disappears and a limit cycle $C$ is generated from the loop (Fig. 1.19c).

This bifurcation, when considered in reverse order, is a bifurcation of cycle $C$ extinction, leading to the creation of a saddle-node on it. The cycle period grows infinitely as $\alpha \to \alpha^*$, and the cycle multipliers tend to zero.

The bifurcation described above preserves the boundaries of an absorbing area (basin of attraction) and is thus an internal bifurcation.

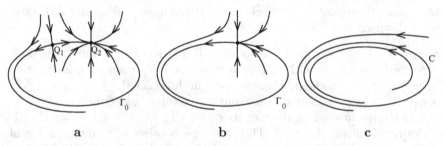

**Fig.1.19a–c.** Bifurcation of saddle-node separatrix loop formation

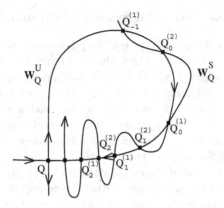

**Fig. 1.20.** Homoclinic intersection of saddle cycle trajectories (graphical representation in a secant plane)

**Homoclinic Trajectory Appearance of a Saddle Limit Cycle.** This bifurcation is realized when $N \geq 3$. In this case there may exist saddle limit cycles with two-dimensional stable $W^{\mathrm{s}}$ and two-dimensional unstable $W^{\mathrm{u}}$ manifolds. Such a cycle corresponds to a saddle fixed point in a secant plane. This fixed point has one-dimensional stable and unstable manifolds. With increasing parameter $\alpha$ the manifolds of the cycle approach each other and touch at $\alpha = \alpha^*$. This bifurcation being of codimension one results in the emergence of a nonrobust double-asymptotic curve $\Gamma_0$, called the *Poincaré homoclinic curve*. When $\alpha > \alpha^*$, the manifolds $W^{\mathrm{s}}$ and $W^{\mathrm{u}}$ intersect, and two robust homoclinic curves $\Gamma_1^0$ and $\Gamma_2^0$ are created. In the secant plane each homoclinic curve corresponds to an infinite double-asymptotic sequence of intersection points of separatrices $Q_n$, $n = 0, \pm 1, \pm 2, \ldots$ (Fig. 1.20). The points $Q_n$ become denser when approaching the saddle, but they tend to it only in the limit as $n \rightarrow \pm\infty$.

In [19, 20] it has been shown that a complicated set of trajectories is generated in the vicinity of a homoclinic curve of a saddle cycle. This set is called a *homoclinic structure* which is similar to the structure arising in the neighborhood of a dangerous saddle-focus separatrix loop and is also connected with the formation of local horseshoe-type maps in the vicinity of the loop. Stable, unstable and saddle periodic orbits are dense everywhere in the vicinity of the homoclinic trajectory. Besides, the homoclinic structure includes a subset of chaotic trajectories, which, under appropriate conditions, may become attracting.

A similar structure is characteristic for the neighborhood of a *heteroclinic trajectory* which emerges when the unstable manifold of one saddle cycle touches and then intersects the stable manifold of another saddle cycle.

### 1.1.5 Attractors of Dynamical Systems. Deterministic Chaos

It is known that in a dissipative system the phase space volume is contracted by the time evolution. Because of this, the final limit set of the system always has zero volume. In addition, the limit set can be a point, or a line, or a surface or a set of surfaces, which has a Cantor structure in its Poincaré section.

The image of dynamical chaos was associated with a *strange attractor* for a long time [21]. More recently, there came the understanding that chaotic self-sustained oscillations may be substantially different in their properties. This fact leads to differences in the structures of the corresponding attractors. It has become clear that a strange attractor is the image of some "ideal" chaos satisfying a number of rigorous mathematical requirements. The regime of the strange attractor in the strict sense of mathematical definition cannot be realized in real systems. What we observe in experiments is more often the regimes of so-called *quasihyperbolic* or *nonhyperbolic* attractors [10, 22]. A distinctive feature of strange, quasihyperbolic and nonhyperbolic chaotic attractors is *exponential instability* of phase trajectories and a fractal (noninteger) dimension. The exponential instability serves as a criterium of chaotic behavior of the system in time. The fractal metric dimension shows that the attractor is a complex geometric object.

The time evolution of the state of a system with a finite number of degrees of freedom is described by either a system of ordinary differential equations

$$\frac{\mathrm{d}x_i}{\mathrm{d}t} = \dot{x}_i = f_i(x_1, \ldots, x_N, \alpha_1, \ldots, \alpha_k) \tag{1.42}$$

or discrete maps

$$x_{n+1}^i = f_i(x_n^1, x_n^2, \ldots, x_n^N, \alpha_1, \ldots, \alpha_k),$$

$$i = 1, 2, \ldots, N.$$

Here $x_i(t)$ (or $x_n^i$) are variables uniquely describing the system state (its phase coordinates) and $\alpha_l$ are system parameters, $f_i(\boldsymbol{x}, \boldsymbol{\alpha})$ are, in general, nonlinear functions.

In what follows, we shall consider only self-oscillatory regimes of the system (1.42) motion. The latter means that the system demonstrates some steady-state oscillations whose characteristics do not depend, to a certain extent, on the choice of initial state. We shall also consider the regime of a stable equilibrium state as a limiting case of the self-oscillatory regime.

Examine the phase space $\mathbf{R}^N$ of system (1.42). All values of the system parameters $\alpha_l$ are fixed. Let $G_1$ be some finite (or infinite) region belonging to $\mathbf{R}^N$ and including a subregion $G_0$. The regions $G_1$ and $G_0$ satisfy the following conditions [6, 23–25]:

- For any initial conditions $x_i(0)$ (or $x_0^i$) from the region $G_1$ all phase trajectories reach as $t \to \infty$ (or $n \to \infty$) the region $G_0$.

- The region $G_0$ is a minimal compact subset in the system phase space.
- If a phase trajectory belongs to the region $G_0$ at the time moment $t = t_1$ $(n = n_1)$, it will always belong to $G_0$, i.e., for any $t \geq t_1$ $(n > n_1)$.

If these conditions are satisfied, the region $G_0$ is called an *attractor* of a DS (1.42). $G_1$ is called the *region (or basin) of attraction* of the attractor $G_0$.

**Regular Attractors.** Before deterministic chaos was discovered only three types of stable steady-state solutions of (1.42) were known: an equilibrium state, when after a transient process the system reaches a stationary (non-changing in time) state; a stable periodic solution; and a stable quasiperiodic solution. The corresponding attractors are a point in the phase space, a limit cycle, and a limit $n$-dimensional torus. The LCE spectrum of a phase trajectory includes zero and negative exponents only.

Non-periodic solutions of (1.42) correspond to strange chaotic attractors of a complex geometric structure. They have at least one positive Lyapunov exponent and, as a consequence, a fractal dimension that can be estimated by using Kaplan–Yorke's definition [26]:

$$D = j + \frac{\sum_{i=1}^{j} \lambda_i}{|\lambda_{j+1}|}, \tag{1.43}$$

where $j$ is the largest integer number for which $\lambda_1 + \lambda_2 + \cdots + \lambda_j \geq 0$. The dimension $D$ calculated from (1.43) is one of the fractal dimensions of the set and is called the Lyapunov dimension. In the general case, it is a lower bound for the metric dimension of the attractor [27]. Applying (1.43) to the three types of the above-listed attractors, we have $D = 0$ for a point-like attractor, $D = 1$ for a limit cycle, and $D = n$ for an $n$-dimensional torus. In all cases the fractal dimension is equal to the metric dimension of the attractors. The aforementioned solutions are asymptotically stable, and the dimension $D$ is defined by an integer number and strictly coincides with the metric dimension. All these facts allow us to say that these attractors are regular. If one of the stated conditions is violated, the attractor is excluded from the group of regular attractors.

In 1971 Ruelle and Takens rigorously proved the existence of nonperiodic solutions of (1.42). They also introduced the notion of the strange attractor as the image of deterministic chaos [21]. Since that time, very often the phenomenon of deterministic chaos and the concept of the strange attractor are related to each other. However, under close examination this is not always correct and needs some explanation.

**Robust Hyperbolic Attractors.** A proof of the strange attractor existence was given under the strong assumption that the DS (1.42) was robust hyperbolic [10,21]. The system is hyperbolic if all of its phase trajectories are saddle ones. A point as the image of a trajectory in the Poincaré section is always a saddle. Robustness means that when the right-hand side of (1.42) is slightly perturbed or the control parameters are slightly varied, all the trajectories remain as saddle ones.

Hyperbolic attractors must satisfy the following conditions [24]:

- A hyperbolic attractor consists of a continuum of "unstable leaves", or curves, which are dense in the attractor and along which close trajectories exponentially diverge.
- A hyperbolic attractor (in the neighborhood of each point) has the same geometry defined as a product of the Cantor set on an interval.
- A hyperbolic attractor has a neighborhood foliated into "stable leaves" along which the close trajectories converge to the attractor.

Robustness means that these properties hold under perturbations.

Due to the presence of the attractor, unstable manifolds $W^u$ of saddle trajectories appear to be concentrated in the region $G_0$ and may intersect with stable manifolds $W^s$ along which the trajectories approach the attractor. This leads to the appearance of homoclinic points (surfaces) and the formation of homoclinic structures which must be robust in robust hyperbolic systems. From the topological viewpoint, the intersection structure of $W^s$ and $W^u$ must correspond to Fig. 1.21a and should not change qualitatively under perturbations. The cases shown in Fig. 1.21b and Fig. 1.21c are excluded because they characterize two nonrobust phenomena, namely, the closure of the manifolds with the loop formation (Fig. 1.21b) and the phenomenon of tangency of the stable and the unstable manifolds (Fig. 1.21c). If the nonlocal properties of manifolds lead to the nonrobust situations pictured in Fig. 1.21b and Fig. 1.21c when the DS is perturbed, bifurcations of system solutions are possible [10]. However, no bifurcations must occur in robust hyperbolic systems.

Strange (according to Ruelle–Takens) attractors are always robust hyperbolic limit sets. The main feature in which strange chaotic attractors differ from regular ones is exponential instability of the phase trajectory on the attractor. In this case the LCE spectrum includes at least one positive exponent. The fractal dimension of an attractor is always more than two and, in general, is not defined by an integer. A minimal dimension of the phase space in which a strange attractor can be "embedded" appears to be equal to 3.

In mathematics at least two examples of robust hyperbolic attractors are known. These are the Smale–Williams attractor [28] and the Plykin attractor [29]. Unfortunately, up to now the regime of rigorously hyperbolic robust chaos has not been found in real systems. "Truly" strange attractors are an ideal but still unattainable model of deterministic chaos.

**Quasihyperbolic Attractors – Lorenz-Type Attractors.** The hyperbolicity conditions stated above are not fulfilled for real DS. Nevertheless, there are DS whose attractors are very close in their structure and properties to hyperbolic ones. Such attractors are chaotic, do not enclose stable regular attractors and preserve these properties under perturbations. Theoretically, quasihyperbolic attractors are not structurally stable. At least one

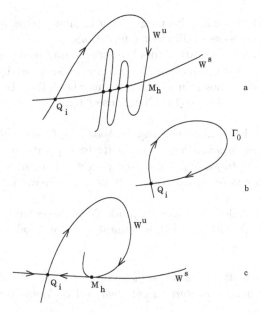

**Fig.1.21a–c.** Possible cases of intersection of stable and unstable separatrices of saddle point $Q_i$ in the Poincaré section

of the three hyperbolicity conditions is violated for these attractors. However, changes in their structures are so minor that they may not be reflected in experimentally observed characteristics.

Following the definition given in [24] we shall call almost hyperbolic attractors *quasihyperbolic*. There are quasihyperbolic Lozi, Belykh and Lorenz-type attractors.

**Nonhyperbolic Attractors.** These attractors are the most typical in experiments and illustrate experimentally observed chaos [10]. Systems with nonhyperbolic attractors exhibit regimes of deterministic chaos, which are characterized by exponential instability of phase trajectories and a fractal structure of the attractor. From this point of view, characteristics of nonhyperbolic attractors are very similar to those of robust hyperbolic and quasihyperbolic attractors. However, there is a very essential difference. A distinctive feature of nonhyperbolic attractors is the coexistence of a countable set of different chaotic and regular attracting subsets in a bounded element of the system phase space volume for fixed system parameters. This collection of all coexisting limit subsets of trajectories in the bounded region $G_0$ of phase space, which all or almost all the trajectories from the region $G_1$ including $G_0$ approach, is called a *nonhyperbolic attractor* of the DS. Hence, nonhyperbolic attractors have a very complicated structure of embedded basins of attraction. But the complexity is greater than this fact. When system parameters are varied in some finite range, both regular and chaotic attractors demonstrate the cascades of different bifurcations.

DS with nonhyperbolic attractors are structurally unstable. The fundamental reason for their existence is the presence of structurally unstable homoclinic trajectories, shown in Fig. 1.21b and Fig. 1.21c. For this situation we have rigorous results proved by Newhouse [20]:

**Theorem 1** (Newhouse). *In any neighborhood of a $C^r$-smooth ($r \geq 2$) two-dimensional diffeomorphism having a saddle fixed point with a structurally unstable homoclinic trajectory, there exist regions where systems with structurally unstable homoclinic trajectories are dense everywhere.*

These regions are called *Newhouse regions*. When a control parameter of a structurally unstable DS is varied, a countable set of Newhouse intervals is registered.

**Theorem 2** (Newhouse). *Given a sufficiently small control parameter $|\alpha_0| > 0$, the interval $(-\alpha_0, \alpha_0)$ contains a countable set of Newhouse intervals.*

$\alpha_0 = 0$ corresponds to the case of homoclinic tangency.

These results are especially important for solving many problems of nonlinear dynamics. Note that the Newhouse results are generalized for the multidimensional case in [30].

One of the methods to experimentally detect a nonhyperbolic attractor is to calculate angles between the stable and the unstable manifolds along a chaotic trajectory [31]. If the angles are different from zero, the attractor is hyperbolic or quasihyperbolic. Zero angles appear if there is a homoclinic tangency between the manifolds along the trajectory. In this case the DS is structurally unstable and the regime being observed is characterized by a nonhyperbolic attractor.

Consider the well-known Henon map [32]

$$x_{n+1} = 1 - a x_n^2 + y_n, \quad y_{n+1} = b x_n. \tag{1.44}$$

This map is dissipative for $0 < b < 1$ and possesses a nonhyperbolic attractor [33].

When following the evolution of angle $\varphi$ between the manifolds along a chaotic trajectory of diffeomorphism (1.44), one can calculate the angle probability distribution $P(\varphi)$. The numerical results are presented in Fig. 1.22a and testify that the probability of close to zero angles is reliably larger than zero, i.e., there is a tangency between the manifolds.

The occurrence of homoclinic tangency in the Henon map is caused by the presence of quadratic nonlinearity $a x_n^2$. Replacing this term in (1.44) by $|x_n|$, we derive the Lozi diffeomorphism [34]:

$$x_{n+1} = 1 - a|x_n| + y_n, \quad y_{n+1} = b x_n. \tag{1.45}$$

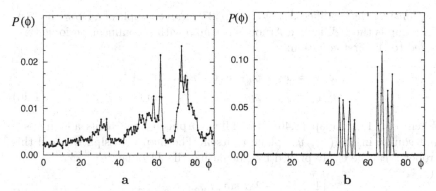

**Fig. 1.22.** Angle probability distribution between the directions of stable and unstable manifolds of a chaotic trajectory (**a**) on the Henon attractor for $a = 1.179$ and $b = 0.3$ and (**b**) on the Lozi attractor for $a = 1.7$ and $b = 0.3$

The results of angle calculation for the Lozi attractor are shown in Fig. 1.22b and clearly indicate the absence of tangency. Thus, the Lozi attractor is quasihyperbolic.

When control parameters are varied, a transition from one type of attractor to another can be observed. Under this transition a robust hyperbolic (or quasihyperbolic) attractor may be transformed into a nonhyperbolic one. This situation can be readily illustrated in the classical Lorenz model.

**Strange Nonchaotic and Chaotic Nonstrange Attractors.** Chaotic attractors of the three types described above have two fundamental properties: complex geometric structure and exponential instability of individual trajectories. It is these properties that are used by researchers as a criterium for diagnostics of the regimes of deterministic chaos.

However, it has been clarified that chaotic behavior in the sense of intermixing and geometric "strangeness" of an attractor may not correspond to each other. Strange attractors in terms of their geometry can be nonchaotic due to the absence of exponential divergence of phase trajectories. On the other hand, there are examples of intermixing dissipative systems whose attractors are not strange in a strict sense, that is, they are not characterized by the fractal structure and the fractal metric dimension [22, 37].

In other words, there are examples of concrete dissipative DS whose attractors are characterized by the following properties:

- An attractor possesses a regular geometric structure from the viewpoint of integer metric dimension, but individual phase trajectories on the attractor are exponentially unstable on average.
- An attractor has a complicated geometric structure, but trajectories on it are asymptotically stable. There is no intermixing.

The first type is called a *chaotic nonstrange attractor* (CNA). The second one is referred to as a *strange nonchaotic attractor* (SNA).

Consider the modified Arnold's map [27] as an example of a DS with CNA. This map is the well-known Arnold's cat map with a nonlinear periodic term added to the first equation:

$$x_{n+1} = x_n + y_n + \delta \cos 2\pi y_n, \quad \text{mod } 1,$$
$$y_{n+1} = x_n + 2y_n, \qquad\qquad \text{mod } 1. \tag{1.46}$$

When $\delta < 1/2\pi$, map (1.46) is a diffeomorphism on a torus and maps a unit square in the $(x_n, y_n)$ plane to itself. The map is dissipative, and this property can be easily proved by calculating the Jacobian

$$J = \begin{vmatrix} 1 & 1 - 2\pi\delta \sin 2\pi y_n \\ 1 & 2 \end{vmatrix} \neq 0, \quad \delta < \frac{1}{2\pi}. \tag{1.47}$$

The time-average value $|J| < 1$, and the LCE spectrum signature is "+" and "–", which suggests the presence of intermixing.

In spite of phase volume contraction, the motion of a representative point of the map is ergodic. As $n \to \infty$, the point visits any element of the unit square. The metric dimension of the attractor (the capacity according to Kolmogorov) is equal to 2. Although the points of the map are not uniformly distributed in the unit square, their density is nowhere equal to zero. Thus, in spite of the contraction, the attractor of map (1.46) is the whole unit square. In this sense Arnold's attractor is not strange, as its geometry is not fractal.

As seen from the phase portrait of Arnold's attractor, shown in Fig. 1.23, although the points cover the square practically entirely, their distribution density is explicitly inhomogeneous. To quantify such inhomogeneity, the information dimension is often used, $1 < D_I < 2$. For example, for $\delta = 0.05$ $D_I \simeq 1.96$, and for $\delta = 0.10$, $D_I \simeq 1.84$. In addition, as we have said, the capacity $D_C = 2.0$ (this is a rigorous result of Y. Sinai). As a consequence of inhomogeneity of the probability distribution density of the points on the

**Fig. 1.23.** Chaotic nonstrange attractor in Arnold's cat map (1.46) for $\delta = 0.15$

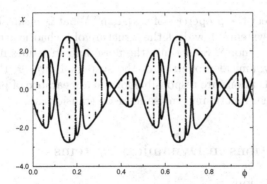

**Fig. 1.24.** Strange nonchaotic attractor in the map (1.48) for $\lambda = 1.5$

attractor, the values of all probability-metric dimensions of Arnold's attractor will lie in the interval $1 \leq D \leq 2$. These dimensions take into account not only geometric but also dynamical properties of the attractor.

Consider the possibility of realizing the opposite situation, when the system demonstrates a complicated nonperiodic oscillatory regime, which is asymptotically stable (without intermixing), but the attractor is not regular from the viewpoint of its geometric structure. SNAs are typical for quasiperiodically driven DS.

SNAs were first found and studied in the following map [37]:

$$x_{n+1} = 2\lambda \tanh(x_n) \cos 2\pi\varphi_n,$$
$$\varphi_{n+1} = \omega + \varphi_n, \mod 1. \tag{1.48}$$

$\omega$ is often set to be the inverse of the so-called *golden mean*: $\omega = 0.5(\sqrt{5} - 1)$. It was analytically shown in [37] that an SNA in system (1.48) exists for $\lambda > 1$ (Fig. 1.24). SNAs have also been observed in other DS, including the quasiperiodically forced circle map, logistic map, Henon map, etc. [38,39].

### 1.1.6 Summary

In the present section we have considered some of the elements of dynamical system theory which are highly important for understanding the bifurcation mechanisms of appearance, structure and properties of chaotic attractors. The occurrence of chaotic attractors in the system phase space is always preceded by soft and hard bifurcations of regular regimes. It is important to determine which kind of chaotic attractor is realized in the system. A deep understanding of bifurcation phenomena in the system becomes more significant if the effect of different fluctuations is taken into account. It is particularly known that noise is inevitably present in dissipative systems. The dynamics of such systems can be more adequately described by treating them as stochastic. Consequently, some questions arise. For example: Will noise significantly affect the properties of a system? What is meant by bifurcation of a stochastic system? How does the structure of a chaotic attractor change in the presence of noise? Sometimes the noise influence does not change the fundamental dynamical properties of the system. However, there are cases when noise fluctuations are responsible for the appearance of principally new regimes of system functioning (noise-induced transitions).

## 1.2 Fluctuations in Dynamical Systems

### 1.2.1 Introduction

A purely deterministic description of DS remains incomplete, as noise inevitably is present in any real dissipative system [40–42]. Therefore, variables of DS in the presence of noise should be considered as stochastic processes which depend in their realization on the particular choice of random numbers or otherwise require a probabilistic approach.

The reason for the inclusion of stochastic sources is manifold. On the one hand, the variables of a DS describe properties of a many-particle system. These variables on a macroscopic scale are permanently subject to thermal fluctuations. Also the always finite number of particles included in a DS as well as the quantum character of the micro-particles lead to noise in the system. This type of noise is usually called *internal*, being inherently present in the system [43].

On the other hand, as a DS describes a real process at a certain level only, one usually distinguishes between the system and its environment. In this case, the interaction of a low-dimensional system with the environment is described as *external* noise affecting the system [44].

Nowadays the inclusion of random sources into the description of DS is well accepted in equilibrated systems within statistical physics. It leads to the consideration of fluctuations of variables as deviations from their means and to the occurrence of dissipative forces counteracting these deviations. Their common appearance is formulated in fluctuation–dissipation relations of a different kind, connecting characteristic time scales of the dissipative and fluctuating forces [45]. However, besides the enormous effort and the success of statistical physics at meso- and macroscopic scales, the microscopic foundation of fluctuation forces is still based on an a priori introduction of probabilistic concepts into the description [46]. In particular, it is necessary to introduce such concepts during a coarse graining of the description, during a decorrelation of modes and events and during the performance of the thermodynamic limit, all involving a loss of precise information and its replacement by probabilistic assumptions [47–51].

Recently many studies have been devoted to stochastic nonequilibrium systems. Steady nonequilibrium states always appear due to the action of external constraints which might be fluctuating. These forces and flows hinder the dissipative fluctuating system from approaching the equilibrium. But due to the large diversity of possible situations, at present there is a lack of a general theory even at the meso- and macroscopic levels. Nevertheless, many possibly different nonequilibrium situations have much in common in their strong interaction with the environment. The majority of studies consider minimal models where the external perturbations act as independent forces. Such a situation implies that the behavior of a fluctuating nonequilibrium DS is controlled by such additional parameters as the intensity or correlation times of external perturbations. Variations in the parameters of the external perturbations may strongly affect the response of the DS. The latter has found its expression in a large number of studies on *noise–induced* behavior in nonequilibrium DS [44, 52–62].

In this section we formulate only the basic concepts of stochastic DS. We emphasize that this chapter is addressed to readers interested in applications of stochastic processes in DS. Most of the mathematical elucidation can be properly presented within the theory of distributions and measures going beyond the scope of this book. The reader interested in questions of a rigorous treatment of stochastic processes is referred to the already large body of mathematical literature [63–70].

## 1.2.2 Basic Concepts of Stochastic Dynamics

A phase trajectory of deterministic DS is uniquely defined by the initial condition. In stochastic dynamics the state $x(t)$ of the system is not uniquely mapped in time. The mathematical formulation can be given by *stochastic differential equations* explicitly including the random sources

$$\dot{x}_\xi = f\left(x_\xi, \xi(t)\right). \tag{1.49}$$

Here $\xi(t)$ is a temporal sequence of subsequent random numbers generated due to some rule. It makes the time mapping dependent on the generated number; $x_\xi(t) \rightarrow x_\xi(t+\mathrm{d}t)$ depends on choice, and, hence, $x_\xi(t)$ is a stochastic process.

Due to the existence of these random forces $\xi(t)$ perturbing the system, different measurements of phase trajectories starting from the same initial condition will give different realizations of $x_\xi(t)$. Therefore, as in statistical physics, we have to consider an *ensemble* of $N$ different phase trajectories, instead of a single one. This ensemble is determined in its turn by an ensemble of realizations of random sources perturbing the system $\xi_N(t)$. The statistical ensemble of $N \rightarrow \infty$ realizations defines a *stochastic process*.

The alternative definition of a stochastic process $x(t)$ can be made in terms of probability distributions or probability densities. In this case we measure realizations of the process $x_1, x_2, \ldots, x_n$ at times $t_1, t_2, \ldots, t_n$ and give the joint probability density $p(x_1, t_1; x_2, t_2; \ldots; x_n, t_n)$ for their occurrence.

In the theory of DS this description is based on deterministic equations for the dynamics of probability distributions $P(x_\xi(t) < x)$ or its densities $p(x, t)$. For problems with continuous time the densities obey evolution operators,

$$\frac{\partial}{\partial t} p = \mathbf{L} p, \qquad (1.50)$$

being linear with respect to the densities. The operators $\mathbf{L}$ contain the information of the dynamics and might include an integration over the past. Later on, we concentrate on the subclass of DS where the operator is memoryless but can depend on time (Markovian processes).

These two definitions of stochastic processes, in terms of a statistical ensemble and of a set of joint probability densities, determine two alternative approaches to the analysis of stochastic DS.

Both kinds of description [(1.49) or (1.50)] can be derived as approximate descriptions of many-particle systems within statistical mechanics. Well established examples in textbooks are Brownian particles in solutions [71, 72] and the dynamics of lasers [73–76].

**Probability Distributions and Densities.** The probability distribution counts the number of realizations $x_\xi$ at time $t$ smaller than a fixed $x$:

$$P_x(x_\xi(t) < x) = \lim_{N \to \infty} \frac{1}{N} \sum_{i=1}^{N} \theta(x - x_{\xi_i}(t)) = \langle \theta(x - x_\xi(t)) \rangle. \quad (1.51)$$

The $\langle \cdot \rangle$ assigns the mean over $N \to \infty$ different realizations of $\xi(t)$, and $\theta$ stands for the Heaviside function. As a parametric function of $x$ the distribution (1.51) varies between 0 and 1. The probability density of the process $p(x, t)$ is determined by the amount of realizations in the interval $[x, x + dx)$ at time $t$: $p_x(x, t) \, dx = P_x(x \leq x_\xi(t) < x + dx)$. It can be obtained taking the derivative of (1.51) with respect to $x$, i.e.,

$$p_x(x, t) = \frac{d}{dx} P_x(x_\xi(t) < x) = \langle \delta(x - x_\xi(t)) \rangle, \qquad (1.52)$$

where $\delta(\cdot)$ is Dirac's delta-function.

Several properties of the probability density can be easily listed. Normalization is given as $\int_X p_x(x, t) \, dx = 1$, and $X$ is the set of possible values of $x$. If $y = f(x)$ one can present it as $y = \int f(x')\delta(x - x') \, dx'$ and average results as $\langle f(x) \rangle = \langle y(t) \rangle = \int f(x')p_x(x', t) \, dx'$, which gives the mean of functions $f$ of $x$.

The density of $y$ is

$$p_y(y, t) = \langle \delta(y - y_\xi(t)) \rangle = \langle \delta(y - f(x_\xi(t))) \rangle = \int_X \delta(y - f(x))p_x(x, t) \, dx. \qquad (1.53)$$

If the inverse of $y$ exists, i.e., $x = f^{-1}(y) = g(y)$ (otherwise one may proceed piecewise), then the $\delta$-function can be integrated, yielding

$$p_y(y,t) = p_x(x,t)\frac{1}{\left|\frac{\mathrm{d}f(x)}{\mathrm{d}x}\right|}; \qquad (1.54)$$

$x = g(y)$ must be inserted on the right-hand side everywhere. A generalization to stochastic dynamics with higher dimensions of $x$ is straightforward. In the last equation (1.54) the derivative should be replaced by the Jacobian of the corresponding variable transformation.

**Joint and Conditional Probability.** If $t_i$ is an arbitrary number of subsequent times such that $i = 1, 2, \ldots, n$, i.e., $t_0 < t_1 < t_2 < \ldots < t_n$, then the complete information on the dynamics of the stochastic process $x(t)$ is given by the infinite sequence of $n$-time joint probability densities:

$$p_n(x_1, t_1; \ldots; x_n, t_n) = \left\langle \delta(x_1 - x_\xi(t_1)) \ldots \delta(x_n - x_\xi(t_n)) \right\rangle. \qquad (1.55)$$

Another useful function is the conditional probability density:

$$p_n(x_n, t_n | x_1, t_1; \ldots; x_{n-1}, t_{n-1}) = \left\langle \delta(x_n - x_\xi(t_n)) \right\rangle_{x_\xi(t_1)=x_1; \ldots; x_\xi(t_{n-1})=x_{n-1}}, \qquad (1.56)$$

defining the density of the stochastic processes $x_\xi$ at time $t_n$ with known history by holding $x_\xi$ fixed at former times $t_i$, $i = 1, \ldots, n-1$. It immediately yields the relation

$$p_n(x_1, t_1; \ldots; x_n, t_n) \qquad (1.57)$$
$$= p_n(x_n, t_n | x_1, t_1; \ldots; x_{n-1}, t_{n-1}) \, p_{n-1}(x_1, t_1; \ldots; x_{n-1}, t_{n-1}).$$

**Markovian Processes.** It is indeed impossible to calculate a complete set of $n$-time joint probability densities. Fortunately, we can extract a special class of stochastic processes which allows a simplified description. These are the Markovian processes, for which only the present state determines the future. A process is called Markovian if the conditional probability density reduces to the transition probability between two subsequent times independent of the states at previous times, i.e.,

$$p_n(x_n, t_n | x_1, t_1; \ldots; x_{n-1}, t_{n-1}) = p_2(x_n, t_n | x_{n-1}, t_{n-1}). \qquad (1.58)$$

For Markovian processes the transition probability density is the central value, since it holds that

$$p_n(x_1, t_1; \ldots; x_n, t_n) = p_2(x_n, t_n | x_{n-1}, t_{n-1}) \ldots p_2(x_2, t_2 | x_1, t_1) p(x_1, t_1), \qquad (1.59)$$

with $p(x_1, t_1)$ given by (1.52). Full information is given by knowledge of the transition probability density and the probability density at the initial time.

An important relation between the transition probabilities can be formulated for Markovian processes. Summing all possible densities at time $t_2$ by integrating as

$$p_2(x_1, t_1; x_3, t_3) = \int_X dx_2 \, p_3(x_1, t_1; x_2, t_2; x_3, t_3) \tag{1.60}$$

and expanding the two joint probability densities at both sides according to (1.59) yields

$$p_2(x_3, t_3 | x_1, t_1) = \int_X dx_2 \, p_2(x_3, t_3 | x_2, t_2) \, p_2(x_2, t_2 | x_1, t_1), \tag{1.61}$$

which is the Chapman–Kolmogorov equation and the basic evolutionary equation for Markovian processes. Equation (1.61) can be treated in several small time limits, which will be shown for the class of diffusion processes in Sect. 1.2.4.

A special form of (1.61) distinguishes between transitions out of the given state and the probability to stay in the state [43]. Let us formally introduce rates of jumps between different states, $\Delta x \neq 0$, by

$$W(x \to x + \Delta x, t) = \lim_{dt \to 0} \frac{1}{dt} p_2(x + \Delta x, t + dt | x, t). \tag{1.62}$$

The full transition probability may be replaced for short times by

$$p_2(x, t + dt | x + \Delta x, t) = \left(1 - dt \int dx' \, W(x + \Delta x \to x', t)\right) \delta(\Delta x)$$
$$+ W(x + \Delta x \to x, t) \, dt + \mathcal{O}(dt^2), \tag{1.63}$$

with the first term being the probability remaining in state $x$ with $\Delta x = 0$. Insertion into the integral of (1.61) by appropriate assignments gives for the limit $dt \to 0$

$$\frac{\partial}{\partial t} p_2(x, t | x_1, t_1) \tag{1.64}$$
$$= \int dx' \left( W(x' \to x, t) \, p_2(x', t | x_1, t_1) - W(x \to x', t) \, p_2(x, t | x_1, t_1) \right).$$

This equation is called the master equation, or Pauli equation in the case of discrete events. It expresses the temporal change of probability at $x$ through the balance of out- and inflowing probability. The rates (1.62) require further definition from the physical model under consideration.

**Stationarity.** A stochastic process is called strictly stationary if the joint probability density depends on the difference between two subsequent times. A change of $t_i \to t_i + \tau$ for all $i$ does not affect the probability density. One may also state that the process is independent on the initial time $t_0$.

As conclusion one gets immediately for the transition probability density

$$p_2(x_2, t + \tau | x_1, t) = p_2(x_2, \tau | x_1). \tag{1.65}$$

It depends on the time difference and is, therefore, time homogeneous.

A minimal but important measure, characterizing stationarity of a process, is its final stationary probability density. If ergodic, the process loses its dependence on the initial state $x_0$. The transition probability density in the limit $\tau \to \infty$ approaches the stationary probability density

$$p^s(x) = \lim_{\tau \to \infty} p_2(x, \tau | x_0). \tag{1.66}$$

We would like to recall stationarity in the wide sense. It requires that the first moment $\langle x_\xi \rangle(t)$ is time independent and the autocorrelation function depends on the time difference only. The requirements are generally less restrictive, except that the second moment is finite in the wide sense case.

*Asymptotically periodic stochastic processes* possess asymptotic probability densities which vary periodically in time: $p(x, t - t_0) = p(x, t + T - t_0)$. Hence, these processes are invariant with respect to time shifts over one period. The usual case describes periodically driven stochastic systems. This type of processes can be coarse grained into stationary ones. In case of harmonic temporal forces, an integration over the initial phase $\varphi_0 = 2\pi t_0/T$ of the driving force assumes randomly equidistributed initial phases. Also averaging over one period $T$ removes the explicit time dependence.

**Moments of Stochastic Processes.** Several stochastic processes can be defined by their time-dependent moments. The number of moments which is sufficient to define the stochastic process gives the number of independent parameters of the process. It is well known, for instance, that the Gaussian processes can be characterized through the first and second moment only, while all higher cumulants vanish. For exponential or Poissonian processes all higher moments are expressed by the first moment. Reduction of the description to the first two moments is known as the correlation theory of stochastic processes [77].

The conditional first moment of the transition probability density is given by

$$\langle x_\xi(t) \rangle_{x_\xi(t_0) = x_0} = \int \mathrm{d}x\, x\, p_2(x, t | x_0, t_0). \tag{1.67}$$

In the asymptotic limit $t_0 \to -\infty$, it loses its dependence on the initial time,

$$\langle x_\xi(t) \rangle_{\mathrm{asy}} = \lim_{t_0 \to -\infty} \langle x_\xi(t) \rangle_{x_\xi(t_0) = x_0}, \tag{1.68}$$

if the transition probability density converges in this limit as $p_2(x, t | x_0, t_0) \to p(x, t)$. For stationary processes this limit is independent of time, and, therefore, without loss of generality, we can assume $\langle x \rangle_{\mathrm{asy}} = 0$.

An important characterization is given by the second moment defined at two different times. Again, conditioned with an initial state, $x_\xi(t_0) = x_0$, it reads

$$\langle x_\xi(t)\, x_\xi(t+\tau) \rangle_{x_\xi(t_0)=x_0} = \int dx_1\, dx_2\, x_1\, x_2\, p_2(x_2, t+\tau|x_1, t)\, p_2(x_1, t|x_0, t_0).$$

(1.69)

It is known as the autocorrelation function of the stochastic process $x$. It may be expressed through the conditional first moment as

$$\int dx_2 \langle x_\xi(t+\tau) \rangle_{x_\xi(t)=x_1}\, x_1\, p_2(x_1, t|x_0, t_0).$$

(1.70)

If $t_0 \to -\infty$ the autocorrelation function approaches its asymptotic limit:

$$\langle x_\xi(t)\, x_\xi(t+\tau) \rangle_{\text{asy}} = \int dx_1 dx_2\, x_1\, x_2 p_2(x_1, t, x_2, t+\tau).$$

(1.71)

With $\tau = 0$ it equals the second moments $\langle x^2(t) \rangle$. On the other hand, if $\tau$ becomes infinitely large (1.71) factorizes in many cases into

$$\langle x_\xi(t)\, x_\xi(t+\tau) \rangle_{\text{asy}} = \langle x_\xi(t+\tau) \rangle_{\text{asy}} \langle x_\xi(t) \rangle_{\text{asy}}.$$

(1.72)

For stationary processes the asymptotic values do not depend explicitly on time, and the stationary autocorrelation function remains a function of the time difference $\tau$ only:

$$c_{x,x}(\tau) = \int dx_1\, dx_2\, x_1\, x_2\, p_2(x_1, \tau|x_2)\, p(x_2).$$

(1.73)

The stationary process is invariant to a shift in time $t \to t - \tau$, and it follows that $c_{x,x}$ is an even function in time, $c_{x,x}(\tau) = c_{x,x}(-\tau)$.

The autocorrelation function gives an important measure of the stochastic process, the correlation time $\tau_c$. It may be defined in several ways, but most often it is defined as

$$\tau_c = \frac{1}{c_{x,x}(0)} \int_0^\infty dt\, |c_{x,x}(t)|.$$

(1.74)

A generalization of correlation functions to multi-dimensional stochastic processes is straightforward. For simplicity, consider the two-dimensional stochastic process $\{x(t), y(t)\}$. Cross-correlations can be similarly introduced on the basis of the joint probability density $p_{x,y}(x, t; y, t+\tau)$. For the stationary processes it immediately gives $c_{x,y}(\tau) = c_{y,x}(-\tau)$. The absolute value of the correlation function obeys $|c_{x,y}(\tau)|^2 \le \langle x^2 \rangle \langle y^2 \rangle$, which is known as Cauchy–Schwartz inequality. Particularly, it yields for the autocorrelation function $|c_{x,x}| \le \langle x^2 \rangle$.

**Fourier Transform of Correlation Functions.** An important measure is the two-sided Fourier transform of the correlation function:

$$G_{x,x}(\omega) = \int_{-\infty}^{\infty} d\tau\, c_{x,x}(\tau)\, e^{-i\omega\tau} = 2 \int_{0}^{\infty} d\tau\, c_{x,x}(\tau) \cos(\omega\tau), \qquad (1.75)$$

with $G_{x,x} \geq 0$. For multi-dimensional stochastic processes we can introduce the Fourier transform of the cross-correlation function:

$$G_{x,y}(\omega) = \int_{-\infty}^{\infty} d\tau\, c_{x,y}(\tau)\, e^{-i\omega\tau}. \qquad (1.76)$$

Both expressions exist if the correlation functions are absolutely integrable with corresponding inverse transformations. From these the following useful expressions are obtained:

$$G_{x,x}(\omega = 0) = \int_{-\infty}^{\infty} d\tau\, c_{x,x}(\tau),$$

$$c_{x,x}(\tau = 0) = \langle x^2 \rangle = \frac{1}{2\pi} \int_{-\infty}^{\infty} d\omega\, G_{x,x}(\omega). \qquad (1.77)$$

**Derivatives and Integrals.** The derivative of any given realization $x(t)$ of a stochastic process is defined as follows:

$$\dot{x}(t) = \frac{dx(t)}{dt} = \lim_{\varepsilon \to 0} \left( \frac{x(t+\varepsilon) - x(t)}{\varepsilon} \right). \qquad (1.78)$$

However, the existence of the limit can be understood in different senses as well as the existence of the derivative. We will restrict ourselves in the definition of the limit to the mean square and, consequently, the derivative is

$$\lim_{\varepsilon \to 0} \left\langle \left| \frac{x(t+\varepsilon) - x(t)}{\varepsilon} - \dot{x}(t) \right|^2 \right\rangle = 0. \qquad (1.79)$$

The necessary condition for convergence is the existence of the mixed second derivative with respect to $t_1$ and $t_2$ of the correlation function or the existence of the second derivative of $c_{x,x}(\tau)$ for the stationary processes. The autocorrelation function for the derivative of $x(t)$,

$$\langle \dot{x}(t_1)\dot{x}(t_2) \rangle = \frac{\partial^2}{\partial t_1 \partial t_2} \langle x(t_1)x(t_2) \rangle, \qquad (1.80)$$

and the following equation connect the spectral densities of the stationary processes, $G_{\dot{x},\dot{x}}(\omega) = \omega^2 G_{x,x}(\omega)$. A similar expression exists for the cross correlation function $\langle x(t_1)\dot{x}(t_2) \rangle = \frac{\partial}{\partial t_2} \langle x(t_1)x(t_2) \rangle$ and for the cross-spectral density $G_{x,\dot{x}}(\omega) = i\omega\, G_{x,x}(\omega)$.

Integrals are also defined in the mean square limit. Let $T_n$ be a partition in the interval $T = [t_0, t]$ which becomes dense in the limit $n \to \infty$ in $T$. Then the integral over a stochastic process multiplied by a time-dependent deterministic function is the stochastic process

$$I_\xi^f(t, t_0) = \int_{t_0}^t ds\, f(s)\, x_\xi(s) = \lim_{n \to \infty} q.m. \sum_{k=0}^{n-1} f(t_{k,n})\, x_\xi(t_{k,n}) \left( t_{k+1}^{(n)} - t_k^{(n)} \right),$$

(1.81)

with $t_0^{(n)} = t_0$, $t_n^{(n)} = t$ and $t_k^{(n)} \leq t_{k,n} \leq t_{k+1}^{(n)}$. The notation $q.m.$ in (1.81) stands for the mean square limit. The precise location does not play a role as long as $x_\xi$ is piecewise continuous. Sufficient and necessary for the existence of the integral is the convergence of

$$\left\langle \left( I_\xi^f(t, t_0) \right)^2 \right\rangle = \int_{t_0}^t \int_{t_0}^t ds_1 ds_2\, f(s_1)\, f(s_2) \langle x_\xi(s_1) x_\xi(s_2) \rangle.$$

(1.82)

A more rigorous approach replaces $x_\xi(t_{k,n}) \left( t_{k+1}^{(n)} - t_k^{(n)} \right)$ with the increments $\Delta I_{t_k} = I_\xi^{f=1}(t_{k+1}, t_k)$, which can be considered as a stochastic process with defined properties. Such a formulation allows the treatment of stochastic integrals even if the increments are independent and, hence, $I_\xi^{f=1}(t, t_0)$ is nondifferentiable. The meaning of a stochastic integral with smooth $f(t)$ was proven by Ito for the case in which $t_{k,n} = t_k^{(n)}$, which is called an Ito integral [78]

$$I_\xi^f(t, t_0) = \int_{t_0}^t dI_s\, f(s) = \lim_{n \to \infty} q.m. \sum_{k=0}^{n-1} \Delta I_{t_k^{(n)}} f(t_k^{(n)}).$$

(1.83)

Further specification will be necessary if $f$ is random and a function of $\xi$, as in the case of multiplicative noise.

**Fourier Transform of Stochastic Processes. Power Spectrum.** Numerous applications of stochastic integrals are possible by using the spectral decomposition of $x_\xi(t)$. But a spectral analysis which gains its importance from a large success in linear dynamics cannot immediately be formulated for stochastic realizations. Fourier transforms in the common sense do not exist for permanently changing $x_\xi(t)$, because the stochastic integrals do not converge in the quadratic mean; the integral (1.82) diverges.

Formally, the Fourier transforms exist only for averaged values. Nevertheless, often the Fourier transforms can still be found, interpreted as an operational algorithm performed in several steps. For a given smooth realization $x_\xi(t)$ one may take $x_\xi^T(t) = x_\xi(t)$ inside $|t| \leq T$ and outside $x_\xi^T(t) = 0$. Further on, one defines

$$x_\xi^T(\omega) = \int_{-T/2}^{T/2} dt\, x_\xi(t)\, e^{-i\omega t}.$$

(1.84)

The limit $T \to \infty$ of this expression formally gives the spectral decomposition of the process and can be taken in expressions averaged over different realizations, if the corresponding integrals exist.

In particular, we look at the power of $x_\xi(t)$ in the band $[\omega, \omega + \Delta\omega)$ of positive frequencies $\omega \geq 0$. With $\Delta\omega = 2\pi/T$ it is defined by

$$P_\xi^T(\omega) = \frac{2}{T} |x_\xi^T(\omega)|^2, \tag{1.85}$$

which is a random variable even in the limit $T \to \infty$. Averaging over different realizations gives

$$P^T(\omega) = \langle P_\xi^T(\omega) \rangle = \frac{2}{T} \int_{-T/2}^{T/2} dt_1 \int_{-T/2}^{T/2} dt_2 \, \langle x_\xi(t_1) x_\xi(t_2) \rangle e^{-i\omega(t_1 - t_2)}. \tag{1.86}$$

Even if $P^T(\omega)$ exists, the random function $P_\xi^T(\omega)$ does not necessarily converge in the mean square limit if $T \to \infty$. It may still remain in the long time limit a random variable with finite variance at arbitrary frequencies [79].

For (sufficiently, in the wide sense) stationary processes the integrals in (1.86) can be simplified, yielding

$$P^T(\omega) = 2 \int_{-T}^{T} d\tau \left(1 - \frac{|\tau|}{T}\right) c_{xx}(\tau) e^{-i\omega\tau}. \tag{1.87}$$

In the long time limit, $P^T(\omega)$ becomes the average power density called the power spectrum of $x_\xi(t)$ and the introduced band of frequencies $\Delta\omega$ vanishes:

$$P(\omega) = \lim_{T \to \infty} P^T(\omega) = 2 \int_{-\infty}^{\infty} d\tau \, c_{x,x}(\tau) e^{-i\omega\tau}. \tag{1.88}$$

As seen for stationary processes, the power spectrum equals the doubled Fourier transform of the correlation function $G_{x,x}(\omega)$, which is the content of a theorem by Wiener and Khinchin. Later on, we will call $G_{x,x}(\omega)$ simply the power spectrum or spectrum of $x_\xi(t)$, omitting the 2. We emphasize that in the operation (1.85) information about the absolute phase of the process is lost.

### 1.2.3 Noise in Dynamical Systems

**Langevin Equations.** Paul Langevin invented stochastic differential equations by adding a fluctuating force in the equation of motion of a Brownian particle. Langevin assumed a zero average and vanishing correlations between the position of a particle and the random force at equal times. Later on he treated the equations in ensemble and obtained ordinary differential equations for the expectations. In the strong damping limit, after integration, he found diffusive behavior, i.e., a linear increase of the mean square displacement of a particle in time. Based on equal partition of energy, he derived the fluctuation dissipation relation, connecting diffusion with the Stokes friction coefficient, which was determined by Einstein three years earlier, starting from an approach based on the kinetic equation for the probability density of Brownian particles.

Let $x(t) = [x_1(t), \ldots, x_n(t)]$ be the vector of state variables of a DS. In its most general form the Langevin equation can be written as follows:

$$\dot{x}_i(t) = f_i(x_1, \ldots, x_n, t) + g_i(x_1, \ldots, x_n, \xi_1(t) \ldots \xi_m(t), t). \qquad (1.89)$$

There are two distinguished parts in this equation, both generally described by nonlinear functions: the deterministic part $f(x, t) = [f_i(x, t), \ldots, f_n(x, t)]$, and the stochastic part, $g(x, \xi, t) = [g_i(x, \xi, t), \ldots, g_n(x, \xi, t)]$ with property $g_i(x, \xi = 0, t) = 0$. The multi-dimensional stochastic process $\xi(t) = [\xi_1(t), \ldots, \xi_n(t)]$ represents noise forces acting on the system. We will assume throughout the book $\langle \xi_i(t) \rangle = 0$.

In the case of *internal noise*, that is, the system is in thermodynamic equilibrium, the intensities of noise forces are in specific fluctuation–dissipation relations with the counteracting dissipative forces. For example, the well-known Einstein relation connects the intensity of the random force with the Stokes friction coefficient, which is included in the deterministic part, $f(x)$, of (1.89). Similarly, the intensity of noise in electronic circuits is given by the Nyquist theorem, including the resistance as a parameter. The situation is simplified in the case of *external noise*, which originates outside the system: the two parts of the Langevin equations can be considered as independent, and the characteristics of random forces become additional independent parameters of the system [46, 47].

In this book we consider situations in which noise enters linearly into the Langevin equations, which simplifies the consideration significantly:

$$\dot{x}_i(t) = f_i(x, t) + \sum_{j=1}^{m} g_{i,j}(x, t) \, \xi_j(t). \qquad (1.90)$$

Additionally, if for a given $j$ all $g_{i,j} = $ const., we speak about *additive noise*, as the intensity of the action of $\xi_j$ on all $x_i(t)$ is independent of the actual state $x_i(t)$. The opposite case, when $g_{i,j}$ is a function of state variables is called *multiplicative noise*.

**Characterization of Noise.** There are many possible ways to introduce random functions $\xi(t)$ with different properties. Generally one distinguishes between a discrete-valued process, which takes only a discrete set of values, and a continuous-valued process, which is defined on a continuous set. The second feature which characterizes the random function is its dependence on time. Again $\xi(t)$ can vary continuously in time or be defined via jumps at discrete times only.

**Gaussian Processes.** These represent a large subclass of stochastic processes. The joint probability is given by the Gaussian distribution

$$p_n(x_1, t_1, \ldots, x_n, t_n)$$

$$= \frac{1}{\sqrt{(2\pi)^n \operatorname{Det}[\sigma]}} \exp\left[ -\frac{1}{2} \sum_{i,j=1}^{n} \sigma_{i,j}^{-1} (x_i - a_i)(x_j - a_j) \right], \qquad (1.91)$$

with

$$a_i \equiv a(t_i) = \langle x(t_i) \rangle, \qquad (1.92)$$

being the time-dependent mean and

$$\sigma_{i,j} \equiv \sigma(t_i, t_j) = \langle x(t_i) \, x(t_j) \rangle - a_i \, a_j, \qquad (1.93)$$

the time-dependent covariance. Thus, only two moments are necessary to fully define a Gaussian process.

**Wiener Process.** Brownian motion can be described as a Gaussian process. If $W_i$ is the erratic position of the Brownian particle at time $t_i$, its joint distribution depends on a single parameter $D$ as

$$p_n(W_1, t_1, \ldots, W_n, t_n) = \prod_{i=1}^{n} \frac{1}{\sqrt{4\pi D(t_i - t_{i-1})}} \exp\left( -\frac{1}{4D} \frac{(W_i - W_{i-1})^2}{t_i - t_{i-1}} \right),$$
$$(1.94)$$

with $W_0 = 0$ and $t_0 = 0$. This process $W(t)$ is named after Norbert Wiener. It generates nondifferentiable but almost always continuous sample paths. It is obviously Markovian, since the factors with $i \geq 2$ in the product (1.94) are the transition probabilities between two subsequent positions:

$$p\left(W_i, t_i \,|\, W_{i-1}, t_{i-1}\right) = \frac{1}{\sqrt{4\pi D \left(t_i - t_{i-1}\right)}} \exp\left( -\frac{\left(W_i - W_{i-1}\right)^2}{4D \left(t_i - t_{i-1}\right)} \right); \quad (1.95)$$

it is independent of the former history and a solution of the Chapman–Kolmogorov equation (1.61).

This process is nonstationary, also not in the wide sense. With the initial condition $p(W_0, t_0 = 0) = \delta(W_0)$, one finds for $t \geq 0$ the probability density:

$$p_1(W, t) = \int dW_0 \, p\left(W, t \,|\, W_0, t_0\right) \delta(W_0) = \frac{1}{\sqrt{4\pi D t}} \exp\left( -\frac{W^2}{4Dt} \right), \quad (1.96)$$

which depends explicitly on time. The first moment of the Wiener process vanishes, $\langle W(t) \rangle = 0$, but the second moment increases linearly in time:

$$\langle W(t_1) \, W(t_2) \rangle = 2D \min(t_1, t_2), \qquad (1.97)$$

and for $t_1 = t_2 = t$: $\langle W(t)^2 \rangle = 2Dt$. Thus, the introduced parameter $D$ governs the rate of increasing variance. In the case of Brownian motion this rate is just the diffusion coefficient.

An important conclusion can be made for the increments $\Delta W_{t_1,t_2} = W(t_2) - W(t_1)$ ; $t_2 > t_1$. The increments are Gaussian-distributed due to (1.95) with zero mean,

$$\langle \Delta W_{t_1,t_2} \rangle = 0. \tag{1.98}$$

Moreover, the increments satisfy the stationarity condition,

$$\left\langle (\Delta W_{t_1,t_2})^2 \right\rangle = 2D (t_2 - t_1), \tag{1.99}$$

and independence of the increments and different times,

$$\langle \Delta W_{t_1,t_2} \Delta W_{t_2,t_3} \rangle = 0 \quad ; \quad t_3 > t_2 > t_1 . \tag{1.100}$$

Thus, the increments form a stationary Markovian process since (1.95) is an exact solution of the Chapman–Kolmogorov equation, hence $\Delta W_{t_1,t2} = \Delta W_{t_2-t_1}$. In its differential form for $t_{i+1} - t_i \to 0$, this equation may be presented as a diffusive one:

$$\frac{\partial}{\partial t} p(\Delta W, t | \Delta W_0, t_0) = D \frac{\partial^2}{\partial \Delta W^2} p(\Delta W, t | \Delta W_0, t_0). \tag{1.101}$$

**White Gaussian Noise.** The Wiener process can be represented as a continuous sum over subsequent independent increments:

$$W(t) = \sum_{k=0}^{n-1} \Delta W_{t_{k+1}-t_k}, \tag{1.102}$$

with $t_n = t$ and $t_0 = 0$, which is a special case of the Ito integral (1.83). The question which we address now is: What would be the properties of a process $\xi(t)$ if one represents the Wiener process as a stochastic integral over time like (1.81),

$$W(t) = \int_0^t ds\, \xi(s)\,? \tag{1.103}$$

Formally, $\xi(t)$ is the temporal derivative of the Wiener process, $\Delta W/\Delta t$. However, this derivative does not exist in the quadratic limit sense if $\Delta t \to 0$ since the Gaussian distribution of $\Delta W/\Delta t$ diverges. The integral of $\xi(t)$ over time achieves sense only if replaced by the sums over the increments as in (1.83) and (1.102) and using the properties of the increment given above.

Nevertheless, usage of $\xi(t)$ in the Langevin equation is rather common in the physical literature. A treatment within the calculus of wide stationary processes gives reasonable properties for $\xi(t)$. This process is Gaussian with a vanishing average. The correlation functions of the Wiener process (1.97) are obtained if $\xi(t)$ is delta-correlated as $c_{\xi,\xi}(t_1 - t_2) = 2D\,\delta(t_1 - t_2)$. In consequence, the process $\xi(t)$ defined by (1.103) is not correlated at different times, which expresses the independence of the increments. The power spectrum $G_{\xi,\xi}(\omega)$ is constant at all the frequencies: $G_{\xi,\xi}(\omega) = 2D$; this is the reason why $\xi(t)$ is called a white noise.

White noise has no physical foundation, for example, the integrated power over all frequencies is infinitely large. It requires complicated mathematics but there is an enormous advantage. As will be seen, only if the random source in the Langevin is white noise (1.89), the resulting process $x(t)$ is Markovian. This important simplification is lost for all other kinds of nonwhite noise.

A possible physical justification to use white noise is based on the separation of the time scales of random perturbations and the characteristic temporal scales of noiseless dynamics. The limit case with a noise correlation time $\tau_c$, vanishingly small compared to the relaxation time $\tau_r$ of the deterministic system, leads to a process driven by white noise.

The single parameter characterizing Gaussian white noise is its intensity $D$. This value changes if the time is rescaled. Transforming $t' = a\,t$ gives $\xi(t') = \frac{1}{\sqrt{a}}\xi(t)$. For the correlation function at the new time scale one finds $c_{\xi,\xi}(t_1' - t_2') = 2\,(D/a)\,\delta\,(t_1' - t_2')$. Hence the intensity scales with the inverse scaling of time.

Throughout later on we will denote the Gaussian white noise again by $\xi(t)$ but with $\langle \xi(t)\,\xi(t+\tau)\rangle = \delta(\tau)$ and, correspondingly, for the increment of the Wiener process we put $\langle \Delta W_{\Delta t}^2 \rangle = \Delta t$. The intensity of the white Gaussian noise as well as the value of the variance of the increment is presented as a multiplicative factor explicitly in front of $\xi(t)$ or $\Delta W_{\Delta t}$.

**Poissonian Process.** Let $N(t)$ be the number of random events occurring in the time interval $[0,t)$ and $N(0) = 0$. It is clear that such a process is discrete-valued and always increases in time. The probability of finding a state $N$ at time $t$ is given by the Poissonian distribution:

$$p_N(t) = \frac{(\gamma t)^N}{N!}\,\exp\,(-\gamma t), \qquad (1.104)$$

with $t \geq 0$. Then $N(t)$ is called the Poissonian process and is defined only by the $\gamma$. It is, obviously, non-Gaussian. Its first as well as all higher moments are expressed through $\gamma$. We list the mean and the second moment

$$\langle N(t)\rangle = \gamma t\,, \quad \langle\,(N(t) - \langle N(t)\rangle)\,(N(t') - \langle N(t')\rangle)\,\rangle = \gamma\min(t,t').$$
$$(1.105)$$

The Poisson distribution (1.104) is a solution of the master equation

$$\dot{P}_N(t) = -\gamma P_N(t) + \gamma P_{N-1}(t)\,; \ N \geq 1, \quad \dot{P}_0(t) = -\gamma P_0(t)\,, \qquad (1.106)$$

describing a birth process $N \to N+1$ with transition probabilities per unit time $W(N \to N+1) = \gamma$. This master equation is a differential form of the Chapman–Kolmogorov equation for discrete events. It describes the time evolution of the probability function. It is important to note that it is possible to obtain $P_N(t+\Delta t)$ if one knows $P_N(t)$. Therefore the Poisson process $N(t)$ is Markovian.

Like the Wiener process, the Poisson process has independent increments. The increment $\Delta N(t) > 0$ for the step $N_0 \to N_0 + \Delta N(t)$ in the interval

$[t_0, t)$ is independent of the previous history. One finds for the transition probabilities from $\Delta N_0 = 0$ at $t = t_0$ to $\Delta N(t)$ at $t > t_0$ again a Poissonian distribution. But this distribution with $\gamma = $ const. defines now a Markovian stationary process $\Delta N(t)$ with

$$P(\Delta N, t | \Delta N_0 = 0, t_0) = \frac{[\gamma(t - t_0)]^{\Delta N}}{\Delta N!} \exp[-\gamma(t - t_0)], \qquad \Delta N \geq 0.$$
(1.107)

Another problem is devoted to waiting times: How long does a process remain in a certain state $N = N_0$ if it is there at $t_0$? For the Poissonian process the answer is easily given. From (1.105) with $\langle N \rangle = 1$ we immediately conclude that $\langle t \rangle = \gamma^{-1}$. Indeed, the probability that no jump occurs in the time period $[0, \tau]$ is $P_0(\tau)$. Therefore

$$D(\tau) = 1 - P_0(\tau)$$
(1.108)

is the probability that $N_0$ will be left in that time interval. The probability that the jump will take place in the period $[\tau, \tau + \Delta\tau]$ follows

$$dD(\tau) = -\frac{dP_0}{d\tau} d\tau = \gamma \exp(-\gamma\tau) d\tau = w_N(\tau) d\tau.$$
(1.109)

Averaging then yields

$$\langle t \rangle = \int_0^\infty \tau \gamma e^{-\gamma\tau} d\tau = \int_0^\infty \tau w_N(\tau) d\tau = \frac{1}{\gamma}.$$
(1.110)

In (1.109) we have introduced $w_N(\tau)$, which is called the density of the waiting time distribution. It is a powerful concept of describing stochastic processes and is widely used in the literature for various problems [80, 81].

**White Shot Noise.** Like the Wiener process we present the Poissonian process as an integral:

$$N(t) = \int_0^t ds \, \xi_{SN}(s).$$
(1.111)

The integrand is called shot noise, and it cannot be Gaussian. It is a sequence of Dirac $\delta$-functions occurring at random times $t_i$:

$$\xi_{SN}(t) = \sum_i \delta(t - t_i).$$
(1.112)

The times $t_i$ are due to a Poissonian distribution with increasing $i$. Each delta peak corresponds to a jump $\Delta N = 1$ of $N(t)$. It was introduced by W. Schottky to describe the impacts of single electrons at cathodes in vacuum tubes. On average, $\xi_{SN}(t)$ possesses a value different from zero, $\langle \xi_{SN}(t) \rangle = \gamma > 0$. Sometime it is more convenient to consider $\xi_{SN}(t) - \gamma = \eta_{SN}(\tau)$ as a white noise with zero average. Due to the independence of increments $\Delta N$, the sequence of $\delta$-peaks at times $t_i$ should be independent at different times. It is expressed by the moments

$$\langle (\xi_{\text{SN}}(t_1) - \gamma)(\xi_{\text{SN}}(t_2) - \gamma) \rangle = \gamma \delta (t_1 - t_2). \qquad (1.113)$$

Therefore, the Poissonian shot noise, also called Campbell's process, is white noise.

**Colored Noise: Ornstein–Uhlenbeck Process.** As we mentioned previously, white noise is a mathematical abstraction. That is why it is important to study the influence of finite correlation times of random forces on noise-driven dynamics. To avoid the problem of dealing with non-Markovian processes, we can introduce a colored noise source (that is, the noise with finite correlation) using some filtration of white Gaussian noise. In the simplest case the filter is linear (like $RC$ or $RCL$ circuits, the word filter obviously is borrowed from electronics).

A simple low-pass filter is described by the following Langevin equation:

$$\dot{y}(t) = -\frac{1}{\tau_{\text{c}}} y + \frac{\sqrt{2D}}{\tau_{\text{c}}} \xi(t), \qquad (1.114)$$

with zero-mean Gaussian white noise $\xi(t)$ and $\langle \xi(t) \xi(t + \tau) \rangle = \delta(\tau)$. The process $y(t)$ governed by (1.114) is called an Ornstein–Uhlenbeck process. It is easy to check that the Ornstein–Uhlenbeck process is a stationary Gaussian process in the asymptotic limit with the stationary probability density

$$P(y) = \sqrt{\frac{\tau_{\text{c}}}{2\pi D}} \exp \left( -\frac{\tau_{\text{c}}}{2D} y^2 \right), \qquad (1.115)$$

and the autocorrelation function

$$c_{y,y}(\tau) = \frac{D}{\tau_{\text{c}}} e^{-\frac{\tau}{\tau_{\text{c}}}}, \qquad (1.116)$$

and with the power spectrum

$$G_{y,y} = \frac{2D}{1 + \omega^2 \tau_{\text{c}}^2}. \qquad (1.117)$$

The Ornstein–Uhlenbeck process as the noise source has been applied to DS in many publications. It has been shown that it may significantly modify the stationary distributions as well as their temporal behavior from the situation where white noise has been applied [54].

**Colored Noise: The Markovian Dichotomic or the Random Telegraph Process.** The discrete counterpart of the Ornstein–Uhlenbeck process with an exponentially decaying correlation function in (1.116) is the dichotomic Markovian process $I_t$. It is the piece-wise constant sequence of two states $\Delta$ and $\Delta'$. Transitions between these states are determined by the exponential waiting time distributions (1.109) with the average time $\tau_\Delta$ for the transition $\Delta \to \Delta'$ and $\tau'_\Delta$ for $\Delta' \to \Delta$. Then $I_t$ is Markovian and its transition probability obeys the master equation

$$\frac{d}{dt} P(\Delta, t \,|\, I_0, t_0) = -\frac{1}{\tau_\Delta} P(\Delta, t \,|\, I_0, t_0) + \frac{1}{\tau_{\Delta'}} P(\Delta', t \,|\, I_0, t_0), \quad (1.118)$$

$$\frac{d}{dt} P(\Delta', t \,|\, I_0, t_0) = \frac{1}{\tau_\Delta} P(\Delta, t \,|\, I_0, t_0) - \frac{1}{\tau_\Delta} P(\Delta', t \,|\, I_0, t_0), \quad (1.119)$$

with initial state $I_0$. The stationary distribution $P^0(I) = P(I, t \to \infty \,|\, I_0, t_0)$ follows immediately:

$$P^0(\Delta) = \frac{\tau_\Delta}{\tau_{\Delta'} + \tau_\Delta}, \qquad P^0(\Delta') = \frac{\tau_{\Delta'}}{\tau_{\Delta'} + \tau_\Delta}. \quad (1.120)$$

After introducing the correlation time $\tau_c$

$$\frac{1}{\tau_c} = \frac{1}{\tau_\Delta} + \frac{1}{\tau_{\Delta'}}, \quad (1.121)$$

one simply finds the time-dependent solution of (1.118):

$$P(\Delta, t \,|\, \Delta, t = 0) = \frac{\tau_c}{2} \left[ \frac{1}{\tau_\Delta} + \frac{1}{\tau_{\Delta'}} \exp\left(-\frac{t}{\tau_c}\right) \right] \quad (1.122)$$

$$P(\Delta', t \,|\, \Delta, t = 0) = \frac{\tau_c}{2} \frac{1}{\tau_{\Delta'}} \left[ 1 - \exp\left(-\frac{t}{\tau_c}\right) \right]. \quad (1.123)$$

If $I_0$ is initially distributed stationary according to (1.120), we obtain a stationary process with the following mean values and correlation function:

$$\langle I_t \rangle = \frac{\Delta \tau_\Delta + \Delta' \tau_{\Delta'}}{\tau_\Delta + \tau_{\Delta'}}, \quad \langle (I_{t_1} - \langle I_{t_1} \rangle)(I_{t_2} - \langle I_{t_2} \rangle) \rangle = \frac{D}{\tau_c} \exp\left(-\frac{|t_1 - t_2|}{\tau_c}\right), \quad (1.124)$$

with

$$D = \frac{\tau_c^3}{\tau_\Delta \tau_{\Delta'}} (\Delta - \Delta')^2. \quad (1.125)$$

White shot noise $\eta_{SN}(t)$ with zero average is recovered if $\tau_{\Delta'} \to 0$ in the limit $\Delta' \tau_{\Delta'} = -\Delta \tau_\Delta = \text{const}$. Then $I_t$ stays at $\Delta < 0$ for mean times $\tau_\Delta$ interrupted by delta-peaks with weights $|\tau_\Delta \Delta|$. The intensity of this shot noise is

$$D = \tau_\Delta \Delta^2. \quad (1.126)$$

From the shot noise une gets Gaussian white noise in the limit of vanishing weights but with constant intensity $D$. Indeed $\tau_\Delta \to 0$, $\Delta \to -\infty$ with $D = \text{const}$. generates a fast sequences of switchings between positive and negative infinite values with zero weight but having the required correlation function. Higher than second order cumulants scaling with $D\,(\tau_\Delta)^{(n-2)}$ vanish; therefore the limiting process is Gaussian. Following van den Broeck, $\gamma = \Delta \tau_\Delta$ can be defined as a "non-Gaussianity" parameter [82].

**Harmonic Noise and Telegraph Signal.** When solving applied problems one often has to deal with some models of random processes such as noisy harmonic oscillations and a telegraph signal. The first model is used to describe the influence of natural and technical fluctuations on spectral and correlation characteristics of oscillations of Van der Pol type oscillators [77,92,151]. The model of telegraph signal serves to outline statistical properties of impulse random processes, for example, random switchings in a bistable system in the presence of noise (the Kramers problem, noise-induced switchings in the Schmitt trigger, etc. [77,87,98]). Experience of the studies of chaotic oscillations in three-dimensional differential systems shows that the aforementioned models of random processes can be used to describe spectral and correlation properties of a certain class of chaotic systems. As we will demonstrate below, the model of narrow band noise represents sufficiently well correlation characteristics of spiral chaos, while the model of telegraph signal is quite suitable for studying statistical properties of attractors of the switching type, such as attractors in the Lorenz system [35] and in the Chua circuit [83].

We consider the basic characteristics of the above-mentioned models of random processes.

*Narrow band noise* $x(t)$ is a stationary zero-mean random process defined by the relation [77,92,151]:

$$x(t) = R_0[1 + \rho(t)] \cos[\omega_0 t + \phi(t)], \tag{1.127}$$

where $R_0$ and $\omega_0$ are constant (mean) values of the amplitude and frequency of oscillations and $\rho(t)$ and $\phi(t)$ are random functions characterizing the amplitude and phase fluctuations, respectively. The process $\rho(t)$ is considered to be stationary. The narrow band noise model assumes that the amplitude and phase fluctuations are slow functions as compared with $\cos(\omega_0 t)$. The most frequently used simplifying assumptions are as follows: (i) the amplitude and phase fluctuations are statistically independent, and (ii) the phase fluctuations $\phi(t)$ represent a Wiener process:

$$\dot{\phi}(t) = \sqrt{2B}\xi(t), \tag{1.128}$$

where $\xi(t)$ is the normalized Gaussian white noise ($\langle \xi(t) \rangle \equiv 0$, $\langle \xi(t+\tau)\xi(t) \rangle = \delta(\tau)$). The constant $B$ is a phase diffusion coefficient. Under the assumptions made, the autocorrelation function (ACF) of the process (1.127) is given by the expression [77,92,151]

$$\psi(\tau) = \frac{R_0^2}{2}[1 + K_\rho(\tau)] \exp(-B|\tau|) \cos \omega_0 \tau, \tag{1.129}$$

where $K_\rho(\tau)$ is the covariance function of the reduced amplitude fluctuations $\rho(t)$. The Wiener–Khinchin theorem yields a corresponding expression for the spectral power density.

The *generalized telegraph signal* is a process that describes random switchings between two possible states $x(t) \pm a$. Two basic types of telegraph signals – random and quasirandom – can be considered [77, 85]. A random telegraph signal has a Poissonian distribution of switching moments $t_k$. For the stationary case, the mean switching frequency will be constant. For the Poissonian distribution, the impulse duration $\theta$ has the exponential distribution

$$p(\theta) = \nu \exp(-\nu\theta), \qquad \theta \geq 0, \tag{1.130}$$

where $\nu$ is the mean switching frequency. The ACF of such a process can be represented as follows [77, 85]:

$$\psi(\tau) = a^2 \exp(-2\nu|\tau|). \tag{1.131}$$

The other type of tepegraph signal (quasirandom) corresponds to random switchings between the two states $x(t) = \pm a$, which can occur only in discrete time moments $t_n = nT_0 + \delta$, $n = 1, 2, 3, \ldots$, where $T_0 = \text{const}$ and $\delta$ is a random quantity uniformly distributed on the interval $[0, T]$. If the probability of switching events is equal to $1/2$, the ACF of this process decays with time according to the linear law

$$\psi(\tau) = a^2 \left(1 - \frac{|\tau|}{T_0}\right), \quad \text{if } |\tau| < T_0; \quad \psi(\tau) = 0, \quad \text{if } |\tau| \geq T_0. \tag{1.132}$$

### 1.2.4 The Fokker–Planck Equation

**Differential Form of the Chapman–Kolmogorov Equation.** An alternative approach to describe Markovian processes is based on evolution equations for the transition probability density $p(x, t|x_0, t_0)$, which guarantees full knowledge of Markovian dynamics. Let us be concerned with the derivation of differential operators for this transition probability density.

Generally, there are two different evolution problems, as $p(x, t|x_0, t_0)$ depends on two tuples of $x, t$ and $x_0, t_0$ giving adjoint evolution operators for the transition probability density. The first "forward" problem is to study the evolution of the probability density varying the state $x$ at time $t > t_0$, when fixing the initial state $x_0$ at time $t_0$. Similarly, we can pose the problem "backwardly", asking for a variable initial state $x_0, t_0$ with constant $x, t$. The usage of both evolution operators depends on the particular physical situation. The backward problem, for example, arises when studying the time difference $t - t_0$ which trajectories of a stochastic process need to reach a fixed value $x$ from some initial point $x_0$.

The derivation of both differential operators starts with the Chapman–Kolmogorov equation and can be found in many books devoted to stochastic processes (see, for example, [71]). The forward differential form is

$$\frac{\partial p(x, t|x_0, t_0)}{\partial t} = \sum_{n=1}^{\infty} (-1)^n \frac{\partial^n}{\partial x^n} K_n(x, t)\, p(x, t|x_0, t_0), \tag{1.133}$$

where the kinetic coefficients $K_n(x,t)$ are defined as rates of the conditional averages

$$K_n(x,t) = \frac{1}{n!} \lim_{dt \to 0} \frac{1}{dt} \int [x' - x]^n \, p(x', t + dt|x, t) \, dx'$$

$$= \frac{1}{n!} \lim_{dt \to 0} \frac{\langle dx^n \rangle}{dt} \bigg|_{x(t)=x}. \tag{1.134}$$

This equation is known as the Kramers–Moyal expansion of the Chapman–Kolmogorov equation and is valid for any Markovian processes with existing coefficients (1.134). The Kramers–Moyal expansion for the backward problem has the form

$$-\frac{\partial p(x,t|x_0,t_0)}{\partial t_0} = \sum_{n=1}^{\infty} K_n(x_0,t_0) \frac{\partial^n p(x,t|x_0,t_0)}{\partial x_0^n}. \tag{1.135}$$

The evolution operators in (1.133) and (1.135) are adjoint to each other and simplify significantly for so-called *diffusion* processes. For these processes the two coefficients (1.134) with $n = 1$ and $n = 2$ are nonzero only, while $K_n = 0$ for $n > 2$ [72]. In this case (1.133) and (1.135) are called the forward and backward Fokker–Planck equations, respectively. Thus the forward Fokker–Planck equation (FPE) is

$$\frac{\partial p(x,t|x_0,t_0)}{\partial t} = \mathbf{L}_x^F \, p(x,t|x_0,t_0) \tag{1.136}$$

$$= -\frac{\partial}{\partial x} K_1(x,t) \, p(x,t|x_0,t_0) + \frac{\partial^2}{\partial x^2} K_2(x,t) \, p(x,t|x_0,t_0),$$

and, analogously, the backward version is

$$-\frac{\partial p(x,t|x_0,t_0)}{\partial t_0} = \mathbf{L}_{x_0}^B \, p(x,t|x_0,t_0) \tag{1.137}$$

$$= K_1(x_0,t_0) \frac{\partial p(x,t|x_0,t_0)}{\partial x_0} + K_2(x_0,t_0) \frac{\partial^2 p(x,t|x_0,t_0)}{\partial x_0^2}.$$

Generalization to the case of many dynamical variables $x_1, \ldots, x_n$ is straightforward. With $p = p(x_1, \ldots, x_n, t|x_{1,0}, \ldots, x_{n,0}, t_0)$ and corresponding conditioned averages

$$K_1^i(x_1, \ldots, x_n, t) = \lim_{dt \to 0} \frac{\langle dx_i \rangle}{dt} \bigg|_{x_1(t)=x_1, \ldots, x_n(t)=x_n}, \tag{1.138}$$

$$K_2^{i,j}(x_1, \ldots, x_n, t) = \frac{1}{2} \lim_{dt \to 0} \frac{\langle dx_i dx_j \rangle}{dt} \bigg|_{x_1(t)=x_1, \ldots, x_n(t)=x_n}, \tag{1.139}$$

one readily obtains

$$\frac{\partial p}{\partial t} = -\sum_{i=1}^{n} \frac{\partial}{\partial x_i} K_1^i(x_1, \ldots, x_n, t) \, p + \sum_{i,j=1}^{n} \frac{\partial^2}{\partial x_i \partial x_j} K_2^{i,j}(x_1, \ldots, x_n, t) \, p; \tag{1.140}$$

and the case is similar for the backward equation.

**Fokker–Planck Versus Langevin Equations.** The kinetic coefficients can be derived from a corresponding Langevin equation. For a given state variable $x(t)$ we need to determine the increment $dx$ in linear order of $dt$ from the Langevin equation only, since higher contributions with $\mathcal{O}(dt^2)$ vanish in the definitions of the kinetic coefficient. Such a procedure gives the required connection between the Langevin and FPE.

The procedure shown below will explain the increment $dx$ formulated in the sense of an Ito integral (1.83). It means that functions of the stochastic process $x$ defining the increment $dx$ will be given at times $t$. Therefore, and this is the purpose, they will be independent on and, hence, not correlated to the actual increment $dW_{dt}$ of the noise source at time $t$. If additionally the corresponding averages for the particular process under consideration exist, the stochastic Ito integral as the sum of the increments is mathematically justified.

For all $x$ at later times $s > t$ this statement is not true and correlations between $x(s)$ and the increment $dW_{dt}$ at $t$ have to be taken into account. The reason is originated by the discontinuous behavior of the increments $dW_{dt}$ which sample path $W(t)$ is not smooth any time. In result a limit from the sum to an integral is not unique if $dt \to 0$ but depends on the choice of the $s$ where $x(s)$ is taken inside $dt$. If $s = t$, as outlined, there are no contributions from the correlations, oppositely at $s = t+dt$ the process is already correlated which results in additional items to the increment $dx$ in $\mathcal{O}(dt)$.

Hence every other choice of $s$ different from $s = t$ needs further explanation. We sketch how an arbitrary $s$ can be traced back to the Ito-formulation. Let us start from a stochastic differential equation

$$\dot{x} = f(x,t) + \sqrt{2D}\,g(x,t)\,\xi(t) \tag{1.141}$$

for an one-dimensional variable $x(t)$ and white Gaussian noise $\xi(t)$ with $\langle\xi(t)\rangle = 0$ and $\langle\xi(t)\,\xi(t+\tau)\rangle = \delta(\tau)$. The increment for a small time interval $dt$ is

$$dx_t = x(t+dt) - x(t) = f(x(s),s)dt + g(x(s),s)\sqrt{2D}\,dW_{dt}, \tag{1.142}$$

where $dW_{dt}$ is the increment of the corresponding Wiener process at time $t$ for the interval $dt$ with $\langle dW_{dt}^2\rangle = dt$.

The explicit time dependence of functions $f$ and $g$ can be treated as a Riemann integral. However, the increment $dx_t$ depends on a particular selection of the grid inside the time interval $[t, t + dt]$ for the sample path $x(s)$. It is due to the noncontinuous behavior of white noise $\xi(t)$ and the resulting nondifferentiable Wiener process with increments $dW_t$. Let $q$ be a constant inside $[0, 1]$ and we put $s = q(t+dt)+(1-q)t$. Insertion into (1.142) and expansion of $f$ and $g$ gives

$$dx_t = f_t\,dt + q\,\frac{\partial f}{\partial x_t}\,dx_t\,dt + g_t\,\sqrt{2D}\,dW_{dt} + q\,\frac{\partial g}{\partial x_t}\,dx_t\,\sqrt{2D}\,dW_{dt}, \tag{1.143}$$

with all functions taken at time $t$. Nevertheless, the increment $dx_t$ still depends on the random $dW_{dt}$. After re-insertion of $dx_t$, into the right hand side of (1.143) and omitting third-order deviations we obtain

$$dx_t = f_t\, dt + g_t\,\sqrt{2D}\, dW_{dt} + q\frac{\partial g}{\partial x_t}\, g_t\, 2\, D\, dW_{dt}^2$$
$$+ \left(q\frac{\partial f}{\partial x_t}\, g_t\, dt + f_t\, g_t + q\frac{\partial g}{\partial x_t}\, f_t\right)\sqrt{2D}\, dW_{dt}\, dt. \quad (1.144)$$

Stratonovich [84] has shown that additionally to the two first items the term with $2\,D d W_{dt}^2$ contributes. It is a higher-order term which modifies $dx_t$. In the short time limit $dt \to 0$ it can be replaced by its expectation $2D\, dt$ since the mean and the second moment of deviations of the stochastic $2\, D\, dW_{dt}^2$ from $2D\, dt$ vanish. Both moments taken of the remaining term in (1.144) in the limit $dt \to 0$ are zero as well. We obtain in $\mathcal{O}(dt)$

$$dx_t = f(x,t)\, dt + 2D\, dt\, q\frac{\partial g(x,t)}{\partial x}\, g(x,t) + \sqrt{2D}\, g(x,t)\, dW_{dt}. \quad (1.145)$$

Several important statements can be made from this expression. First, the increment is independent of the process $x(t')$ at former times $t' < t$. Thus, a white-noise-driven process is Markovian. Second, the increment is linear with respect to $dW_{dt}$ and therefore erratic in its time evolution as $dW_{dt}$. For example, one may ask for the number of intersections of $x(t)$ with a given boundary value $x_0$ if the process is in the vicinity of $x_0$. In the limit $dt \to 0$, the linear dependence on $dW_{dt}$ guarantees unbounded growth of the number of intersections.

The increment (1.145) is a Ito-formulation of the increment $dx$. Summing up the increments between $0 \to t$ gives the stochastic Ito-integral as defined in (1.83) with the integration grid as shown [85]:

$$x(t) = x(0) + \int_0^t ds\left(f(x,s) + 2D\, q\frac{\partial g(x,s)}{\partial x}\, g(x,s)\right) + \sqrt{2D}\int_0^t dW_s\, g(x,s).$$
$$(1.146)$$

Since $W_t$ is continuous, we obtain continuity also for $x(t)$.

The kinetic coefficients of the FPE that are conditional moments of $x(t)$ per unit time can be obtained immediately from (1.146) as

$$K_1(x,t) = \frac{d\langle x\rangle}{dt} = f(x,t) + 2\, D\, q\frac{\partial g(x,t)}{\partial x}\, g(x,t) \quad (1.147)$$

and

$$K_2(x,t) = \frac{1}{2}\frac{\langle dx^2\rangle}{dt} = D\, g^2(x,t). \quad (1.148)$$

The higher-order coefficient vanishes due to the properties of the Wiener process (1.100) [84]. Hence, processes defined by the Langevin equation (1.89) with Gaussian white noise are Markovian diffusion processes.

Thus, the FPE for the transition probability density $p_2 = p_2(x, t | x_0, t_0)$ corresponding to the Langevin equation (1.141) reads

$$\frac{\partial}{\partial t} p_2 = -\frac{\partial}{\partial x} \left( f(x,t) + 2 D q \frac{\partial g(x,t)}{\partial x} g(x,t) \right) p_2 + D \frac{\partial^2}{\partial x^2} g^2(x,t) p_2.$$

(1.149)

In the same way we can obtain kinetic coefficients in the multi-dimensional case,

$$\dot{x}_i = f_i(x_1, \ldots, x_n) + \sum_{j=1}^{m} g_{i,j}(x_1, \ldots, x_n) \xi_j(t),$$

(1.150)

with $m$ white noise sources vanishing in the mean and correlated as

$$\langle \xi_i(t) \xi_j(t + \tau) \rangle = 2 D_{i,j} \delta(\tau).$$

(1.151)

From expressions of the conditioned moments for the increments $dx_i$ one obtains

$$K_1^i = f_i + 2q \sum_{j,k,l} D_{j,l} \frac{\partial g_{i,j}}{\partial x_k} g_{k,l},$$

(1.152)

$$K_2^{i,j} = \sum_{k,l} D_{k,l} \, g_{i,k} \, g_{j,l}.$$

(1.153)

Again insertion of these kinetic coefficients into (1.140) gives the forward time evolution operator which fully defines the stochastic dynamics of the higher dimensional process.

This type of evolution operator with a drift and a diffusion term was first formulated for the motion of a Brownian particle in the phase space by Fokker, and later the description was supplemented by Planck. In the case of a single variable first Rayleigh and later Smoluchowski made pioneering contributions. Klein and Kramers formulated operators for Brownian particles affected by external forces. A first mathematical foundation goes back to Kolmogorov. All the listed names can be found as a label for (1.140) for different cases; a collection of early manuscripts can be found in [86] and a distinguished description in [72, 87]. Later on, we will use FPE (Fokker–Planck equation) throughout to abbreviate for the evolution operator for the Markovian diffusion processes.

We point out that the last term in (1.146) is always defined as the Ito integral (1.83). But so far the relation between the kinetic coefficients and the Langevin equation in the case of multiplicative noise depends on the choice of the grid, i.e., on the constant $q$. There is no universal approach to selecting this parameter. It depends how the limit to white noise is taken compared to $dt \to 0$.

Next we will see that $q = 1/2$ if the white noise is assumed to be the small time limit of colored noise with vanishing temporal correlations. It corresponds to the Stratonovich calculus for treating stochastic Ito-integrals.

This choice also possesses the advantage that the relation between the conditional moments and the Langevin equation is invariant with respect to a variable transformation [66,85]. For all other values of $q$ the transformation rules have to be modified, even for the case $q = 0$. Recently Klimontovich has shown [50] that the case of $q = 1$ provides an agreement with results from statistical mechanics for the equilibrium distribution. In mathematical literature the preference is given to the case $q = 0$, which is called the Ito calculus [66, 70]. In this case the increment $dx_t$ relies on the random impact of $dW_t$ at $t$ only. It can be justified if the continuous $x(t)$ stands for a discrete-valued process and the white noise limit is taken prior the continuous approximation.

Later on we use the Stratonovich calculus with $q = 1/2$.

**FPE from Dichotomic Processes.** Let us consider a DS driven by Markovian dichotomic noise $I_t$, which is a symmetric two-state process, $I_t = \pm\Delta$. Transitions between both states occur with the mean time $1/\gamma$. The Langevin equation in the one-dimensional case is

$$\dot{x} = f(x) + g(x) I_t. \tag{1.154}$$

As long as $I_t = $ const. in one of the possible states, the kinetic coefficients are

$$K_1(x, \pm\Delta) = f(x) \pm g(x) \Delta, \quad K_2 = 0. \tag{1.155}$$

Inclusion of transitions in $I$ leads to the balance of probability $p_\Delta(x)$ and, respectively, $p_{-\Delta}(x)$

$$\frac{\partial}{\partial t} p_{\pm\Delta}(x) = -\frac{\partial}{\partial x} \left( f(x) \pm \Delta g(x) \right) p_{\pm\Delta}(x) + \gamma p_{\mp\Delta}(x) - \gamma p_{\pm\Delta}(x). \tag{1.156}$$

Since the two equations are linear in $p_{\pm\Delta}$, one can derive equations for the summed probability $p(x,t) = p_\Delta(x,t) + p_{-\Delta}(x,t)$:

$$\frac{\partial}{\partial t} p(x,t) = -\frac{\partial}{\partial x} \left( f p(x,t) + \Delta g(x) Q(x,t) \right), \tag{1.157}$$

and the difference of both $Q(x,t) = p_\Delta(x,t) - p_{-\Delta}(x,t)$. The solution giving $Q$ with respect to $p$ was obtained in [44, 88]:

$$Q(t) = -\Delta \int_{-\infty}^{t} ds \exp\left[-\left(2\gamma + \frac{\partial}{\partial x} f\right)(t - s)\right] \frac{\partial}{\partial x} g(x) p(x, s). \tag{1.158}$$

Therein $Q(-\infty) = 0$ was selected and the derivative in the exponential acts also on $g(x)$ and $p(x, s)$. It finishes the description if inserted into (1.157). In the Gaussian white noise limit (see 1.126) the exponential function in (1.158) becomes a $\delta$-function, and the evolution operator for the Markovian process becomes

$$\frac{\partial}{\partial t} p(x,t) = -\frac{\partial}{\partial x} f(x) p(x,t) + D \frac{\partial}{\partial x} \left( g(x) \frac{\partial}{\partial x} g(x) p(x,t) \right), \quad (1.159)$$

with $D = \Delta^2/(2\gamma)$.

From this equation we obtain the grid point for the considered limit. To obtain Gaussian white noise from the limit of a colored noise, we have to take $q = 1/2$ [44, 66], which gives agreement between (1.159) and (1.149).

The stationary probability density for the problem (1.154) can be easily obtained by taking $\partial p/\partial t = 0$ in (1.156):

$$p^s(x) = N \frac{g(x)}{\Delta^2 g^2(x) - f^2(x)} \exp \left( -2\gamma \int^x dx' \frac{f}{f^2 - \Delta^2 g^2} \right). \quad (1.160)$$

**One-Dimensional Markovian Diffusion Systems.** Although the FPE is linear with respect to transition probability density, the nonlinearity of a DS is reflected in the dependence of kinetic coefficients on the state variables. This complicates the analytical treatment in most cases. In fact, only few model systems allow exact analytical solution of the FPE. Even in the one-dimensional case, an exact analytical time-dependent solution is only available for linear systems.

The main information one can extract from the FPE is the stationary probability density $p^s(x)$ by setting $\partial p_2/\partial t = 0$. For this purpose it will be of advantage to introduce the probability flux or current

$$j(x,t) = f(x) p(x,t|x_0,t_0) - D g(x) \frac{\partial}{\partial x} (g(x) p(x,t|x_0,t_0)) \quad (1.161)$$

which presents the FPE as a continuity equation. In the one-dimensional case and for additive noise, if $g = 1$ is in (1.159), a first integration gives a constant flux of probability

$$j^s = f(x) p^s - D \frac{\partial}{\partial x} p^s(x) = \text{const.} \quad (1.162)$$

For the highly interesting case with $j^s \neq 0$ we refer the reader to Sections 1.2.6, 1.3.4 and Chapter 3.4. Here we consider the case that boundary conditions allow to put $j^s = 0$ which is valid if probability and derivatives vanish at some position (for example, if $x \to \infty$) or in case of periodic systems without bias. Then the stationary density is easily obtained and reads

$$p^s(x) = N \exp \left( -\frac{1}{D} U(x) \right). \quad (1.163)$$

In this Boltzmann-like distribution $U(x) = -\int^x dx' f(x')$ is called the potential, and $N$ is the normalization constant,

$$N^{-1} = \int_X dx' \exp \left( -\frac{1}{D} U(x') \right). \quad (1.164)$$

with $X$ being the set of possible values. The extrema $x_e$ of the stationary density given by $\partial p^s / \partial x = 0$ converge with the states of equilibrium of the corresponding deterministic system, given by the roots of the equation $f(x_0) = 0$. Stable states with $f'(x_0) < 0$ correspond to the maxima of $p^s$, and unstable ones with $f'(x_0) > 0$ to the minima, respectively.

The Boltzmann-like structure of the solution

$$p^s(x_1, \ldots, x_n) = N \exp\left(-\frac{1}{D} U(x_1, \ldots, x_n)\right). \tag{1.165}$$

is also obtained in higher dimensions for the class of so called gradient dynamics with symmetric noise which also allows solutions with vanishing probability flow. That are stochastic DS as

$$\dot{x}_i = -\frac{\partial}{\partial x_i} U(x_1, \ldots, x_n) + \sqrt{2D}\, \xi_i(t), \tag{1.166}$$

where all Gaussian white noise sources are independent. Again fixed points correspond to extremal states of the probability. The large class of overdamped reaction diffusion systems approximated in a box description under additive noise belongs to this class. Other situations which are solved by vanishing probability flows were presented in [76]. There a condition of integrability which connects the kinetic coefficients (1.138) and (1.139) was derived for this case.

Modification of the correspondence between the state of equilibrium and extrema of the stationary probability density arises in the case of multiplicative noise. With $q = 1/2$ the stationary solution also in the case that the flow $j^s$ disappears yields

$$p^s(x) = N \frac{1}{|g(x)|} \exp\left(\frac{1}{D} \int^x dx' \frac{f(x')}{g^2(x')}\right). \tag{1.167}$$

This solution is also the Gaussian white noise limit of (1.160). The extrema $x_e$ of $p^s(x)$ now depend on the considered multiplicative part, $g(x)$, obeying [89]

$$h(x_e, D) = f(x_e) - D\, g(x_e) \frac{\partial g(x_e)}{\partial x_e} = 0. \tag{1.168}$$

Thus, the noise intensity $D$ appears to be a new parameter of the system. The stationary probability density exhibits maxima for $h'(x_e) < 0$ and minima for $h'(x_e) > 0$. Regions in the parameter space with $h'(x_e) = 0$ define structurally unstable situations. They are called $P$-bifurcations if the sign of $f'(x_0)$ changes in the parameter space in the vicinity of $x_e$ [70]. As a result of $P$-bifurcations the correspondence between attractors of the deterministic dynamics and maxima of the stationary density is lost. Horsthemke and Lefever [44] introduced the notion of *noise-induced transitions* for the appearance of maxima after a $P$-bifurcation by variation of noise intensity $D$. In their book [44], several examples and models were investigated and expressions for the critical noise intensities $D_c$ were derived. Therein the concepts of noise-induced states are generalized to situations with dichotomic noise (see (1.156) and when an Ornstein–Uhlenbeck process (1.114) replaces white noise in the Langevin equation [54, 100].

It is obvious from (1.168) that the noise-induced states appear in the neighborhood of bifurcations of the deterministic dynamics. If $\Delta a = a - a_c$ is the distance from a bifurcation point of $\dot{x} = f(x, a)$ the critical intensity is $D_c \propto \Delta a$. Hence, with assumed small $D$ the appearance of noise-induced maxima is generally restricted to systems in the vicinity of critical points of the deterministic dynamics.

The concept of noise-induced transitions was criticized because the stationary mean $\langle x \rangle^s = x_m$ as a solution of

$$K_1(x_m) = f(x_m) + D\,g(x_m)\,\frac{\partial g(x_m)}{\partial x_m} = 0 \qquad (1.169)$$

differs in its localization from the roots of (1.168). The influence of noise acts oppositely as expressed by the different size of the second term in (1.169). Thus the stationary first moment does not reflect a noise-induced transition defined by the maxima of the stationary density. Additionally, there are a few other measures for characterizing the influence of multiplicative noise, as, for example, higher moments or stochastic Lyapunov exponents [70].

### 1.2.5 Stochastic Oscillators

In this section we will study two-dimensional oscillatory systems driven by additive noise. Typical phenomena taking place in these systems are bi- and multi-stability, excitability and self-sustained oscillations [90]. We intend to summarize the principal effects of additive noise on these phenomena by looking at minimal models. These models are of high relevance in mechanics, electronics, chemistry and biology. Mainly we discuss the stationary solutions of the FPE, while dynamical properties will be discussed in the next chapter and throughout the book.

We start with a mechanical system, described by the following Langevin equations:

$$\dot{x} = v\,, \quad \dot{v} = -\gamma(x, v)\,v - \frac{\partial}{\partial x}U(x) + \sqrt{2D}\xi(t), \qquad (1.170)$$

where $x$ and $v$ are the position and the velocity of a particle, respectively, which moves in the potential $U(x)$ with the friction $\gamma(x, v)$. In the case, $\gamma = $ const. we speak about damped conservative oscillators with a nonlinear potential. The second kind of mechanical oscillator assumes the damping term to be nonlinear with possible zeros and refers to dissipative nonlinear systems, with $U$ being a harmonic potential, in most cases [50, 90]. The FPE for the transition probability density $p_2 = p_2(x, v, t|x_0, v_0, t_0)$ in the form of a Boltzmann equation is

$$\frac{\partial}{\partial t} p_2 + v \frac{\partial}{\partial x} p_2 - \frac{\partial U}{\partial x} \frac{\partial}{\partial v} p_2 = \frac{\partial}{\partial v} \gamma(x, v) \, v \, p_2 + D \frac{\partial^2}{\partial v^2} p_2. \qquad (1.171)$$

In the stationary limit $p_2$ tends to its stationary density, $p_2 \to p^s(x, v)$, with $\partial p^s / \partial t = 0$ and does not depend on the initial state.

**Amplitude and Phase Description.** In the case of harmonic potential and weak dissipative forces, a successful analytical approach was developed by Bogolyubov, Krylov and Mitropolsky (BKM) [91]. They proposed to search for solutions of the deterministic problem by transforming to an amplitude and phase description. In the regime of small dissipation one may assume a slow variation of the amplitude $A$ of the oscillations and a shift $\varphi$ of the phase $\Phi = \Phi_0 + \varphi$ added to the fast motion $\Phi_0 = \omega_0 t$, where $\omega_0$ is the natural frequency of the oscillator.

This approach can be applied to many stochastic oscillators [92–97]. Let us consider the harmonic potential $U(x) = \omega_0^2 x^2 / 2$ and the following transformation of variables:

$$x = A \cos(\omega_0 t + \varphi), \quad v = -\omega_0 A \sin(\omega_0 t + \varphi). \qquad (1.172)$$

The Langevin equations for the amplitude $A$ and the phase shift $\varphi$ are

$$\dot{A} = \frac{\sin(\omega_0 t + \varphi)}{\omega_0} \gamma(x, v) \, v + y_A, \quad \dot{\varphi} = \frac{\cos(\omega_0 t + \varphi)}{A \omega_0} \gamma(x, v) \, v + y_\varphi, \qquad (1.173)$$

where $y_A(t)$ and $y_\varphi(t)$ are new noise sources:

$$y_A = -\frac{\sin(\omega_0 t + \varphi)}{\omega_0} \sqrt{2D}\xi(t), \quad y_\varphi = -\frac{\cos(\omega_0 t + \varphi)}{A \omega_0} \sqrt{2D}\xi(t). \qquad (1.174)$$

The first term in both equations in (1.173) can be treated by averaging over one period $T = 2\pi/\omega_0$ under the assumption of constant $A$ and $\varphi$. This procedure gives the first-order expansions, $f_A(a, \varphi)$ and $f_\varphi(a, \varphi)$, of the BKM theory. The noise in (1.173) is multiplicative. In the first kinetic coefficient (1.152) one obtains from the noise

$$K_1^A \propto D \frac{\cos^2(\omega_0 t + \varphi)}{A \omega_0^2}, \quad K_1^\varphi \propto -D \frac{\cos(\omega_0 t + \varphi) \sin(\omega_0 t + \varphi)}{A^2 \omega_0^2}. \qquad (1.175)$$

After averaging over one period only one nonvanishing contribution $D/(2A\omega_0^2)$ remains in $K_1^A$. The second moments can be calculated following (1.153) and after averaging over one period the cross-correlations vanish, i.e., $K_2^{A,\varphi} = 0$. The noise sources of the amplitude and the phase shift are therefore independent Gaussian white noises with the intensities

$$K_2^{A,A} = \frac{D}{2\omega_0^2}, \qquad K_2^{\varphi,\varphi} = \frac{D}{2\omega_0^2 A^2}. \tag{1.176}$$

Thus, the FPE in the amplitude–phase variables follows

$$\frac{\partial}{\partial t} p_2 = -\frac{\partial}{\partial A}\left(f_A + \frac{D}{2 A \omega_0^2}\right) p_2 - \frac{\partial}{\partial \varphi} f_\varphi \, p_2 + \frac{D}{2\omega_0^2}\frac{\partial^2 p_2}{\partial A^2} + \frac{D}{2\omega_0^2 A^2}\frac{\partial^2 p_2}{\partial \varphi^2} \tag{1.177}$$

for the transition probability density $p_2(A, \varphi, t | A_0, \varphi_0, t_0)$. The Rayleigh distribution solves in the stationary case this equation with $\gamma = \text{const.}$ and a harmonic potential with frequency $\omega_0$

$$p^s(A, \varphi) = \frac{1}{2\pi}\frac{A}{\sigma^2}\exp\left(-\frac{A^2}{2\sigma^2}\right) \tag{1.178}$$

and $\sigma^2 = D/\omega_0^2\gamma$.

**Energy Representation.** In mechanics integrals of motion are widely used, one being the mechanical energy. Therefore, another important representation of the FPE uses the energy as a variable [98]. This representation is preferable for the cases of slow energy variations [72].

Let the friction be linear, that is $\gamma = \text{const.}$ Then the velocity can be replaced by the energy as

$$v(x, E) = \pm \sqrt{2[E - U(x)]}, \tag{1.179}$$

using the definition of mechanical energy. As a result we obtain the Langevin equations

$$\dot{x} = v(x, E), \quad \dot{E} = -\gamma v^2(x, E) + \sqrt{2D}\, v(x, E)\, \xi(t), \tag{1.180}$$

with the insertion of $v$ from (1.179). The equation for the energy contains multiplicative noise, $y_{E,x} = \sqrt{2D}\, v(x, E)\, \xi(t)$, being the source of fluctuations in this representation. With the Stratonovich interpretation we find the kinetic coefficients to be

$$K_1^x = v(x, E), \quad K_1^E = -\gamma v^2(x, E) + D, \quad K_2^{E,E} = D v^2(x, E). \tag{1.181}$$

Thus, the FPE in a coordinate and energy representation for the transition probability density $p_2 = p_2(x, E, t | x_0, E_0, t_0)$ is

$$\frac{\partial p_2}{\partial t} = -\frac{\partial}{\partial x} v(x, E)\, p_2 + \frac{\partial}{\partial E}\left(\gamma v^2(x, E) - D\right) p_2 + D\frac{\partial^2}{\partial E^2} v^2(x, E) p_2. \tag{1.182}$$

Its stationary solution reads

$$p^s(x, E) = \frac{N}{|v(x, E)|}\exp\left(-\frac{\gamma E}{D}\right). \tag{1.183}$$

Integration over $x$ in regions where $E \geq U(x)$ removes the dependence of the stationary probability density on the coordinate:

$$p^s(E) \propto T(E) \exp\left(-\frac{\gamma E}{D}\right), \qquad (1.184)$$

with $T(E)$ being the period of a closed trajectory for a given energy $E$. In particular, for a Hamilton function containing $d$ quadratic terms, the stationary probability density is

$$p^s(E) \propto E^{\frac{d}{2}-1} \exp\left(-\frac{\gamma E}{D}\right). \qquad (1.185)$$

Both latter formulas have to be supplemented by a normalization.

**Stochastic Bistable Oscillator.** One of the best investigated nonlinear stochastic DS is the bistable oscillator [87]. It is a damped conservative oscillator with two stable equilibria. Without noise its equations of motion are

$$\frac{dx}{dt} = v, \qquad m\frac{dv}{dt} = -\gamma v - \frac{\partial U}{\partial x}, \qquad (1.186)$$

where $U(x)$ is the potential of a bistable external force. Without loss of generality we consider the symmetric double-well potential

$$U(x) = -\frac{a}{2}x^2 + \frac{b}{4}x^4, \qquad (1.187)$$

which gives the force

$$f(x) = -\frac{\partial U(x)}{\partial x} = ax - bx^3. \qquad (1.188)$$

The simple analysis yields two stable equilibria at

$$v_0 = 0, \qquad x_{1,3} = \pm\sqrt{\frac{a}{b}}, \qquad a, b > 0; \qquad (1.189)$$

these basins of attraction are divided by two separatrices merging at the saddle point

$$v_0 = 0, \qquad x_2 = 0. \qquad (1.190)$$

For a low damping if $\gamma < 8a$, the stable equilibria possess complex eigenvalues and thus are stable foci. They transform into centers if $\gamma \to 0$.

In the overdamped case $\gamma \to \infty$ or for light particles $m \to 0$, the trajectories quickly approach the $v = 0$ axes. In such a case it is sufficient to eliminate $v(t)$ and to restrict the study to the dynamics of the coordinate $x(t)$ only:

$$\frac{dx}{dt} = \frac{a}{\gamma}x - \frac{b}{\gamma}x^3. \qquad (1.191)$$

The nonlinearity is conservative and part of the external force. The mechanical system dissipates its energy permanently. A derivation of the mechanical energy with respect to time gives

$$\frac{\mathrm{d}}{\mathrm{d}t} E = \frac{\mathrm{d}}{\mathrm{d}t} \left( \frac{m}{2} v^2 + U(x) \right) = -\gamma v^2 \leq 0, \tag{1.192}$$

which vanishes at $v_0 = 0$ in one of the stable equilibria (if it does not start from the saddle point).

The bistable oscillator is one of the standard examples for describing complex behavior in low-dimensional systems. Results which will be obtained later on can be applied in a good approximation for bistable nonequilibrium systems such as electronic flip-flop devices, bistable optical systems, bistable nonlinear semiconductors and bistable reaction schemes [99, 100].

We add white Gaussian noise to the velocity dynamics as a random force:

$$\frac{\mathrm{d}x}{\mathrm{d}t} = v, \quad m\frac{\mathrm{d}v}{\mathrm{d}t} = -\gamma v - \frac{\partial U(x)}{\partial x} + \sqrt{2D}\,\xi(t). \tag{1.193}$$

The corresponding FPE, first formulated for this problem by Kramers and Klein, is as follows:

$$\frac{\partial}{\partial t} p_2(x, v, t | x_0, v_0, t_0) + v\frac{\partial p_2}{\partial x} - \frac{1}{m}\frac{\partial U(x)}{\partial x}\frac{\partial p_2}{\partial v} = \frac{\gamma}{m}\frac{\partial}{\partial v} v\, p_2 + \frac{D}{m^2}\frac{\partial^2 p_2}{\partial v^2}. \tag{1.194}$$

The stationary probability density $p^s(x, v)$ describes the long-term behavior after the stochastic system has equilibrated over the whole phase space and can be found easily. Assuming that $p^s = p^s\big(E(x, v)\big)$ gives the unique solution

$$p^s(x, v) = N^{-1} \exp\left( -\frac{\gamma m}{2D} v^2 - \frac{\gamma U(x)}{D} \right). \tag{1.195}$$

It is the Maxwell–Boltzmann distribution which is obtained in the equilibrium if the noise relates to the temperature as $D = \gamma k_B T$ by the fluctuation dissipation theorem.

We mention that the peaks of the probability density coincide with the stable fixed points, whereas the point $x_2 = v_0 = 0$ is shaped like a saddle, density decreasing in both directions of velocity and, oppositely, increasing in the coordinate. But the random excitations make the separatrice permeable. The trajectories are able to perform transitions from one potential well to another and back. Thus we can speak about noise-induced oscillations between the two attractors.

A first insight into the transition phenomena can be achieved by looking at the ratio of the probabilities of being in the lowest state, at the saddle point $x_2$, to the probability of being in one of the potential minima. It yields the Arrhenius law, determined by the activation energy $\Delta U = U(x = x_2) - U(x = x_1) > 0$,

$$\frac{p^s(x = x_2, v = v_0)}{p^s(x = x_1, v = v_0)} = \exp\left(-\frac{\gamma \Delta U}{D}\right). \tag{1.196}$$

It contributes to all characteristic values describing the transition behavior combined with a time scale.

**Stochastic Self-sustained Oscillations.** Noise in damped harmonic oscillators pumps energy into the system and enhances damped oscillations, with the mean energy being proportional to the noise intensity $D$. The second and higher moments of energy increase with powers of $D$. Thereby the relative energy fluctuations given by the variance always remain equal to 1. The stationary probability density in the velocity and coordinate representation is Gaussian.

This picture changes in the case of oscillators which possess a limit cycle solution. As an example we consider the stochastic Van der Pol oscillator:

$$\dot{x} = v, \quad \dot{v} = \varepsilon v (\alpha - \beta x^2 - \delta v^2) - \omega_0^2 x + \sqrt{2D}\, \xi(t), \tag{1.197}$$

with $\beta; \delta = \mathcal{O}(1) > 0$. Two different scenarios to excite a limit cycle are known, which are hard and soft excitations. A hard excitation occurs if $\varepsilon$ passes zero with $\alpha = 1$. The oscillator exhibits a limit cycle with amplitude $\propto \mathcal{O}(1)$. In contrast, with $\varepsilon = 1$ and $\alpha$ vanishing a soft generation of a limit cycle with amplitude $\propto \sqrt{\alpha}$ takes place. In both cases for sufficiently small distances from the bifurcation point, i.e., $\varepsilon, \alpha \ll 1$, respectively, the system oscillates with a constant frequency, $\omega_0$.

If the noise is also assumed to be a small perturbation the BKM method can be applied, giving for slowly varying amplitude and phase

$$\dot{A} = \varepsilon \frac{A}{2}\left(\alpha - b A^2\right) + y_A, \quad \dot{\varphi} = y_\varphi, \tag{1.198}$$

where the noise sources for the amplitude $y_A$ and phase $y_\varphi$ are given by (1.174). With $b = (\beta + 3\delta\omega_0^2)/4$, where the noise sources for amplitude and phase are given by (1.174). Solving the corresponding stationary FPE we obtain,

$$p^s(A, \varphi) = N A \exp\left[\frac{2\varepsilon\omega_0^2}{D}\left(\frac{\alpha}{2} A^2 - \frac{b}{4} A^4\right)\right]. \tag{1.199}$$

Beyond the Hopf bifurcation the stationary density exhibits in the phase space a crater-like distribution, where the maximal values correspond to the attracting limit cycle (see Fig. 1.25).

The density (1.199) possesses a maximum for $\alpha, \varepsilon > 0$. On the other hand, noise only enhances a diffusive motion of the phase near the bifurcation point. We note that in the case of canonical dissipative systems, where $\gamma(x, v) = \gamma(E)$, if $\beta = \delta\omega_0^2$, the presented stationary density (1.199) is an exact solution of the original FPE transformed to amplitude and phase variables.

It is interesting to look at the energy representation. The probability density can be easily transformed to the energy representation since $\omega_0$ is constant for arbitrary values of energies. Hence with

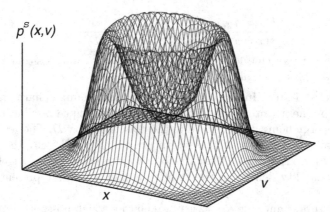

$p^s(x,v)$

$x$     $v$

**Fig. 1.25.** Stationary probability density $p^s(x, v)$ of a stochastic self-sustained oscillator (1.197)

$$p_2(E, t|E_0, t_0) = \int_0^{2\pi} d\varphi \int_0^\infty dA\, \delta\left(E - \frac{A^2 \omega_0^2}{2}\right) p_2(A, \varphi|A_0, \varphi_0), \quad (1.200)$$

the FPE is

$$\frac{\partial}{\partial t} p_2(E, t|E_0, t_0) = -\varepsilon \frac{\partial}{\partial E} 2\, E \left(\alpha - \frac{b}{2\omega_0^2} E\right) p_2 + D \frac{\partial}{\partial E} E \frac{\partial}{\partial E} p_2. \quad (1.201)$$

The stationary solution follows:

$$p^s(E) = N \exp\left\{\frac{2\varepsilon}{D} E \left(\alpha - \frac{bE}{\omega_0^2}\right)\right\}; \quad (1.202)$$

and differs from a canonical form. Important properties can be obtained from the moments of energy in certain cases. The case of $\varepsilon < 0$ and $E_s = A_s^2 \omega_0^2/2 \to \infty$ gives a canonical-like density with the first moment $D/(2|\varepsilon|\,\alpha)$ and higher moments $\langle E^n \rangle = n!\langle E \rangle^n$. In particular, the relative energy fluctuations are

$$\frac{\langle \Delta E^2 \rangle}{\langle E \rangle^2} = 1. \quad (1.203)$$

Near the bifurcation point the critical energy fluctuations with standard deviation

$$\frac{\langle \Delta E^2 \rangle}{\langle E \rangle^2} = \frac{\pi}{2} - 1 < 1 \quad (1.204)$$

occur independently on all other parameters. In the case of a strongly nonlinear behavior the density becomes sharply Gaussian around $E_s = \alpha \omega_0^2/2b$ as

$$p^s(E) = N \exp\left\{-\frac{2b\varepsilon}{D\omega_0^2} \left(E - \frac{\omega_0^2 \alpha}{2b}\right)^2\right\} \quad (1.205)$$

with relative variance

$$\frac{\left\langle \Delta E^2 \right\rangle}{\left\langle E \right\rangle^2} = \frac{D\,b}{\varepsilon\,\omega_0^2\,\alpha^2}. \tag{1.206}$$

The quantity $\varepsilon\omega_0^2\alpha^2/Db$ can be considered to be a parameter of strong non-linearity. If it increases, then the energy distribution around the attracting manifolds $E_{\mathrm{s}} = \omega_0^2\,\alpha/2\,b$ becomes vanishingly small and collapses eventually in a $\delta$-function as, for example, $D \to 0$. This behavior should be generic for small noise and strong forcing in the case of extended attracting manifolds [50]. Otherwise, as for a single state of equilibrium, a lot of probability is concentrated above the deterministic attracting manifold, reducing the amount of probability for possible deviations.

### 1.2.6 The Escape Problem

One can generically claim that additionally to diffusion the qualitatively new impact of noise in DS consists in the problem of escaping from basins of attraction of stable manifolds. The question at which times and with which statistics an attractor will be left or, correspondingly, what the rates of transitions out of a single attractor are is the central topic in the physical literature where stochastic tools have been applied.

This question was first formulated in nucleation theory [101, 102] and in the theory of chemical reactions [98]. Nucleation implies the stochastic growth of a piece of the new phase to a critical size impossible without fluctuations. In the energetic landscape the critical nucleus forms a saddle-point-like configuration unstable in directions of smaller and larger sizes of the droplet or bubble. Once the critical size is surpassed, the nucleus grows with high probability. Similarly, chemical reactions implicate a surmounting of potential energy across the reaction path to leave a bound state with another particle. Many other processes include such escapes over barriers, firing of neurons, transport through many types of ion channels, the enforcement of populations, and electronic and optical relays, to mention a few only.

Problems of the stochastic escape from a region of attraction can be investigated with the aid of mechanical oscillators. It was first proposed by Kramers in his pioneering paper in 1940. For this purpose Kramers calculated the stationary flux $j^{\mathrm{s}}$ through certain boundaries in the phase space $x, v$, generalizing earlier attempts in nucleation theory to the case of two variables. He defined the rate $r$ of transitions as

$$r = \frac{j^{\mathrm{s}}}{n_0}, \tag{1.207}$$

where $n_0$ denotes the probability near the attractor which will be left. Kramers succeeded in giving rates over an energy barrier $\Delta U$ for low, moderate and strong damping $\gamma$.

Earlier, in 1933, Pontryagin, Andronov and Vitt [103] extended the method of the first passage to the escape problem. They defined the probability distribution

$$W_\Omega(t|x_0, t_0) = \int_\Omega dx\, p_2(x, t|x_0, t_0), \tag{1.208}$$

for the time $t$ after a start at $t_0$ in $x_0$ when trajectories escape for the first time out of a region $\Omega$. For stationary processes $W_\Omega$ as well as its density $w_\Omega = dW_\Omega/dt$ obviously depend on $t - t_0$. Starting from a given evolution operator of $p_2(x, t|x_0, t_0)$ Pontryagin et al. and, later on, Weiss [104] were able to derive ordinary differential equations for moments $\langle T^n \rangle$ of $w_\Omega(t - t_0|x_0)$.

Both methods complement each other and their equivalence has been proven for many problems [87, 105], recently for homogeneous stochastic processes [106]. It holds

$$\langle T \rangle_\Omega \propto \frac{1}{r}. \tag{1.209}$$

Both methods will be elucidated in this section. We will specify the bistable oscillator as presented above (1.193) and will be interested in the escape from the left well near $x_1$. In the flux method we assume an oscillator in equilibrium fixing $D = \gamma k_B T$ due to the fluctuation-dissipation theorem which gives the Maxwell–Boltzmann density (1.195)

$$p^{\mathrm{eq}} = N \exp\left(-\frac{v^2}{2\,k_B\,T} - \frac{U(x)}{k_B\,T}\right). \tag{1.210}$$

**Rates in Equilibrium.** A first estimate for the rates assumes the establishment of the time-independent $P^s(x, v)$ throughout the phase space without additional constraints. Even in this case the oscillator still has a permanent flux of particles from left to right through values $x = 0$, if $v > 0$, which is compensated by a back flux with negative velocities.

This stationary flux of particles with positive velocities from left to right can be calculated as follows:

$$j^{\mathrm{eq}} = \int_0^\infty dv\, v\, p^{\mathrm{eq}}(x = 0, v). \tag{1.211}$$

The rate $k^{\mathrm{eq}}$ of leaving the left site is defined by (1.207). The probability distribution under the integral

$$n_0 = \int_{-\infty}^0 dx \int_{-\infty}^\infty dv\, p^{\mathrm{eq}}(x, v) \tag{1.212}$$

gives its main contribution around the maximum at $x_1$. For low thermal energy $k_B T$ compared to the potential difference $\Delta U$ in (1.196) the density can be approximated as Gaussian around $x_1$, ranging from $-\infty$ to $\infty$. Therefore (1.212) can be approximated by taking

$$U(x) = U(x_1) + \frac{1}{2}\frac{\partial^2 U(x)}{\partial x^2}\Big|_{x=x_1}(x - x_1)^2 + o\left((x - x_1)^3\right). \tag{1.213}$$

By assigning the second derivative of the potential in (1.213) as $\omega_0^2 = 2a$, both integrals in (1.211) and (1.212) can be taken and we find that

$$r^{\text{eq}} = k^{\text{TST}} = \frac{\omega_0}{2\pi} \exp\left(-\frac{\Delta U}{k_{\text{B}}T}\right). \qquad (1.214)$$

The obtained expression is often used as first approximation. It results from several approaches in statistical mechanics and is known as the rate of the transition state theory (TST) [87]. The prefactor contains information on the outgoing attractor, whereas the energy state at the barrier enters via the Arrhenius factor only. In general it overestimates the rate and may serve as the upper boundary for the rate.

Rice [107] extended this result for arbitrary Gaussian-distributed processes with given spectra or correlation functions of the velocity and coordinate related by $\sigma_{v,v}(t) = \mathrm{d}^2\sigma_{x,x}(t)/\mathrm{d}t^2$. He found the rates of exceeding a potential barrier from the ground state $\Delta U$ to be as follows:

$$r^{\text{eq}}_{\text{Rice}} = \frac{1}{2\pi} \sqrt{\frac{\sigma_{v,v}(0)}{\sigma_{x,x}(0)}} \exp\left(-\frac{\Delta U}{\sigma_{v,v}(0)}\right). \qquad (1.215)$$

The values of the correlation functions at $t = 0$ can be expressed by integrals of the corresponding spectra. In the case of white noise this expression converges to (1.214).

**Rates in Nonequilibrium.** The question of how rates are modified if the mechanical system is not fully equilibrated was first addressed by Kramers [98]. Kramers succeeded in finding the solution in the mechanical model explicitly including $\gamma$, which is the main parameter estimating the relaxation of the density.

For that purpose one has to assume a time-scale separation between several processes: first the density relaxes in the neighborhood of the next attractor; at a longer time scale it escapes realizing a transition from the basin of attraction to a possible second existing attractor; eventually, at a third time scale the weights between several attractors become equilibrated.

The escape problem concerns the second step, the surmounting of an energetic barrier which binds trajectories in the neighborhood of the attractor. A quasistationary regime is assumed. It is mathematically formulated by imposing special boundary conditions with respect to the kinetic equation of the probability density. Near the attractor the equilibrium density has built up. All realizations which have surpassed the energetic barrier are absorbed and reinserted into the attractor region. Hence, the averaged surmounting realizations give a stationary flux out of the attracting basin before full equilibrium has been established.

Kramers considered three different regimes of damping: the overdamped case, the moderate to strong case and the low-damping case. Here we restrict ourselves to the first case if $\gamma \gg \omega_b$ where $\omega_b^2$ is related to the curvature of the potential as $\omega_b = (1/m)|\partial^2 U(x)/\partial x^2|$ at the barrier. In the case of (1.187) at $x_b = x_2 = 0$, it follows that $\omega_b^2 = a/m$. In the following we consider overdamped motion. With regard to low friction, the reader will find an excellent presentation in [87].

In the large-damping case the instantaneous velocities abruptly change their values without inertia, which is described by neglecting $dv/dt$ in the Langevin equations (1.193) which then reads

$$\dot{x} = v = -\frac{1}{\gamma}\frac{\partial U(x)}{\partial x} + \sqrt{\frac{2k_{\mathrm{B}}T}{\gamma}}\,\xi(t)\,, \tag{1.216}$$

and again we made use of the fluctuation–dissipation relation. The evolution of the probability density $p_2(x, t|x_0, t_0)$ follows the Smoluchowski equation for a one-dimensional Markovian process:

$$\frac{\partial p_2(x, t|x_0, t_0)}{\partial t} = -\frac{\partial}{\partial x}\left(-\frac{1}{\gamma}\frac{\partial U}{\partial x}p_2\right) + \frac{k_{\mathrm{B}}T}{\gamma}\frac{\partial^2 p_2}{\partial x^2}\,. \tag{1.217}$$

The initial value problem of this equation cannot be solved for anharmonic potentials. Therefore, we look for a stationary approximation $p^{\mathrm{s}}$ describing escapes with a constant circulating flow $j^{\mathrm{s}}$ as outlined above. Escaped probability will be absorbed at a position larger than the barrier value $x = x_A > x_2 = 0$, requiring $p^{\mathrm{s}}(x_A) = 0$ which is a sink of probability. The absorbed flow of probability $j^{\mathrm{s}}$ is reinjected at the stable fixed point $x_1$. For this purpose a source term is added to (1.217)

$$0 = -\frac{\mathrm{d}}{\mathrm{d}x}\left(-\frac{1}{\gamma}\frac{\partial U}{\partial x}p^{\mathrm{s}}\right) + \frac{k_{\mathrm{B}}T}{\gamma}\frac{\mathrm{d}^2 p^{\mathrm{s}}}{\mathrm{d}x^2} + j^{\mathrm{s}}\delta(x - x_1)\,. \tag{1.218}$$

Left from $x_1$ the probability flow vanishes again and the equilibrium density will be established $p^{\mathrm{s}}(x < x_1) = p^{\mathrm{eq}}(x)$. Continuity at $x_1$ gives the second necessary boundary condition $\lim_{\varepsilon \to +0} p^{\mathrm{s}}(x_1 + \varepsilon) = p^{\mathrm{eq}}(x_1)$.

In this way probability permanently flows through the system between $x_1$ and $x_A$ and surpasses the unprobable state $x_2$. The solution of (1.218) is the Green's function of the forward FPE-operator. The first integration gives in the interval $[x_1, x_A]$

$$-\frac{1}{\gamma}\frac{\partial U}{\partial x}p^{\mathrm{s}} - \frac{k_{\mathrm{B}}T}{\gamma}\frac{\mathrm{d}}{\mathrm{d}x}p^{\mathrm{s}} = j^{\mathrm{s}}\,\Theta(x - x_1)\,. \tag{1.219}$$

Equation (1.219) can be integrated the second time using the multiplicative factor $1/p^{\mathrm{eq}}(x)$, which yields

$$\frac{\mathrm{d}}{\mathrm{d}x}\left(\frac{p^{\mathrm{s}}(x)}{p^{\mathrm{eq}}(x)}\right) = -j^{\mathrm{s}}\frac{1}{p^{\mathrm{eq}}(x)}\frac{\gamma}{k_{\mathrm{B}}T}\Theta(x - x_1)\,, \tag{1.220}$$

and eventually

$$p^{\mathrm{s}}(x) = -j^{\mathrm{s}}\frac{\gamma}{k_{\mathrm{B}}T}p^{\mathrm{eq}}(x)\int^{x}\mathrm{d}x'\frac{1}{p^{\mathrm{eq}}(x')}\,\Theta(x' - x_1)\,. \tag{1.221}$$

Usage of the formulated boundary conditions of the stationary nonequilibrium density $p^s$ defines the value of the constant flow by the relation

$$1 = j^s \frac{\gamma}{k_B T} \int_{x_1}^{x_A} dx' \frac{1}{p^{eq}(x')}. \tag{1.222}$$

Further on we use the definition of the rate (1.207) and find the integral formulation

$$r_K = \frac{k_B T}{\gamma \int_{-\infty}^{0} dx' \, p^s(x')} \frac{1}{\int_{x_1}^{x_A} dx' \, (p^{eq}(x'))^{-1}}. \tag{1.223}$$

The remaining two integrals have to be estimated, and one proceeds in a way quite similar to the earlier presented case of $k^{TST}$. The first one gives its main contribution around $x = x_1$. In this region $p^s$ can be replaced by the equilibrium distribution $p^{eq}$; as a result of this, the normalization constant under both integrals drops out. Further on, one uses the Gaussian approximation of the potential around $x_1$ and extends the integration over the whole range of coordinates. Oppositely, the function under the second integral approaches maximal values near the unstable fixed point $x_2 = 0$, and other regions of integration give vanishingly small contributions. Near $x_2$ one may expand as (1.213), with $U(x) = U(x_b) - (m/2)\omega_b^2(x - x_b)^2$. Again extending the limits of the second integral to $\pm\infty$, the integral results eventually in

$$r_K = \frac{\omega_b}{\gamma} \frac{\omega_0}{2\pi} \exp\left(-\frac{\Delta U}{k_B T}\right). \tag{1.224}$$

Thus the rate of transition decreases for large damping as $\propto 1/\gamma$. As a consequence of the condition $\gamma \gg \omega_b$ the escape rate is always smaller than in the case of full equilibrium.

**Pontryagin Equation.** There are several tools in the theory of stochastic processes which characterize the dynamical behavior of a stochastic system without explicitly calculating the time-dependent probability density [63,67,80]. Two of them, the most popular ones, will be considered here: the probability to reach a certain boundary from a given initial state for the first time, and the mean time which is necessary for this event.

Let $x(t)$ be a one-dimensional Markovian diffusion process given either by its Langevin equation or by the corresponding FPE. We will assume stationarity of this process. Further, let $x_0$ be the initial state at $t_0 = 0$ which is located inside the interval $x_0 \in (a, b)$. Then

$$P_{a,b}(t, x_0) = \int_a^b p_2(x, t | x_0) \, dx \tag{1.225}$$

is the probability that $x(t)$ is still inside the region $(a, b)$ at time $t > 0$.

Now the problem is modified in such a way that we can find the probability of reaching the boundaries of the region for the first time. It implies that trajectories which have reached the boundaries have to be excluded from returning back into the interval to avoid multiple crossings. The simple way to do this is to remove them from consideration by applying special boundary conditions. In the case of Markovian diffusion processes every trajectory which leaves the region has to approach either the value $x = a$ or the value $x = b$. This follows from the continuity of the sample paths we have mentioned.

Exclusion of these trajectories is realized if we require that the probability of a transition from the boundaries back into the interval is exactly zero, or explicitly written, if

$$p_2(x, t|a) = p_2(x, t|b) = 0, \quad x \in (a, b). \tag{1.226}$$

Insertion into (1.225) gives

$$P_{a,b}(t, a) = P_{a,b}(t, b) = 0, \tag{1.227}$$

and the boundary conditions of this type are called absorbing.

Another boundary condition causes reflection of particles. It means that the flow of probability $j(x, t)$ (1.161) vanishes at this boundary $j(x = a, t) = 0$ which gives a condition for the forward problem, the probability density at $x$ for times $t > t_0$. Application of the property that the forward and backward FPE-operators are adjoint operators

$$\int dx\, p_2(y, t_y|x, t_x)\, \mathbf{L}_x^{\mathrm{F}}\, p_2(x, t_x|x_0, t_0) = \int dx\, p_2(x, t_x|x_0, t_0)\, \mathbf{L}_x^{\mathrm{B}}\, p_2(y, t_y|x, t_x) \tag{1.228}$$

leads to the formulation of a reflecting boundary condition for the backward problem [87]. Indeed, taking the integration in (1.228) from $-\infty$, where the density and its derivative should vanish, to the reflecting boundary $b$ gives the equality of

$$p_2(y, t_y|x, t_x)\, j(x, t) = p_2(x, t_x|x_0, t_0)\, \frac{\partial}{\partial x}\, p_2(y, t_y, |x, t_x) \tag{1.229}$$

with $x$ taken at $a$. Hence, $j(a, t) = 0$ requires vanishing of the right hand side and if the density at the boundary does not vanish

$$\left. \frac{\partial p_2(x, t|x_0, t_0)}{\partial x_0} \right|_{x_0 = a} = 0. \tag{1.230}$$

After integration one obtains

$$\left. \frac{\partial P_{a,b}(t, x_0)}{\partial x_0} \right|_{x_0 = a} = 0. \tag{1.231}$$

Henceforth let $P_{a,b}(t, x_0)$ label the probability that all escaped trajectories have been eliminated. To find a differential equation for $P_{a,b}$, we make use of the assumption that the dynamics inside the region is not modified due to the application of the boundary conditions [92] and given by the rates of the conditioned moments (1.134). It will be valid if trajectories need some finite time to reach the boundaries, i.e., the probability to be out of the interval after vanishingly small times $\tau$ decreases faster than linear

$$\lim_{\tau \to 0} \frac{1}{\tau}(1 - P_{a,b}(\tau, x_0)) = 0, \quad x_0 \neq a, b. \tag{1.232}$$

Then the moments of the increments (1.142) of the considered process with the corresponding kinetic coefficients (1.147) and (1.148) can be applied and are not modified inside the interval. Hence, the Kolmogorov backward and forward equations can be used inside for the first passage problem.

Taking the backward equation (1.137) and integrating it over the considered interval, we readily obtain

$$-\frac{\partial}{\partial t_0} P_{a,b} = \frac{\partial}{\partial t} P_{a,b}(t, x_0) = K_1(x_0) \frac{\partial}{\partial x_0} P_{a,b} + K_2(x_0) \frac{\partial^2}{\partial x_0^2} P_{a,b}. \tag{1.233}$$

The change in the time derivation to $t$ can be performed with the assumed stationarity of the stochastic process (comp.(1.65)).

Equation (1.233) is the evolution operator for the probability of staying inside the considered interval without having contact with the boundaries. It was first derived by Pontryagin, Andronov and Witt in 1933 [103]. Boundary conditions and initial conditions have to be added to find solutions for special physical problems. One then solves (1.233) with respect to $P_{a,b}(t, x_0)$ and averages over the initial values.

**Survival Probability.** Several phenomena require one to look for the asymptotic solutions of (1.233), addressing the problem of survival probability. If at least one absorbing boundary exists which the process can reach in finite times $P_{a,b}(t \to \infty, x_0) = 0$, then the interval will be depleted with certainty. Obviously, the function

$$W_{a,b}(t, x_0) = 1 - P_{a,b}(t, x_0) \tag{1.234}$$

is the probability of escaping the region for the first time. It obeys the same evolution operator with corresponding boundary conditions following from the definition (1.227) and (1.231). With an absorbing state and by waiting for an infinitely long time under ergodic conditions, being outside is a certainty:

$$W_{a,b}(t \to \infty, x_0) = W_{a,b}^{as}(x_0) = 1, \tag{1.235}$$

for arbitrary $x_0$.

One may ask for the asymptotic probability of leaving the interval by crossing $b$ without reaching before $a$. Then, $b$ absorbs probability, i.e., $W_{a,b}^{as}(b) = 1$, but a trajectory reaching $a$ has to be bound within the region forever, i.e., it should be unable to leave by reaching $b$, i.e., $W_{a,b}^{as}(a) = 0$. The asymptotic solution by setting $\partial W_{a,b}^{as}(x_0)/\partial t = 0$ then reads

$$W_{a,b}^{as}(x_0) = \frac{\int\limits_a^{x_0} dx \exp\left(-\int^x \frac{K_1}{K_2} dx'\right)}{\int\limits_a^b dx \exp\left(-\int^x \frac{K_1}{K_2} dx'\right)}. \tag{1.236}$$

Obviously it differs from the certainty due to the implication of formulated constraint.

**Waiting-Time Density.** More detailed information will be obtained by investigating the probability density at time $t$ when the region will be left for the first time. It is called the density of the waiting time distribution. If $W_{a,b}(t, x_0)$ is the probability distribution that the interval $[a, b]$ is left for the first time during the time period $t_0$ and $t$, then its density is

$$w_{a,b}(x_0, t)\, \Delta t = \frac{\partial W_{a,b}(t, x_0)}{\partial t} \Delta t = -\frac{\partial P_{a,b}}{\partial t} \Delta t. \tag{1.237}$$

We immediately obtain the following equation for the waiting-time probability density:

$$\frac{\partial}{\partial t} w_{a,b} = K_1(x_0) \frac{\partial}{\partial x_0} w_{a,b} + K_2(x_0) \frac{\partial^2}{\partial x_0^2} w_{a,b}, \tag{1.238}$$

supplemented by the initial and boundary conditions

$$w_{a,b}(x_0, t = 0) = 0, \quad x_0 \in (a, b), \quad w_{a,b}(a, t) = w_{a,b}(b, t) = \delta(t), \tag{1.239}$$

if both boundaries are absorbing. If $a$ is a reflecting boundary, the corresponding boundary condition is replaced by

$$\left.\frac{\partial w_{a,b}(x_0, t)}{\partial x_0}\right|_{x_0 = a} = 0. \tag{1.240}$$

**Moments of First Escape.** The first moment of the waiting time distribution is called the mean first passage time. It is

$$T_{a,b}(x_0) = \int_0^\infty t w_{a,b}(x_0, t)\, dt, \tag{1.241}$$

and by definition it simply follows

$$T_{a,b}(x_0) = \int_0^\infty P_{a,b}(t, x_0)\, dt. \tag{1.242}$$

Averaging over a density of initial states inside $[a, b]$ yields

$$\langle T_{a,b} \rangle_{x_0} = \int_0^\infty P_{a,b}(t, x_0) P(x_0) \, dt \, dx_0 \ . \tag{1.243}$$

An ordinary differential equation for $T_{a,b}(x_0)$ can be obtained by multiplying with $t$ and integrating (1.238) with respect to time. It gives on the left-hand side

$$\int_0^\infty \frac{\partial}{\partial t} P_{a,b} \, dt = P_{a,b}(\infty, x_0) - P_{a,b}(0, x_0) = -1 \ . \tag{1.244}$$

and integrating the right-hand side we find eventually

$$-1 = K_1(x_0) \frac{\partial}{\partial x_0} T_{a,b}(x_0) + K_2(x_0) \frac{\partial^2}{\partial x_0^2} T_{a,b}(x_0) \ . \tag{1.245}$$

This equation [103] solves the problem of after which mean time the region $[a, b]$ will be left according to the boundary conditions. Absorption in $1D$ implies that no time is required to leave if we are initially at the boundary, i.e.,

$$T_{a,b}(a) = T_{a,b}(b) = 0. \tag{1.246}$$

In the case when $x = a$ is a reflectory boundary

$$\left. \frac{\partial T_{a,b}(x_0)}{\partial x_0} \right|_{x_0 = a} = 0. \tag{1.247}$$

It is of interest to introduce higher moments of the mean first passage problem as [104]

$$T_{a,b}^{(n)} = -\int_0^\infty t^n \frac{\partial}{\partial t} W_{a,b}(t, x_0, t_0) \, dt. \tag{1.248}$$

Again, for these moments one is able to find a chain of coupled differential equations from (1.238). By denoting $T_{a,b}^{(0)} = 1$ and $T_{a,b} = T_{a,b}^{(1)}$, one obtains

$$-n T_{a,b}^{(n-1)} = K_1(x_0) \frac{\partial}{\partial x_0} T_{a,b}^{(n)}(x_0) + K_2(x_0) \frac{\partial^2}{\partial x_0^2} T_{a,b}^{(n)}(x_0) \tag{1.249}$$

with boundary conditions as for $T_{a,b}^{(1)}$.

The solution of (1.249) is as follows: Denoting

$$u_{a,b}^{(n)} = \frac{dT_{a,b}^{(n)}}{dx_0}, \tag{1.250}$$

it follows from (1.249) with $a < b$ that

$$u_{a,b}^{(n)}(x_0) = \exp\left(-\Phi(x_0)\right)\left(C_n - n\int_a^x \frac{T_{a,b}^{(n-1)}(x')}{K_2(x')}\exp\left(\Phi(x')\right)dx'\right), \quad (1.251)$$

where $C_n$ are constants and

$$\Phi(x) = \int^x \frac{K_1(y)}{K_2(y)}dy. \quad (1.252)$$

In addition, we can apply boundary conditions and consider special cases: In the case $x = a$ is a reflectory state and $b$ absorbing, $u_{a,b}^{(n)}(a) = 0$, $T_{a,b}^{(n)}(b) = 0$, one gets

$$T_{a,b}^{(n)}(x_0) = n\int_{x_0}^b \exp\left(-\Phi(x)\right)\left(\int_a^x \frac{T_{a,b}^{(n-1)}}{K_2(x')}\exp\left(\Phi(x')\right)dx'\right)dx. \quad (1.253)$$

Particularly, for the first moment we find the expression

$$T_{a,b}^{(1)}(x_0) = \int_{x_0}^b dx \exp\left(-\Phi(x)\right)\left(\frac{\int_a^x dx' \exp\left(\Phi(x')\right)}{K_2(x')}\right). \quad (1.254)$$

If both $a$ and $b$ are absorbing states, integration of (1.251) gives

$$T_{a,b}^{(n)}(x_0) = C_n\int_a^{x_0}\exp\left(-\Phi(x)\right)dx -$$

$$n\int_a^{x_0}\exp\left(-\Phi(x)\right)\left(\int_a^x \frac{T_{a,b}^{(n-1)}}{K_2(x')}\exp\left(\Phi(x')\right)dx'\right)dx, \quad (1.255)$$

with $C_n$ following from $T_{a,b}^{(n)}(b) = 0$.

The most common situation corresponding to the rate problem in the overdamped oscillator concerns the probability of leaving a region $(-\infty, b)$. It means that $x = b$ is an absorbing boundary and $a$ is shifted to minus infinity. The physically relevant situation in that case is that there is no probability at $-\infty$ and also its derivative vanishes. The flux of probability is exactly zero, which is called a natural boundary condition of a physical problem.

Evaluation of the integrals in (1.254) with the kinetic coefficients specified to the bistable oscillator are performed again by local Gaussian approximations if $\Delta U \gg k_B T$. The internal integral in (1.241) is maximal near the attractor $x = x_1$. In contrast, the first integrand contributes significantly near the barrier only. If the initial state $x_0$ is sufficiently far left of the barrier $x_2$ the mean first passage time is simply the inverse of the overdamped rate (1.224),

$$T_{-\infty,b}(x_0) = Q(b)\frac{1}{r_K}. \quad (1.256)$$

$Q(b)$ is a form factor which grows abruptly from zero to one if the final state $b$ approaches and crosses the position of the barrier $x_2$. At $b = x_2$, $Q = 1/2$.

A generalization to higher-dimensional cases can be found in [108] and more recent development is included in [52–62, 87, 88, 106, 109–112].

### 1.2.7 Summary

We have introduced elements of stochastic processes and applied them to dynamical systems. Despite the fact that the basic knowledge and application of stochastic processes in physics goes back to the beginning of the 20th century, the usage of stochastic tools in physics is still a modern, agile and attracting field.

Nowadays inspiring investigations with stochastic methods come from research related to biophysics, which inherently combines nonlinearities, nonequilibrium, signals and noise or fluctuations. Some of the methods developed in statistical physics are helpful in the consideration of the biological sphere. But, most important, the new area of physical research impacts on the formation of novel stochastic techniques at a qualitatively higher level. The interested reader is referred to the large amount of published literature and to Chap. 3.

## 1.3 Synchronization of Periodic Systems

### 1.3.1 Introduction

One of the fundamental nonlinear phenomena observed in nature and, particularly, one of the basic mechanisms of self-organization of complex systems [7] is synchronization. From the most general point of view, synchronization is understood as an adjustment of some relations between characteristic times, frequencies or phases of two or more DS during their interaction. Synchronization has attracted much attention in different fields of natural sciences. For instance, applications of synchronization in engineering sciences [113] have achieved great practical importance and are widely employed. Moreover, specifying to biophysics, several kinds of synchronization have been observed in biological systems. We would like to mention here the behavior of cultured cells [114] and of neurons [115–117] and, indeed, synchronization of biological populations [118]. More complex types of synchronization have been reported recently for the human cardio-respiratory system [119] as well as in magnetoencephalography [120]. Synchronization of regular, chaotic and stochastic oscillations has been reported for ensembles of interacting oscillators [121–131] and for extended systems [132–135].

The classical theory of synchronization distinguishes between forced synchronization by an external periodic driving force and mutual synchronization between coupled oscillators. In both cases manifestations of synchronization are the same. They are determined by an interplay of time scales by phase locking or, respectively, natural frequency entrainment or due to suppression of inherent frequencies.

From the mathematical point of view, the theory of phase synchronization of periodic self-sustained oscillators is well established [1, 113, 136, 137]. If $\Phi(t)$ is the phase of a periodic oscillator and $\Psi(t)$ is the phase of an external periodic force or, otherwise, the phase of another periodic oscillator coupled with the first one, then the condition of synchronization can be formulated as

$$|m\Phi(t) - n\Psi(t)| = \text{const.} , \qquad (1.257)$$

where $m$ and $n$ are integers. This condition defines the locking of two phases $\Phi(t)$ and $\Psi(t)$ and requires that the phase difference should be constant. Synchronization is also defined as frequency entrainment, provided that the frequencies of the oscillator and the driving force are in rational relation. In this section we mostly consider the simplest case of 1:1 synchronization ($m = n = 1$).

Synchronization of periodic self-sustained oscillators in the presence of noise has been studied in detail by Stratonovich [92]. The theory shows that noise counteracts synchronization in a sense that under the influence of noise synchronization occurs only for a limited time interval.

Recent developments in nonlinear dynamics have opened up new perspectives for the theory of synchronization. From the modern point of view, regular (e.g., periodic or quasiperiodic) oscillatory regimes are only a small fraction of all possible types of dynamical behavior. With an increase in the degree of nonlinearity and the dimension of the phase space of the forced or the mutually interacting DS nonperiodic or chaotic behavior is more typical. The effect of noise on nonlinear systems far from equilibrium is also a nontrivial problem. Nonequilibrium noise might even change the qualitative behavior of a DS, inducing new regimes which are absent in the noiseless system [44]. Recently, synchronization-like phenomena have been also found in stochastic systems, such as systems with noise-induced switchings [138–140] and excitable systems of the neuron model type [131, 141, 142]. Rigorously speaking, these systems are not self-sustained. Their signals are not periodic, and the power spectrum may not contain peaks at any distinct frequencies. However, it has been proved that the concept of phase and frequency locking can be applied to this case as well. That is why the problem of an extension of the concept of synchronization to chaotic and stochastic motion is of great interest and importance.

In this section we describe the results of the classical theory of synchronization of periodic oscillations. Using as an example the Van der Pol oscillator, we consider forced and mutual synchronization, including the analysis of noise effects. The generalization of the classical theory of synchronization to chaotic and stochastic oscillations is made and discussed in Sects. 2.2 and 2.3 and in Chap. 3.

## 1.3.2 Resonance in Periodically Driven
## Linear Dissipative Oscillators

To study synchronization one has to assign a phase to the dynamical variable(s) of the system. First we discuss the notion of a phase for periodic motions. Later, in the following sections, this notion will be generalized for more complicated types of dynamics.

The term "phase" was originally introduced for harmonic processes with $x(t) = A \cos(\omega_0 t + \varphi_0)$ (see, for instance, [1, 91]). As long as $x(t)$ is strictly periodic with a constant amplitude, $A$, one defines $\Phi(t) = \omega_0 t + \varphi_0$ as the instantaneous phase. The amplitude and phase can be found from $x(t)$ and a second conjugated variable, e.g., its time derivative $\dot{x}(t)$. This definition of amplitude and phase corresponds to a transition to the system of polar coordinates with the radius vector $A$ and the angle $\Phi$ given by

$$A = \sqrt{x^2 + \frac{\dot{x}^2}{\omega_0^2}}, \quad \tan \Phi = -\frac{\dot{x}}{x\,\omega_0}, \tag{1.258}$$

if $\omega_0$ is the circular frequency of the harmonic process which can be found, for example, by comparing the maximal elongation of the coordinate and the maximal velocity.

A similar approach is the analytic continuation of a periodic signal into the complex plane. Instead of the real signal $x(t)$, we introduce

$$z(y) = x(t) + \mathrm{i}y(t) = A \exp \mathrm{i}\,(\omega_0 t + \varphi_0) = A\left[\cos(\omega_0 t + \varphi_0) + \mathrm{i}\sin(\omega_0 t + \varphi_0)\right], \tag{1.259}$$

which is now complex. The harmonic oscillation is pictured as a rotation of the vector $A$ with constant angular velocity $\omega_0$ in the complex plane. The phase of oscillations corresponds to the angle of the vector $A$, and the angle $\varphi_0$ at $t = 0$ is the initial phase.

However, these definitions cannot be directly used in nonlinear dissipative systems, since the oscillations in such systems are not harmonic. How to proceed in more complex situations? Several approaches will be discussed below for different examples. The particular procedure depends on both the type of process generated by the DS and the structure of the driving force.

Let us consider a linear dissipative oscillator driven by a periodic harmonic force:

$$\ddot{x} + \gamma\dot{x} + \omega_0^2 x = a \cos(\Omega t), \tag{1.260}$$

where $\gamma$ is the damping coefficient and $\omega_0$ is the natural frequency of the oscillator. Without the periodic force it has a single stable attractor at the origin, $x_0 = \dot{x}_0 = 0$. Let us discuss the dynamical properties of this linear system in the amplitude–phase description. In the limit of small friction $\gamma \ll 2\omega_0$ and small amplitude of the driving force, the nonlinear transformation to the amplitude $A(t)$ and phase $\Phi(t)$ by

$$x(t) = A(t) \cos \Phi(t) , \quad \dot{x}(t) = v(t) = -A(t) \, \Omega \, \sin \Phi(t) \qquad (1.261)$$

will give us a well-known asymptotic solution. It is reasonable to decompose the phase variable $\Phi(t)$ into two parts of fast and slow motion: $\Phi(t) = \Omega t + \varphi(t)$. The new function $\varphi(t)$ represents the phase difference between the driving force and the response of the system and is a slowly changing function in comparison with the external force.

We proceed with the usual technique employed in the theory of oscillations, developed by Bogolyubov and Mitropolski [91]. Inserting (1.261) into (1.260) gives a DS which describes the time evolution of $A(t)$ and $\varphi(t)$. With the assumption made about the smallness of the friction coefficient and the applied force, we can suppose that the amplitude and the phase difference change on time scales larger than the period of the applied force. This gives the possibility to average the dynamics of $A$ and $\varphi$ over one period, fixing their values during this period. As a result we derive the following approximate, so-called reduced amplitude–phase equations:

$$\dot{A} = -\frac{\gamma}{2}A - \mu \sin \varphi, \quad \dot{\varphi} = \Delta - \frac{\mu}{A} \cos \varphi . \qquad (1.262)$$

Therein $\mu = a/2\Omega$ is the normalized amplitude of the force and $\Delta = (\omega_0^2 - \Omega^2)/2\Omega \approx \omega_0 - \Omega$ is the frequency mismatch or the detuning between the natural frequency of the autonomous system (1.260) and the frequency of the external signal.

The long-term asymptotics of the amplitude and the phase are given by the stationary solutions of (1.262):

$$A_0 = \frac{a}{\sqrt{(\omega_0^2 - \Omega^2)^2 + \gamma^2 \Omega^2}} , \quad \varphi_0 = \arctan \frac{\gamma \Omega}{\Omega^2 - \omega_0^2} , \qquad (1.263)$$

which fully characterize the response of a linear dissipative oscillator to a harmonic force. We note that expressions for the stationary amplitude and the phase shift (1.263) obtained in the framework of the amplitude–phase approximation are in full agreement with the exact theory discussed in textbooks.

The phenomenon of resonance manifests itself in the abrupt increment of the amplitude of the forced oscillations when the external frequency $\Omega$ nearly coincides with the natural frequency $\omega_0$ of the oscillator. In detail, the peak is located at $\Omega = \omega_p = \sqrt{\omega_0^2 + \gamma^2/2} \approx \omega_0$. Resonance is characterized by this typical dependence of the amplitude $A_0(\Omega)$, which is called the resonance curve.

It is important to note, that the phase of the periodically driven damped harmonic oscillators is completely defined by the external force. In contrast, in the next section we consider so-called self-sustained oscillators with periodic behavior located on a limit cycle. Their natural frequency and amplitude without external forcing are given by the internal parameters of the oscillator. The phase gives the geometrical position on the limit cycle. Contrary to the linear driven oscillator with fixed phase as considered above, the phase variable of self-sustained oscillators possesses a zero Lyapunov exponent along the limit cycle.

### 1.3.3 Synchronization of the Van der Pol Oscillator. Classical Theory

The Van der Pol model is the simplest example of a Thomson-type generator. It is the prototype for many mechanical, electronic and biological systems exhibiting self-sustained oscillations. It is a generic model for an investigation of the different types of synchronization which can occur when the oscillator is periodically driven by or coupled with another oscillator.

First we discuss the unperturbed regime of the autonomous Van der Pol oscillator. Afterwards we elucidate different types of bifurcations and elaborate when synchronization to an external periodic force appears. Finally, we consider mutual synchronization of two coupled Van der Pol oscillators.

The autonomous Van der Pol oscillator is described by the second order differential equation

$$\ddot{x} - \varepsilon(1 - x^2)\dot{x} + \omega_0^2 x = 0 \,, \tag{1.264}$$

where $\varepsilon$ is a small parameter characterizing the degree of nonlinearity and corresponding to the feedback strength in an electronic realization of the oscillator and $\omega_0$ is the frequency of the oscillator. For $-2\omega_0 < \varepsilon < 0$ this system possesses a single stable state of equilibrium of the focus type at the origin with the eigenvalues

$$s_{1,2} = \frac{\varepsilon}{2} \pm i\omega_0 \sqrt{1 - \left(\frac{\varepsilon}{2\omega_0}\right)^2}\,. \tag{1.265}$$

For $\varepsilon > 0$ the limit cycle is the single stable limit set of the system, and its basin of attraction is the whole phase plane.

From the physical point of view, the notion stable "limit cycle" corresponds to self-sustained oscillations. The properties of this regime, e.g., amplitude and frequency, do not depend on initial conditions and are fully determined by the internal properties of the system. For $0 < \varepsilon < 1$ the period of self-sustained oscillations is

$$T = \frac{2\pi}{|\text{Im } s_{1,2}|}\,, \quad \text{Im } s_{1,2} = \omega_0 \sqrt{1 - \left(\frac{\varepsilon}{2\omega_0}\right)^2}\,, \tag{1.266}$$

i.e., the frequency of oscillations is close to the natural frequency $\omega_0$ of the resonance circuit.

To discuss synchronization we add a periodic force to (1.264) and consider the dynamics of the periodically driven Van der Pol oscillator whose equation is written in the form

$$\ddot{x} - \varepsilon(1 - x^2)\dot{x} + \omega_0^2 x = a\cos(\Omega t). \tag{1.267}$$

We consider the case of small nonlinearities $0 < \varepsilon < 1$ and introduce instantaneous phase and amplitude according to the nonlinear transformation (1.261) with $\Phi(t) = \Omega t + \varphi(t)$. Sufficient conditions for slow amplitude and phase evolution are $\omega_0 \simeq \Omega$, small feedback $\varepsilon$ and that the amplitude of the periodic force also scales with $\varepsilon$.

The exact equations for the phase and the amplitude contain fast oscillating terms which can be neglected, taking into account frequency selectivity of the Van der Pol oscillator, by averaging over the period of the external force. Omitting this procedure of straightforward but tedious transformations [91], we further discuss the slow dynamics of the first approximation for the instantaneous amplitude $A(t)$ and phase $\varphi(t)$:

$$\dot{A} = \frac{\varepsilon A}{2}\left(1 - \frac{A^2}{A_0^2}\right) - \mu\sin\varphi, \quad \dot{\varphi} = \Delta - \frac{\mu}{A}\cos\varphi, \tag{1.268}$$

again with $\mu = a/2\Omega$, the frequency mismatch $\Delta = (\omega_0^2 - \Omega^2)/2\Omega \approx \omega_0 - \Omega$ and $A_0 = 2$ being the amplitude of the stable cycle of the autonomous case.

A stationary fixed point of the system (1.268) corresponds to the periodic solution of the initial system (1.267), and a periodic solution of (1.268) corresponds to a quasiperiodic one of (1.267).

Now we turn to the bifurcation analysis of the mechanism of synchronization. Assume that the system (1.268) has a fixed point $\dot{A} = 0$, $\dot{\varphi} = 0$ as a solution which is stable. The condition $\dot{\varphi} = 0$ means that $\dot{\Phi} = \Omega$, that is, the frequency of the forced oscillations coincides with the frequency of the external force. Thus, the generator is tuned to the external signal frequency and the effect of forced synchronization takes place.

This phenomenon can be studied in more detail. The phase space of the DS (1.268) is a two-dimensional cylinder. Setting the right-hand sides of the system (1.268) equal to zero, we find the coordinates of the stationary fixed points pictured in Fig. 1.26a. One easily finds three different points with amplitudes $A > 0$: a stable node at $O_1$, a saddle at $O_2$ and an unstable node at $O_3$.

The presence of a stable point $O_1$ corresponds to the regime of 1:1 synchronization ($A = \text{const.}$, $\varphi = \text{const.}$). Increasing the detuning parameter $\Delta$, the points $O_1$ and $O_2$ approach each other, merge and disappear through a saddle-node bifurcation at a certain critical value of $\Delta$ (see Fig. 1.26b). As a result, a limit cycle $C_1$ of the second kind (that is, surrounding the whole cylinder) is born. This bifurcation corresponds to a loss of synchronization, and a regime with two frequencies appears in the original system (1.267). Indeed, if $\varphi(t)$ changes periodically, then $\dot{\Phi} \neq \Omega$. The region in the space of control parameters $\Delta$ and $\mu$, in which the fixed point $O_1$ remains stable, represents a synchronization region with frequency relation 1:1.

In order to find the boundaries of this region one solves the linearized equations of (1.268) in the vicinity of the fixed point $O_1$ for its eigenvalues.

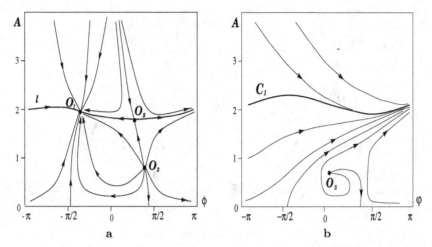

**Fig. 1.26.** Structure of the phase space of system (1.268) for different values of the parameters: **(a)** $\varepsilon = 0.1$, $\mu = 0.042$, $\Delta = 0.02$; **(b)** $\varepsilon = 0.1$, $\mu = 0.042$, $\Delta = 0.032$

A vanishing real part of the eigenvalues indicates the values of the control parameters where the stability of this point is lost or $O_1$ disappears. This problem can be solved analytically, and the results are shown in Fig. 1.27.

Inside the first synchronization region (region I) the fixed point $O_1$ is a stable node. On the lines $l_a$ a saddle-node bifurcation of $O_1$ and $O_2$ takes place. Moving from I to III by crossing the lines $l_a$ corresponds to the birth of a limit cycle $C_1$. Thus, the synchronous regime is destroyed by increasing

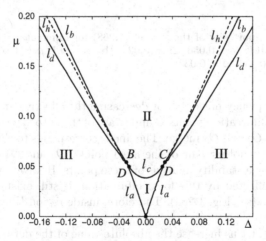

**Fig. 1.27.** Bifurcation diagram for systems (1.267) and (1.268). Lines $l_a$ and $l_c$ correspond to the saddle-node bifurcation in (1.268); $l_b$ denotes the line of Andronov–Hopf bifurcation in (1.268); $l_d$ is the bifurcation line of a cycle $C_2$ crisis; $l_h$ is the bifurcation line of torus birth from the resonant cycle in the original system (1.267)

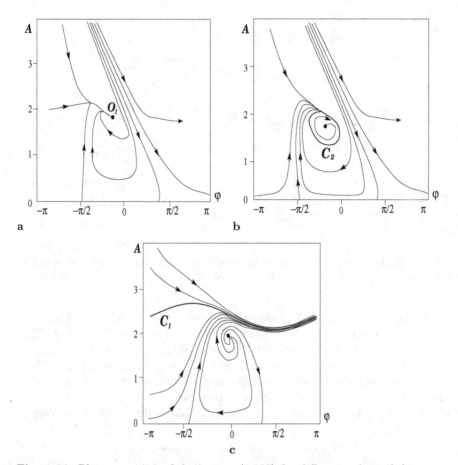

**Fig. 1.28.** Phase portraits of the system (1.268) for different values of the parameters: (a) $\varepsilon = 0.1$, $\mu = 0.056$, $\Delta = 0.028$; (b) $\varepsilon = 0.1$, $\mu = 0.056$, $\Delta = 0.031$; (c) $\varepsilon = 0.1$, $\mu = 0.056$, $\Delta = 0.033$

the absolute frequency mismatch or decreasing the effective amplitude of the external force. Bifurcation points $B$ and $C$ are of the cusp type where all three fixed points $O_1$, $O_2$ and $O_3$ merge. The line $l_c$ corresponds to the merging and disappearance of another pair of the fixed points, $O_2$ and $O_3$. This line can be determined by a stability analysis of these points. However, since the fixed point $O_1$ is unaffected by this local bifurcation, it still exists and is stable above the line $l_c$ (see Fig. 1.28a). Therefore, inside region II synchronization still takes place.

Furthermore, let us increase the absolute value of the detuning with fixed $\mu$ and $\varepsilon$ above $l_c$ in the region II. Starting from a certain critical value of $|\Delta|$ the point $O_1$ looses its stability through the Andronov-Hopf bifurcation indicated in Fig. 1.27 by the lines $l_b$. Outside the region bounded by $l_b$ a stable limit cycle $C_2$ of the first kind (lying fully within the surface between $-\pi$ and $\pi$) appears. This bifurcation again destroys synchronous oscillations as the original system (1.267) performs quasiperiodic oscillations [see Fig. 1.28b].

As the detuning is increased further, the re-building of quasiperiodic regime $C_1$ into $C_2$ can be observed (see Fig. 1.28c). Above the line $l_c$ the Andronov–Hopf bifurcation of equilibrium takes place at lines $l_b$ shown in Fig. 1.27. At lines $l_d$ $C_2$ is transformed into $C_1$.

Thus, regions of 1:1 synchronization for the Van der Pol oscillator are bounded by the bifurcation lines $l_a$ below points $B$, $C$ and by the lines $l_b$ beyond these points. Inside the regions of synchronization the original system (1.267) exhibits a stable limit cycle whose frequency coincides with the frequency of the external force. This fact means that both the frequency and the phase are locked by the external force.

Let us summarize the bifurcations described above. Assume that the parameters of the oscillator are inside region I of Fig. 1.27. As seen from Fig. 1.26a, inside region I the separatrices of the saddle $O_2$ are pointing into the stable node $O_1$. Both points lie on the closed invariant curve $l$ going around the full cylinder. This curve is the image of a two-dimensional resonant torus in the original system (1.267). The stable point corresponds to the synchronous regime of the oscillator. At the onset of the saddle-node bifurcation on lines $l_a$ of Fig. 1.27, curve $l$ takes the form of an ergodic curve, being the image of a two-dimensional ergodic torus in the original phase space. In region III the motion is quasiperiodic and is represented by an ergodic torus in the phase space of the system.

On line $l_c$ the torus is destroyed and the invariant curve disappears. However, the point $O_1$ exists and is stable. Therefore, the synchronization regime in region II is no longer related to a resonance on the torus, but to a stable limit cycle of the original system (1.267). At the transition from region III to region II, the synchronous regime is realized by passing through the bifurcation lines $l_b$. Quasiperiodic oscillations in system (1.267) disappear softly, and the regime of the stable limit cycle arises. This mechanism is called synchronization via asynchronous suppression of oscillations. In terms of the reduced equations (1.268) this mechanism corresponds to a suppression of periodic oscillations with an amplitude $A(t)$ and to the appearance of the regime with $A = $ const. The bifurcation lines $l_a$ and $l_b$ converge at points $D$, which are called Bogdanov–Takens points [137].

It is clear that the phenomena for higher values of the parameter $\mu$, especially the crisis, cannot be described exactly by the reduced equations (1.268) derived under the assumption of small forcing. In this case, numerical methods should be applied to the original system (1.267) to build the bifurcation lines. The results of a numerical study are incorporated in Fig. 1.27. For the case of weak external driving ($\mu \leq 0.05$ in region I) numerical and analytical results coincide completely. Also lines of the Andronov–Hopf bifurcation (lines $l_h$ into Fig. 1.27) and Bogdanov–Takens points $D$ were confirmed numerically. However, this good quantitative agreement disappears above and left and right, respectively, from the points $D$ (torus birth lines). Nevertheless, it is important to emphasize that the numerically obtained bifurcations inside region I coincide with results of the analytical study of the reduced dynamics. Moreover, these bifurcations are typical for any periodic self-sustained oscillator synchronized by an external periodic force in the case of small detuning.

**Synchronization as Phase and Frequency Locking.** Though the bifurcation scenario significantly depends on the dynamics of the instantaneous amplitude, synchronization can be well represented by the long-term behavior of the instantaneous frequency of the oscillator (1.267). The instantaneous frequency is given by the time derivative of the instantaneous phase

$$\omega(t) = \frac{\mathrm{d}}{\mathrm{d}t}\,\Phi(t). \tag{1.269}$$

Obviously, this additional reduction of the dynamics, allowing a consideration of the phase dynamics only, requires further assumptions on the amplitude dynamics. It will be valid for the interesting situation of a small amplitude of the external periodic force: $\mu/(A_0\varepsilon) \ll 1$ or respectively, $\varepsilon \gg a/(2\Omega A_0)$. This condition can be established in region I of the bifurcation diagram Fig. 1.27. It is possible to show that in this case the amplitude changes much faster than the phase (to do so we compared coefficients of the linear terms, which define relaxational time scales). That is why we can substitute $\varphi = $ const. into the first equation of the system (1.268) and use the unperturbed amplitude $A_0$ in the equation for the phase:

$$\frac{\mathrm{d}}{\mathrm{d}t}\,\varphi = \Delta - \frac{\mu}{A_0}\cos\varphi. \tag{1.270}$$

This equation is one of the canonical ones in the theory of phase synchronization [143]. It can be re-written in a potential form: $\dot{\varphi} = -\mathrm{d}U(\varphi)/\mathrm{d}\varphi$ with the potential $U(\varphi) = -\Delta \cdot \varphi + \frac{\mu}{A_0}\sin\varphi$. Therefore, the dynamics of the phase difference $\varphi$ can be viewed as the motion of an overdamped particle in the tilted potential $U(\varphi)$ (see Fig. 1.29). The detuning parameter $\Delta$ determines the slope of the potential, and $\mu/A_0$ gives the height of the potential barriers. For $\Delta < \mu/A_0$ the minima of the potential $\varphi_k = \arccos{(\Delta \cdot A_0/\mu)} + 2\pi k$, $k = 0, \pm 1, \pm 2, \ldots$, correspond to synchronization as the instantaneous phase difference remains constant in time.

The instantaneous frequency is constant and matches the driving frequency in the regime of synchronization. Otherwise it changes in time and we have to calculate the mean frequency as $\langle\omega\rangle = \lim_{T\to\infty} 1/T \int_0^T \omega(t)\mathrm{d}t$. The dependence of the frequency difference $\langle\omega\rangle - \Omega$ on detuning is shown in Fig. 1.30 by the solid line. As clearly seen from this figure, the mean frequency coincides with the external frequency $\Omega$ in a finite range of $\Delta$. The plateau in Fig. 1.30 corresponds to the synchronization region. Outside this

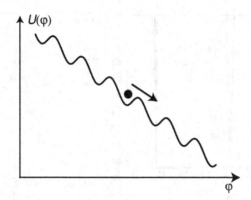

**Fig. 1.29.** Potential profile $U(\varphi)$ in the case of phase locking for $\Delta \neq 0$

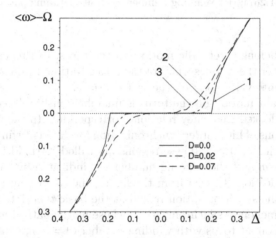

**Fig. 1.30.** Dependence of the difference between the mean frequency of oscillations in system (1.270) and the frequency of the external signal versus the detuning parameter for different values of the noise intensity $D$

region, $\langle \omega \rangle$ differs from the external frequency and two-frequency oscillations occur.

If we increase the detuning further, then higher-order regimes of synchronization can occur. To study these regimes, we introduce the ratio of the driving frequency to the mean frequency of the oscillator, $\theta = \Omega / \langle \omega \rangle$, also called the *winding number*. This ratio tells us how many periods of external force are within one period of the oscillator. Up to now we have studied the regime of 1:1 synchronization when $\theta = 1$. As we already know, this situation corresponds to the existence of the resonance stable limit cycle on a two-dimensional torus. However, with an increase of the driving frequency $\Omega$, $\theta$ also increases and can take both rational and irrational values. The structure of the phase trajectories on the torus will undergo bifurcations. Irrational

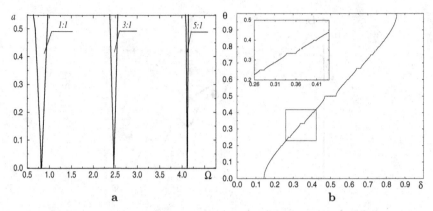

**Fig. 1.31.** (a) Typical resonance regions for the indicated values of winding number for the system (1.267); (b) Winding number $\Theta$ versus detuning parameter $\delta$ for the circle map

numbers of $\theta$ belong to ergodic motion on the torus. In this case the phase trajectories cover the whole surface of the torus. Rational values of $\theta$ conform to resonant limit cycles lying on the torus surface.

Such resonant motion on the torus is unambiguously related to synchronization, with locked frequency relations corresponding to the winding number. Some regions of high-order synchronization for different winding numbers are presented in Fig. 1.31a. These regions are called "Arnold tongues". The rational values $m{:}n$ of the winding number are indicated by the numbers in the plot. As additionally seen from the figure, the tongues are topologically equivalent to the synchronization region at the basic tone 1:1.

The phenomenon of synchronization, whose mathematical image is represented by a resonant torus with winding number $\Theta = m{:}n$, can be described using the circle map. The Poincaré section along the small torus circle gives rise to a one-dimensional map of the circle onto itself. It has the following form:

$$\varphi_{n+1} = \varphi_n + f(\varphi_n), \quad f(\varphi_n) \equiv f(\varphi_n + 2\pi k). \tag{1.271}$$

Each iteration of the map corresponds to one turn of a phase trajectory along the large torus circle, and in general leads to a shift of a representative point on the circle at a certain angle $\varphi$. If a finite number of points is fixed on the circle as $n \to \infty$, we can observe the image of a resonant torus. If the number of points is infinite and they cover the circle densely, we deal with the image of an ergodic torus in the form of an invariant circle.

The circle map is governed by the following difference equation:

$$x_{k+1} = x_k + \delta - \frac{K}{2\pi} \sin(2\pi x_k), \quad \text{mod 1.} \tag{1.272}$$

For $K = 0$ the parameter $\delta$ represents the winding number, which characterizes the ratio of two frequencies of uncoupled oscillators. If $0 < K < 1$, the $\delta$ is the frequency detuning. The map (1.272) may have a period $n$ cycle ($n = 1, 2, \ldots$). In this case we deal with the effect of synchronization. To illustrate this, let us calculate the winding number $\Theta$,

$$\Theta = \lim_{k \to \infty} \frac{x_k - x_0}{k}, \tag{1.273}$$

as a function of the parameter $\delta$ (in addition one needs to exclude the operation mod 1). The results are shown in Fig. 1.31b and indicate the presence of plateaus, which correspond to the synchronization regions with different rational winding numbers $\Theta = m : n$. The graph in Fig. 1.31b also demonstrates the property of self-similarity. The self-similarity manifests itself through the fact that between any two plateaus with winding numbers $\Theta_1 = r : s$ and $\Theta_2 = p : q$ there always exists one more region of synchronization with winding number $\Theta = \frac{r+p}{s+q}$. For this reason, the dependence $\Theta(\delta)$ in Fig. 1.31b is called the "devil's staircase". On the parameter plane $(K, \delta)$ synchronization regions, inside which the winding number $\Theta = m : n$ is rational, form Arnold tongues.

**Mutual Synchronization: Two Coupled Van der Pol Oscillators.** So far we have been concerned with forced synchronization when the driving influence on the oscillator is unidirectional without a feedback to the force. However, let us imagine the periodic force originates from a second Van der Pol oscillator and both generators interact but have different natural frequencies, $\omega_{01}$ and $\omega_{02}$. The interaction is symmetric, and we assume that one oscillator is driven by the second one with strength $\gamma$ and in proportion to the difference of their coordinates. From the physical point of view it may be realized by a spring with constant $\gamma$ which tries to synchronize the motion of the two oscillators. The particular equations are

$$\ddot{x}_1 - \varepsilon(1 - x_1^2)\dot{x}_1 + \omega_{01}^2 x_1 = \gamma(x_2 - x_1),$$
$$\ddot{x}_2 - \varepsilon(1 - x_2^2)\dot{x}_2 + \omega_{02}^2 x_2 = \gamma(x_1 - x_2), \tag{1.274}$$

and starting from here we call $\gamma$ the coupling parameter.

The question to be sketched here is: Is it possible to observe the effect of synchronization in this case and what are its peculiarities?

The answer can be given qualitatively by means of an analysis of the bifurcations in system (1.274). The structure of the bifurcation diagram for the system (1.274) is pictured in Fig. 1.32. As seen from the figure, the bifurcation diagram for the case of the two coupled oscillators is topologically equivalent to the situation in Fig. 1.31a. From the viewpoint of a bifurcation analysis the case of mutual synchronization is completely equivalent to the earlier studied case of forced synchronization.

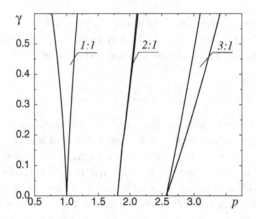

**Fig. 1.32.** Resonance regions for the system (1.274) in the "detuning–coupling" parameter plane. The parameters are $p = \omega_{02}/\omega_{01}$ and $\varepsilon = 2.0$

### 1.3.4 Synchronization in the Presence of Noise. Effective Synchronization

The above considered problems do not take into account the presence of random perturbations. Noise is inevitably present in any real system in the form of natural (or internal) fluctuations caused by the presence of dissipation as well as in the form of random perturbations of the environment.

The introduction of a phase in a noisy oscillating system requires a probabilistic approach. With the transformation (1.261), the instantaneous amplitude and phase change into stochastic variables since $x(t)$ and $\dot{x}(t)$ are stochastic. With noise taken into account, the amplitude and phase dynamics is described by stochastic differential equations including a noise term, $\xi(t)$.

To extract information from the stochastic dynamics we have to calculate the moments of $A(t)$, $\Phi(t)$ and $\omega(t) = \dot{\Phi}(t)$ or consider the transition probability density $p(A, \Phi, t \mid A^*, \Phi^*, t_0)$ which is sufficient for Markovian approximations. It gives the conditional probability to observe the amplitude $A$ and the phase $\Phi$ at time $t$ if started at time $t_0$ with $A^*$ and $\Phi^*$, respectively.

In noisy systems the phase $\Phi(t)$ as well as the difference with respect to the external driving $\varphi(t) = \Phi(t) - \Omega t$ performs motion similar to a Brownian particle in the potential $U(\varphi)$ (see Fig. 1.29). The stochastic process $\varphi(t)$ can be decomposed into two parts: a deterministic part given by its mean value or the mean value of the instantaneous frequency and a fluctuating part characterized, for example, by the diffusion coefficient around its mean value. Synchronization as a fixed relation between two phases is always interrupted by randomly occurring abrupt changes in the phase difference, also known as phase slips. Therefore, in noisy oscillating systems the notion of synchronization must be mathematically expressed by relations and conditions between the moments of the fluctuating phase or its corresponding probability density.

**Langevin Equation Description.** The stochastic force $\xi(t)$ is added to the deterministic differential equation of a periodically driven Van der Pol oscillator as

$$\ddot{x} - \varepsilon(1 - x^2)\dot{x} + \omega_0^2 x = a \, \cos \, \Omega t + \sqrt{2D_0} \, \xi(t) \,. \qquad (1.275)$$

For simplicity we argue that the noise is part of the external driving force. We assume $\xi(t)$ to be normalized Gaussian white noise with zero mean, and the new parameter, $D_0$, is the noise intensity.

Following [92] with the ansatz (1.261) we can obtain reduced equations for the stochastic amplitude and phase difference (comp. (1.173) and the following elucidation and (1.198)) :

$$\dot{A} = \frac{\varepsilon A}{2} \left(1 - \frac{A^2}{A_0^2}\right) - \mu \sin \varphi + \frac{D}{A} + \sqrt{2D}\xi_1(t) \,,$$

$$\dot{\varphi} = \Delta - \frac{\mu}{A} \cos \varphi + \frac{\sqrt{2D}}{A}\xi_2(t) \,, \qquad (1.276)$$

where (1.276) is an Ito equation and $\xi_i$ are statistically independent Gaussian noise sources obeying $\langle \xi_i(t)\xi_j(t+\tau)\rangle = \delta_{i,j}\delta(\tau)$ and $\langle \xi_i(t)\rangle = 0$ for both $i, j = 1, 2$. $D = D_0/(2\Omega^2)$ denotes the common intensity of the two transformed noise sources $\xi_{1,2}$.

Again let us consider the most interesting situation corresponding to region I in Fig. 1.27. For small noise, $D \ll \varepsilon A_0^2/2$ and a weak external signal this probability distribution is narrowly centered at $A \approx A_0$, i.e., the amplitude will be very close to its unperturbed value $A_0$. This gives us the possibility to consider the second equation of (1.276) separately by substitution of $A_0$ instead of $A$:

$$\dot{\varphi} = \Delta - \frac{\mu}{A_0} \cos \varphi + \frac{\sqrt{2D}}{A_0} \xi_2(t) \,. \qquad (1.277)$$

Therefore, the dynamics of the phase difference $\varphi$ can be viewed as the motion of an overdamped Brownian particle in the tilted potential $U(\varphi)$ (see Fig. 1.29), with the slope defined by the detuning. The parameter $\mu/A_0 = \Delta_s$ gives the height of the potential barriers. The presence of noise leads to the diffusion of the instantaneous phase difference in the potential $U(\varphi)$, that is, $\varphi(t)$ fluctuates for a long time inside a potential well (which means phase locking) and rarely makes jumps from one potential well to another (i.e., displays phase slips) changing its value by $2\pi$.

Time series of the phase difference for different values of the noise intensity $D$ obtained numerically by integrating (1.277) are shown in Fig. 1.33. As clearly seen from this figure, for small noise intensities ($D = 0.02$) the instantaneous phase difference remains bounded during a long observation time. The increase of the noise intensity leads to a decrease of the average

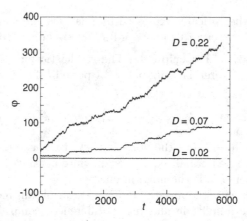

**Fig. 1.33.** Time dependence of the instantaneous phase difference for the indicated values of noise intensity. Other parameters are $\Delta = 0.06$, $\mu = 0.15$ and $A_0 = 1$

duration of residence times inside a potential well and causes the hopping dynamics of the phase difference ($D = 0.07$). Although the phase locking epochs, $\varphi \approx$ const., are clearly seen, the mean value of the phase difference increases in time. Evidently, for a large slope (detuning) and for a small value of the periodic force amplitude, the jumps from one metastable state to another become very frequent, and the duration of phase locking segments becomes very short. This leads to the growth of phase difference (see the dependences of $\varphi(t)$ for $D = 0.22$), causing a change of the mean frequency of oscillations.

**Fokker–Planck Equation Description.** The FPE corresponding to the stochastic differential equation (1.277) is

$$\frac{\partial p(\varphi, t)}{\partial t} = -\frac{\partial}{\partial \varphi}\left[(\Delta - \Delta_s \cos \varphi)\, p(\varphi, t) - Q\frac{\partial p(\varphi, t)}{\partial \varphi}\right], \qquad (1.278)$$

where $Q = D/A_0^2$. The phase difference $\varphi$ is an unbounded variable and the stochastic process defined by (1.278) is nonstationary. However, since coefficients of the FPE are periodic with respect to $\varphi$, we can introduce the probability distribution $P(\varphi, t)$ of the wrapped phase, which is bounded in $[-\pi, \pi]$:

$$P(\varphi, t) = \sum_{n=-\infty}^{\infty} p\left(\varphi + 2\pi n, t\right). \qquad (1.279)$$

The FPE for $P(\varphi, t)$ has the same structure as (1.278), but now we can find the stationary probability density $P_{st}(\varphi)$, taking into account the periodic boundary conditions $P(-\pi, t) = P(\pi, t)$ and the normalization condition $\int_{-\pi}^{\pi} P(\varphi, t)\mathrm{d}\varphi = 1$ [92]:

$$P_{st}(\varphi) = N \exp\left(\frac{\Delta \cdot \varphi - \Delta_s \sin\varphi}{Q}\right) \int_{\varphi}^{\varphi+2\pi} \exp\left(-\frac{\Delta \cdot \psi - \Delta_s \sin\psi}{Q}\right) d\psi,$$
$$-\pi \le \varphi \le \pi, \qquad (1.280)$$

where $N$ is the normalization constant. In the particular case when $\Delta = 0$, e.g., when the natural frequency of the oscillator exactly matches the driving frequency, the stationary probability density of the wrapped phase difference takes a simple form:

$$P_{st}(\varphi) = \frac{1}{2\pi I_0(\Delta_s/Q)} \exp\left(\frac{\Delta_s}{Q}\cos(\varphi + \pi/2)\right), \quad -\pi \le \varphi \le \pi, \quad (1.281)$$

where $I_0(z)$ is the modified Bessel function. For a large noise intensity, $I_0(\Delta_s/Q) \approx 1$ and $\exp\left[(\Delta_s/Q)\cos(\varphi + \pi/2)\right] \approx 1$; thus the stationary probability density tends to the uniform one, $P_{st}(\varphi) = 1/2\pi$. This situation indeed corresponds to the absence of synchronization. Otherwise, for very weak noise, $\cos(\varphi + \pi/2) \approx 1 - (\varphi + \pi/2)^2/2$, $I_0(\Delta_s/Q) \approx \exp(\Delta_s/Q)/\sqrt{2\pi\Delta_s/Q}$, and the stationary probability density has a Gaussian shape: $P_{st}(\varphi) = \exp\left(-\Delta_s(\varphi + \pi/2)^2/2Q\right)/\sqrt{2\pi\Delta_s/Q}$ centered at $\varphi_0 = -\pi/2$. The pronounced Gaussian peak in the stationary probability density of the phase difference indicates phase locking. In the limit $Q \to 0$ the probability density becomes a $\delta$-function, i.e., $\lim_{Q\to 0} P_{st}(\varphi) = \delta(\varphi + \pi/2)$.

The mean frequency of oscillations $\langle\omega\rangle$ can be found via the stationary probability density $P_{st}(\varphi)$ of the wrapped phase difference:

$$\langle\omega\rangle = \langle\dot\varphi\rangle + \Omega = \int_{-\pi}^{\pi} (\Delta - \Delta_s\cos\varphi)\, P_{st}(\varphi) d\varphi + \Omega, \qquad (1.282)$$

where $\Omega$ is the frequency of a synchronizing signal. The difference between the mean frequency and the external frequency versus detuning parameter is presented in Fig. 1.30 (curves 2 and 3) for different values of the noise intensity. With an increase in noise intensity the region of frequency locking shrinks, which is another manifestation of noise-induced breakdown of synchronization in the Van der Pol oscillator.

Let us now go back to Fig. 1.33, where the noise-induced motion of the unwrapped phase difference is shown. Let the distribution of the phase difference be initially concentrated at some initial value $\varphi_0$, i.e., $p(\varphi, t = 0) = \delta(\varphi - \varphi_0)$, so that $\langle\varphi^2(t = 0)\rangle = 0$. Due to noise the phase difference diffuses according to the law $\langle\varphi^2(t)\rangle \propto D_{eff} \cdot t$ [92], where $D_{eff}$ is the effective diffusion constant which measures the rate of diffusion:

$$D_{eff} = \frac{1}{2}\frac{d}{dt}\left[\langle\varphi^2(t)\rangle - \langle\varphi(t)\rangle^2\right]. \qquad (1.283)$$

In the absence of noise there is no phase diffusion, $D_{eff} = 0$. With the increase in noise intensity the effective diffusion constant also increases, so that the diffusion is enhanced. In Fig. 1.33 this situation corresponds to very frequent phase slips. The effective diffusion constant is therefore connected to the mean duration of the phase locking epochs: the longer the phase locking segments, the slower the spreading of the phase difference, and, thus, the smaller the effective diffusion constant. The effective diffusion constant can be estimated analytically for the case of (1.277) by solving the Kramers problem [87] regarding the escape from a well of the potential $U(\varphi)$ [92]:

$$
D_{\text{eff}} = \frac{\sqrt{\Delta_s^2 - \Delta^2}}{2\pi} \left[ 1 + \exp\left( -\frac{2\pi\Delta}{Q} \right) \right]
$$

$$
\times \exp\left[ -\frac{2}{Q} \left( \sqrt{\Delta_s^2 - \Delta^2} - \Delta \cdot \arcsin\frac{\Delta}{\Delta_s} \right) \right]. \tag{1.284}
$$

Thus, the effective diffusion measures the average number of $2\pi$-jumps of the phase difference per unit time and grows exponentially with an increase of the noise intensity.

### 1.3.5 Phase Description

To discuss synchronization in complex systems as an entrainment of frequencies and phases an *instantaneous phase* $\Phi(t)$ has to be assigned to aperiodic processes. Compared with harmonic and quasiharmonic oscillations this will require new concepts. Therefore, we first look for possible generalizations of the notion of an *instantaneous phase* as introduced in Sect. 1.3.2. The proposed approaches do not account for generality. A lot of other alternatives exist. The usage of various definitions strongly depends on the DS and the signals under consideration.

Later on, we show their applicability for the purpose to find synchronization in stochastic systems, specifically for examples of stochastic resonance (SR). With a given stochastic $\Phi(t)$, stochastic synchronization will be defined by making use of the mean velocity or mean angular frequency,

$$
\langle \omega \rangle = \lim_{T \to \infty} \frac{1}{T} \int_{t_0}^{t_0+T} \frac{d\Phi(t)}{dt} \, dt = \lim_{T \to \infty} \frac{1}{T} (\Phi(t_0 + T) - \Phi(t_0)), \tag{1.285}
$$

and the effective diffusion coefficient defined by (1.283).

**Phases in the Analytic Signal Representation.** One new concept is found from a generalization of the analytic expansion of oscillations and was successfully applied to a description of chaos synchronization. For a signal $x(t)$ one constructs an analytic signal $w(t)$ in the complex plane by

$$
w(t) = x(t) + iy(t) = A(t)e^{i\Phi(t)}. \tag{1.286}
$$

Then by definition the instantaneous amplitude and phase straightforwardly follow:

$$
A(t) = \sqrt{x^2(t) + y^2(t)}, \quad \Phi(t) = \arctan\left(\frac{y}{x}\right) + \pi k, \quad k = 0, \pm 1, \ldots \tag{1.287}
$$

Immediately, one can define the instantaneous frequency as

$$\omega(t) = \frac{d}{dt}\Phi(t) = \frac{1}{A^2(t)}[x(t)\dot{y}(t) - y(t)\dot{x}(t)]. \tag{1.288}$$

Until now the choice of $y(t)$ has not been specified. The key point in the analytic signal representation [144–146] is the usage of the Hilbert transform of $x(t)$ in the definition of $y(t)$. It is not a unique choice of selecting the conjugated variable. Therefore, let a short motivation support this selection showing that it generally complies with well-known definitions.

In case of harmonic processes one adds the signal $x(t) = A\cos(\omega t)$ to the conjugated signal $y(t) = A\sin(\omega t)$ to obtain the correct phase $\Phi = \omega t$. In more detail, this map $x \to y$ can be found by $x$ taking $y = -\dot{x}\,\text{sign}(\omega)/\omega$. Hence, it differs for positive and negative frequencies, but the map effects a phase shift, only. For positive frequencies $y$ advances $x$ by $+\pi/2$; in contrast to the case of $\omega < 0$ where it is delayed by $-\pi/2$.

Additionally one requires that $y(t)$ results from a convolution of $x(t)$, i.e.,

$$y(t) = \int_{-\infty}^{\infty} K(t-\tau)x(\tau)d\tau. \tag{1.289}$$

What does the convolution kernel look like for the harmonic process? Convoluting $x(t)$ means, in the spectral representation, multiplication of the Fourier transforms of $x$ and $K$ to obtain that of $y$. According to the convolution theorem of Fourier transforms:

$$y_\omega = K_\omega x_\omega. \tag{1.290}$$

The necessary phase shift is obtained by multiplying $x_\omega$ by $\pm i$. Hence, the Fourier transform of the kernel is

$$K_\omega = -i\,\text{sign}(\omega) \tag{1.291}$$

with $\text{sign}(0) = 0$. Performing the inverse Fourier transform gives

$$K(t) = \frac{1}{\pi t}. \tag{1.292}$$

The described procedure for harmonic signals can be generalized to complex processes and gives an advice how to construct the analytic signal. With the kernel $y(t)$ is given by the Hilbert transform as

$$y(t) = H[x] = \frac{1}{\pi}\int_{-\infty}^{\infty}\frac{x(\tau)}{t-\tau}d\tau = \frac{1}{\pi}\int_{0}^{\infty}\frac{x(t-\tau) - x(t+\tau)}{\tau}d\tau. \tag{1.293}$$

The integral in the latter expression is taken in the sense of the Cauchy principal value.

As a linear transformation $H[x]$ obeys several useful properties. Every Hilbert transform of a linear superposition of two signals is the superposition of the separate Hilbert transforms. If the time of the signal is shifted by some amount, it also shifts the argument of the Hilbert transform. The Hilbert transform of a Hilbert transform gives the negative original signal. Even functions give odd Hilbert transforms and vice versa. The original signal and the Hilbert transform are orthogonal. The full energy of the original signal, the integral of $x^2(t)$ over all times, is equal to the energy of the transformed one.

Hilbert transforms can also be performed for stochastic variables. In the case of a stochastic signal $x(t)$, the convergence of this integral should be understood in the sense of the mean square [147]. One finds that the transformed signal is correlated in the same manner as the original signal. But both are anticorrelated, with the correlation function being the Hilbert transform of the autocorrelation function.

In some special cases a simpler definition can be used to introduce the instantaneous amplitude and phase of oscillations [150]. This approach may be understood as a generalization of the classical phase definition to the case when the phase trajectory rotates irregularly about an unstable equilibrium point with some time-dependent velocity. In this case the instantaneous phase can be defined as the rotation angle of a radius vector projection on some plane of variables $(x, z)$. The radius vector specifies the phase point location and usually originates from the equilibrium point. One can introduce the following change of variables:

$$x(t) = A(t) \cos \Phi(t),$$
$$z(t) = A(t) \sin \Phi(t). \tag{1.294}$$

Instantaneous phase $\Phi(t)$ and amplitude $A(t)$ are determined by a formula similar to (1.287), but the Hilbert transform $y(t)$ is replaced by the dynamical variable $z(t)$. Again, the instantaneous frequency $\omega$ is introduced as a derivative of the instantaneous phase and the mean frequency $\langle \omega \rangle$ is given by (1.285).

**Phases for Discrete Events.** Stochastic bistable dynamics at the global scale can be well approximated as a discrete process. We will interpret changes of the phase as a hopping between attractors. These events occur at random times $t_k$ given by a distribution function or a dynamical process. Thus the stochastic bistable system is reduced to a simple point process.

The time between two successive crossings is $\tau(t) = t_{k+1} - t_k$, $t_k < t < t_{k+1}$. The instantaneous phase of $x(t)$ can be defined as

$$\Phi(t)_{\text{lin}} = \pi \frac{t - t_k}{t_{k+1} - t_k} + \pi k, \quad t_k < t < t_{k+1}, \tag{1.295}$$

which is a piecewise linear function of time. The mean frequency is

$$\langle\omega\rangle = \lim_{M\to\infty} \frac{1}{M} \sum_{k=1}^{M} \frac{\pi}{t_{k+1} - t_k} . \tag{1.296}$$

Another definition neglects the linear interpolation between two subsequent events. One takes

$$\Phi(t)_{\text{discr}} = \pi k(t) = \pi \sum_k \theta(t - t_k), \tag{1.297}$$

with $k(t)$ being a sequence of increasing integers and $t_k$ is again subject to some dynamics or distribution. $\theta(x)$ stands for the Heavyside function. The mean frequency of this definition takes the form of an average of a sequence of delta pulses:

$$\langle\omega\rangle = \pi \left\langle \sum_k \delta(t_k - t) \right\rangle. \tag{1.298}$$

Correspondingly, the continuous stochastic process is mapped on a dichotomic process

$$x(t) = \exp[i\,\Phi(t)_{\text{discr}}]. \tag{1.299}$$

It represents a time sequence of successive states $+1$ and $-1$, as in a two-state approximation. The monotonous increase in the phase $\Phi(t)$ characterizes the temporal structure of the time sequence. Generally, it may be periodic or random or somewhere in between as in the case of SR. Thus, we are speaking about dichotomic periodic and random sequences, respectively.

Let us compare the definitions of the phase (1.295) and (1.297) with that of the analytic signal concept [148]. For this purpose we analytically calculate the Hilbert transform of a dichotomic process $x(t)$ (even $k$ stands for a transition $-1 \to +1$)

$$y(t) = \frac{1}{\pi} \sum_k (-1)^k \ln \left| \frac{t - t_k}{t - t_{k-1}} \right| = \frac{2}{\pi} \sum_k \ln \left| \frac{t - t_{2k}}{t - t_{2k+1}} \right|. \tag{1.300}$$

For $t < t_k$, $x(t) = -1$, and the Hilbert transform $y(t)$ decreases monotonously and reaches $-\infty$ at $t = t_k$. At this moment $x$ switches to $+1$, and $y(t)$ starts to grow, becoming $+\infty$ at $t = t_{k+1}$. Thus an arrow following $x$ and $y$ in phase space completes a circle during two subsequent transitions, returning to the initial state $x = -1$. According to definition (1.287) the instantaneous phase from the analytic signal reads

$$\Phi(t)_{\text{Hilbert}} = \pi k(t) + \arctan \frac{y(t)}{x(t)}, \tag{1.301}$$

with $y(t)$ given by (1.300).

Figure 1.34 compares the three different phases graphically. Both, the piecewise linear and the piecewise constant definition perfectly shadow $\Phi_{\text{Hilbert}}$ from the analytic signal. Hence, we find that the phase concept in stochastic point processes well approximates the findings from the analytic signal [148].

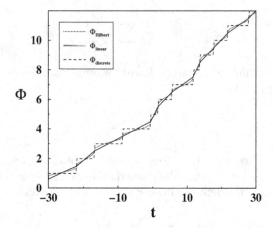

**Fig. 1.34.** Comparison of the three different definitions of the instantaneous phase. Both the linear interpolation and the definition by piecewise constant segments approximate the phase from the analytic signal well

The above-described definitions of the instantaneous phase and the mean frequency can be used to develop a concept of phase–frequency synchronization of both chaotic oscillations and stochastic processes in bistable and excitable systems. A generalized definition of phase synchronization, which can also be applied to these cases, is based on the restriction [149, 150]:

$$\lim_{t \to \infty} |n\Phi_1(t) - m\Phi_2(t)| < \text{const.} \tag{1.302}$$

Numerical experiments indicate that, although the precise values of instantaneous phases and frequencies defined in different approaches may be slightly different, the behavior of the mean frequencies practically proves to be the same. Hence, any of the above-listed definitions of an instantaneous phase can be used to identify the phase–frequency synchronization. The mean frequency is related to a characteristic time of the system. For the regime of spiral chaos, the characteristic time is represented by the average return period $\langle T \rangle$ of a phase trajectory to some secant surface. The following equality is valid:

$$\langle T \rangle = \frac{2\pi}{\langle \omega \rangle}. \tag{1.303}$$

Moreover, in this case the mean frequency of oscillations $\langle \omega \rangle$ practically, i.e., within the calculation accuracy, coincides with the basic frequency $\omega_0$ occurring in the power spectrum.

### 1.3.6 Summary

Concluding this section we point out that the analysis of different types of synchronization allows us to formulate some fundamental properties and criteria of synchronization. The fact that we first considered the case of forced synchronization is not of principal importance. It was important that the external force is periodic. This follows, in particular, from the qualitative consideration of the dynamics of two symmetrically coupled generators. It is known that oscillations in real generators are periodic but not harmonic. The basic indication of both forced and mutual synchronization is the appearance of the oscillatory regime with a constant and rational winding number $\theta = m{:}n$ which holds in some finite region of the system's parameter space. This region is called the synchronization region and is characterized by the effects of phase and frequency locking. Frequency locking means a rational ratio of two initially independent frequencies $\omega_1/\omega_2 = m{:}n$ everywhere in the synchronization region. Phase locking means that the instantaneous phase difference is constant in the synchronization region ($\dot{\varphi} = 0$, $\varphi_{st} = \text{const.}$).

The influence of noise on an oscillator leads to destruction of the synchronization regime in the sense of the above-given definition. However, if the noise intensity is relatively small, the main physical properties of synchronization may survive.

Due to phase diffusion, the definition of synchronization in the presence of noise appears to be "blurred". That is why the conditions of synchronization should be defined in a statistical way, by using the notion of effective synchronization [151]. This can be done by imposing a certain restriction on some statistical measures of corresponding stochastic processes. In particular, a definition of effective synchronization can be based on the following items:

- The stationary probability density of the wrapped phase difference. In this case a peak in $P_{st}(\varphi)$ should be well expressed in comparison to the uniform distribution.
- The mean frequency. This should match the driving frequency (up to some small statistical error).
- The effective diffusion constant. This measure should be small enough so that the phase locking segments are much longer than one period of the external force. In other words, this restriction requires that the oscillator phase is locked during a considerable number of periods of the external signal and can be expressed as

$$D_{\text{eff}} \leqslant \frac{\Omega}{n},$$
(1.304)

where $n \gg 1$ is the number of periods of the external force.

Utilizing these definitions of effective synchronization, we can define synchronization regions in the parameter space.

# References

1. A.A. Andronov, E.A. Vitt, S.E. Khaikin, *Theory of Oscillations* (Pergamon Press, Oxford 1966).

2. V.S. Anishchenko, *Dynamical Chaos – Models and Experiments* (World Scientific, Singapore 1995).
3. Yu.I. Neimark, *Method of Return Mappings in the Theory of Nonlinear Oscillations* (Nauka, Moscow 1972) (in Russian).
4. A.M. Lyapunov, *Collection of Works*, Vols. 1 and 2 (Academic, Moscow 1954–1956) (in Russian).
5. A.A. Andronov, E.A. Leontovich, I.E. Gordon, A.G. Maier, *The Theory of Dynamical Systems on a Plane* (Israel Program of Scientific Translations, Israel 1973).
6. V.S. Afraimovich, V.I. Arnold, Yu.S. Il'yashenko, L.P. Shilnikov, In: *Dynamical Systems V, Encyclopedia of Mathematical Sciences* (Springer, Berlin, Heidelberg 1989).
7. H. Haken, *Advanced Synergetics* (Springer, Berlin, Heidelberg 1985).
8. B.P. Demidovich, *Lectures on Mathematical Theory of Stability* (Nauka, Moscow 1967) (in Russian).
9. S. Smale, Bull. Am. Math. Soc. **73**, 747 (1967).
10. L.P. Shilnikov, Int. Bifurc. Chaos **7**, 1953 (1997).
11. C. Grebogi, E. Ott, J.A. Yorke, Phys. Rev. Lett. **48**, 1507 (1982).
12. T. Poston, I. Steward, *Catastrophe Theory and its Applications* (Pitman, London 1978).
13. R. Thom, *Structural Stability and Morphogenesis* (Benjamin-Addison Wesley, 1975).
14. H. Whitney, Ann. Math. **62**, 247 (1955).
15. V.I. Arnold, A.N. Varchenko, S.M. Gusein-Zade, *Singularities of Differentiable Mappings* (Nauka, Moscow 1968).
16. L.E. Marsden, M. McCraken, *The Hopf Bifurcation and its Applications* (Springer, New York 1976).
17. L.P. Shilnikov, Math. USSR Sb. **61**, 443 (1963) (in Russian).
18. L.P. Shilnikov, Math. USSR Sb. **10**, 91 (1970) (in Russian).
19. N.K. Gavrilov, L.P. Shilnikov, Math. USSR Sb. **17**, 467 (1972); Math. USSR Sb. **19**, 139 (1973) (in Russian).
20. S.E. Hewhouse, Publ. Math. IHES **50**, 101 (1979).
21. D. Ruelle, F. Takens, Commun. Math. Phys. **20**, 167 (1971).
22. V.S. Anishchenko, G.I. Strelkova, Discrete Dyn. Nat. Soc. **2**, 53 (1998).
23. V.S. Afraimovich, L.P. Shilnikov, "Strange Attractors and Quasiattractors", In: *Nonlinear Dynamics and Turbulence*, ed. by G.I. Barenblatt, G. Iooss, D.D. Joseph (Pitman, Boston 1983) p. 1.
24. V.S. Afraimovich, "Attractors", In: *Nonlinear Waves*, ed. by A.V. Gaponov, M.I. Rabinovich, J. Engelbrechet (Springer, Berlin, Heidelberg 1989) p. 6.
25. L.P. Shilnikov, J. Circuit Syst. Comput. **3**, 1 (1993).
26. J.L. Kaplan, J.A. Yorke, Lect. Notes Math. **730**, 204 (1971).
27. J.D. Farmer, E. Ott, J.A. Yorke, Physica D **7**, 153 (1983).
28. R. Williams, Publ. Math. IHES **43**, 169 (1974).
29. R.V. Plykin, Uspekhi Math. Nauk. **35**, 94 (1980) (in Russian).
30. S.V. Gonchenko, L.P. Shilnikov, D.V. Turaev, Sov. Math. Dokl. **44**, 422 (1992) (in Russian).
31. Y.-C. Lai, C. Grebogi, J.A. Yorke, I. Kan, Nonlinearity **6**, 779 (1993).
32. M.A. Henon, Commun. Math. Phys. **50**, 69 (1976).
33. V.I. Arnold, *Additional Chapters of the Theory of Ordinary Differential Equations* (Nauka, Moscow 1978) (in Russian).
34. R. Lozi, J. Phys. **39**, 9 (1978).
35. E.N. Lorenz, J. Atmos. Sci. **20**, 130 (1963).

36. V.S. Anishchenko, T.E. Vadivasova, J. Kurths, A.S. Kopeikin, G.I. Strelkova, Phys. Lett. A **270**, 301 (2000).
37. C. Grebogi, E. Ott, S. Pelikan, J.A. Yorke, Physica D **13**, 261 (1984).
38. V.S. Anishchenko, T.E. Vadivasova, O.V. Sosnovtseva, Phys. Rev. E **53**, 4451 (1996).
39. J.F. Heagy, S.M. Hammel, Physica D **70**, 140 (1994).
40. L.D. Landau, E.M. Lifshitz, *Lehrbuch der Theoretischen Physik*, Band 5, Statistische Physik, 3rd ed. (Akademie, Berlin 1971).
41. Sh. Kogan, *Electronic Noise and Fluctuations in Solids* (Cambridge University Press, Cambridge 1996).
42. W.T. Coffey, Yu.P. Kalmykov, J.T. Waldron, *The Langevin equation with Applications in Physics, Chemistry and Electrical Engineering* (World Scientific, Singapore 1996).
43. N.G. Van Kampen, *Stochastic Processes in Physics and Chemistry*, 2nd ed. (North Holland, Amsterdam 1992).
44. W. Horsthemke, R. Lefever, *Noise Induced Transitions, Theory and Applications in Physics, Chemistry and Biology*, Springer Series in Synergetics, Vol. 15 (Springer, Berlin 1983).
45. R. Kubo, M. Toda, N. Hashitsume, *Statistical Physics II*, Springer Series in Solid-State Science, Vol. 31 (Springer, Berlin, Heidelberg 1985).
46. P. Hänggi, H. Thomas, Phys. Rep. **88**, 207 (1982).
47. R.L. Stratonovich, *Nonlinear Nonequilibrium Thermodynamics* (Nauka, Moscow 1985) (in Russian); [Springer Series in Synergetics, Vols. 57 and 59 (Springer, Berlin, Heidelberg 1992, 1994)].
48. G. Nicolis, I. Prigogine, *Self-Organization in Non-Equilibruium Systems* (Wiley, New York 1977).
49. R. Feistel, W. Ebeling, *Evolution of Complex Systems* (Kluwer, Dordrecht 1989).
50. Yu.L. Klimontovich, *Statistical Physics*, Vol. 1 (Kluwer, Dordrecht 1995)
51. G. Nicolis, *Introduction to Nonlinear Science* (Cambridge University Press, Cambridge 1995).
52. F. Moss, P.V.E. McClintock (eds.), *Noise in Nonlinear Dynamical Systems*, 3 Vol. (Cambridge University Press, Cambridge 1990).
53. A. Bulsara, S. Chillemi, L. Kiss, P.V.E. McClintock, R. Mannella, F. Marchesoni, C. Nicolis, K. Wiesenfeld (eds.), *Stochastic Resonance, Signal Processing and Related Phenomena*, Nuovo Cimento D **17**, 653 (1995).
54. P. Hänggi, P. Jung, Adv. Chem. Phys. **89**, 239 (1995).
55. M. Milonas (ed.), *Fluctuations and Order, The New Synthesis* (Springer, New York 1996).
56. L. Schimansky-Geier, T. Pöschel (eds.), *Stochastic Dynamics*, Lecture Notes in Physics, Vol. 484 (Springer, Berlin 1997).
57. J.B. Kadtke, A. Bulsara (eds.), *Applied Nonlinear Dynamics and Stochastic Systems Near the Millennium*, APS Conf. Proc., Vol. 411 (AIP, Woodbury, N. Y. 1997).
58. D.G. Luchinsky, P.V.E. McClintock, M.I. Dykman, Rep. Prog. Phys. **61**, 889 (1998).
59. J. García-Ojalvo, J.M. Sancho, *Noise in Spatially Extended Systems* (Springer, New York 1999).
60. D.S. Broomhead, E.A. Luchinskaya, P.V.E. McClintock, T. Mullin (eds.), *Stochastic and Chaotic Dynamics in the Lakes*, APS Conf. Proc., Vol. 502 (AIP, Melville 2000).

61. J.A. Freund, T. Pöschel T. (eds.), *Stochastic Processes in Physics, Chemistry and Biology*, Lecture Notes in Physics, Vol. 557 (Springer, Berlin, Heidelberg 2000).
62. M. San Miguel, R. Toral, In: *Instabilities and Nonequilibrium Structures VI*, ed. by E. Tirapegui, J. Martnez, R. Tiemann (Kluwer, Dordrecht 2000).
63. A.T. Bharucha-Reid, *Elements in the Theory of Markovian Processes and their Applications* (McGraw-Hill, New York 1960).
64. A. Papoulis, *Probability, Random Variables, and Stochastic Processes* (McGraw-Hill, New York 1965).
65. Yu.A. Rusanov, *Stochastic Processes* (Nauka, Moscow 1971) (in Russian).
66. E. Wong, *Stochastic Processes in Information and Dynamical Systems*, Mc-Graw Series in System Science (McGraw-Hill, New York 1971).
67. B. Karlin, M.M. Taylor, *A First Course in Stochastic Processes*, 2nd ed. (Academic, New York 1975).
68. W. Feller, *An Introduction to Probability Theory and Applications*, Vol. 2 (Wiley, New York 1971).
69. M.I. Freidlin, A.D. Wencel, *Random Perturbations in Dynamical Systems* (Springer, New York 1984).
70. L. Arnold, *Stochastic Differential Equations, Theory and Applications* (Krieger, Malabar 1992).
71. C.W. Gardiner, *Handbook of Stochastic Methods*, Springer Series in Synergetics, Vol. 13 (Springer, Berlin, Heidelberg 1982).
72. H. Risken, *The Fokker–Planck Equation*, Springer Series in Synergetics, Vol. 18 (Springer, Berlin, Heidelberg 1984).
73. M. Lax, Rev. Mod. Phys. **38**, 541 (1966).
74. M. Lax, *Fluctuations and Coherence in Classical and Quantum Physics* (Gordon and Breach, New York 1968).
75. H. Haken, Rev. Mod. Phys. **47**, 67 (1975).
76. R. Graham, *Springer Tracts in Mod. Phys.* 66 (Springer, Berlin, Heidelberg 1973) p. 1.
77. S.M. Rytov, *Introduction in Statistical Radiophysics* (Nauka, Moscow 1966) (in Russian).
78. H.M. Ito, J. Stat. Phys. **35**, 151 (1984).
79. B.P. Levin, *Theoretical Foundations of Radio Engineering*, Vol. 1 (Sov. Radio, Moscow 1974) (in Russian).
80. E.W. Montroll, K.E. Shuler, Adv. Chem. Phys. **1**, 361 (1969).
81. E. W. Montroll and J. L. Lebowitz, (eds.), *Fluctuation Phenomena*, 2nd ed. (North Holland., Amsterdam, 1987).
82. C. van den Broeck, J. Stat Phys. **31**, 467 (1983).
83. *Chua's circuit: A paradigm for chaos*, ed. by R.N. Madan (World Scientific, Singapore, 1993).
84. R.L. Stratonovich, "Some Markov Methods in the Theory of Stochastic Processes in Nonlinear Dynamical Systems", In: *Noise in Nonlinear Dynamical Systems*, ed. by F. Moss, P.V.E. McClintock (Cambridge University Press, Cambridge 1990), p. 16.
85. V.I. Tikhonov, M.A. Mironov, *Markovian Processes* (Sov. Radio, Moscow 1979) (in Russian).
86. N. Wax (ed.), *Selected Papers on Noise* (Dover, New York 1954).
87. P. Hänggi, M. Borkovec, P. Talkner, Rev. Mod. Phys. **62**, 251 (1990).
88. C. Van den Broeck, P. Hänggi, Phys. Rev. A **30**, 2730 (1984).
89. W. Ebeling, Phys. Lett. A **68**, 430 (1978).
90. W. Ebeling, H. Engel, H. Herzel, *Selbstorganisation in der Zeit* (Akademie, Berlin 1990).

91. N.N. Bogolyubov, Yu.A. Mitroplski, *Asymtotic Methods in the Theory of Non-linear Oscillations* (Fismatgiz, Moscow 1963) (in Russian).
92. R.L. Stratonovich, *Selected Problems of Fluctuation Theory in Radiotechnics* (Sov. Radio, Moscow 1961) (in Russian); *Selected Topics in the Theory of Random Noise*, Vols. 1 and 2 (Gordon and Breach, New York 1963, 1967).
93. P.I. Kuznetsov, R.L. Stratonovich, V.I. Tikhonov, *Non-Linear Transformations of Stochastic Processes* (Pergamon, Oxford 1965).
94. R.L. Stratonovich, P.S. Landa, In: *Non-Linear Transformation of Stochastic Processes*, ed. by P.I. Kuznetsov, R.L. Stratonovich, V.I. Tikhonov (Pergamon, Oxford 1965) p. 259
95. R.L. Stratonovich, Yu.M. Romanovski, In: *Non-Linear Transformation of Stochastic Processes*, ed. by P.I. Kuznetsov, R.L. Stratonovich, V.I. Tikhonov (Pergamon, Oxford 1965).
96. E. Shidlovskaya, L. Schimansky-Geier, Yu.M. Romanovski, Z. Phys. Chemie **214**, 65 (2000).
97. P. Hänggi, P. Riseborough, Am. J. Phys. **51**, 347 (1983).
98. H.A. Kramers, Physica **7**, 284 (1940).
99. W. Ebeling, *Strukturbildung bei irreversiblen Prozessen* (Teubner, Leipzig 1976).
100. H. Malchow, L. Schimansky-Geier, *Noise and Diffusion in Bistable Nonequilibrium Systems* (Teubner, Leipzig 1985).
101. L. Farkas, Z. Phys. Chem. **125**, 236 (1927).
102. R. Becker, W. Döring, Ann. Phys. **24**, 719 (1935).
103. L.S. Pontryagin, A.A. Andronov, A.A. Vitt, J. Exp. Theor. Phys. **3**, 165 (1933) (in Russian); ibid., *Non-Linear Transformation of Stochastic Processes,* ed. by P.I. Kuznetsov, R.L. Stratonovich, V.I. Tikhonov (Pergamon, Oxford 1965), Vol. 1.
104. G.H. Weiss, Adv. Chem. Phys. **13**, 1 (1967).
105. B.J. Matkowsky, Z. Schuss, SIAM J. Appl. Math. **33**, 365 (1977).
106. P. Reimann, G.J. Schmid, P. Hänggi, Phys. Rev. E **60**, 1 (1999).
107. S.O. Rice, Bell Syst. Technol. J. **23**, 1 (1944); in: *Selcted Papers on Noise*, ed. by N. Wax (Dover, New York 1954).
108. J.S. Langer, Ann. Phys.(N.Y.) **54**, 258 (1969).
109. V.I. Mel'nikov, Phys. Rep. **209**, 1 (1991).
110. P. Jung, Phys. Rep. **234**, 175 (1993).
111. P. Hänggi, P. Talkner (eds.), *New Trends in Kramers Reaction Rate Theory* (Kluwer, Boston 1995).
112. H. Frauenfelder, P.G. Wolynes, R.H. Austin, Rev. Mod. Phys. **71**, 149 (1999).
113. I. Blekhman, *Synchronization of Dynamical Systems* (Nauka, Moscow 1971) (in Russian); I. Blekhman, *Synchronization in Science and Technology* (Nauka, Moscow 1981) (in Russian); English translation, (ASME, New York 1988).
114. Y. Soen, N. Cohen, D. Lipson, E. Braun, Phys. Rev. Lett. **82**, 3556 (1999).
115. A. Neiman, X. Pei, D. Russell, W. Wojtenek, L. Wilkens, F. Moss, H. Braun, M. Huber, K. Voigt, Phys. Rev. Lett. **82**, 660 (1999).
116. R.C. Elson, A.I. Selverston, R. Huerta, N.F. Rulkov, M.I. Rabinovich, H.D.I. Abarbanel, Phys. Rev. Lett. **81**, 5692 (1998).
117. See, for instance, H.D.I. Abarbanel, M. I. Rabinovich, A. Selverston, M.V. Bazhenov, R. Huerta, M.M. Sushchik, L.L. Rubchinskii, Phys.–Uspekhi **39**, 337 (1996).
118. A.T. Winfree, J. Theor. Biol. **16**, 15; A.T. Winfree, *The Geometry of Biological Time* (Springer, New York 1980); J. Buck, Q. Rev. Biol. **63**, 265 (1988); S.H. Strogatz, I. Stewart, Sci. Am. **269**, 102 (1993).

119. C. Schäfer, M. Rosenblum, J. Kurths, H. Abel, Nature **392**, 239 (1998); Phys. Rev. E **60**, 857 (1998).
120. P. Tass, M. Rosenblum, J. Weule, J. Kurths, A. Pikovsky, J. Volkmann, A. Schnitzler, H. Freund, Phys. Rev. Lett. **81**, 3291 (1998).
121. V.A. Vasilev, Yu.M. Romanovsky, D.S. Chernavsky, V.G. Yakhno, *Autowave Processes in Kinetic Systems* (VEB Deutscher Verlag der Wissenschaften, Berlin 1987).
122. V. Anishchenko, I. Aranson, D. Postnov, M. Rabinovich, Dokl. Acad. Nauk. SSSR **286**, 1120 (1986) (in Russian).
123. V. Belykh, N. Verichev, L. Kocarev, L. Chua, J. Circuit Syst. Comput. **50**, 1874 (1993).
124. L. Pecora, T. Carroll, Phys. Rev. Lett. **64**, 821 (1990).
125. G.V. Osipov, A.S. Pikovsky, M.G. Rosenblum, J. Kurths, Phys. Rev. E **55**, 2353 (1997).
126. L. Kocarev, U. Parlitz, Phys. Rev. Lett. **74**, 5028 (1995).
127. V.S. Afraimovich, V.I. Nekorkin, G.V. Osipov, V.D. Shalfeev, *Stability, Structures and Chaos in Nonlinear Synchronization Networks* (World Scientific, Singapore 1994).
128. J. Jamaguchy, H. Shimizu, Physica D **11**, 212 (1984).
129. S.H. Strogatz, R.E. Mirollo, J. Phys. A **21**, L699 (1988).
130. P.C. Matthews, R.E. Mirollo, S.H. Strogatz, Physica D **51**, 293 (1991).
131. B. Hu, Ch. Zhou, Phys. Rev. E **61**, R1001 (2000).
132. E. Mosekilde, F. Larsen, G. Dewell, P. Borcmans, Int. J. Bifurc. Chaos **8**, 1003 (1998).
133. A. Amengual, E. Fernandez-Garcia, R. Montagne, M. San Miguel, Phys. Rev. Lett. **78**, 4379 (1997).
134. S. Boccaletti, J. Bragard, F.T. Arecchi, Phys. Rev. E **59**, 6574 (1999).
135. S. Boccaletti, J. Bragard, F.T. Arecchi, H. Mancini, Phys. Rev. Lett. **83**, 536 (1999).
136. C. Hayashi, *Nonlinear Oscillations in Physical Systems* (McGraw-Hill, NY 1964).
137. J. Guckenheimer, P. Holmes, *Nonlinear Oscillations, Dynamical Systems, and Bifurcations of Vector Fields* (Springer, New York 1983).
138. A.B. Neiman, Phys. Rev. E **49**, 3484 (1994).
139. B.V. Shulgin, A.B. Neiman, V.S. Anishchenko, Phys. Rev. Lett. **75**, 4157 (1995).
140. V.S. Anishchenko, A.B. Neiman, F. Moss, L. Schimansky-Geier, Uspekhi Fiz. Nauk. **69**, 7 (1999) (in Russian); ibid., Phys.–Uspekhi **42**, 7 (1999).
141. S.K. Han, T.G. Yim, D.E. Postnov, O.V. Sosnovtseva, Phys. Rev. Lett. **83**, 1771 (1999).
142. S. Tanabe, T. Shimokawa, Sh. Sato, K. Pakdaman, Phys. Rev. E **60**, 2182 (1999).
143. Y. Kuramoto, *Chemical Oscillations, Waves, and Turbulence*, Springer Series in Synergetics, Vol. 19 (Springer, Berlin 1984).
144. P. Panter, *Modulation, Noise and Spectral Analysis* (McGraw-Hill, New York 1965).
145. D. Gabor, J. IEE Lond. **93**, 429 (1946).
146. L. Vainshtein, D. Vakman, *Frequency Separation in Theory of Oscillations and Waves* (Nauka, Moscow, 1983) (in Russian).
147. D. Middleton, *An Introduction to Statistical Communication Theory* (McGraw-Hill, New York 1960).

148. J.A. Freund, A. Neiman, L. Schimansky-Geier, In: *Stochastic Climate Models. Progress in Probability*, ed. by P. Imkeller and J. von Storch (Birkhuser, Boston 2001).
149. L. Schimansky-Geier, V. Anishchenko, A. Neiman, In: *Neuro-informatics*, ed. by S. Gielen, F. Moss, Handbook of Biological Physics, Vol. 4, Series Editor A.J. Hoff (Elsevier, Amsterdam 2000).
150. M. Rosenblum, A. Pikovsky, J. Kurths, Phys. Rev. Lett. **76**, 1804 (1996).
151. A.N. Malakhov, *Fluctuations in Auto-oscillation Systems* (Nauka, Moscow 1968) (in Russian).

# 2. Dynamical Chaos

## 2.1 Routes to Chaos

### 2.1.1 Introduction

A dynamical system (DS) displays its nonlinear properties in different ways with variation of system control parameters. An increase in the influence of nonlinearity causes the dynamical regime to become complicated. Simple attractors in the phase space of a dissipative system are replaced by more complicated ones. Under certain conditions, nonlinearity can lead to the onset of dynamical chaos. Moving along a relevant direction in parameter space, one can observe a set of bifurcations resulting in the appearance of a chaotic attractor. Such typical bifurcation sequences are called the *bifurcation mechanisms*, or the *scenarios of the transition to chaos*.

The first scenario of transition to nonregular behavior was proposed by L.D. Landau in 1944 [1] and independently by E. Hopf [2] when attempting to explain the onset of turbulent behavior in fluid flow with increasing Reynolds number. The corresponding bifurcation mechanism was called the *Landau–Hopf scenario*. The latter involves a sequence of bifurcations (of the Andronov–Hopf bifurcation type), each giving rise to a new incommensurate frequency. As a result, a multi-frequency quasiperiodic regime appears following a multi-dimensional torus in the phase space of a DS. If the number $k$ of bifurcations is large enough and if one takes into account fluctuations inevitably present in real systems, the power spectrum of the process becomes practically continuous. However, such multi-frequency oscillations, even in the presence of noise, are not chaotic, since the Landau–Hopf scenario does not involve the development of instability and intermixing in the system. Besides, this scenario is not able to explain the onset of oscillations with continuous spectrum in low-dimensional systems.

In the early 1970s the idea of developing turbulence through quasiperiodic oscillations was revised by D. Ruelle, F. Takens and S. Newhouse [3,4]. They connected turbulent behavior with chaotic dynamics and were the first to introduce the notion of a *strange attractor* (SA) as the mathematical image of chaos in a DS. It was also shown that a strange attractor can arise in low-dimensional systems ($N \geq 3$).

By now there are three known major routes to chaotic attractors: (i) the period-doubling cascade route; (ii) the crisis and the intermittency route; and (iii) the route to chaos through quasiperiodicity destruction. These bifurcation mechanisms can already be realized in three-dimensional phase space [5–9]. Each transition to chaos has *universality* properties, i.e., certain generalities, independent of a particular form of the evolution operator. In the present section we describe these three routes to chaos in detail. In the framework of the quasiperiodic route attention will be given to the peculiarities of the development of nonregular dynamics in systems with a robust ergodic two-dimensional torus.

## 2.1.2 Period-Doubling Cascade Route. Feigenbaum Universality

A large number of various DS, ranging from the simplest maps to distributed systems demonstrate the transition to chaos through *a cascade of period-doubling bifurcations* [10–17]. Since the period-doubling is a codimension one bifurcation, this route assumes a one-parametric analysis and is as follows: Let $\alpha$ be the control parameter of a DS, and at some $\alpha = \alpha_0$ the system has a stable limit cycle $C$ with period $T_0(\alpha)$. As $\alpha$ is increased to $\alpha_1$, the supercritical period-doubling bifurcation takes place, and a period-2 cycle ($2C$) appears. At $\alpha = \alpha_k$, $k = 1, 2, 3, \ldots$, the system exhibits an infinite sequence of period doublings of $2^k C$. The power spectrum shows new components that appear at subharmonics of the natural frequency $\omega_0 = 2\pi/T_0$, and thus the period-doubling bifurcation sequence is sometimes called the *subharmonic cascade*. When $k \to \infty$, the bifurcation points accumulate to some critical value $\alpha = \alpha_{cr}$, at which the period becomes infinite and the power spectrum continuous. For $\alpha > \alpha_{cr}$, there occur aperiodic oscillations which are unstable according to Lyapunov. These oscillations follow a chaotic attractor in the system phase space. Figure 2.1 exemplifies the changes which happen in the generator with inertial nonlinearity (GIN) of Anishchenko–Astakhov [14] during the period-doubling route to chaos. The generator is governed by the following system of equations:

$$\dot{x} = mx + y - xz,$$
$$\dot{y} = -x, \qquad\qquad\qquad (2.1)$$
$$\dot{z} = -gz + g\Phi(x),$$

where $\Phi(x) = x^2$ for $x \geq 0$ and $\Phi(x) \equiv 0$ for $x < 0$.

It has been experimentally established that in all continuous-time systems, a chaotic attractor arising through the period-doubling route has fractal dimension $2 < d < 3$, and its section resembles a horseshoe in shape. In this case a Poincaré surface of section map can be modeled, for example, by the simplest and well-known Henon map [18]:

$$x_{n+1} = 1 + y_n - ax_n^2,$$
$$y_{n+1} = bx_n. \qquad\qquad\qquad (2.2)$$

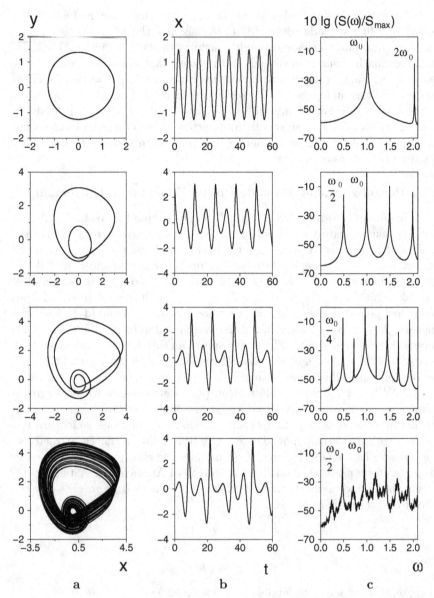

**Fig. 2.1.** Period-doubling cascade route in a GIN (2.1). (**a**) Projections of phase trajectories, (**b**) waveforms $x(t)$, and (**c**) power spectra $S(\omega)$ for period $2^k T_0$ cycles, $k = 0, 1, 2$, and for a chaotic attractor

The map (2.2) is invertible and contracts a square element for $b < 1$. As one of the system control parameters $a$ or $b$ is varied, the Henon map demonstrates the period-doubling route to chaos. If the square element is strongly contracted, then the transverse Cantor structure of a horseshoe can be neglected and points in the map can be considered to lie on one smooth bent curve. Introducing a new coordinate along this curve one can arrive at a non-invertible model map of an interval to itself, which is defined by a smooth first return function with a single extremum. The map stretches an interval element and then "folds" it into the same interval. Since the first return function is assumed to be smooth everywhere, then, as in the case of horseshoe, there exists a region near the extremum for which stretching is absent. The existence of such a region causes the birth of stable periodic orbits inside a chaotic zone, so-called *stability windows*. Periodic windows can be eliminated provided that the map stretches everywhere. An example is the tent map. However, in such models periodic windows disappear together with a period-doubling cascade. The theory of the period-doubling route to chaos was developed by M. Feigenbaum on the basis of one-dimensional model maps [19–21], and thus, this bifurcation mechanism was named the *Feigenbaum scenario*.

The simplest model for studying the Feigenbaum scenario is the *logistic map*

$$x_{n+1} = f(x_n) = r - x_n^2, \qquad (2.3)$$

where $r$ is a parameter.

The logistic map may also be re-written in another form which can be reduced to (2.3) by a linear change of variables, e.g.,

$$x_{n+1} = \alpha x_n (1 - x_n), \quad x'_{n+1} = 1 - \beta x_n'^2. \qquad (2.4)$$

Consider how the map (2.3) evolves as the parameter $r$ increases. The map has a fixed point $x_0 > 0$, also called a period-1 cycle, or 1-cycle, with coordinate $x_0 = -1/2 + \sqrt{r + 1/4}$. The fixed point $x_0$ is stable in the interval $r \in [-1/4, 3/4]$, and its multiplier $\mu_1$ is equal to $-2x_0$. At $r = r_1 = 3/4$, we have $\mu_1 = -1$, and the first period-doubling bifurcation takes place, giving rise to a stable period-2 cycle. The latter consists of two points $x_{1,2} = 1/2 \pm \sqrt{r - 3/4}$. The 2-cycle has a multiplier $\mu_2 = f'_x(x_1) f'_x(x_2) = 4(1 - r)$ and is stable in the range $r \in [3/4; 5/4]$. At $r = r_2 = 5/4$, the second period-doubling bifurcation occurs, and a period-4 cycle is created, etc. By this means, map (2.3) generates the sequence of period-doubling bifurcation values of the parameter $r$, namely, $r_1 = 3/4$, $r_2 = 5/4$, $r_3 \approx 1.368099$, $r_4 \approx 1.394046$, $r_5 \approx 1.399637, \ldots$, with the accumulating point $r_{\mathrm{cr}} \approx 1.40115\ldots$. When $k \to \infty$, the convergence rate of the bifurcation values tends to some finite limit

$$\delta = \lim_{k \to \infty} \frac{r_{k+1} - r_k}{r_{k+2} - r_{k+1}} = 4.669201\ldots \qquad (2.5)$$

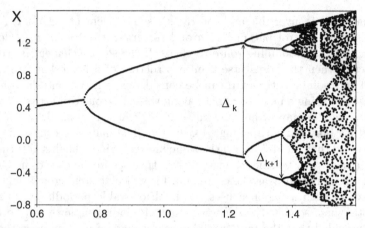

**Fig. 2.2.** Phase-parametric diagram of dynamical regimes for the map (2.3)

Figure 2.2 shows a phase-parametric diagram for the logistic map (2.3), which appears to be typical for systems exhibiting the period-doubling route to chaos. Such a diagram is called *Feigenbaum's tree*. The bifurcation diagram readily illustrates the scale splitting of the dynamical variable as well as *scaling* properties, i.e., *scale invariance*, when the same image element reproduces itself under an arbitrary small change in scale. Denoting the distances between similar points of the tree branches by $\Delta_k$ (see Fig. 2.2), we can introduce the scaling factors $a_k = \Delta_k/\Delta_{k+1}$, which in the limit as $k \to \infty$ converge to

$$a = \lim_{k\to\infty} \frac{\Delta_k}{\Delta_{k+1}} = -2.5029\ldots^{[1]} \qquad (2.6)$$

Numerical investigations have shown that the values of $\delta$ and $a$ do not depend on the particular form of a map. However, it is necessary for the map to be unimodal, i.e., it must have a single extremum, and for the extremum to be quadratic.

Feigenbaum explained the universal character of quantitative regularities of the period-doubling route and created the *universality theory* [19–21]. For analyzing maps of the logistic parabola type he applied a *renormalization-group* (RG) method, which is as follows: Assume that at the critical point $r = r_{\mathrm{cr}}$ we have the map

$$x_{n+1} = f_0(x_n), \qquad (2.7)$$

---

[1] The negative sign means that the vector that connects similar points reverses its direction each time.

where $f_0$ is an arbitrary unimodal function with a quadratic extremum at $x_n = 0$, and $f_0(0) = 1$. Applying the map (2.7) twice we obtain a map $x_{n+1} = f_0\big(f_0(x_n)\big)$. Re-scale the variable $x \to x/a_0$ so that the new map is also normalized to unity at the origin, i.e., $a_0 = 1/f_0\big(f_0(0)\big)$. Denote the new map as $x_{n+1} = f_1(x_n) = a_0 f_0\big(f_0(x_n/a_0)\big)$. Repeating this procedure many times we derive the *renormalization-group equation*

$$f_{i+1}(x) = a_i f_i\big(f_i(x/a_i)\big), \tag{2.8}$$

where $a_i = 1/f_i\big(f_i(0)\big)$. Due to the self-similarity, the following limits exist at the critical point:

$$\lim_{i\to\infty} f_i(x) = g(x), \quad \lim_{i\to\infty} a_i = a. \tag{2.9}$$

Function $g(x)$ represents a fixed point of the functional *Feigenbaum–Cvitanovic equation*:

$$\hat{T}g(x) = ag\big(g(x/a)\big) = g(x), \tag{2.10}$$

where $\hat{T}$ is a doubling operator and $a = 1/g\big(g(0)\big)$. For the critical point corresponding to the period-doubling transition to chaos, the boundary conditions for (2.10) are $g(0) = 1$ and $g'_x(0) = 0$. The function $g(x)$ is *universal* because it does not depend on a particular form of the original map and is defined by the order of extremum only. Taking into account the variable $x$ renormalization, this function yields an asymptotic form of $2^i$-times the applied evolution operator at the critical point as $i \to \infty$. The constant $a$ involved in the fixed point equation is also universal. The fixed point solution of (2.10) was numerically found by Feigenbaum under the assumption of a quadratic extremum and of the above-stated boundary conditions. It is given by

$$g(x) = \; 1 - 1.5276330x^2 + 0.1048152x^4 + 0.0267057x^6 - 0.0035274x^8$$
$$+ 0.0000816x^{10} + 0.0000254x^{12} - 0.0000027x^{14}. \tag{2.11}$$

The *universal Feigenbaum constant, a* appears to be equal to $-2.502907876\ldots$

If the evolution operator $f(x_n)$ is slightly perturbed due to a small deviation of the parameter from its critical value, the doubling operator $\hat{T}$ and the function $g(x)$ turn out to be perturbed as well. Having linearized $\hat{T}$ at the point $g(x)$ for $r = r_{\mathrm{cr}}$, one can obtain an operator $\hat{L}_g$ which defines the behavior of perturbations as well as the equation for eigenfunctions $h(x)$ and eigenvalues $\rho$ of the linearized operator:

$$\hat{L}_g h(x) = a\left[ g'\big(g(x/a)\big)h(x/a) + h\big(g(x/a)\big)\right] = \rho h(x). \tag{2.12}$$

The behavior of perturbations will be mainly determined by the eigenvalues exceeding unity in modulus. In the case of quadratic extremum, there is one such value corresponding to an unremovable component of the perturbations. This value defines a second *universal Feigenbaum constant*, $\rho_1 = \delta = 4.6692016091\ldots$.

Maps with a nonquadratic extremum are characterized by different values of the universal constants [22–25]. However, computer and physical experiments performed for a variety of continuous-time systems, including distributed ones, with the Feigenbaum scenario to chaos have shown that the scaling factor $a$ and the convergence rate of the bifurcation sequence $\delta$ coincide within the experimental error with the theoretical values found for maps with a quadratic extremum. Evidently, the typical case is when a map generated by the evolution operator of a continuous-time system in the neighborhood of the critical point may be approximated by a one-dimensional map with a quadratic extremum. Other cases are considered to be atypical.

The Feigenbaum scenario is universal and its universality manifests itself in the behavior of spectral amplitudes of subharmonics which appear with each period doubling. The ratio of amplitudes of subharmonics $\omega_0/k$ and $\omega_0/(k+1)$ in the limit as $k \to \infty$ is a universal constant [6, 26].

At $r = r_{cr}$, the logistic map (2.3) generates a limit set of points, which is called the *Feigenbaum attractor*. It is strange since its capacity dimension is fractal but nonchaotic because the Lyapunov exponent $\lambda$ is zero at the critical point. The universal properties also hold in a supercritical region $r > r_{cr}$ both for model maps and for continuous-time systems. The Lyapunov exponent having become positive beyond the critical point grows according to the universal law [27]

$$\lambda \sim \varepsilon^\gamma, \quad \gamma = \frac{\ln 2}{\ln \delta} \approx 0.4498, \tag{2.13}$$

where $\varepsilon = r - r_{cr}$ is the supercriticality parameter. By analogy with the theory of second-type phase transitions, the coefficient $\gamma$ is called the *critical index* of the transition to chaos. Figure 2.3 shows numerically computed Lyapunov exponents as a function of the control parameter for the logistic map and for GIN.

Beyond the critical point, systems with period doubling demonstrate a cascade of *merging* bifurcations. This kind of bifurcation consists in merging parts, or bands, of a chaotic attractor, which are visited by a phase point in a certain order. Each attractor-band-merging bifurcation is accompanied by the disappearance of appropriate subharmonics in the power spectrum. A fragment of the band-merging bifurcation cascade observed in system (2.1) [8, 28] is illustrated in Fig. 2.4 with phase portraits and corresponding power spectra. For a one-dimensional map, the band-merging bifurcation looks like the merging of neighboring intervals which are filled with points of a chaotic set. Denote by $\bar{r}_k$ the parameter values corresponding to the merging bifurcations, where index $k = 1, 2, \ldots$ increases when approaching the critical point from right to left. $2^i$-periodic orbits exist on intervals to the left of the critical point, whereas $2^i$-band chaotic attractors are realized in the parameter $r$

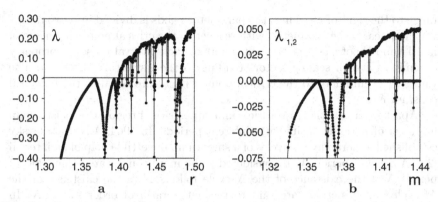

**Fig. 2.3.** Lyapunov exponents versus the control parameter (**a**) for the logistic map and (**b**) for GIN with $g = 0.3$ ($\lambda_2(m) \equiv 0$)

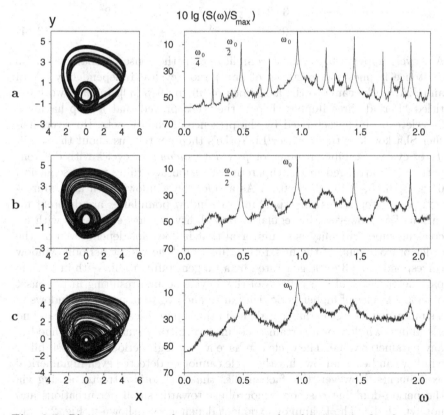

**Fig. 2.4.** Phase portraits and power spectra under the attractor-band-merging bifurcations in GIN (2.1): (**a**) a four-band chaotic attractor, (**b**) a two-band chaotic attractor, (**c**) a single-band chaotic attractor. $\omega_0$ is the natural spectrum frequency corresponding to the period $T_0$ of generating cycle

range to the right of $r_{\mathrm{cr}}$. Thus, the parameter $r$-axis is divided into two ranges which appear to be symmetric with respect to the critical point. Segments of multi-band chaotic sets at relevant points of each interval possess a property of similarity, with scaling factors tending to the universal constant $a$. The values $\bar{r}_k$ accumulate to the critical point with a rate equal to the universal constant $\delta$.

Apart from chaotic trajectories, in a supercritical region the logistic map has a set of periodic orbits with different periods. In 1964, A.N. Sharkovsky established a hierarchy for cycles of a smooth noninvertible map of an interval [29]. A cycle of period $M$ is considered to be more complicated than that of period $N$ if the existence of the $N$-cycle follows from the existence of the $M$-cycle. Their periods are said to be in the ratio of order $M \to N$. In correspondence with Sharkovsky's theorem, this ratio arranges cycles in a certain order, the so-called *Sharkovsky series*:

$$3 \to 5 \to 7 \to 3 \cdot 2 \to 5 \cdot 2 \to 7 \cdot 2 \to \ldots \to 3 \cdot 2^2$$
$$\to 5 \cdot 2^2 \to 7 \cdot 2^2 \to \ldots \to 2^3 \to 2^2 \to 2 \to 1. \tag{2.14}$$

A 3-cycle appears to be most complicated in the sense of Sharkovsky. Its existence implies the existence of any period cycles. Independently, T. Li and J.A. Yorke came to the analogous result in 1975 [30]. In their work entitled "Period three implies chaos", they also proved that a map having a 3-cycle can be characterized by the presence of chaotic sets. However, neither Sharkovsky's theorem nor Li–Yorke's theorem tells us about the stability of cycles. Stability regions, or *periodic windows*, of cycles with different periods are arranged in the supercritical region according to the sequence: 6, 5, 3, 6, 5, 6, 4, 6, 5, 6, ... As a rule, the window width and bifurcation parameter values corresponding to window boundaries are different for various DS. However, the regularity of stability window emergence with increasing supercriticality is so universal that it does not depend even on the order of extremum of the first return function. The widest periodic window corresponds to a 3-cycle generated by a tangent bifurcation. With increasing parameter the doubling process of the 3-cycle occurs, resulting in the onset of chaos. Cycles of higher periodicity arise and evolve in periodic windows in a similar manner. Generally speaking, in the supercritical region one can find a stability window of a certain cycle in an arbitrary small neighborhood of any parameter value. The cycle can have a very high period, and its stability window can be so narrow that the cycle cannot be detected even in numerical experiments. However, this fact implies that a chaotic attractor arising via the period-doubling cascade is nonrobust towards small perturbations and nonhyperbolic. The Lyapunov exponent dependences, shown in Fig. 2.3, are also typical for nonhyperbolic attractors.

### 2.1.3 Crisis and Intermittency

As knowledge of dynamical chaos developed, it has been established that a transition from regular oscillations to chaos may suddenly occur as a result of only one bifurcation. Such a mechanism for chaos onset is said to be *hard* and is accompanied by the phenomenon of *intermittency*. The latter is a regime when chaotic behavior (*turbulent phase*), immediately appearing after passing the border of chaos, is interspersed with periodic-like behavior (*laminar phase*) in an intermittent fashion. Figure 2.5 illustrates a time realization of oscillation process $x(t)$ typical for intermittency. The abrupt transition to chaos and the intermittency phenomenon were first considered in the works of I. Pomeau and P. Manneville [31,32]. The corresponding bifurcation mechanism of the chaos onset was thus called the *Pomeau–Manneville scenario*.

In the intermittency route, the sole bifurcation of a periodic regime may cause drastic qualitative changes of the structure of phase space as well as of the structure of the basin of attraction of an attractor. Such attractor bifurcations are called *crises* [33,34]. Typical crises of periodic regimes (limit cycles) are related to certain kinds of local codimension one bifurcations, namely, tangent (saddle-node), subcritical period-doubling and subcritical torus birth bifurcations (the Andronov–Hopf bifurcation in the Poincaré map). In the case of tangent bifurcation a stable limit cycle disappears through merging with a saddle cycle. In the two other situations, the limit cycle still exists after a bifurcation but becomes unstable, i.e., a saddle one.

Suppose that for $\alpha < \alpha_{cr}$ a system has an attractor in the form of a limit cycle $C$. When any of the aforementioned bifurcations takes place at $\alpha = \alpha_{cr}$, the attractor $C$ disappears. For $\alpha > \alpha_{cr}$, phase trajectories coming from the local vicinity of the former cycle $C$ must fall on another attractor

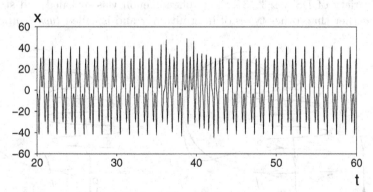

**Fig. 2.5.** Intermittency in the Lorenz system ($\dot{x} = \sigma(y - x)$, $\dot{y} = -xy + rx - y$, $\dot{z} = xy - bz$) for $r = 166.1$, $\sigma = 10$, $b = 8/3$

which either already exists in the system for $\alpha < \alpha_{cr}$ or is generated by the bifurcation. Assume that the system already possesses one more attractor. In this case the bifurcation results in a simple transition from one attractor to another. Intermittency does not arise even if the new regime is chaotic. This fact can be explained as follows: The chaotic attractor is not created through a limit cycle crisis and does not capture the local neighborhood of cycle $C$. Phase trajectories leave this region and never return. The question is: What are the conditions under which the limit cycle crisis results in the appearance of intermittent chaos? Evidently, this situation may happen when at the bifurcation point $\alpha = \alpha_{cr}$ there already exists a chaotic set which becomes attracting for $\alpha > \alpha_{cr}$ and encloses the local vicinity of cycle $C$, so that a phase trajectory moving on the chaotic attractor returns to this vicinity from time to time. Such system behavior is realized provided a saddle limit set which is involved in the cycle $C$ crisis possesses a homoclinic structure. As an example, Fig. 2.6 illustrates the tangent bifurcation of cycles, leading to chaotic intermittency. The saddle cycle has a pair of robust homoclinic orbits. At the bifurcation point $\alpha = \alpha_{cr}$, a nonrobust saddle-node orbit is created which possesses a homoclinic structure in its vicinity. Phase trajectories move away from it and approach it along double-asymptotic homoclinic curves (they correspond to the intersection points of the manifolds in the section shown in Fig. 2.6). For $\alpha > \alpha_{cr}$ the nonrobust closed orbit disappears and the non-attracting homoclinic structure becomes attracting. As a result, a chaotic attractor emerges in the system phase space. Trajectories lying on this attractor are concentrated in the region of the former saddle-node orbit and repeat the motion on it for a long period of time. This motion testifies to a laminar phase of intermittent chaos.

Intermittency related to the tangent bifurcation of cycles is most typical for a variety of DS [31, 32, 35]. This phenomenon was revealed and studied much earlier than other types of intermittency and is called *type-I intermit-*

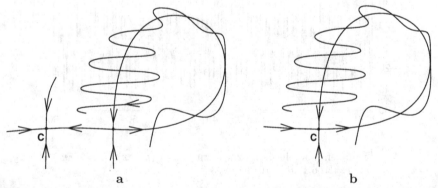

**Fig. 2.6.** Qualitative picture of the Poincaré section for the tangent bifurcation of stable and saddle cycles, leading to the emergence of chaos via intermittency: **(a)** before the bifurcation and **(b)** at the bifurcation point

*tency.* To analyze its properties the following one-dimensional model map is usually used:

$$x_{n+1} = f(x_n) = \varepsilon + x_n + \beta |x_n|^p + \text{``return''}. \tag{2.15}$$

Here, the parameter $\varepsilon$ denotes the supercriticality $(\alpha - \alpha_{cr})$ of the system since in (2.15) a tangent bifurcation occurs at $\varepsilon = 0$. $p$ is an integer number and defines the order of an extremum of the first return function. The return of a phase point to a bounded interval of $x$ values can proceed in different ways. For example, for the map presented in Fig. 2.7 the branch of the return function graph on the interval $AB$ serves to return the phase point. Figure 2.7a shows the map at the moment of tangent bifurcation, $\varepsilon = 0$. The dashed lines constructed by using Lamerei's diagram indicate a double-asymptotic trajectory of a saddle-node. The map displayed in Fig. 2.7b corresponds to the case $\varepsilon \gtrsim 0$. In the neighborhood of the former fixed point, a so-called *channel* opens up along which the phase point moves for a long period of time. This motion reflects the laminar phase of intermittency. When the point leaves the channel, a turbulent phase is observed in the course of which the point must reach the interval $AB$, the latter providing its return to the channel again.

The study of maps in the form of (2.15) reveals certain quantitative features of the type-I intermittency, e.g., how the average duration of laminar phase depends on the supercriticality. These regularities are universal in the sense that they do not depend on a particular form of a map and are determined by the order of an extremum $p$. For a typical case $p = 2$, these features are in good agreement with results of numerical and experimental investigations of the type-I intermittency in continuous-time systems. The RG method has been applied to study this type of intermittency [7]. Let us consider a map at the critical point and restrict ourselves to the interval

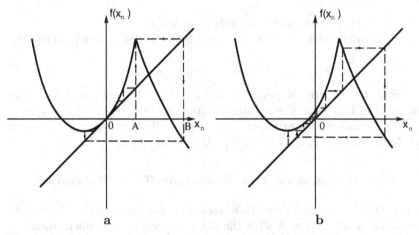

**Fig. 2.7.** The map modeling type-I intermittency (**a**) at the bifurcation point and (**b**) immediately after the bifurcation

$x_n \in [0, 1]$, where the map is defined by a monotonic function of the form $x_{n+1} = f_0(x_n)$, with $f_0(0) = 0$ and $f_0'(0) = 1$. Repeating all the arguments as in the case of Feigenbaum scenario one can derive the same Feigenbaum Cvitanovic equation (2.10) but with different boundary conditions: $g(0) = 0$ and $g'(0) = 1$. The RG method enables us to theoretically determine the asymptotic behavior of the average duration of laminar phase, which reads

$$T_1 \sim \varepsilon^{-\nu}, \quad \nu = \frac{p-1}{p}. \tag{2.16}$$

For $p = 2$, we have $T_1 \sim 1/\sqrt{\varepsilon}$, which concurs well with results of numerous experiments.

As already mentioned, other types of intermittency are associated with a subcritical Andronov–Hopf bifurcation in the Poincaré section and to a sub-critical period doubling. They are called *type-II* and *type-III intermittency*, respectively [9]. Type-II intermittency is modeled by the map of the plane, defined in polar coordinates:

$$r_{n+1} = (1 + \varepsilon)r_n + \beta r_n^3 + \text{``return''},$$
$$\phi_{n+1} = \phi_n + \Omega, \mod 1. \tag{2.17}$$

For $\varepsilon = 0$, the map displays the subcritical Andronov–Hopf bifurcation in which an unstable invariant circle and a stable focus merge. The unstable invariant circle corresponds to a saddle torus in a continuous-time system with dimension $N \geq 4$. For type-II intermittency the asymptotic behavior of the average duration of laminar phase is defined by

$$T_1 \sim \frac{1}{\varepsilon}, \tag{2.18}$$

where $\varepsilon = \alpha - \alpha_{\text{cr}}$ is the supercriticality parameter.

*Type-III intermittency* can be described by a one-dimensional model map in the form

$$x_{n+1} = -(1 + \varepsilon)x_n - \beta x_n^2 + \text{``return''}. \tag{2.19}$$

For $\varepsilon = 0$, the subcritical period-doubling bifurcation of a period-1 cycle occurs in the map. This type of intermittency has been observed experimentally [36]. The average duration of laminar phase is evaluated approximately by the same ratio as in the case of type-II intermittency.

### 2.1.4 Route to Chaos via Two-Dimensional Torus Destruction

According to the Ruelle–Takens–Newhouse scenario, the transition from quasiperiodicity to chaos occurs after the third frequency birth when unstable, according to Lyapunov chaotic trajectories appear on a three-dimensional torus [3,4]. However, the study of particular DS has shown that the appearance of chaos following the destruction of two-frequency quasiperiodic motion is also a typical scenario of the transition to chaos. According to this route, a two-dimensional (2-D) torus, $T^2$, in phase space is destroyed, and trajectories fall in a set with fractal dimension $2 + d$, $d < 1$. This set is created in the vicinity of $T^2$ and thus called a *torus-chaos* [37]. Such a route may be thought of as a special case of the quasiperiodic transition to chaos.

Unlike the Feigenbaum scenario, the transition $T^2 \rightarrow strange\,attractor$ (SA) requires a two-parameter analysis. This is associated with the fact that the character of quasiperiodic motion depends on a winding number $\theta$, which determines the ratio of basic frequencies of oscillations. If the frequencies are rationally related, i.e., the winding number $\theta$ has a rational value, resonance on a torus takes place and, consequently, periodic oscillations are realized. If the frequencies become irrationally related, the motion on the torus will be ergodic. The transition from the torus $T^2$ to chaos at a fixed winding number can be realized only by controlling simultaneously at least two system parameters. On the torus birth line, specified by the bifurcation condition $\mu_{1,2} = \exp(\pm j\phi)$, where $\mu_{1,2}$ is a pair of complex conjugate multipliers of a limit cycle, the winding number is defined as $\theta = \phi/2\pi$. Resonance regions on a two control parameter plane have a tongue-like shape and originate from relevant points on the torus birth line. Such regions are called *Arnold tongues* (in honor of V.I. Arnold, who studied the structure of resonance regions [38,39]). Each direction of the paths on the parameter plane is characterized by its own sequence of bifurcations associated with the appearance and disappearance of different resonances on the torus.

**Two-Dimensional Torus Breakdown.** The results obtained by mathematicians in the framework of the qualitative theory of DS play an important role in understanding the mechanisms which lead to the destruction of $T^2$ and to the birth of torus-chaos. One of the basic results was described in [40,41] and is concerned with the breakdown of $T^2$ as well as possible routes to chaos from $T^2$.

Consider an $N$-dimensional DS ($N \geq 3$):

$$\dot{x} = F(x, \alpha), \qquad (2.20)$$

where components of the vector-function $F$ belong to the smoothness class $C^k$, $k \geq 3$, and $\alpha$ is the system parameter vector.

(i)  Suppose that at $\alpha = \alpha_0$ a smooth attracting torus $T^2(\alpha_0)$ exists in some region of the phase space of system (2.20) and has a robust structure consisting of stable and saddle cycles. This case corresponds to a resonance region. The surface of the resonance torus is formed by the closure of the unstable manifold $W^u$ of the saddle cycle $C^{sd}(\alpha_0)$ running to the stable cycle $C^{st}(\alpha_0)$, i.e., $T^2(\alpha_0) = W^u(\alpha_0) \bigcup C^{st}(\alpha_0)$. The Poincaré section of a resonance torus is shown in Fig. 2.8a. Assume that

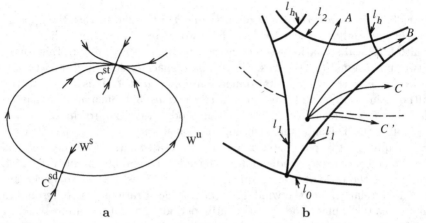

**Fig. 2.8.** (a) The section of resonance torus $T^2$ and (b) the qualitative bifurcation diagram for the breakdown of $T^2$. Routes of torus destruction are indicated by $A, B$ and $C$

the invariant torus does not exist at $\alpha = \alpha_1$. Then, for the continuous curve $\alpha(s)$, where $s \in [0, 1]$, $\alpha(0) = \alpha_0$, $\alpha(1) = \alpha_1$, there exists a value $s = s^*$ such that at $\alpha(s^*)$ the torus $T^2$ is destroyed and no longer exists in system (2.20) at least for some arbitrary values of $s$, close to $s^*$, $s > s^*$;

(ii) Let for all $0 \le s < s^*$ the attracting set of (2.20) coincides with the torus $T^2(\alpha(s))$;

(iii) Suppose that in the limit $s \to s^*$ the unstable manifold $W^u(\alpha(s))$ of the saddle cycle does not enclose periodic orbits, different from $C^{st}$ and $C^{sd}$.

Under the above assumptions, the *theorem on two-dimensional torus breakdown* which asserts the following three possible mechanisms of the $T^2$ destruction is valid: (i) due to the loss of cycle $C^{st}$ stability; (ii) as a result of the emergence of a homoclinic tangency between $W^u$ and $W^s$ of $C^{sd}$; and (iii) due to a tangent bifurcation of $C^{st}$ and $C^{sd}$ on $T^2$. Before being destroyed, $T^2$ loses its smoothness for $s > s^{**}$, i.e., $T^2(\alpha(s^{**} < s < s^*))$ is *homeomorphic* but not *diffeomorphic* to the torus.

Figure 2.8b shows a sketch of the qualitative bifurcation diagram in the two parameter $(\alpha_1 - \alpha_2)$ plane. The directions indicated in Fig. 2.8b by $A, B, C$ correspond to the three routes of the resonance torus destruction. In the diagram the curve $l_0$ corresponds to the bifurcation of $T^2$ birth. The phase-locking region is bounded by the two bifurcation curves $l_1$ related to a tangent bifurcation of $C^{st}$ and $C^{sd}$ on $T^2$. On the curve $l_2$ the resonance cycle $C^{st}$ loses its stability inside the synchronization region. The bifurcation lines $l_h$ correspond to a tangency of $W^s$ and $W^u$ of $C^{sd}$. Outside the phase-locking region, $T^2$ is destroyed on the curves shown in Fig. 2.8b as dashed lines. Actually this boundary has a complicated fractal structure. Moving along the route $C'$ results in a tangent bifurcation on the curve $l_1$ without torus destruction. In this case a transition from the phase-locked $T^2$ to an ergodic $T^2$ takes place.

**Route A.** On the bifurcation curve $l_2$, $C^{\mathrm{st}}$ on $T^2$ becomes unstable due to either period doubling or a 2-D torus birth. Before this, the resonance torus $T^2$ loses its smoothness (in the cycle $C^{\mathrm{st}}$ points) when a pair of multipliers of the cycle becomes complex conjugate or one of the multipliers is negative. At the moment of bifurcation, the length of an invariant curve in the Poincaré map becomes infinite (Fig. 2.9a), and the torus is destroyed. The transition to chaos along route $A$ can come either from a period-doubling bifurcation cascade or via the breakdown of a torus occurring from $C^{\mathrm{st}}$ on the curve $l_2$.

**Route B.** When moving along route $B$ $W^{\mathrm{u}}$, which forms the torus surface, of $C^{\mathrm{sd}}$ is distorted and on the bifurcation curve $l_{\mathrm{h}}$ a tangency of $W^{\mathrm{u}}$ and $W^{\mathrm{s}}$ occurs. This is illustrated in Fig. 2.9b. At this moment $(s = s^*)$ a structurally unstable homoclinic trajectory arises and $T^2$ does not exist above the bifurcation curve $l_{\mathrm{h}}$. For $s > s^*$ two robust homoclinic orbits appear, and a homoclinic structure of the cycles and of chaotic trajectories is formed in their vicinity. However, $C^{\mathrm{st}}$ is still stable and remains as an attractor. A chaotic attractor may arise either through the disappearance of $C^{\mathrm{st}}$ or when it loses its stability. When we cross the synchronization region above the curve $l_{\mathrm{h}}$, a transition to chaos appears accompanied by type-I intermittency.

**Route C.** In this case $W^{\mathrm{u}}$ is also distorted when approaching $C^{\mathrm{st}}$. $T^2$ is destroyed as our route intersects the curve $l_1$ corresponding to a tangent bifurcation. Suppose that an invariant curve in the Poincaré section of the torus becomes unsmooth at the bifurcation point (Fig. 2.9c). This means that when applying the Poincaré map successively, the image of a small segment of the unstable separatrix of a saddle-node bends in the form of a horseshoe. The disappearance of the saddle-node leads to the emergence of a Smale horseshoe map in its vicinity. This map generates a countable set of saddle cycles as well as a continuous set of aperiodic hyperbolic trajectories. Under certain additional conditions these trajectories can be transformed into a chaotic attractor.

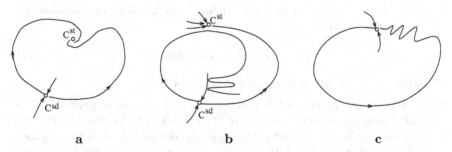

**Fig. 2.9.** Qualitative illustration of the invariant curve in the Poincaré section at the moment of torus destruction when moving along the routes (**a**) $A$, (**b**) $B$ and (**c**) $C$ indicated in Fig. 2.8b

Thus, the mechanisms of resonance torus destruction leads to the emergence of a chaotic set in the vicinity of the torus, and this set may become attracting. The chaotic attractor originates from a horseshoe-type map with a smooth bend and is nonhyperbolic [42, 43]. The bifurcation scenarios of the transition to chaos described are concerned with the bifurcations of resonance cycles on $T^2$. They do not cause the absorbing area to change abruptly and thus represent the bifurcation mechanism of a soft transition to chaos. All conclusions of the torus breakdown theorem are generic, which has been verified by physical experiments and computer simulations for a variety of discrete and continuous-time systems [44–51]. If we consider the evolution of an invariant curve in the Poincaré section of a torus by changing system parameters in such a way that the winding number is kept irrational, the following phenomena can be observed: First, the invariant curve in the Poincaré map of the ergodic torus is distorted in shape and repeats the unstable manifold of a resonance saddle cycle [50]. Then the ergodic torus loses its smoothness and is destroyed. However, a chaotic motion does not yet appear, since in the neighborhood (in parameter space) of the former torus with an irrational winding number there always exist other resonance tori, remaining as attractors of the system. Hence, a resonance on $T^2$ always precedes the transition to chaos. The torus destruction line in a two control parameter plane is characterized by a complex structure. It consists of a countable set of intervals, on which the resonance torus is destroyed according to the scenarios indicated by the theorem, and of a set of points corresponding to the breakdown of an ergodic torus and having a joint zero measure.

**Soft Transition from Quasiperiodicity to Chaos** In the general case, the motion on a two-dimensional torus can be modeled by an isomorphic dissipative map of the ring $Q$ into itself. For different parameter values of the map the following cases are possible [52]:

(i) Inside the ring there is a closed contour $L$ (Fig. 2.10a) which is transformed into itself. In other words, there exists a closed invariant curve corresponding to a two-dimensional torus in a continuous-time system. On the contour $L$ a new map can be defined which will be one-dimensional and homeomorphic to the circle map:

$$\phi_{n+1} = \Phi(\phi_n, \boldsymbol{\alpha}), \quad \mathrm{mod}\ 1, \tag{2.21}$$

where $\boldsymbol{\alpha}$ is the vector parameter of the circle map.

(ii) A horseshoe-type map arises generating a countable set of periodic orbits and a continuous set of aperiodic hyperbolic trajectories. Such a structure is created when some part of the region $Q$, denoted by $\sigma$, is transformed into $\tilde{\sigma}$ as shown in Fig. 2.10b. In this case the closed contour $L$ no longer exists and the model map (2.21) becomes noninvertible.

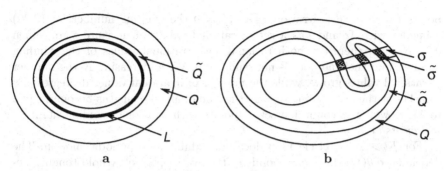

**Fig. 2.10.** Ring map (**a**) in the case of the existence of invariant closed curve $L$ and (**b**) in the case of formation of a local horseshoe-type map

The circle map is often defined as follows:

$$\phi_{n+1} = \Phi(\phi_n, \Omega, K) = \phi_n + \Omega - \frac{K}{2\pi} \sin(2\pi\phi_n), \quad \text{mod } 1. \tag{2.22}$$

The angle $\phi$ is determined in the interval $[0,1]$; $K \geq 0$ and $\Omega \in [0,1]$ are considered as parameters of the map. Generally speaking, the form of function $\Phi(\phi)$ is not so important (as in the case of the logistic map) but the following conditions must be satisfied: (i) $\Phi(\phi + 1) = 1 + \Phi(\phi)$; (ii) for $K < K_{\text{cr}}$, $\Phi(\phi)$ and its inverse function $\Phi^{-1}(\phi)$ exist and are differentiable, i.e., the map is a diffeomorphism of the circle; (iii) at $K = K_{\text{cr}}$ the inverse function $\Phi^{-1}(\phi)$ becomes nondifferentiable at the point $\phi = 0$, and the single-valued inverse function no longer exists for $K > K_{\text{cr}}$. All these conditions are fulfilled for (2.22) with $K_{\text{cr}} = 1$.

Dynamics of a point in the circle map is characterized by the winding number $\theta$, which is given by

$$\theta = \lim_{n \to \infty} \frac{\Phi^n(\phi_0) - \phi_0}{n}. \tag{2.23}$$

This number represents the mean angle about which a phase point rotates on a circle in one iteration. For a smooth one-to-one map, i.e., when $0 \leq K < 1$, the limit (2.23) exists and does not depend on the initial point $\phi_0$. From the above statement it follows that for $K < 1$ the map (2.22) has no fixed points if the winding number is irrational. When $\theta$ takes a rational number $\theta = p/q$ with $p$, $q$ being integers, the circle map possesses an even number of stable and unstable fixed points of $q$ multiplicity, i.e., at least one stable and one unstable $q$-cycle of the map. The numerator $p$ determines the number of full turns along a circle in $q$ iterations. The resonance structure corresponding to a rational value of the winding number is robust [53]. Each rational value of $\theta$ remains unchanged within a certain range of parameter variation, the so-called Arnold tongue. The dependence of the winding number on the parameter $\Omega$ is known as the *devil's staircase* and represents a fractal curve consisting of an infinite number of "steps" corresponding to rational numbers of $\theta$ and a set of isolated points for which $\theta$ is irrational. At $K = 0$ the winding number for (2.22) coincides with $\Omega$ and has a set of rational values of measure zero. When $0 < K < 1$, there are both rational and irrational values of the winding number, and their number is not equal to 0. With increasing $K$ the number of rational values grows, while the number of irrational values decreases and vanishes at the critical line $K = 1$ (the sum of all step lengths becomes equal to 1). However, a countable set of points with irrational winding numbers still exists at $K = 1$.

For $K > 1$ the circle map does not exhibit quasiperiodic motion. The dependence $\theta(\Omega)$ that corresponds to the overlapping of Arnold tongues becomes ambiguous. In a supercritical region the circle map describes resonances on a torus as well as chaotic motions in the vicinity of the former torus $T^2$. The map demonstrates the scenarios of torus breakdown and appearance of chaos, indicated in the Afraimovich–Shilnikov theorem. Inside the Arnold tongues, the stable resonance cycle loses its stability in a period-doubling bifurcation and a transition to chaos via the Feigenbaum scenario occurs. In the regions where the resonance tongues are overlapped, crises may take place, leading to the merging of chaotic attractors which appear from different resonance cycles. As a result, torus-chaos is created. Figure 2.11 shows the diagram of dynamical regimes for (2.22) in the $\Omega - K$ plane [54] and reflects the complicated self-similar structure of Arnold tongues.

As ergodic quasiperiodic motions are destroyed, map (2.22) exhibits certain quantitative universal properties, independent of a particular form of $\Phi(\phi)$, provided that $\Phi(\phi)$ satisfies the above-mentioned conditions [55, 56]. However, these features depend on $\theta$.

An irrational number can be represented in the form of a *continued fraction* [38]:

$$\theta = \cfrac{1}{m_1 + \cfrac{1}{m_2 + \cfrac{1}{\cdots}}} = \langle m_1, m_2, \ldots, m_k, \ldots \rangle. \tag{2.24}$$

If only the first $k$ expansion terms are taken into consideration, one obtains a rational number $\theta_k = p_k/q_k$, which is called the *rational approximation* of $\theta$ of order $k$. In this case the irrational number can be defined as the limit of a sequence of rational numbers:

$$\theta = \lim_{k \to \infty} \theta_k. \tag{2.25}$$

The irrational number called *the golden mean*[2], $\theta_g = 0.5(\sqrt{5} - 1) = \langle 1, 1, 1, \ldots \rangle$, has the simplest expansion into a periodic continued fraction.

Certain regularities are characteristic for irrational numbers of $\theta$, having a periodic (at least from some $m_k$) expansion into a continued fraction. Assume $\Omega_k(K)$ to be the value of $\Omega$ at some fixed $K$, for which $\theta = \theta_k$ and the point

---

[2] For the golden mean, $p_k$ and $q_k$ are the sequential terms of the main Fibonacci series, namely, $p_k = F_k$, $q_k = F_{k+1}$, and, consequently, $\theta_g = \lim_{k \to \infty} F_k/F_{k+1}$. The Fibonacci series are determined by the recurrent formula $F_{k+1} = F_{k-1} + F_k$, where $(F_0, F_1)$ are the base of the series. The main series has the base $(0, 1)$.

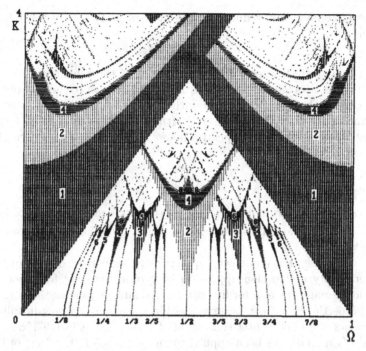

**Fig. 2.11.** Diagram of dynamical regimes of the circle map. Periodic regimes are realized in shaded regions and their periods are indicated by numbers [54]

$\phi = 0$ belongs to a stable limit cycle with period $q_k$. In other words, $\Omega_k$ is defined by the relation $\Phi^{q_k}(0, \Omega_k, K) = p_k$, where $\Phi^{q_k}$ is a function applied $q_k$ times. The values of $\Omega_k$ converge to some value $\Omega_\infty(\theta, K)$ according to a geometric progression law with the rate $\delta$:

$$\delta = \lim_{n \to \infty} \frac{\Omega_k - \Omega_{k-1}}{\Omega_{k+1} - \Omega_k}. \qquad (2.26)$$

The magnitude $\delta$ is a universal constant and depends on $\theta$ and $K$ only. For the golden mean, $\delta = -2.6180339\ldots = -\theta_\mathrm{g}^{-2}$ for $K < K_\mathrm{cr}$ and $\delta = -2.83362\ldots$ at $K = K_\mathrm{cr}$.

The scale given by $d_k = \Phi_{\Omega_k}^{q_{k-1}}(0, \Omega_k, K) - p_{k-1}$ is characterized by the limit

$$\lim_{k \to \infty} \frac{d_k}{d_{k+1}} = a, \qquad (2.27)$$

where $a$ is a universal constant. For $\theta = \theta_\mathrm{g}$, $a = -1.618\ldots = -\theta_g^{-1}$ for $K < K_\mathrm{cr}$ and $a = -1.28857\ldots$ at $K = K_\mathrm{cr}$.

The spectrum of the circle map at $K = K_\mathrm{cr}$ also demonstrates a number of universal properties. If $\theta$ can be presented in the form of a periodic continued fraction, the spectrum has the property of *scaling*. For $\theta = \theta_\mathrm{g}$, the frequencies of spectral components, reduced to the interval $[0, 1]$, satisfy the following relation:

$$\nu = |F_{k+1}\theta_g - F_k|, \qquad (2.28)$$

where $F_k, F_{k+1}$ are the sequential terms of one of the Fibonacci series. The spectral series, arranged in decreasing order with respect to their amplitudes, correspond to the Fibonacci series with the following bases: main series – (0,1); 2nd one – (2,2); 3rd one – (1,3); 4th one – (3,3); 5th one – (1,4); 6th one – (2,5), etc. The reduced spectral power for the lines of each series has a limit as $j \to \infty$:

$$S_i = \lim_{j \to \infty} \frac{S_i^j}{\nu^2(j)} = \text{const.} \qquad (2.29)$$

The normalized spectrum $a_i^j = S_i^j / \left( S_1^j \nu^2(j) \right)$, represented in the coordinates $\log a_i^j - \log \nu$, is divided into identical intervals located between relevant lines of each series.

The above-mentioned and other quantitative regularities of two-frequency quasiperiodicity destruction appear to be typical not only for model one-dimensional maps but also for invertible maps and continuous-time systems. These properties have been observed within the given accuracy in full-scale experiments and in computer simulation of various DS [49, 51, 57, 58].

The RG method has been applied to analyze maps in the form of (2.22) [56, 59–63]. For $\theta = \theta_g$, one can derive a functional equation for the fixed point, which reads

$$\Phi^*(\phi) = a\Phi^* \left( a\Phi^*(\phi/a^2) \right), \qquad (2.30)$$

where $\Phi^*(\phi + 1) = \Phi^*(\phi) + 1$. Its solution $\Phi^*(\phi)$ is a universal function and the scaling multiplier $a$ represents a universal constant. Equation (2.30) has the linear solution $\Phi^*(\phi) = \phi - 1$, for which $a_{1,2} = 0.5(\pm\sqrt{5} - 1)$. The value of $a$, found numerically for $K < K_{cr}$, coincides with the solution $a_2 = 0.5(-\sqrt{5} - 1) = -\theta_g^{-1} \approx -1.618$. At $K = K_{cr}$ the linear solution does not obey (2.30). Since $\tilde{\Phi}(\phi)$ has a cubic inflection point at zero, the universal function $\Phi^*(\phi)$ must contain a cubic term $\phi^3$. A nontrivial function of this sort has been deduced numerically in the form

$$\Phi^*(\phi) = 1 + c_1\phi^3 + c_2\phi^6 + \ldots \qquad (2.31)$$

The value of $a$ found agrees with numerically obtained results. One of eigenvalues of the linearized equation at the fixed point coincides with the constant derived numerically from (2.26). This fact explains the universal character of the constant $\delta$. Results obtained for $\theta = \theta_g$ by means of the RG method have been generalized to the case of an arbitrary irrational winding number, which can be presented in the form of a periodic continued fraction. With this, the form of $\Phi^*(\phi)$ as well as the values of $a$ and $\delta$ naturally depend on $\theta$ [62].

## 2.1.5 Route to Chaos via a Three-Dimensional Torus.
## Chaos on $T^3$. Chaotic Nonstrange Attractors

In the previous section we considered the quasiperiodic route to chaos, which did not assume the emergence of a third independent oscillation frequency. Consequently, the route to chaos described may already occur in continuous-time systems with the phase space dimension $N = 3$. However, this scenario does not rule out the appearance of third, fourth, etc., independent frequencies, provided that the system has a phase space of appropriate dimension. Multi-frequency quasiperiodic regimes have been observed and studied in a large number of numerical and experimental works [5, 64–66]. Thus, the Ruelle–Takens–Newhouse scenario can be confidently realized in systems with dimension $N \geq 4$. However, the theorems proved in the works of Ruelle, Takens, and Newhouse do not explain the bifurcation sequences which lead to chaos. They only give evidence that motions on $T^3$ are structurally unstable and, therefore, will be converted into chaotic motion if the system is perturbed. Besides, the strange attractor that appears belongs to the flow on $T^3$, i.e., the torus is not destroyed.

Numerical investigations have shown that the transition to chaos from three-frequency quasiperiodicity is quite similar to the route to chaos via quasiperiodic motion on $T^2$ [67–72]. To visualize the evolution of $T^3$ in a system with continuous time the double Poincaré section is usually used. In discrete-time systems, three-frequency quasiperiodic motion is represented by an attractor in the form of an invariant $T^2$. Applying the Poincaré section in a map we obtain a one-dimensional closed curve corresponding to a $T^3$ in a continuous-time system. Investigations have shown that an invariant curve in the double Poincaré section of $T^3$ behaves like an invariant curve in the Poincaré section of $T^2$ [71]. Distortion of the invariant curve and resonances on $T^3$ precede the emergence of a chaotic attractor.

The motion on $T^3$ is often modeled using invertible maps of a unit square. The Poincaré section of $T^3$ gives birth to a map of $T^2$ to itself, which may be written as follows:

$$x_{n+1} = \Phi(x_n, y_n), \quad \text{mod } 1,$$
$$y_{n+1} = \Psi(y_n, x_n), \quad \text{mod } 1. \tag{2.32}$$

The phase space of such a map is a square with sides equal to 1, which represents a development of $T^2$. If the map is invertible, one can associate it with a flow on $T^3$. If the map becomes non-invertible for certain parameter values, this implies that $T^3$ is destroyed. Regular regimes on $T^3$ are characterized by two winding numbers which define the relation of three basic frequencies. For map (2.32) the winding numbers are introduced as follows:

$$\theta_x = \lim_{n \to \infty} \frac{\Phi^n(x_n, y_n)}{n},$$
$$\theta_y = \lim_{n \to \infty} \frac{\Psi^n(x_n, y_n)}{n}. \tag{2.33}$$

*Partial* and *full* resonances on $T^3$ are distinguished. Partial resonances are observed when only one of the winding numbers becomes rational or when both winding numbers are irrational but a multiple relation is kept between them. In this case ergodic two-dimensional tori exist on $T^3$. If both winding numbers take rational values, a full resonance occurs and periodic orbits appear on $T^3$. However, the limits (2.33) may not always exist and be unambiguous. Unlike the circle map, the torus map (2.32) can exhibit chaotic dynamics and remain invertible. This fact gives evidence that a chaotic motion can occur on $T^3$.

The motion on $T^3$ is often described by a system of two coupled circle maps:

$$x_{n+1} = x_n + \Omega_x - \frac{K}{2\pi} \sin 2\pi y_n \,, \ \text{mod } 1,$$

$$y_{n+1} = y_n + \Omega_y - \frac{K}{2\pi} \sin 2\pi x_n \,, \ \text{mod } 1. \tag{2.34}$$

This map may be thought of as representing the phase dynamics of three interacting oscillators. The parameters $\Omega_x$ and $\Omega_y$ denote the frequency shifts and govern the winding numbers $\theta_x$ and $\theta_y$. $K$ is the strength of nonlinearity. For $K < 1$ the map (2.34) is a diffeomorphism. For $K > 1$ the map is noninvertible and $T^3$ is destroyed. As $K$ increases, the measure of ergodic three-frequency motions decreases, while the measure of partial and full resonances increases. Different kinds of chaotic behavior can also be observed in map (2.34), even when $K < 1$. Bifurcations of the map (2.34) for $K < 1$, which are related to resonance fixed points and invariant curves, have been studied theoretically and numerically in [73]. Three different kinds of chaos on $T^3$ have been distinguished, namely, *contractible*, *rotational*, and *toroidal*. The distinction is made basing on different behavior of manifolds of saddle fixed points inside the complete synchronization regions and, correspondingly, on different behavior of the limits (2.33) [73]. Figure 2.12 displays phase portraits of the different chaotic attractors observed in map (2.34).

Toroidal chaos can develop into an ergodic chaotic attractor whose points densely fill the torus surface everywhere. Such an attractor is characterized by an integer capacity dimension and is thus called a *chaotic nonstrange attractor* [74]. Such regimes have been observed numerically in a system of coupled circle maps [75].

Many conservative, i.e., preserving phase space volumes, diffeomorphisms of a torus exhibit chaotic dynamics. For example, the toroidal chaos (in this case it is impossible to speak of an attractor) filling all the torus surface can be realized in the well-known *Arnold's map*, often called *Arnold's cat map* [76]:

$$x_{n+1} = x_n + y_n, \ \text{mod } 1,$$

$$y_{n+1} = x_n + 2y_n, \ \text{mod } 1. \tag{2.35}$$

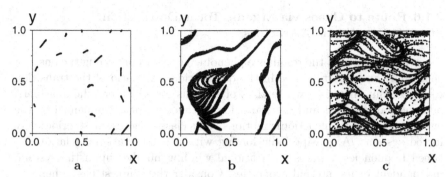

**Fig. 2.12.** Chaos in map (2.34): (a) contractible; (b) rotational; (c) toroidal

Arnold's cat map is a *robust hyperbolic system* (the *Anosov cascade*). Adding the nonlinear term $\delta \sin 2\pi x_n$ to the first equation of (2.35), we arrive at the dissipative map

$$x_{n+1} = x_n + y_n + \delta \sin 2\pi x_n, \ \ \mathrm{mod}\ 1,$$
$$y_{n+1} = x_n + 2y_n, \ \ \mathrm{mod}\ 1, \tag{2.36}$$

which is invertible for $\delta < 1/2\pi$. As was shown in [77], for small $\delta$ the map (2.36) has a chaotic attractor with integer capacity dimension $D_C = 2$. The stable and the unstable manifolds of saddle fixed points of Arnold's map bend around the torus surface, intersecting transversally so that no tangencies occur. Evidently, such behavior of manifolds holds for the perturbed map (2.36) when $\delta$ is small. In this case a chaotic attractor in (2.36) may appear to be an example of a robust hyperbolic attractor.

Thus, different kinds of chaotic motion including such unusual ones as chaotic nonstrange attractors are possible on $T^3$. However, the results presented are concerned with maps on a two-dimensional torus. The question of to what extent such maps can reflect the behavior of a flow on $T^3$ is still open. Nevertheless, in [78] the authors tried to detect chaotic attractors on $T^3$ when simulating a continuous-time system in the form of coupled oscillators. Unfortunately, their attempts failed. Up to now, a chaotic motion on $T^3$ has not been found in flow systems. The most typical case is when $T^3$ is broken down before chaos sets in. But the possibility of the existence of chaos on $T^3$ cannot be excluded.

The quasiperiodic route to chaos in high-dimensional systems allows the emergence of fourth, fifth, etc., independent frequencies of oscillations. However, due to nonlinear interaction between the oscillatory modes of the system, a measure of ergodic multi-frequency regimes is associated, as a rule, with the appearance of resonance structures on high-dimensional tori.

## 2.1.6 Route to Chaos via Ergodic Torus Destruction. Strange Nonchaotic Attractors

As mentioned above, the resonance phenomena always preceding the onset of chaos play an important role in the quasiperiodic scenario of the transition to chaos. But there is a whole class of systems for which ergodic quasiperiodic motion is robust and the transition to chaos is not accompanied by the emergence of resonant periodic motions. This class encloses quasiperiodically forced systems. A quasiperiodic forcing with a fixed irrational relation between frequencies imposes an irrational winding number upon the system, independent of its internal properties. Consider the simplest case when one of the forcing frequencies coincides with the natural frequency of periodic oscillations of the system. With this, a robust two-frequency quasiperiodic regime with a fixed winding number can be observed in a synchronization region of natural oscillations. This winding number is defined from the outside and supposed to be irrational. Varying system parameters and the amplitude of the external forcing one can achieve $T^2$ destruction and the transition to chaos. The transition from an ergodic torus to chaos is characterized by its own peculiarities as compared with the above-considered case when the torus destroyed had an arbitrary varying winding number. Numerous studies of flow systems and maps with quasiperiodic forcing [74, 79–84] have shown that the appearance of a special class of attractors, the so-called strange nonchaotic attractors (SNA), is typical for such systems. An SNA is defined as an attracting limit set of a DS, which is not a manifold and for which there is no exponential divergence of phase trajectories.

In order to study the mechanisms of transition from an ergodic torus to chaos, it is convenient to use quasiperiodically forced maps presented in the following form:

$$\boldsymbol{x}_{n+1} = \boldsymbol{F}(\boldsymbol{x}_n, \phi_n, \boldsymbol{\alpha}),$$
$$\phi_{n+1} = \phi_n + \theta, \ \text{mod} \ 1. \tag{2.37}$$

Here, $\boldsymbol{x} \in \mathbf{R}^N$ is the state vector of an autonomous system, $\boldsymbol{F} \in \mathbf{R}^N$ is periodic in argument $\phi$ with period 1, $\boldsymbol{\alpha}$ is the parameter vector of the system. $\phi$ denotes the phase of the forcing, and $\theta$ is the winding number. The forcing is quasiperiodic if $\theta$ is irrational. The winding number is usually chosen to be the golden mean, $\theta = 0.5(\sqrt{5} - 1)$. One-dimensional maps ($N = 1$) are most easy to use. A two-dimensional torus in map (2.37) corresponds to an invariant closed curve. As $\theta$ is irrational, this curve is densely covered everywhere by points of phase trajectories. Since no resonance structure arises on the torus and there is no need to control the winding number, the bifurcation mechanisms of $T^2$ destruction and the appearance of chaos allow in this case a one-dimensional analysis. Hence, we consider $\alpha$ to be a scalar.

Suppose that at $\alpha = \alpha_0$ there exists an ergodic $T^2$ with a strictly fixed irrational winding number, whereas at $\alpha = \alpha_1$ a chaotic attractor (CA) arises.

What is the scenario of transition to chaos in this case? Investigations have shown that the destruction of a robust ergodic $T^2$ leads initially to the appearance of an SNA that is then transformed into a CA [80, 82–85]. In the map, the invariant curve is first deformed and then loses its smoothness. According to the theorem [41], a resonance torus also loses its smoothness before it is destroyed. This happens on a finite set of fixed points of the invariant curve, corresponding to the points of the stable resonance cycle. Such an "nonsmooth torus" can exist for some time in the system phase space before its destruction takes place.

In the case of an ergodic torus, the invariant curve has no fixed points, and at some $\alpha = \alpha_{cr1}$ it loses its smoothness simultaneously on an everywhere dense set of points. As a result, the invariant curve is destroyed, and a set that is not a manifold appears. However, the torus destruction does not automatically lead to the emergence of exponential instability of the motion. The dynamics becomes chaotic later when $\alpha = \alpha_{cr2} > \alpha_{cr1}$. Thus, there exists a finite range of parameter $\alpha$ values, $\alpha_{cr1} < \alpha < \alpha_{cr2}$, where an SNA is observed. The regime of the SNA possesses the properties of being intermediate between quasiperiodicity and chaos. In order to establish that an SNA is really observed, one has: (i) to calculate the Lyapunov exponents and (ii) to check that the attractor is not of regular type. The first task is relatively simple, because there are reliable methods for calculating the Lyapunov exponents, and if there is no positive exponent, then one can be sure that the attractor is nonchaotic. The second problem is much more difficult. The numerical criteria available do not allow one to confidently detect whether the set under study is an SNA or a strongly deformed but still smooth torus.

The most reliable numerical methods for diagnosing the regime of an SNA have been proposed in works by Pikovsky et. al. [86, 87]. The first approach is related to the rational approximation of the winding number. The second method is based on the property of phase sensitivity of the SNA. The latter deals with the criterium of local Lyapunov exponents. Consider (2.37) for the case of $N = 1$. The map attains the following form:

$$x_{n+1} = f(x_n, \phi_n, \alpha),$$
$$\phi_{n+1} = \phi_n + \theta, \ \text{mod} \ 1. \tag{2.38}$$

Using this map, we explore the destruction of an invariant curve in a phase plane. The *rational approximation method* is based on the bifurcational analysis of cycles arising in the map under the rational approximation of the winding number $\theta_k = p_k/q_k$, $\lim_{k \to \infty} \theta_k = \theta$. In this case the behavior of the map strongly depends on the choice of the initial phase $\phi_0$. If for sufficiently large (theoretically for arbitrary large) $k$ the cycles exhibit bifurcations as $\phi_0$ changes, one can conclude that an SNA exists in (2.38). When using this method, a phase-parametric diagram $x(\phi_0)$, $\phi_0 \in [0; 1/q_k]$, is constructed for $\theta = \theta_k$. This diagram was called the *approximating attracting set* in [86]. Indeed, at $\theta = \theta_k$, the dependence of attractors' coordinates of the map on

$\phi_0$ approximates a small segment of an ergodic attractor. The fact that the approximating set is smooth testifies to the quasiperiodic regime of the map, whereas the presence of points of nonsmoothness corresponds to an SNA.

Another approach is based on the sensitivity of the dynamical variable to the phase of the external forcing. The map is now considered for an irrational value of the winding number. The derivative $\partial x_n/\partial \phi_0$ is calculated along a trajectory, and its maximum is estimated. For (2.38) it is easy to obtain

$$\frac{\partial x_n}{\partial \phi_0} = \sum_{k=1}^{n} f_\phi \mu_{n-k}(x_k, \phi_k) + \mu_n(x_0, \phi_0)\frac{\partial x_0}{\partial \phi_0}, \tag{2.39}$$

where

$$\mu_m(x_k, \phi_k) = \prod_{i=0}^{m-1} f_x(x_{k+i}, \phi_{k+i}), \quad \mu_0 = 1, \tag{2.40}$$

is a "local multiplier" of the phase trajectory. Map (2.38) has one Lyapunov exponent

$$\lambda = \lim_{n \to \infty} \frac{1}{n} \ln |\mu_n|, \tag{2.41}$$

which is not equal to zero. Since it is negative in the regime of the SNA, the local multiplier must tend to zero with $n$. Taking this into account, the derivative (2.39) may be re-written as follows:

$$\frac{\partial x_n}{\partial \phi_0} = \sum_{k=1}^{n} f_\phi \, \mu_{n-k}(x_k, \phi_k). \tag{2.42}$$

Introduce the quantity $\Gamma_n$:

$$\Gamma_n = \min_{x_0, \phi_0} \max_{0 \le i \le n} \left| \frac{\partial x_i}{\partial \phi_0} \right|, \tag{2.43}$$

where the maximum is sought in all points of a single trajectory, and the minimum is determined with respect to randomly chosen initial points. The fact that the value of $\Gamma_n$ grows infinitely with $n$ means that the derivative $\partial x_n/\partial \phi_0$ does not exist, i.e., the invariant curve loses its smoothness and an SNA appears. Figure 2.13 shows dependences of $\Gamma_n$ on $n$ obtained in [86] for the map

$$x_{n+1} = 2\sigma(\tanh x_n)\cos(2\pi\phi_n) + \alpha\cos\big(2\pi(\phi_n + \beta)\big),$$
$$\phi_{n+1} = \phi_n + \theta, \text{ mod } 1, \tag{2.44}$$

for $\theta = \theta_g = 0.5(\sqrt{5} - 1)$ and for different $\alpha$. As seen from the graphs, the SNA exists only for $\alpha = 0$. $\Gamma_n$ may be presented as

$$\Gamma_n \sim n^\eta. \tag{2.45}$$

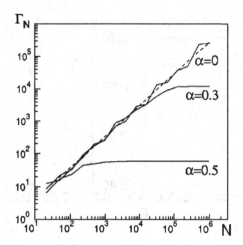

**Fig. 2.13.** Values of $\Gamma_n$ versus $n$ for the system (2.44) for $\theta = \theta_g$, $\sigma = 1.5$, $\beta = 1/8$ [86]

Here, $\eta$ is called a *phase sensitivity exponent*. In [86] it has been proved that $\eta = 1$ in the regime of the SNA.

Phase sensitivity is connected to the existence of a nonzero measure of positive local Lyapunov exponents. The *local Lyapunov exponent* is a Lyapunov exponent of a trajectory which is calculated on a finite time interval [86]. For (2.38) the local Lyapunov exponent reads

$$\Lambda_n(x, \phi) = \frac{1}{n} \ln |\mu_n(x, \phi)|, \tag{2.46}$$

where $\mu_n(x, \phi)$ is defined by (2.40). It is clear that both the value and the sign of the local Lyapunov exponent depend on the initial point $(x_0, \phi_0)$. Since the SNA is nonsmooth and the local multiplier $\mu_n$ is unbounded, a measure of positive values $\Lambda_n$ must be different from zero even for sufficiently large $n$. With this, $\lim_{n \to \infty} \Lambda_n = \lambda < 0$, because the attractor is nonchaotic.

Unfortunately, the criteria considered can be mainly applied to one-dimensional discrete models with quasiperiodic forcing and are little suited to systems with continuous time. The exception may be the criterium of positive local Lyapunov exponents. But its application is associated with the problem concerning a minimal time interval for which the local exponents of a quasiperiodic regime of the system under study cannot be already positive. Other numerical criteria for the existence of SNA have also been proposed. They are related to the properties of the spectrum and autocorrelation function [79, 87]. However, in most cases they also do not provide unambiguous results. Therefore, the mechanism of ergodic torus destruction should be explored with caution by using as many available criteria of SNA as possible.

The reasons why an invariant curve in model maps loses its smoothness and then is destroyed have been analyzed, and two mechanisms leading to

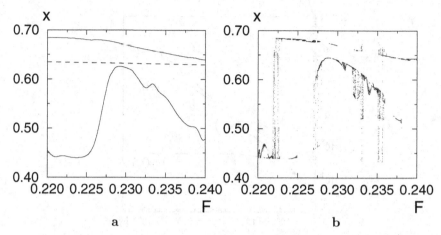

**Fig. 2.14.** Band merging crisis of the invariant curve in map (2.47) for $\varepsilon = 0.1$, $\theta = \theta_{\mathrm{g}}$. A segment of the invariant curve before tangency ($\alpha = 3.271$) (**a**) and after tangency ($\alpha = 3.272$) (**b**) with the repeller (*dashed line*)

the appearance of SNA have been revealed: (i) a crisis of an ergodic torus at $\alpha = \alpha_{\mathrm{cr1}}$ through a nonlocal bifurcation and (ii) a gradual evolution of the torus, leading to its loss of smoothness and destruction at $\alpha = \alpha_{\mathrm{cr1}}$ [85]. The ergodic torus crisis occurs when a stable torus touches an unstable torus or its stable manifold, the latter playing the role of a separatrix surface. In one-dimensional models with quasiperiodic forcing (2.38) a saddle torus corresponds to an unstable invariant curve (repeller). The crisis may be connected with the merging of bands of a quasiperiodic attractor.[3] The point is that as a result of torus doubling bifurcation in one of the periods, the invariant curve in the torus section will consist of $2^k$ bands. These bands are visited by a representative point in a strictly defined order and separated by a separatrix surface. The latter can be represented by a surface in the section of the stable manifold of a saddle torus. In terms of one-dimensional maps, a separatrix corresponds to an unstable invariant curve. Figure 2.14 illustrates the band merging crisis of an invariant curve in the logistic map with quasiperiodic parameter modulation [82]:

$$x_{n+1} = \alpha(1 + \varepsilon \cos 2\pi\phi_n)\, x(1 - x),$$
$$\phi_{n+1} = \phi_n + \theta, \ \mathrm{mod}\ 1. \tag{2.47}$$

The invariant curve exists for $\alpha < \alpha_{\mathrm{cr1}}$ and consists of two bands separated by a repeller. A representative point visits each band in one iteration of the map. The distortion of the invariant curve leads to the situation that at $\alpha = \alpha_{\mathrm{cr1}}$

---

[3] From the viewpoint of crisis definition, such a merge is not a crisis since it does not cause the absorbing area to change. But in this case the term "crisis" is commonly accepted.

the invariant curve touches the repeller separating the attractor bands. At the moment of tangency, the stable invariant curve loses its smoothness and is destroyed. For $\alpha > \alpha_{cr1}$ there exists an SNA uniting both bands of the curve destroyed, which are now visited by the representative point randomly. This mechanism of torus destruction restricts the sequence of ergodic torus doublings, which is typically finite.

Torus crisis can also be related to the merging of different quasiperiodic attractors [84, 85]. The situation is possible when the crisis occurs without collision of quasiperiodic attractors or bands of a single attractor. This case is especially typical for a resonance on $T^3$ whose one winding number has a fixed irrational value and whose other is varied arbitrarily. The distortion of the shape of invariant curves in the section of stable and unstable two-dimensional tori on $T^3$ results in a crisis of tangency at separate points instead of a tangency bifurcation of the tori on the boundary of the synchronization region. Such a crisis has been found and studied in [83].

Smoothness loss and destruction of an ergodic torus can also take place without bifurcations related to a separatrix tangency. In this case, as the control parameter $\alpha$ is varied, the shape of an invariant curve in a torus section is gradually distorted, which causes the phase sensitivity to increase. The derivative with respect to the initial phase is no longer bounded at the critical point $\alpha = \alpha_{cr1}$. If the winding number is replaced by rational approximations, the dynamics of the map turns out to be chaotic at some $\phi_0$ values and remains regular at other $\phi_0$ values. However, the Lyapunov exponent still converges to a negative value when $n \to \infty$. In this case the approximating set cannot be smooth and an SNA is observed within some interval $\alpha_{cr1} \leq \alpha \leq \alpha_{cr2}$. Such an evolutionary mechanism for the appearance of SNA has been studied in [84, 85].

It is assumed that an SNA arising from a crisis or by evolution exists on a nonzero measure in parameter space. But this statement is difficult to prove. It is known that on the interval $\alpha \in [\alpha_{cr1}, \alpha_{cr2}]$ an SNA can degenerate into an invariant curve which possesses a finite number of points of discontinuity and then appear again [85]. The bifurcation mechanism of transition from SNA to chaos is still not understood. The well-studied scenarios of chaos formation assume the presence of homoclinic trajectories of saddle cycles or dangerous separatrix loops of a saddle-focus. In their neighborhood a horseshoe-type map arises and a chaotic set of trajectories can be created. When the winding number $\theta$ is kept constant and has an irrational value, a system possesses neither equilibrium states nor limit cycles. In this case a homoclinic structure must be connected to a tangency and intersection of manifolds of saddle tori. At the present time, homoclinics of this sort have not been adequately explored.

### 2.1.7 Summary

In the present section we have described three typical routes to chaos: (i) the period-doubling cascade route (Feigenbaum scenario); (ii) the crisis of periodic oscillations and the intermittency transition route (Pomeau–Manneville scenario); and (iii) different kinds of quasiperiodic routes to chaos (Ruelle–Takens–Newhouse scenario). It is worth noting that for the same system different routes to chaos can be observed corresponding to different regions and directions in parameter space. Moreover, the bifurcational sequences being observed may be combined in a complex way. Thus, to imagine the full picture of the appearance of chaotic motion, one should not restrict oneself to a one-parametric analysis. It is necessary to have an idea, at least in general terms, of the bifurcation diagram of dynamical regimes of the system in parameter space, what happens on its different "leaves", where they are "sewed", etc.

The scenarios considered are typical in the sense that they are observed for a wide class of DS with both low- and high-dimensional phase space, as well as for distributed systems for arbitrarily chosen control parameters and for different types of variation. Evidently, other scenarios for chaos formation may also be possible, but they are not typical. They may be related to some peculiarities (degenerations) of DS or to a special choice of directions of motion in parameter space which pass through high-codimension critical points [54, 88].

## 2.2 Statistical Properties of Dynamical Chaos

### 2.2.1 Introduction

Dynamical chaos, as a random process, requires a statistical description. As chaotic systems are studied in laboratory experiments or simulated numerically, some probabilistic characteristics are normally calculated or measured, such as the stationary probability distribution over the attractor, correlation functions, power spectra, etc. Chaotic oscillations, which can be mathematically represented by chaotic attractors of various types, differ in their statistical properties and their sensitivity to the influence of noise.

From the standpoint of rigorous theory, hyperbolic chaos is frequently said to be the "ideal" chaos. Its structure is topologically homogeneous and stable against perturbations [3, 89–91]. However, as a rule, the strange chaotic attractors of dynamical systems do not behave as structurally stable hyperbolic systems. Nearly hyperbolic (quasi-hyperbolic) attractors include unstable orbits of the separatrix-loop type. The generation and disappearance of such orbits do not affect the characteristics of chaos such as the phase portrait of the attractor, the power spectrum, the Lyapunov exponents, etc. Dynamical systems in chaotic regimes can be characterized by an invariant measure that is independent of the initial distribution and completely specifies the statistical properties of the attractor. The existence of an invariant measure has been proven theoretically for structurally stable hyperbolic and quasi-hyperbolic systems [92–97].

However, most chaotic attractors that have been studied numerically and/or experimentally are not hyperbolic [42,98,99]. The problem of the existence of an invariant measure on a nonhyperbolic chaotic attractor encounters serious difficulties because a stationary probability distribution independent of the initial distribution cannot be introduced in general. The nonhyperbolic attractor is the maximal attractor of the dynamical system; it includes a countable set of regular and chaotic attracting subsets [42,98]. Therefore, one can consider the invariant measure of a nonhyperbolic attractor only if the influence of external noise is present [100]. As a rule, nonhyperbolic attractors change their properties dramatically under the action of noise [101–104], whereas hyperbolic and quasi-hyperbolic attractors are stable against noise perturbations [101,102,105,106].

The statistical description of noise-affected nonhyperbolic chaotic attractors is an important problem of the theory of dynamical chaos that still remains unresolved. Among other things, the time of relaxation to a stationary distribution should be investigated. A number of fundamental questions that have not yet been clearly answered arise. What is the actual relaxation time for a stationary distribution? What are the factors controlling this time? What characteristics could be used to quantitatively estimate the relaxation time to the stationary measure? How do the noise statistics and noise intensity affect the law of establishment of the stationary distribution? Is the process of relaxation related to the dynamics of the system? These questions were partly answered in [107,108] by means of computer simulation.

The process of establishing a stationary distribution can be described by Fokker–Planck-type or Frobenius–Perron-type evolutionary equations. The eigenvalues and eigenfunctions of the evolution operator specify the relaxation process and the characteristics of mixing, which are linked to the relaxation to the invariant probability measure. However, if the dimension of the DS is high ($N \geq 3$), the Fokker–Planck and Frobenius–Perron equations are virtually unsolvable, even in a numerical form. For this reason, the studies described in [107,108] used the technique of stochastic differential equations.

By definition, chaotic dynamics implies mixing and, therefore, a positive Kolmogorov entropy. As a result of mixing, autocorrelation functions (ACF) decrease to zero (correlation splitting). States of the system separated by a sufficiently long time interval become statistically independent [93,95,109–111]. It is important to note that any system with mixing is ergodic. The temporal splitting of correlations in chaotic DS is related to the exponential instability of chaotic trajectories and to the system's property of generating a positive Kolmogorov entropy [93,95,109–113]. Although the correlation properties of chaotic processes are very important, they have not yet been adequately explored. It is commonly assumed that the ACF of chaotic systems decay exponentially at a rate determined by the Kolmogorov entropy [109]. It is then assumed that the Kolmogorov entropy $H_K$ is bounded from above by the sum of positive Lyapunov exponents [95,113,114]. Regrettably, this estimate proves to be wrong in the general case of hyperbolic systems.

It has been proven for certain classes of discrete maps with mixing (expanding maps with a continuous measure and the Anosov diffeomorphism) that the time decrease in the correlations is bounded from above by an exponential function [96, 115–117]. Various estimates have been obtained for the rate of this exponential decay, which are not always related to the Lyapunov exponents [107, 118–120]. As regards systems with continuous time, no theoretical estimates for the rate of correlation splitting are available as yet [121].

Experimental studies of some particular chaotic systems testify to a complex behavior of the correlation functions, which is controlled not only by the positive Lyapunov exponents but also by the properties of the chaotic dynamics of the system [107, 118, 120, 122]. It is important to reveal specific parameters of the chaotic dynamics that are responsible for the decay rate of the autocorrelations and for the spectral-line width of the fundamental frequency of the chaotic attractor. In this section we summarize our results [107, 108, 123–125] obtained recently in studying classical systems with nonhyperbolic and quasi-hyperbolic attractors [126–129]. These results include a technique for diagnosing nonhyperbolic chaos, noise effects on nonhyperbolic attractors, some probabilistic aspects of chaotic dynamics (such as the features of the relaxation to the stationary probability distribution and the mixing rate), and a spectral-correlation analysis of various types of chaotic oscillation regimes. Particular attention is given in this section to the influence of external noise on the statistical properties of chaos.

### 2.2.2 Diagnosis of Hyperbolicity in Chaotic Systems

Strange attractors in finite-dimensional systems can be divided into three basic classes: structurally stable (robust) hyperbolic, nearly hyperbolic (quasi-hyperbolic), and nonhyperbolic [42,98,99]. The property of robust hyperbolicity of a chaotic attractor implies that all its trajectories are of the same saddle-point type, and their stable and unstable manifolds are everywhere transverse, i.e., the structure of the hyperbolic attractor is homogeneous at any point of the attractor. Furthermore, small perturbations of the parameters of the system preserve these properties. But structurally stable hyperbolic attractors are rather typical of idealized objects such as the Smale–Williams solenoid [130] or the Plykin attractor [131]. The existence of a robust hyperbolic attractor has not been proven for dynamical systems specified by differential equations or discrete maps. Nevertheless, several examples of nearly hyperbolic attractors are known for such systems. These are the Lorenz attractor [132] and the Shimizu–Morioka attractor [133] in flow systems and the Lozi attractor [134] and the Belykh attractor [135] in discrete maps. Singular phase trajectories are characteristic of these systems. For example, the Lorenz attractor is characterized by the presence of a set of separatrix loops of the saddle-point-type equilibrium state, whereas the Lozi attractor includes non-robust homoclinic curves without contacts between stable and unstable manifolds. However, these peculiar trajectories do not generate stable motions, and the quasi-hyperbolic and hyperbolic attractors are similar from the viewpoint of numerical simulation.

Most chaotic attractors of DS are nonhyperbolic [42, 98, 99]. Nonhyperbolic attractors include chaotic limit sets and stable periodic orbits that, as a rule, are difficult to detect in numerical simulations, because they have extremely small attracting basins. If the whole collection of properties is considered, nonhyperbolic attractors differ substantially from hyperbolic ones [99, 136, 137]. Therefore, the diagnosis of the attractor type is of paramount importance for the analysis of nonlinear systems in both the theoretical and practical contexts [103, 138–143].

The direct technique of determining the conditions of hyperbolicity includes the calculation of the angles $\phi$ between the stable and unstable manifolds along a phase trajectory. A numerical procedure of computing these angles was proposed by Lai et al. [144] as a tool for diagnosing the hyperbolicity of chaotic saddle points in two-dimensional systems. This technique consists of transforming an arbitrary vector by the evolution operator in both direct and reverse time, which makes it possible to find the angle between the directions of stability and instability for various points of chaotic sets.

Manifolds are one-dimensional in two-dimensional systems, and hence diagnosing the effect of homoclinic contact does not present major difficulties. The problem is more complex for three-dimensional systems, because the manifolds are two-dimensional in this case. We have suggested a method for diagnosing hyperbolicity in three-dimensional differential systems [145]. It was found that systems such as the Rössler system, the Chua circuit [146], and the Anishchenko–Astakhov oscillator are typically nonhyperbolic, i.e., structurally unstable systems [98]. The Lorenz system can be considered an exception. The Lorenz attractor is nearly hyperbolic in a certain parameter range. The stable and unstable manifolds of the attractor trajectories intersect transversally [145]. However, as the parameters are varied, the Lorenz system exhibits a bifurcational transition to a nonhyperbolic attractor [147]. Figure 2.15a shows the probability distribution of angles $p(\phi)$ for the Lorenz attractor in the system [127]

$$\dot{x} = -\sigma(x - y), \qquad \dot{y} = rx - y - xz, \qquad \dot{z} = -\beta z + xy. \qquad (2.48)$$

It can be seen from the upper graph that the probability of homoclinic tangency is exactly zero $[p(\phi) = 0]$. As the region where the Lorenz attractor exists recedes, the effect of homoclinic tangency emerges (Fig. 2.15b). This

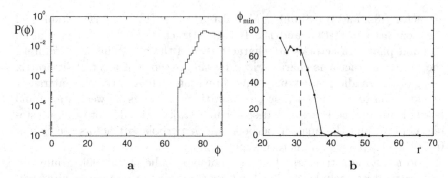

**Fig. 2.15.** Calculation results for Lorenz system (2.48) at $\sigma = 10$ and $\beta = 8/3$. (a) Probability distribution of angles for the Lorenz attractor at $r = 27$; (b) the minimum angle $\phi_{\min}$ as a function of the parameter $r$. The vertical line marks the theoretically determined onset of the transition from the Lorenz attractor to nonhyperbolic attractors

effect largely accounts for the properties of nonhyperbolic chaos, which are considered in subsequent parts of this section.

### 2.2.3 Chaos in the Presence of Noise

Nonlinear stochastic problems are of fundamental and practical importance. Two basic approaches to the analysis of stochastic systems are known [148–151]. The first is based on solving *stochastic equations* and is called the *Langevin method*. Any particular solution to stochastic equations, even for the same initial state, generates a new realization of the random process. This method allows obtaining an ensemble of a large number of realizations and finding a statistical characterization of the process. Averaging can be done over one, sufficiently long, realization, because the chaotic process is ergodic. The second approach consists in solving the *evolutionary equations for the probability measure*, such as the Chapman–Kolmogorov equation, the kinetic equation, or the Fokker–Planck equation. This requires that the random process in the system be at least Markovian, which poses some constraints on the noise sources. For the process to be Markovian, random actions must be independent. In this case, the Chapman–Kolmogorov equation is valid. If the noise is Gaussian, the process is diffusive, and the probability density obeys the Fokker–Planck equation. Of the noise sources satisfy the corresponding requirements, the method of stochastic equations and the method of evolutionary equations yield equivalent results [148–150, 152].

The problem of statistical characterization of dynamical chaos and the role of fluctuations in chaotic systems is of particular interest [93, 94, 105, 128, 152–156]. For systems with hyperbolic-type chaotic dynamics, statistical description is possible even in purely deterministic cases, i.e., in the absence of noise [93, 94, 105, 156]. This means that the stationary solution of the evolutionary equation for the probability density allows the existence of the limit as d $\rightarrow$ 0, where $D$ is the noise intensity; therefore, a solution for the probability measure can be obtained even in a purely deterministic case. As shown in [93, 105], small fluctuations ($D \ll 1$) in hyperbolic systems give rise to small variations in the structure of the probability measure. From this viewpoint, so-called quasi-hyperbolic (nearly hyperbolic) attractors, such as the Lozi attractor and the Lorenz attractor [42, 98], do not virtually differ from hyperbolic attractors. This is because neither quasi-hyperbolic nor hyperbolic attractors contain stable periodic orbits. A rigorous proof of the existence of a probability measure in the Lorenz attractor in the absence of noise, was given in [94].

The effects of noise are important for nonhyperbolic systems. As shown in [102], the mean distance between an orbit with noise and a nonhyperbolic attractor without noise is much larger than in the hyperbolic case and depends on the information dimension of the attractor. It is well known that noise can induce various phase transitions in systems with nonhyperbolic attractors [128, 129, 157, 158]. As sources of Gaussian noise are added to the system, the attraction basins of the coexisting attractors merge. This results in the establishment of a stationary probability density, which is independent of the initial state [100]. A statistical description of nonhyperbolic chaos encounters fundamental difficulties. Strictly speaking, a stationary probability measure independent of the initial distribution does not exist in nonhyperbolic chaotic attractors without noise. The continuum limit as $D \rightarrow 0$ cannot be implemented in those case [100]. Moreover, probabilistic characteristics of nonhyperbolic chaos are highly sensitive even to tiny variations in the parameters of the system [99, 128, 145, 159]. Thus, the existence of a stationary probability measure in a nonhyperbolic attractor is only possible if the system is affected by noise.

### 2.2.4 Relaxation to a Stationary Probability Distribution for Chaotic Attractors in the Presence of Noise

**Models and numerical techniques.** We consider the chaotic attractors of such well-known systems as the Rössler oscillator [126]

$$\dot{x} = -y - z + \sqrt{2D}\xi(t), \qquad \dot{y} = x + ay, \qquad \dot{z} = b - z(m - x) \qquad (2.49)$$

and the noise-affected Lorenz system (2.48) [127]

$$\dot{x} = -\sigma(x - y) + \sqrt{2D}\xi(t), \qquad \dot{x} = rx - y - xz, \qquad \dot{z} = -\beta z + xy. \qquad (2.50)$$

In both models, $\xi(t)$ is the source of Gaussian white noise with the mean value $\langle \xi(t) \rangle \equiv 0$ and correlation $\langle \xi(t)\xi(t+\tau) \rangle \equiv \delta(\tau)$, where $\delta(\tau)$ is the Dirac delta-function. The parameter $D$ denotes the noise intensity. For the Rössler system, we fix the parameters $a = 0.2$ and $b = 0.2$., varying the parameter

$m$ within the range $[4.25, 13]$. For the Lorenz system, we choose two different regimes – a quasihyperbolic attractor ($\sigma = 10$, $\beta = 8/3$, and $r = 28$) and a nonhyperbolic attractor ($\sigma = 10$, $\beta = 8/3$, and $r = 210$).

The chaotic attractors of systems (2.49) and (2.50) have been studied in detail and are classical examples of quasihyperbolic and nonhyperbolic chaos, respectively [129, 160]. Thus, the results obtained for systems (2.49) and (2.50) can be generalized to a broad class of dynamical systems.

To study the processes of relaxation to a stationary distribution in these systems, we analyze the time evolution of a collection of points initially located in a cube of a small size $\delta$ around an arbitrary point of a trajectory that belongs to the attractor of the system. We choose the size of this cube to be $\delta = 0.09$ and fill the cube uniformly with points whose number is $n = 9000$. In due course, they spread over the entire attractor. To characterize the convergence to a stationary distribution, we trace the time evolution of the set of points and calculate the ensemble average

$$\bar{x}(t) = \int_W p(x,t)x\mathrm{d}x = \frac{1}{n}\sum_{i=1}^{n} x_i(t), \qquad (2.51)$$

where $x$ is one of the dynamical variables of the system and $p(x,t)$ is the probability density of the variable $x$ at time $t$, which corresponds to the initial distribution. We introduce the function

$$\gamma(t_k) = |\bar{x}_m(t_{k+1}) - \bar{x}_m(t_k)|, \qquad (2.52)$$

where $\bar{x}_m(t_k)$ and $\bar{x}_m(t_{k+1})$ are successive extrema of $\bar{x}(t)$. The function $\gamma(t_k)$ characterizes the amplitude of fluctuations in the mean of $\bar{x}_m(t)$. The successive time moments $t_k$ and $t_{k+1}$ in (2.52) correspond to the extrema of $\bar{x}$. The temporal behavior of $\gamma(t_k)$ gives an idea of the regularities and rate of relaxation to the probability measure on the attractor. We calculated the largest Lyapunov exponent $\lambda_1$ of a chaotic trajectory on the attractor and the normalized autocorrelation function of the well-established oscillations $x(t)$:

$$\Psi(\tau) = \frac{\psi(\tau)}{\psi(0)}, \qquad \psi(\tau) = \langle x(t)x(t+\tau)\rangle - \langle x(t)\rangle\langle x(t+\tau)\rangle. \qquad (2.53)$$

Here, the angular brackets $\langle \ldots \rangle$ denote time averaging.

Instead of $\gamma(t_k)$ and $\Psi(\tau)$, we plot their respective envelopes $\gamma_0(t_k)$ and $\Psi_0(\tau)$, to make the figures more informative and visual.

**Relaxation to a Stationary Distribution in the Rössler system. Effects of Noise on the Mixing Rate.** The nonhyperbolic chaotic attractor that appears in the Rössler system (2.49) at fixed $a = b = 0.2$ and at $m$ values within the interval $[4.25, 8.5]$ is a well-known example of a spiral (or phase-coherent) attractor. The phase trajectory in a spiral attractor winds

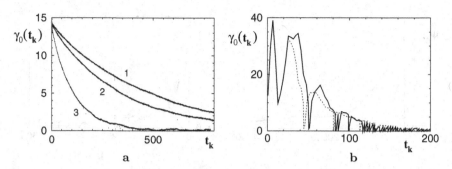

**Fig. 2.16.** Behavior of the function $\gamma_0(t_k)$ for the attractors in Rössler system (2.49): **(a)** for the spiral attractor ($a = b = 0.2$, $m = 6.1$) at $D = 0$ (curve 1), $D = 0.001$ (curve 2), and $D = 0.1$ (curve 3); **(b)** for the funnel attractor ($a = b = 0.2$,, $m = 13$) at $D = 0$ (solid curve) and $D = 0.01$ (dashed curve)

around one or several saddle foci. The ACF is oscillatory, and the narrow-band peaks stand out in its power spectrum; they correspond to the mean winding frequency and its harmonics and subharmonics [129, 161–163].

As the parameter $m$ is increased, the attractor of system (2.49) changes qualitatively. An incoherent attractor arises within the interval $8.5 < m \leq 13$, which is called a funnel attractor [128, 162]. The phase trajectories behave in a more complex way in it. As a result, the ACF of the funnel attractor decays much more rapidly than in the case of spiral chaos, and the power spectrum does not contain pronounced peaks.

Calculations carried out for $m \in [4.25, 7.5]$ (spiral chaos) and for $m \in [8.5, 13]$ (funnel chaos) without including noise suggest that an invariant probability measure exists at these parameter values[4]. Qualitatively, all effects observed in system (2.49) for each attractor type are preserved as the parameter $m$ is varied.

Figure 2.16 shows a typical behavior of the function $\gamma_0(t)$ for the spiral and the funnel attractor of system (2.49). It has been found that noise strongly affects the mixing rate in the spiral-attractor regime. The relaxation time considerably decreases with the increase in the noise intensity (Fig. 2.16a). The situation is radically different in the case of the funnel attractor. Incoherent chaos is virtually insensitive to noise influences. The behavior of $\gamma_0(t)$ does not change substantially as noise is added (Fig. 2.16b). Numerical simulations show that the correlation times are also quite different for these two chaotic regimes: in the absence of noise effects, they differ by two orders of magnitude. For spiral chaos, the correlation time is considerably shorter in the presence of noise (Fig. 2.17a), while the ACF for the funnel attractor in the deterministic case virtually coincides with the ACF in the presence of noise (Fig. 2.17b). Therefore, nonhyperbolic incoherent chaos exhibits some

---

[4] Stable trajectories have vanishingly small basins of attraction and do not manifest themselves, since the accuracy of computations is finite.

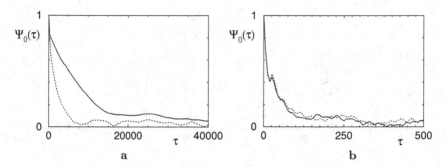

**Fig. 2.17.** Envelopes of the normalized ACF $\Psi_0(\tau)$ for the attractors in system (2.49): (**a**) at $m = 6.1$, for $D = 0$ (solid curve) and $D = 0.01$ (dashed curve); (**b**) at $m = 13$, for $D = 0$ (solid curve) and $D = 0.01$ (dashed curve)

properties of hyperbolic chaos, i.e., as Sinai notes [93], "dynamical stochasticity" proves to be stronger than the stochasticity enforced from outside.

We also note another result. It has been found that a positive Lyapunov exponent is weakly sensitive to fluctuations effects for both spiral and funnel chaos (Fig. 2.18) and slightly decreases as noise is intensified. The correlation time can, however, vary significantly under the influence of noise in this case. Thus, mixing in the regime of spiral chaos is determined not only and not so much by the degree of exponential instability. Other, weightier factors are present. We analyze them below.

For this, we use the notions of instantaneous amplitude and phase of oscillations [163]. Unfortunately, neither is universal. For spiral attractors, these quantities are quite reasonably introduced as

$$x(t) = A(t) \cos \Phi(t), \qquad y(t) = A(t) \sin \Phi(t). \tag{2.54}$$

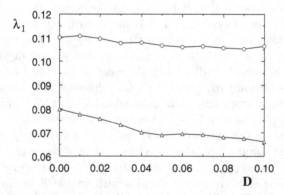

**Fig. 2.18.** Largest Lyapunov exponent $\lambda_1$ for the spiral (triangles) and funnel (circles) attractors as a function of the noise intensity $D$ for the Rössler system

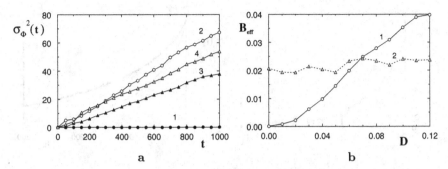

**Fig. 2.19.** (a) Time dependence of the instantaneous-phase variance $\sigma_\Phi^2$ for spiral chaos ($m = 6.1$) at $D = 0$ (curve 1) and $D = 0.1$ (curve 2) and for incoherent chaos ($m = 13$) at $D = 0$ (curve 3) and $D = 0.1$ (curve 4). (b) The effective phase diffusion coefficient $B_{\text{eff}}$ as a function of the noise intensity $D$ for spiral (curve 1) and incoherent (curve 2) chaos

The instantaneous phase, as it follows from (2.54), is determined by the relation

$$\Phi(t) = \arctan \frac{y(t)}{x(t)} + \pi n(t), \tag{2.55}$$

where $n(t) = 0, 1, 2, \ldots$ are the numbers of windings of the phase trajectory around the equilibrium state.

We have found that the component of mixing in the direction of the flux of trajectories is associated with the instantaneous-phase variance $\sigma_\Phi^2$, which controls phase diffusion. Figure 2.19a illustrates the time dependence of the instantaneous-phase variance $\sigma_\Phi^2$ on the ensemble of initially close trajectories for the spiral and funnel attractors of system (2.49). It can be seen that in the time intervals considered, the variance grows in a virtually linear manner both with and without the presence of noise. The assumption that the time dependence of the instantaneous-phase variance for chaotic oscillations in the Rössler system is linear was made in [161]; however, it was justified neither theoretically nor numerically. For spiral chaos without noise (curve 1), $\sigma_\Phi^2$ is small and grows much more slowly than in the other cases under study. The linear growth of the variance allows determining the effective phase-diffusion coefficient (first introduced by Stratonovich [164])

$$B_{\text{eff}} = \frac{1}{2} \left\langle \frac{d\sigma_\Phi^2}{dt} \right\rangle, \tag{2.56}$$

where the angular brackets denote time averaging of fast oscillations.

The diffusion coefficient $B_{\text{eff}}$ as a function of the noise intensity is presented in Fig. 2.19b for the spiral and funnel attractors in Rössler system (2.49). It can be seen that $B_{\text{eff}}$ grows with $D$ in both cases, but this growth is stronger for spiral chaos.

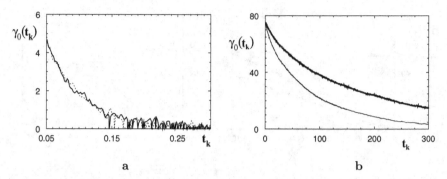

**Fig. 2.20.** Behavior of the function $\gamma_0(t_k)$ for chaotic attractors in Lorenz system (2.50): (**a**) for $r = 28$, $D = 0$ (solid curve) and $D = 0.01$ (dashed curve); (**b**) for $r = 210$, $D = 0$ (heavy curve) and $D = 0.01$ (light curve). The other parameters are $\sigma = 10$ and $\beta = 8/3$

**Relaxation to the probability measure in the Lorenz system.** The well-known quasihyperbolic attractors in three-dimensional differential systems, such as the Lorenz attractor and the Shimizu-Morioka attractor [133], are switching-type attractors. The phase trajectory switches chaotically from the vicinity of one saddle equilibrium state to the vicinity of another. Such switching involves random phase changes even in the absence of noise. Adding noise does not substantially modify the phase dynamics and does not therefore affect the rate of relaxation to the stationary distribution.

Figure 2.20 shows the behavior of the function $\gamma_0(t_k)$ for the quasi-hyperbolic and nonhyperbolic chaotic attractors of system (2.50) with and without noise influences. It has been discovered that noise has virtually no effect on the relaxation rate for the Lorenz attractor (Fig. 2.20a). The situation is radically different for the nonhyperbolic attractor in the Lorenz system. In this case, noise strongly affects the settling rate of the probability measure (Fig. 2.20b).

We now assess the dependence of the Lyapunov exponent and correlation time on the level of noise influence. For the same chaotic attractors in the Lorenz system, the highest Lyapunov exponent $\lambda_1$ and the normalized autocorrelation function $\Psi(\tau)$ ($\tau = t_2 - t_1$)) of the dynamical variable $x(t)$ were calculated for various noise intensities $D$. It was found that within the accuracy of computations, $\lambda_1$ is independent of the noise intensity for either type of chaotic attractor. Similarly, noise has virtually no effect on the ACF of the quasihyperbolic attractor (Fig. 2.21a). However, in the regime of the nonhyperbolic attractor, the ACF declines more rapidly with the presence of noise (see the curves in Fig. 2.21b).

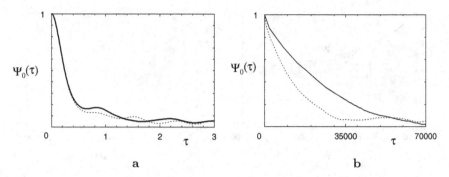

**Fig. 2.21.** Envelopes of the normalized autocorrelation function $\Psi_0(\tau)$ for the attractors in system (2.50): (a) $r = 28$, $D = 0$ (solid curve) and $D = 0.01$ (dashed curve); (b) $r = 210$, $D = 0$ (solid curve) and $D = 0.01$ (dashed curve)

### 2.2.5 Spectral-Correlation Analysis of Dynamical Chaos

**Spectral-correlation analysis of spiral chaos.** From the physical point of view, spiral-type chaotic attractors largely resemble noisy limit cycles. It should be kept in mind in this context that spiral attractors are present in completely deterministic systems, i.e., without fluctuation sources. We consider the regime of spiral chaos in Rössler system (2.49) at $a = b = 0.2$ and $m = 6.5$. With this aim in view, we introduce the instantaneous amplitude $A(t)$ and phase $\Phi(t)$ according to (2.54) and, by means of numerical simulation, we determine the normalized ACF of the chaotic oscillation $x(t)$ (Fig. 2.22, points in shaded region 1), the covariance function of amplitude fluctuations $K_A(\tau)$, and the effective phase-diffusion coefficient $B_{\text{eff}}$. Figure 2.22 shows the results for $\Psi_x(\tau)$ in system (2.49) with and without the presence of noise. The decay of the ACF is virtually exponential both in the absence (Fig. 2.22a) and in the presence of noise (Fig. 2.22b). Furthermore, as can be seen from Fig. 2.22c, an interval exists for $\tau < 20$ where the ACF decreases much more rapidly. The envelope of the computed ACF, $\Psi_x(\tau)$, can be approximated using equation (1.129). For this, we substitute the calculated characteristics $K_A(\tau)$ and $B = B_{\text{eff}}$ into the expression for the normalized envelope $\Psi_0(\tau)$:

$$\Psi_0(\tau) = \frac{K_A(\tau)}{K_A(0)} \exp(-B_{\text{eff}}|\tau|). \tag{2.57}$$

The calculation results for $\Psi_0(\tau)$ are represented by the points of curves 2 in Figs. 2.22a and 2.22b. It can be seen that the behavior of the envelope of the ACF, $\Psi_x(\tau)$, is described well by formula (2.57). We note that taking the factor $K_A(\tau)/K_A(0)$ into account yields a good approximation for all $\tau \geqslant 0$. This means that the amplitude fluctuations are important in short time intervals ($\tau < \tau_{\text{cor}}$), whereas the slow decrease in the correlation is mainly determined by phase diffusion. The surprisingly good agreement between the

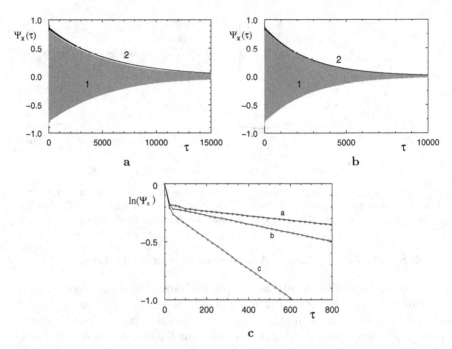

**Fig. 2.22.** The normalized ACF of the oscillation $x(t)$ in system (2.49) at $m = 6.5$ (points in the shaded region 1) and its approximation (2.57) (points on curve 2) for $D = 0$ (**a**) and $D = 10^{-3}$ (**b**); (**c**) envelopes of the ACF on a logarithmic scale for $D = 0$ (curve a), $D = 0.001$ (curve b), and $D = 0.01$ (curve c)

numerical results for spiral chaos and the data for the classical model of harmonic noise is noteworthy. At the same time, this good agreement is quite difficult to explain. First, relation (1.129) was obtained under the assumption that the amplitude and phase of fluctuations are statistically independent. It is absolutely clear that this assumption is not applicable to a chaotic regime. Second, formula (1.129) was derived taking into account that the phase fluctuations can be described in terms of a Wiener process. In the case of chaotic oscillations, $\Phi(t)$ is a more complex process, with unknown statistical properties. It is especially important to emphasize that the results in Fig. 2.22a were obtained for the regime of purely deterministic chaos (without noise), which additionally confirmed the similarity between chaotic self-sustained oscillations and a random process.

It follows from the results presented in Fig. 2.22 that the envelope of the ACF for chaotic oscillations at $\tau > \tau_{\text{cor}}$ can be approximated by the exponential factor $\exp(-B_{\text{eff}}|\tau|)$. According to the Wiener–Khinchin theorem, the spectral peak at the mean frequency $\omega_0$ must have a Lorentzian profile with the width determined by the effective phase diffusion coefficient $B_{\text{eff}}$:

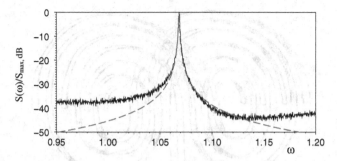

**Fig. 2.23.** A fragment of the normalized power spectrum of the oscillation $x(t)$ in system (2.49) at $a = b = 0.2$, $m = 6.5$ (solid curve) and this spectrum approximation by expression (2.58) (dashed curve) for the noise intensity $D = 10^{-3}$

$$S(\omega) = C\frac{B_{\text{eff}}}{B_{\text{eff}}^2 + (\omega - \omega_0)^2}, \qquad C = \text{const.} \qquad (2.58)$$

The calculation results shown in Fig. 2.23 confirm this assertion. The fundamental spectral line can be approximated by expression (2.58), which is supported by the numerical results for the power spectrum of the oscillation $x(t)$. The results presented in Figs. 2.22 and 2.23 for the noise intensity $D = 10^{-3}$ were reproduced for different $D$ in the interval $0 < D < 10^{-2}$ and for the range of the parameter $m$ that corresponds to the spiral-chaos regime. We note that the above-presented approximation results for the ACF and the profile of the fundamental spectral line of the spiral attractor in the Rössler system were completely confirmed by studies of spiral attractors in other dynamical systems [124, 125].

**Correlation Parameters of the Lorenz Attractor.** The narrow-band-noise model cannot be used to analyze the ACFs of switching-type chaotic oscillations, which have a continuous spectrum without pronounced peaks at any distinguished frequencies. Such attractors are quite complex in their structure [168]. The Lorenz attractor is a classical example of a switching-type attractor [127]. We consider the Lorenz system in the quasihyperbolic-attractor regime at $r = 28$, $\sigma = 10$, and $\beta = 8/3$.

There are two saddle foci in the phase space of the Lorenz system, which are located symmetrically about the $z$ axis and are separated by the stable manifold of the saddle point at the coordinate origin. The stable manifold has a complex structure, which is responsible for random switching between the saddle foci in peculiar paths [98, 168] (Fig. 2.24). The phase trajectory, spiraling around a saddle focus, approaches the stable manifold and, with a certain probability, can subsequently enter the vicinity of the other saddle focus. The winding about the saddle foci does not make a significant contribution to the time dependence of the ACF, while the random switching substantially affects the correlation time.

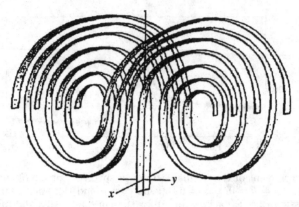

**Fig. 2.24.** A qualitative illustration of the structure of the manifold in the Lorenz system

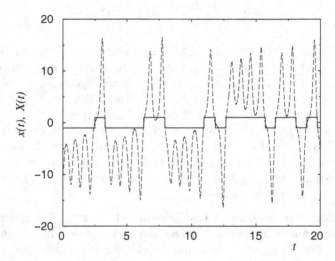

**Fig. 2.25.** Telegraph signal (solid curve) obtained for the oscillation $x(t)$ (dashed curve) in the Lorenz system at $\sigma = 10$, $\beta = 8/3$, and $r = 28$

We consider the time dependence of the $x$ coordinate illustrated in Fig. 2.25. If the winding about the saddle foci is eliminated using the symbolic-dynamics method, we can obtain a signal similar to the telegraph signal [123, 125].

Figure 2.26 shows the ACF of the oscillation $x(t)$ for the Lorenz attractor and the ACF of the corresponding telegraph signal. A comparison between these two graphs indicates that the correlation-decay time and the behavior of the ACF on this time scale are mainly determined by the switching, while the winding about the saddle foci does not contribute considerably to the ACF decay. It is important to note that the ACF decay law is virtually linear at

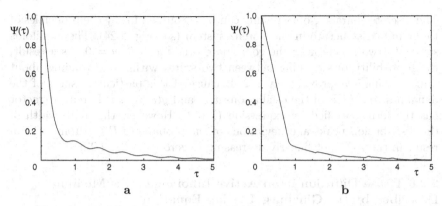

**Fig. 2.26.** ACF of the oscillation $x(t)$ (**a**) and the telegraph signal (**b**)

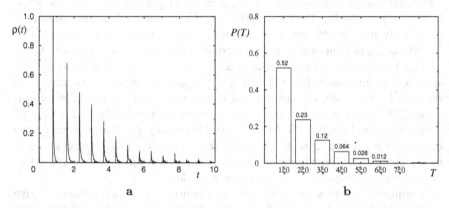

**Fig. 2.27.** The distribution of pulse duration for the telegraph signal (**a**) and the probability distribution of switching within times that are multiples of $\xi_0$ (**b**)

short times. This fact is remarkable, because a linear decline in the ACF corresponds to a discrete equidistant probability distribution of residence times in the form of a set of delta peaks, and the probability of switching between two states should be $1/2$ [166,167].

Figure 2.27 shows the distribution of residence times computed for the telegraph signal presented in Fig. 2.25. As can be seen from Fig. 2.27a, the distribution of residence times indeed has a structure close to an equidistant discrete distribution. At the same time, the peaks are not $\delta$-like spikes but have finite widths. Figure 2.27b shows the probability distribution of switching that occurs at values that are multiples of $\xi_0$, the minimum residence time for one state[5]. This dependence demonstrates that the probability of a transition within the time $\xi_0$ (within one trajectory winding) is close to $1/2$.

---

[5] The time $\xi_0$ corresponds to the duration of one trajectory winding about the saddle focus in the Lorenz attractor.

The discrete character of switching can be accounted for by the properties of the manifold structure in the Lorenz system (see Fig. 2.24). The manifolds split into two sheets near the coordinate origin $x = 0$, $y = 0$. As a result, the probability of switching between two states within one winding about a fixed point is approximately $1/2$. Because of this particular aspect of the dynamics, the ACF of the oscillations $x(t)$ and $y(t)$ on the Lorenz attractor has the form specified by expression (1.132). However, the finite width of the distribution peaks and deviations of the probability $P(\xi_0)$ from $1/2$ can result in the ACF not linearly decreasing to zero (see Fig. 2.27).

### 2.2.6 Phase Diffusion in an Active Inhomogeneous Medium Described by the Ginzburg–Landau Equation

Extended systems are among the most interesting subjects of investigation in theoretical physics. First and foremost, this is due to wave processes that occur only in extended systems. Numerous studies have been dedicated to the dynamics of continuous media, including the onset of turbulence. An irregular behavior of the medium in space and time can develop because of its spatial nonuniformity [169–171]. Effects of spatial nonuniformity have been studied, for example, in ensembles of coupled self-sustained oscillatory systems [172, 173] that can be regarded as models approximating an extended active medium. In ensembles with a spatial frequency gradient, the emergence of frequency clusters – groups of oscillators with equal or close mean frequencies – is typical. Accordingly, perfect (with equal frequencies) or imperfect (with differing frequencies) clusters are considered.

Frequency clusters can also form in a continuous inhomogeneous active medium [174, 175]. In contrast to ensembles, which consist of discrete sets of oscillators, a regime with a continuous coordinate dependence of the frequency is possible in a medium with imperfect clusters. This corresponds to the effect of emergence of imperfect clusters with a continuous power spectrum of oscillations. Because this phenomenon can be observed in a purely deterministic medium in the absence of fluctuations, it implies the onset of deterministic chaos in an extended medium. We here consider the onset of chaotic temporal behavior of a continuous inhomogeneous medium and compare the details of the dynamics of the inhomogeneous medium with the above-described emergence of phase diffusion in finite-dimensional systems.

As an example, we study a one-dimensional self-sustained oscillatory medium that obeys the Ginzburg–Landau equation with a coordinate-dependent frequency,

$$a_t = i\omega(x)a + \frac{1}{2}(1 - |a|^2)a + ga_{xx}, \qquad (2.59)$$

where i is the imaginary unit, $a(x,t)$ is the complex amplitude of oscillations, $t$ is the time, $x \in [0, l]$ ($l = 50$) is the spatial coordinate, and $g$ is the diffusion coefficient.

As $g \to 0$, oscillations at different points of the medium have different frequencies specified by the function $\omega(x)$. We consider the case where the frequency depends linearly on the spatial coordinate, $\omega(x) = x\Delta_{\max}$; in experiments, $\Delta_{\max}$ is set to the fixed value 0.2. The boundary conditions have the form

$$a_x(x,t)|_{x=0;l} \equiv 0. \tag{2.60}$$

The initial state of the medium is chosen at random near some uniform distribution $a_0 = \text{const}$. Equation (2.59) is integrated numerically using an implicit finite-difference technique with a forward–backward marching procedure [176]. We calculate the real oscillation amplitude

$$A(x,t) = |a(x,t)| \tag{2.61}$$

and the phase

$$\phi(x,t) = \arg a(x,t). \tag{2.62}$$

The mean oscillation frequency is computed as the mean time derivative of the phase,

$$\Omega(x) = \langle \phi_t(x,t) \rangle. \tag{2.63}$$

If no mismatch is present ($\Delta_{\max} = 0$), only uniform self-sustained oscillation regimes are possible in medium (2.59): $a(x,t) \equiv a(t)$. At a given mismatch $\Delta_{\max}$, the formation of perfect and imperfect frequency clusters can be observed in a certain range of diffusion-coefficient values (Fig. 2.28). Time-periodic oscillations correspond to perfect clusters. In the regime of imperfect clusters, the time variation in the oscillation amplitude $A$ at any fixed point in the medium $x$ is quite complex and resembles a chaotic process.

This effect can be illustrated by calculating the power spectra for the regimes of perfect and imperfect clusters shown in Fig. 2.29. As the regime evolves from perfect to imperfect clusters, a transition from multifrequency regular oscillations to complex oscillations with a continuous spectrum are observed in the medium at any spatial point.

We also calculate the temporal ACFs of the process $A(x,t)$ for different points in the medium,

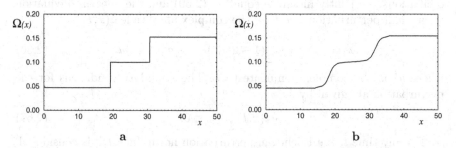

**Fig. 2.28.** Variation in the mean oscillation frequency $\Omega$ along the medium for a perfect cluster structure at $g = 1.0$ (**a**) and for an imperfect structure at $g = 0.85$ (**b**)

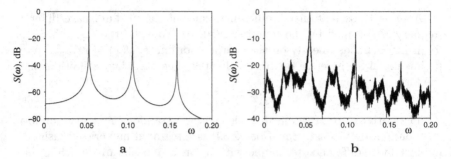

**Fig. 2.29.** Normalized spectral power densities of the process $A(x_{1,2}, t)$: (**a**) in the regime of perfect clusters ($g = 1.0$) at the point $x_1 = 25$ (cluster center); (**b**) in the regime of imperfect clusters ($g = 0.85$) at the point $x_1 = 25$ (cluster center)

$$\psi_A(x, \tau) = \langle A(x, t) A(x, t + \tau) \rangle - \langle A(x, t) \rangle^2, \qquad (2.64)$$

where the angular brackets denote time averaging and $\langle A(x, t) \rangle$ is $t$-independent. We consider the normalized ACF (correlation coefficient)

$$\Psi_A(x, \tau) = \frac{\psi_A(x, \tau)}{\psi_A(x, 0)}. \qquad (2.65)$$

An ACF is exemplified in Fig. 2.30a. Our calculations demonstrate that in the imperfect-cluster regime, $\Psi_A(x, \tau)$ decreases with time at any point in the medium $x$, ultimately approaching zero (Fig. 2.30a). This testifies to the presence of mixing. Two time scales can be distinguished in the ACF-envelope decay law. At small $\tau$ (of several oscillation periods), the correlation declines rapidly. At longer times, an exponential decrease with a certain damping rate $\alpha$ is a fairly good approximation. The damping rate varies within the range $\alpha = (0.15 - 0.4) \times 10^{-3}$, depending on the point in the medium. If the cluster structure is perfect, periodic or quasi-periodic processes occur in the medium, with corresponding correlation functions.

Because no noise sources are present in the model under study, only the onset of dynamical chaos – an absolute exponential instability of oscillations in the medium – can be responsible for mixing. To analyze the stability of the oscillations, we jointly integrate equation (2.59) and the linearized equation for a small perturbation $u(x, t)$ of the complex amplitude $a(x, t)$:

$$u_t = i\omega(x)u + \frac{1}{2}(1 - 2|a|^2)u - \frac{1}{2}a^2 u^* + g u_{xx}, \qquad (2.66)$$

where $u^*$ is the complex conjugate to $u$. The boundary conditions for the perturbation are given by

$$u_x(x, t)|_{x=0;l} \equiv 0. \qquad (2.67)$$

For any time $t$, the Euclidean perturbation norm $||u(x, t)||$ is considered, which reduced to the sum of a finite number of terms because of the discretization of the spatial coordinate. Our calculations have shown that the decay

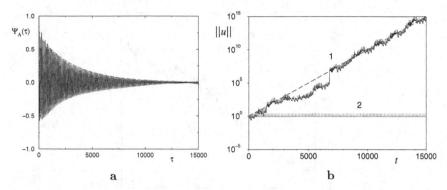

**Fig. 2.30.** (a) Normalized autocorrelation function of the process $A(x_{1,2}, t)$ in the regime of imperfect clusters ($g = 0.85$) at the point $x_1 = 25$ (cluster center). The dashed curve represents the exponential approximation of the envelope of the ACF: $C \exp(-\alpha\tau)$, $\alpha = 0.0003$, $C = $ const. (b) Time dependence of the perturbation norm $\|u(x, t)\|$ for the oscillation of the medium (2.59) in the regime of imperfect clusters at $g = 0.85$ (curve 1) and in the regime of perfect frequency clusters at $g = 1.0$ (curve 2). The dashed straight line corresponds to the exponential function $\exp(0.0023t)$

of the ACF in the regime of imperfect clusters is accompanied by an exponential temporal growth (on average) in the perturbation norm (Fig. 2.30b). The rate of the exponential growth $\lambda_{\max}$ obtained for $g = 0.85$ has the value $\lambda_{\max} \approx 0.002$. We note that this $\lambda_{\max}$ exceeds the damping rate of the ACF by an order of magnitude.

To check the presence of exponential instability in the medium, we calculate the maximal Lyapunov exponent $\lambda_{\max}$, based on the time series of data, using the algorithm suggested in [177]. The calculations yield a positive value of the maximal Lyapunov exponent, which weakly depends on the parameters of the numerical scheme. The results corresponding to different points in the medium differ to a certain extent but are all of the order $10^{-3}$. For instance, at the optimum parameters of the numerical scheme, the reconstruction technique yields

$$\lambda_{\max} = 0.002 \pm 0.0002 \tag{2.68}$$

for the point $x_1 = 25$, which agrees very well with the results of the linear stability analysis. Thus, it can be safely said that the regime of imperfect frequency clusters corresponds to chaotic oscillations in time.

The estimates of the maximal Lyapunov exponent obtained suing two different methods agree well with each other, but differ substantially (by an order of magnitude) from the estimate for the exponential-damping rate of the correlations in the corresponding regime. According to the above discussion, for a broad class of chaotic systems with lumped parameters, the rate of correlation splitting at long time intervals and the width of the basic spectral line are determined by the effective diffusion coefficient for the instantaneous

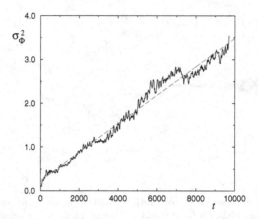

**Fig. 2.31.** Variance of the instantaneous phase computed for the regime of imperfect clusters ($g = 0.85$) at the point $x = 25$ (cluster center), $B_{\text{eff}} \approx 0.00016$. The approximating straight line is shown dashed

phase of the fluctuations,

$$y(t) = A(t) - \langle A(t) \rangle. \tag{2.69}$$

To verify this assertion, we study the dynamics of the instantaneous phase defined as

$$\Phi(t) = \arctan\left(\frac{y_{\text{h}}(t)}{y(t)}\right) \pm \pi k, \qquad k = 0, 1, 2, \ldots, \tag{2.70}$$

where $y_{\text{h}}(t)$ is the Hilbert-conjugate process. The choice of an integer $k$ in expression (2.70) is dictated by the continuity condition for the function $\Phi(t)$.

For an ensemble of segments of a sufficiently long realization $\Phi(t)$, we calculate the variance

$$\sigma_{\Phi}^2(t) = \langle \Phi^2(t) \rangle - \langle \Phi(t) \rangle^2, \tag{2.71}$$

where the angular brackets denote ensemble averaging. The variance of the instantaneous phase is plotted as a function of time in Fig. 2.31. The variance grows with time almost linearly in the interval $t \in [0, 10000]$. A least-square estimate of the angular growth factor makes it possible to determine the effective diffusion coefficient for the phase $\Phi(t)$ (2.56).

In the regime of imperfect clusters, the obtained $B_{\text{eff}}$ values range within the interval $[0.000\,16, 0.000\,38]$, depending on the spatial coordinate $x$. A more accurate, direct calculation of $B_{\text{eff}}$ based on formula (2.56) proved to be quite complicated because of the need of averaging over a vast dataset.

Our numerical investigation reveals a number of important new facts.

(1) The development of chaos and turbulence in a continuous self-sustained oscillatory medium can result from the inhomogeneity of the medium,

which specifies a continuous coordinate dependence of the self-sustained oscillation frequency.

(2) The self-sustained oscillations of the medium in the regime of imperfect, partial (cluster) synchronization are mixable, i.e., they are exponentially unstable, with splitted temporal correlations.

(3) The damping rate of the correlation functions at long times is not directly determined by the Lyapunov exponent but is related to the diffusion of the instantaneous oscillation phase. This testifies to the generality of the correlation splitting laws in finite-dimensional and distributed chaotic systems.

### 2.2.7 The Autocorrelation Function and Power Spectrum of Spiral Chaos in Physical Experiments

Our experiments were carried out using a setup that included a radio generator with inertial nonlinearity (Anishchenko–Astakhov generator [129]) and the fundamental frequency 18.5 kHz, a fast-ADC computer with the discretization frequency 694.5 kHz, and a generator of broadband Gaussian noise in the frequency 0 to 100 kHz [178]. The behavior of the ACF was also analyzed in the presence of external noise. With this aim in view, a signal from an external noise generator with controlled noise intensity was fed to the system. The generator with inertial nonlinearity can be described by the three-dimensional dissipative dynamical system

$$\dot{x} = mx + y - xz - \delta x^3, \qquad \dot{y} = -x, \qquad \dot{z} = -gz + gI(x)x^2, \quad (2.72)$$
$$I(x) = \begin{cases} 1, & x > 0, \\ 0, & x \leq 0. \end{cases}$$

At certain $m$ and $g$, the system realizes spiral-chaos regimes [129].

The first important question to be unambiguously resolved in the experiment is whether the Wiener-process approximation can be used to describe the statistical parameters of the instantaneous phase, as assumed in [107, 123, 125, 161]. The instantaneous phase used to determine the diffusion coefficient $B_{\text{eff}}$ is based on the concept of analytical signal with the application of the Hilbert transform of experimental realizations of $x(t)$ [163]. The phase variance $\sigma_\Phi^2(t)$ is then computed by averaging over an ensemble of $N$ realizations. The effective phase-diffusion coefficient is determined by the temporal-growth rate of the variance.

The time dependence of the phase variance shown in Fig. 2.32a is not strictly linear, as should be expected for a Wiener process. However, the linear growth dominates over small fluctuations in the phase variance. Therefore, the process under consideration can be associated with a Wiener process whose diffusion coefficient is $B_{\text{eff}}$.

The next stage of the experiment is the measurement of the ACF for the chaotic oscillations of the generator with inertial nonlinearity. Several dozen

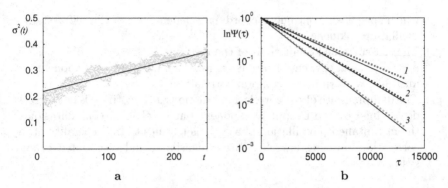

**Fig. 2.32.** (a) Time dependence of the phase variance in the presence of noise with the intensity $D = 0.001$ mV and the linear least-squares approximation of this dependence (the nondimensional time $t$ is equal to the number of oscillation periods). (b) The ACF envelopes (solid lines) obtained experimentally for different values of the noise intensity: $1 - D = 0$, $2 - D = 0.0005$ mV, $3 - D = 0.001$ mV; with their experimental approximations (straight lines) for the respective damping rates $B_{\mathrm{eff}} = 0.00024$, $B_{\mathrm{eff}} = 0.00033$, and $B_{\mathrm{eff}} = 0.000439$. The other parameters of the numerical calculations are $N = 100$, $n = 262144$, and $p = 1/(2n)$

realizations of the signal $x(t)$, with a duration of 10 s each, were recorded by the fast ADC. The total length of the realization was $(3-5) \times 10^5$ oscillation periods with a discretization step $\Delta t$ corresponding to 37 points per period. The ACF was calculated as follows. First, we computed the time-averaged value of the $x$ variable for each of $N$ realizations of the process $x(t)$:

$$\overline{x} = \frac{1}{n} \sum_{i=1}^{n} x(t_i). \tag{2.73}$$

Next, time averaging was used to obtain the mean product $\langle x(t)x(t+\tau) \rangle$,

$$K_l(\tau) = \frac{1}{p} \sum_{i=1}^{p} x(t_i)x(t_i + k\Delta t), \qquad \tau = k\Delta t_i, \qquad k = 0, 1, \ldots, n - p, \tag{2.74}$$

where $l = 1, \ldots, N$ is the realization number. Since the correlation-decay rate is not high in the regime under consideration, the ACF should be calculated for a very long time interval. To achieve high accuracy in the calculation of the ACF, the data obtained were averaged over $N$ realizations:

$$\psi(\tau) = \frac{1}{N} \sum_{i=1}^{N} K_l(\tau) - \overline{x}^2. \tag{2.75}$$

The ACF was normalized to the maximum value at $\tau = 0$, i.e.,

$$\Psi(\tau) = \frac{\psi(\tau)}{\psi(0)}. \tag{2.76}$$

Experimental graphs of the envelopes of the normalized ACF for various external-noise intensities are shown in Fig. 2.32b. The dependences obtained were approximated by the exponential law

$$\Psi_{app}(\tau) = \exp(-B_{eff}\tau), \qquad (2.77)$$

where $B_{eff}$ is the experimentally determined effective instantaneous-phase diffusion coefficient. The approximations are shown by symbols in Fig. 2.32b.

We now analyze the results of the power-spectrum measurements. The power spectrum of a diffusive process has a Lorentzian profile whose width is determined by the effective phase-diffusion coefficient. For a normalized spectrum, the Lorentzian is given by formula (2.58). Experimentally, the diffusion coefficient can be independently determined by measuring the width of the spectral peak. To obtain a more accurate value of the diffusion coefficient, we approximated the spectral peak using formula (2.58) and varying $B_{eff}$. We choose the $B_{eff}$ value at which the approximation error was minimum (Fig. 2.33a). Figure 2.33 presents experimental power spectra of the generator with inertial nonlinearity. The spectrum was computed using the standart technique of the fast Fourier transform (FFT) with averaging. The principal result is that the values of the effective phase-diffusion coefficient based on the power-spectrum measurements agree well with the $B_{eff}$ values obtained from the linear approximation of the growth of the instantaneous-phase variance. The corresponding values of the effective phase-diffusion coefficient are given in the table for three values of the external-noise intensity.

Thus, we have found experimentally that in the spiral-chaos regime, the instantaneous-phase variance of chaotic oscillations grows on average linearly with the diffusion coefficient $B_{eff}$. In the absence of noise, this coefficient is controlled by the chaotic dynamics of the system. If noise is present, the

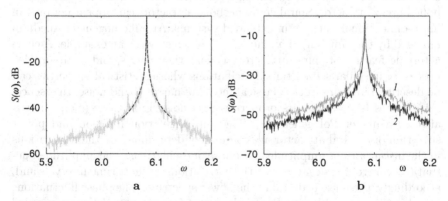

**Fig. 2.33.** (a) Experimentally obtained power spectrum of the oscillation $x(t)$ in system (2.72) and its theoretical approximation (2.58) at $B_{eff} = 0.00033$ in the presence of noise with $D = 0.0005$; (b) power spectra for $D = 0.001$ (curve 1) and $D = 0$ (curve 2)

**Table 2.1.** Comparison of the phase-diffusion coefficients obtained using various techniques with the inclusion of noise of different intensity

| $D$, mV | $B_{eff}$ (Hilbert) | $B_{eff}$ (spectrum) |
|---------|---------------------|----------------------|
| 0       | 0.000244            | 0.000266             |
| 0.0005  | 0.000330            | 0.000342             |
| 0.001   | 0.000439            | 0.000443             |

growth of the phase variance is also linear, but $B_{eff}$ increases. The ACF of spiral chaos decreases exponentially with time, as $\exp(-B_{eff}\tau)$. The spectral-line width of oscillations at the fundamental frequency $\omega_0$ is determined by the phase-diffusion coefficient according to expression (2.58). The corresponding values of the effective phase-diffusion coefficient are given in Table 2.1 for three values of the external noise intensity.

## 2.2.8 Summary

Our results have shown that there is a class of spiral-type nonhyperbolic attractors for which noise has a pronounced effect on the rate of relaxation to the stationary distribution and on the correlation time but, virtually, does not influence the value of the positive Lyapunov exponent. The rate of mixing on nonhyperbolic attractors is determined not only and not very much by the exponential instability but depends on the complex dynamics of the instantaneous phase of chaotic oscillations. In the spiral-chaos regime, noise substantially increases the rate of relaxation to the stationary distribution. For chaotic attractors with an irregular behavior of the instantaneous phase, noise has virtually no effect on the mixing rate. This is the case for nonhyperbolic funnel and switching-type attractors such as the quasihyperbolic Lorenz attractor. Spiral nonhyperbolic attractors can appear not only in finite-dimensional but also in extended systems. An inhomogeneous medium modeled by the Ginzburg–Landau equation can serve as an example. A characteristic feature of spiral attractors is that they correspond to a complex process of irregular self-sustained oscillations whose statistical properties can be described in terms of the classical model of narrow-band noise. In essence, spiral chaos is similar in its properties to a noisy limit cycle (e.g., a noise-affected Van der Pol generator). The autocorrelation function and power spectrum of a spiral attractor are completely determined by the fluctuations in the instantaneous amplitude and phase of oscillations. The amplitude fluctuations control the decay rate of the correlations at short time intervals and, accordingly, the noise pedestal in the power spectrum. The phase fluctuations broaden the spectral line at the fundamental frequency in the spectrum and result in an exponential decay in the autocorrelation function, which is determined by the effective diffusion coefficient $B_{eff}$. The phase-diffusion coefficient in a noise-free system is determined by its chaotic dynamics and does not directly depend on the positive Lyapunov exponent. The following important conclusion can be deduced: in dynamical systems with spiral chaos, the Kolmogorov entropy as a quantitative characteristic of the mixing rate is mainly controlled by the growth rate $B_{eff}$ of the instantaneous-phase variance rather than by the positive Lyapunov exponent, as is generally assumed. Analyses of the statistical properties of the Lorenz attractor have shown that the properties of the ACF are mainly determined by the random-switching process, depending weakly on winding about the saddle foci. The classical model of the telegraph signal can be used to describe the statistics of the Lorenz attractor. In particular, this model provides a good approximation of the interval of linear decrease in the ACF, which enables us to theoretically calculate the correlation time. The fact that the ACF decay rate for the Lorenz attractor is virtually constant both in and without the presence of noise results from the statistics of the switching process. The probability of switching in the Lorenz attractor is nearly 1/2 and virtually independent of the level of noise influence.

## 2.3 Synchronization of Chaos

### 2.3.1 Introduction

Accumulation of the knowledge on chaotic dynamics of nonlinear systems has led to the necessity of extending a classical concept of synchronization to the case of chaotic oscillations. Earlier synchronization and deterministic chaos were thought of as quite opposite tendencies in the behavior of DS [179]. In a series of works based on the above opinion, synchronization of chaos was understood as a transition from chaotic oscillations to periodic ones under external periodic force on the system [180, 181]. However, numerous recent studies have shown that the effect of synchronization can also manifest itself in systems with chaotic dynamics. In the contemporary literature three major kinds of chaos synchronization are well described, namely, complete synchronization, phase–frequency synchronization (or synchronization in the sense of Huygens [182]), and generalized synchronization. The complete synchronization phenomenon is most often discussed in the literature [183]– [188]. It is observed when identical chaotic oscillators interact and manifests itself in that time dependences of the dynamical variables of the interacting systems completely coincide as the coupling strength increases. In this case the systems oscillate completely "in-phase". The concept of complete synchronization can be applied not only to self-sustained systems but also to nonautonomous nonlinear oscillators, e.g., periodically driven Duffing oscillators [187], as well as to discrete time systems, i.e., coupled maps [183, 184]. When interacting systems have a slight parameter mismatch in the presence of sufficiently strong coupling, one can observe a phenomenon close to the complete synchronization effect [189]. The ideas on phase–frequency chaos synchronization in self-sustained systems are developed in [190–197]. They are based on a generalization of the classical notion of synchronization as natural frequency entrainment or suppression and the occurrence of certain phase relations between oscillations of interacting oscillators. Another kind of chaos synchronization, namely, generalized synchronization, is introduced in [189, 198–200]. By generalized synchronization of chaotic oscillations we mean the appearance of a functional dependence between the instantaneous states of partial systems. In slightly nonidentical chaotic self-sustained systems a regime can appear where the states of the subsystems are completely identical but shifted in time. This effect is called the lag synchronization [201]. Different aspects of chaos synchronization, discussed in recent works, are well described in the reviews [202, 203].

Synchronization of chaos has been found for ensembles consisting of a large number of interacting chaotic oscillators [204–213]. It has resulted from a saturation of the growth of the attractor dimension in arrays of oscillators and leads to the formation of stable spatial structures. The possibility of realizing synchronous regimes with different phase relations is closely related to the phenomenon of multistability, i.e., the co-existence of a set of attractors, both regular and chaotic, in the phase space of interacting systems [208]. In turn, the multistability leads to attractor crises, fractalization of their basins of attraction, etc. All these phenomena explain a great variety and complexity of the cooperative dynamics of ensembles of nonlinear systems. The study of distributed systems described by partial differential equations shows that spatio-temporal structures can also be synchronized [214–216].

The comprehensive study of synchronization of complex oscillations evokes great interest in connection with the development of mathematical models and the wide application of nonlinear dynamics methods in biophysics (neural networks, interacting populations, etc.) [217–222]. Besides, experimental investigations of biological systems, for which there is still no satisfactory mathematical description, show phenomena very similar to the well-known effects of synchronization [223, 224]. Due to the variety and complexity of synchronization manifestations in various interacting systems, this effect gains a more rich and universal content. The problem of extending the basic notions underlying the classical approach to synchronization has been considered in order to develop a unified concept reflecting all the aspects of this important phenomenon.

## 2.3.2 Phase–Frequency Synchronization of Chaos. The Classical Approach

The most consistent extension of the synchronization concept is the development of a classical concept which is based on the adjustment of a relation between characteristic frequencies and time scales of DS. The classical concept of synchronization can easily be extended to the case of oscillators in the regime of spiral (phase-coherent) chaos [8, 162]. The power spectrum of spiral chaos contains a well-pronounced peak at the frequency close to that of periodic oscillations (a limit cycle producing a chaotic attractor through a subharmonic bifurcation sequence). This frequency is referred to as the *basic oscillation frequency* $\omega_0$, and one can investigate the effect of entrainment or suppression of the basic frequency as well as introduce a certain analogue for phase locking of chaotic oscillations. Such effects are called *synchronization in the sense of Huygens*, or *phase–frequency synchronization*. This kind of synchronization can be realized in systems which are considerably different in their mathematical description as well as in their behavior. One can also observe forced chaos synchronization in the sense of Huygens, including that which occurs under periodic forcings.

The first attempt to extend the classical notion of synchronization as frequency locking or frequency suppression to the case of interacting chaotic oscillators was made in [190–194,225]. It was found that in a plane of parameters which govern the strength of coupling and the frequency mismatch, one is able to identify regions of chaos synchronization which resemble Arnold's tongues. Inside these regions chaotic oscillations (synchronous chaos) are qualitatively different from the chaos outside them (nonsynchronous chaos).

In [195, 196] several methods were proposed to introduce an instantaneous phase of chaotic oscillations. It has also been shown that in case of interacting chaotic oscillators with a certain frequency mismatch an increasing coupling can bound the instantaneous phase difference of the oscillators. This fact testifies to the effect of phase locking. Obviously, the frequency and phase locking effects in the case of chaotic oscillations, as well as for periodic oscillations, are closely interrelated.

We exemplify frequency synchronization of chaotic oscillations with experiments performed for two interacting radio-technical chaos generators with basic frequency mismatch [191,192]. A block diagram is shown in Fig. 2.34.

The interacting self-sustained generators being considered are described by the following system of differential equations:

$$
\begin{aligned}
\dot{x}_1 &= (m_1 - z_1)x_1 + y_1 + \gamma_1(x_2 - x_1 + y_1 - y_2/p),\\
\dot{y}_1 &= -x_1,\\
\dot{z}_1 &= g_1\big(f(x_1) - z_1\big),\\
\dot{x}_2/p &= (m_2 - z_2)x_2 + y_2 + \gamma_2(Bx_1 - x_2 + y_2 - Bpy_1),\\
\dot{y}_2/p &= -x_2,\\
\dot{z}_1/p &= g_2(f(x_2) - z_2).
\end{aligned} \tag{2.78}
$$

Parameters $m_{1,2}$ and $g_{1,2}$ govern the dynamics of partial generators; $p = C_1/C_2$ is the resonance frequency mismatch of the Wien bridges and determines the frequency mismatch of the partial systems. Parameters $\gamma_{1,2}$ characterize the strength of coupling, and $B$ is the transfer coefficient of the buffer. A choice of $\gamma_1 = 0$ and $B = 3$ corresponds to unidirectional coupling, and

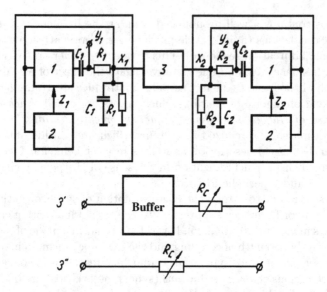

**Fig. 2.34.** Block diagram of the system of two coupled Anishchenko–Astakhov generators. 1 – linear amplifiers with controlled amplification coefficients, 2 – inertial nonlinear transformers, and 3 – the coupling block for unidirectional $(3^{'})$ and mutual $(3^{''})$ coupling

in this case the first generator is acting on the second one. When $\gamma_1 = \gamma_2$ and $B = 1$, the generators are mutually and symmetrically coupled. Function $f(x)$ determines the nonlinearity of inertial transformers and is equal to $x^2$ for $x \geq 0$ and 0 for $x < 0$.

We start by considering forced synchronization of chaotic oscillations with the basic frequency ratio $f_1 : f_2 = 1 : 1$, which was observed in full-scale experiments with unidirectionally coupled generators. Figures 2.35 and 2.36 illustrate the evolution of power spectra corresponding to two mechanisms of synchronization, namely, via locking and suppression of the basic frequency, respectively.

A similar spectrum evolution has also been observed for mutual synchronization of symmetrically coupled generators. A portion of the experimentally constructed bifurcation diagram in this case is presented in Fig. 2.37. Without coupling, each generator demonstrates the regime of spiral chaos which is produced by a period-3 cycle. The diagram illustrates the left half of the major synchronization region inside which the basic frequencies are in the 1 : 1 ratio. On the line $l_1^2$, corresponding to the region boundary of synchronous oscillations with period $2T_0$, the basic frequencies are mutually locked. Lines $l_2^k$, $k = 1, 2, 4$, denote the period-doubling bifurcations of synchronous cycles with period $kT_0$, respectively. On lines $l_0^k, k = 1, 2$, one of the basic frequencies is suppressed ($f_1$ for $p < 1$ and $f_2$ for $p > 1$), and periodic oscillations with period $kT_0$ emerge. In the diagram the regions of periodic oscillations

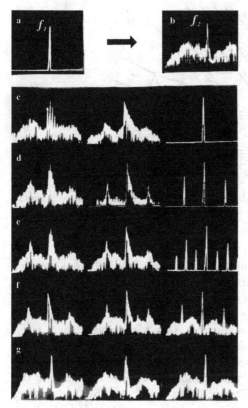

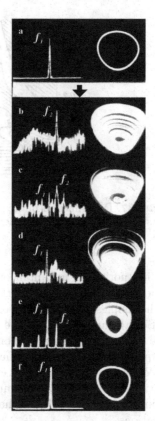

**Fig. 2.35.** Spectra of $x_2(t)$ oscillations in the case of forced synchronization through the basic frequency locking (full-scale experiment). (**a**) Forcing signal spectrum; (**b**) spectrum of autonomous oscillations of the second oscillator; (**c–g**) spectra of oscillations of the second oscillator for different values of the frequency mismatch. The coupling parameter is increased from the left to the right [192]

**Fig. 2.36.** Spectra and phase portraits in the case of forced synchronization via the basic frequency suppression of chaotic oscillations (full-scale experiment). (**a**) Forcing signal; (**b–f**) oscillations of the second oscillator for a fixed frequency mismatch and for different coupling strengths. The coupling parameter grows from the top to the bottom [192]

are unshaded and labeled by $kT_0, k = 2, 3, 4, 8, \ldots$, where $kT_0$ is the period of oscillations and $T_0$ denotes the region of periodic oscillations with period $T_0$ which is determined by the basic frequency $f_1 = f_2 = f_0$. Besides, in the diagram one can see synchronization zones of higher order with the 5:4 and 4:3 frequency ratios. Three regions of synchronous chaos, $CA_0$, $CA_0'$, and $CA_3$, and the region of nonsynchronous chaos $CA_2$ are distinguished in the diagram. The results of numerical studies of model (2.78) are in full agreement with the experimental results [192].

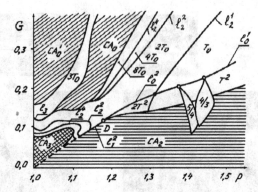

**Fig. 2.37.** Experimental bifurcation diagram for the system of two coupled generators (2.78) in the plane "frequency mismatch – coupling strength" ($G = \gamma_1 = \gamma_2$) [192]

The notions of instantaneous amplitude and phase of oscillations are very important for a deeper understanding of the effect of synchronization. Even in the simplest case of Thomson's generator we need to introduce into consideration the amplitude and the phase as functions of time. For a more general case when oscillations are nonharmonic and even nonperiodic, there is no unique definition of instantaneous amplitude and phase (see Sect. 1.3.5).

We examine how different characteristics of phase–frequency synchronization behave using as an example a nonautonomous Rössler oscillator in the regime of weakly developed chaos. The oscillator is described by the following equations:

$$
\begin{aligned}
\dot{x} &= -\Omega y - z + \gamma \sin \omega_s t, \\
\dot{y} &= \Omega x + \alpha y, \\
\dot{z} &= \beta + z(x - \mu).
\end{aligned}
\tag{2.79}
$$

The frequency of the external signal is fixed at $\omega_s = 1$. The parameter $\Omega$ governs the frequency mismatch between the self-sustained oscillations and the external signal. Figure 2.38a,b shows the winding number $\theta$ as a function of parameter $\Omega$. In Fig. 2.38a, the winding number is calculated as the ratio of the characteristic times, i.e., $\langle T \rangle / 2\pi$, and in Fig. 2.38b it is computed as the ratio of the basic frequencies $\omega_s / \omega_0 = 1/\omega_0$. The dependence of $D_{\text{eff}}$ for the phase difference versus parameter $\Omega$ is plotted in Fig. 2.38c. The instantaneous phase is introduced according to the third definition (1.294). For the parameter values chosen, the transition "synchronous chaos – nonsynchronous chaos" can be observed at the boundaries of the main synchronization region. The parameter values corresponding to the synchronization region boundaries, $\Omega = \Omega_{1,2}$, are determined from the dependences shown in Fig. 2.38a–c and are very close to one another. A minor difference may be related to numeric errors. Within the accuracy of numeric calculations,

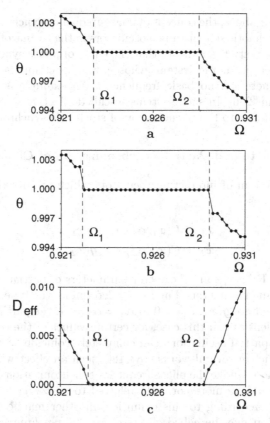

**Fig. 2.38.** Winding number as a function of parameter $\Omega$, calculated for (2.79) when (a) $\theta = \langle T \rangle / 2\pi$ and (b) $\theta = \omega_0^{-1}$. (c) The effective diffusion coefficient versus $\Omega$. The system parameters are $\alpha = 0.2$, $\beta = 0.2$, $\mu = 4$, and $\gamma = 0.02$

all the considered characteristics are found to yield the same values of $\Omega_{1,2}$, corresponding to the boundaries of the synchronization region.

The change of a chaotic attractor at the synchronization region boundary can also be distinguished "by eye" when projections of phase trajectories and their Poincaré sections[6] are analyzed. Chaotic self-sustained oscillations are suppressed for larger values of the external force amplitude and the frequency mismatch. In this case one can observe periodic oscillations inside the synchronization zone, while its boundary corresponds to the torus-birth bifurcation from a limit cycle (as in the classical case of suppression of periodic self-sustained oscillations).

---

[6] In the case of forced synchronization it is more reasonable to project phase trajectories onto the plane of dynamical variables, one of which is the external force.

As outlined above, the concept of phase-frequency synchronization is more applicable to chaotic self-sustained oscillators in the regime of coherent chaos. However, recent studies indicate that the effect of phase synchronization can also be realized in chaotic systems whose power spectrum has no pronounced peaks and, therefore, no basic frequencies. An example is the well-known Lorenz system [226]. In such systems so-called effective synchronization occurs, being similar to the phenomenon of stochastic synchronization.

### 2.3.3 Complete and Partial Synchronization of Chaos

Consider a system of interacting oscillators which are described by the following equations:

$$
\begin{aligned}
\dot{\boldsymbol{x}}_1 &= \boldsymbol{F}(\boldsymbol{x}_1, \boldsymbol{\alpha}_1) + \gamma_1 \boldsymbol{g}(\boldsymbol{x}_1, \boldsymbol{x}_2), \\
\dot{\boldsymbol{x}}_2 &= \boldsymbol{F}(\boldsymbol{x}_2, \boldsymbol{\alpha}_2) + \gamma_2 \boldsymbol{g}(\boldsymbol{x}_2, \boldsymbol{x}_1),
\end{aligned}
\tag{2.80}
$$

where $\boldsymbol{x}_{1,2} \in \mathbf{R}^N$; $\boldsymbol{\alpha}_{1,2}$ are the vector parameters of partial systems and $\gamma_{1,2}$ are the coupling parameters. Function $\boldsymbol{g}$ determines the type of coupling, and if $\boldsymbol{x}_1 = \boldsymbol{x}_2$, we have $\boldsymbol{g}(\boldsymbol{x}_1, \boldsymbol{x}_2) = 0$. If $\boldsymbol{\alpha}_1 = \boldsymbol{\alpha}_2 = \boldsymbol{\alpha}$, the partial oscillators are completely identical. In this case, for certain values of the coupling parameters $\gamma_{1,2}$ the partial oscillations can completely coincide, $\boldsymbol{x}_1(t) \equiv \boldsymbol{x}_2(t)$, but remain chaotic. In several works [183, 186] such an effect was called chaotic synchronization. Unlike the phase-frequency synchronization discussed above, this effect would be more correctly referred to as *complete (in-phase) synchronization*. According to this definition all other manifestations of chaos synchronization may be called effects of *partial synchronization*. The complete synchronization in systems of type (2.80) can be observed both in the case of symmetric ($\gamma_1 = \gamma_2 = \gamma$) and nonsymmetric ($\gamma_1 \neq \gamma_2$) coupling, as well as for unidirectional coupling ($\gamma_1 = 0$) of partial systems.

For $\boldsymbol{\alpha}_1 = \boldsymbol{\alpha}_2 = \boldsymbol{\alpha}$, in the phase space of system (2.80) there exists an invariant manifold $U$: $\boldsymbol{x}_1 = \boldsymbol{x}_2$, which is referred to as a *symmetric subspace*. Phase trajectories lying in $U$ correspond to completely synchronized oscillations. The complete synchronization effect is realized if a limit set belonging to $U$ attracts phase trajectories not only from $U$ but also from some neighborhood of the symmetric subspace. The complete synchronization manifests itself for interacting systems, each demonstrating not only phase-coherent chaos but also other kinds of chaotic dynamics (e.g., the Lorenz attractor [227] and Chua's "double scroll" attractor [188]).

Complete chaos synchronization can also be observed in a case when an active controlling system acts on a passive controlled system [186]. The controlling system is decomposed into two subsystems,

$$
\begin{aligned}
\dot{\boldsymbol{v}} &= \boldsymbol{p}(\boldsymbol{v}, \boldsymbol{y}_1), \\
\dot{\boldsymbol{y}}_1 &= \boldsymbol{q}(\boldsymbol{v}, \boldsymbol{y}_1),
\end{aligned}
\tag{2.81}
$$

and the controlled system is chosen to be identical to one of them. In this case two interacting systems (drive and response) are described by the following equations:

$$\dot{v} = p(v, y_1),$$
$$\dot{y}_1 = q(v, y_1),$$
$$\dot{y}_2 = q(v, y_2),$$

(2.82)

where $v \in \mathbf{R}^m$ and $y_{1,2} \in \mathbf{R}^{N-m}$. In the regime of complete synchronization $y_1(t) \equiv y_2(t)$.

A growing interest in the problem of complete synchronization is related to its possible application in secure communication [228–232]. The information signal is transmitted by using a broadband carrier which can be represented by a chaotic signal of some DS. At the receiver the information signal can be recovered in the regime of complete synchronization. With this, it is assumed that the receiver knows exactly the parameter values of the chaotic system used and that the regime of complete synchronization itself is robust with respect to different kinds of perturbations.

When there is a parameter mismatch of partial systems, the symmetric subspace no longer exists and complete synchronization cannot be realized at some finite value of $\gamma$. However, if the parameter mismatch does not cause the structure of a chaotic attractor to change and only the basic frequency of chaotic oscillations is varied, then, starting from a certain value of the coupling, the so-called *lag synchronization* effect can be observed [201]. It consists in coinciding *shifted in time* states of two systems, i.e., $x_1(t) \equiv x_2(t + \tau_d)$, where $\tau_d$ is the time delay. In the case of lag synchronization the chaotic attractor is topologically equivalent to the "in-phase" attractor in the regime of complete synchronization. Hence, the lag synchronization can be treated as an extension of the complete synchronization notion to systems with a slight parameter mismatch.

Thus, for the case of interaction of chaotic oscillators with frequency mismatch three types of chaos synchronization can be distinguished, namely, phase–frequency synchronization, lag synchronization, and complete synchronization. The border of a synchronization region is determined by synchronization in the sense of Huygens, i.e., it corresponds to locking of instantaneous phases and basic frequencies of chaotic oscillations of partial systems or to suppression of self-sustained oscillations of one of the oscillators (for large values of the mismatch). In the last case, although each of the partial autonomous systems behaves chaotically, periodic oscillations occur inside the synchronization region. With decreasing mismatch and increasing strength of the coupling there may appear a stronger effect of synchronization, namely, lag synchronization. The transition from phase–frequency synchronization to lag synchronization is a complicated process whose bifurcation mechanism is not sufficiently understood yet but is likely to be similar to the mechanism of destruction of complete synchronization [233]. Without detuning and starting from a certain value of the coupling strength, one can observe the complete synchronization effect when chaotic trajectories lying in the symmetric subspace become stable in the full phase space of the system.

Different quantitative characteristics are used to evaluate the synchrony of partial oscillations for two near-identical interacting oscillators. To characterize synchronization of the processes $x_1(t)$ and $x_2(t)$ of partial systems, it is proposed in [201] to use the minimum of a similarity function

$$\kappa = \min_\tau S(\tau), \tag{2.83}$$

where $S(\tau)$ is the similarity function derived from the relation

$$S^2(\tau) = \frac{\langle (x_2(t+\tau) - x_1(t))^2 \rangle}{\sqrt{(\langle x_1^2(t) \rangle \langle x_2^2(t) \rangle)}}. \tag{2.84}$$

The angle brackets mean time averaging. $\kappa$ is equal to zero in the case of complete and lag synchronization. $\kappa$ grows as the mismatch increases and the coupling strength decreases.

Another characteristic for identifying synchronization of chaotic oscillators is the mutual normalized autocorrelation function

$$R_{x_1 x_2}(\tau) = \frac{\langle x_1(t) x_2(t+\tau) \rangle - \langle x_1(t) \rangle \langle x_2(t+\tau) \rangle}{\sqrt{\left( \langle x_1^2(t) \rangle - \langle x_1(t) \rangle^2 \right) \left( \langle x_2^2(t+\tau) \rangle - \langle x_2(t+\tau) \rangle^2 \right)}}. \tag{2.85}$$

The quantity $\eta = \max_\tau R_{x_1 x_2}(\tau)$ is equal to unity in the case of lag synchronization (and, clearly, complete synchronization) and vanishes when $x_1(t)$ and $x_2(t)$ become statistically independent. The degree of synchronous motion can be revealed by analyzing the probability density of the instantaneous phase difference, $\Delta\Phi(t) = \Phi_1(t) - \Phi_2(t)$. The instantaneous phases in this case are defined on the interval $[-\pi, \pi]$. Synchronization can also be quantified by using the effective diffusion coefficient, $D_{\mathrm{eff}}$, of the instantaneous phase difference, the variance of the instantaneous phase difference, defined in the interval $[-\pi, \pi]$, or the entropy of its distribution, which corresponds to a fixed sampling step of $\Delta\Phi$ values.

The degree of synchrony can also be estimated in the context of the spectral approach. For this purpose the coherence function is used [192]:

$$r_{x_1 x_2}(\omega) = \frac{|W_{x_1 x_2}(\omega)|}{\sqrt{W_{x_1}(\omega) W_{x_2}(\omega)}}, \tag{2.86}$$

where $W_{x_1}$ and $W_{x_2}$ are the power spectra of fluctuations $x_1(t) - \langle x_1 \rangle$ and $x_2(t) - \langle x_2 \rangle$; $W_{x_1 x_2}$ is the mutual fluctuation spectrum. If the processes $x_1(t)$ and $x_2(t)$ are statistically independent, then $r_{x_1 x_2} \equiv 0$, and when they are linearly interrelated, we have $r_{x_1 x_2} \equiv 1$. A quantitative frequency-independent characteristic can be introduced by considering the mean value of the coherence coefficient in the frequency interval being studied, i.e., $\bar{r} = \left(1/(\omega_2 - \omega_1)\right) \int_{\omega_1}^{\omega_2} r_{x_1 x_2}(\omega)\mathrm{d}\omega$.

In [192] it is suggested to characterize the degree of phase coherence of different spectral components of the signals $x_1(t)$ and $x_2(t)$. The phase spectra $\phi_{1,2}(\omega) = \arg\left[\int_{-T/2}^{T/2} x_{1,2}(t)\exp(-\mathrm{j}\omega t)\,\mathrm{d}t\right]$ of $x_1(t)$ and $x_2(t)$ are considered on finite time $T$, and the current phase difference is then introduced for each spectral frequency:

$$\Delta\phi(\omega) = \phi_1(\omega) - \phi_2(\omega), \quad \Delta\phi \in [-2\pi, 2\pi], \tag{2.87}$$

A set of functions $\Delta\phi(\omega)$ obtained for different initial conditions on an attractor is used for estimating a probability density $p(\Delta\phi, \omega)$. In the case of complete and lag synchronization the probability density looks like a frequency-independent $\delta$-function. When the lag synchronization is destroyed, the distribution has a finite width and shape which are different at distinct frequencies. To quantify synchronization of oscillators one can utilize the variance averaged over all frequencies as well as the entropy of the phase difference distribution.

As an example consider a system of two coupled Rössler oscillators, which is written in the form:

$$\begin{aligned}
\dot{x}_1 &= -\Omega_1 y_1 - z_1 + \gamma(x_2 - x_1), \\
\dot{y}_1 &= \Omega_1 x_1 + \alpha y_1, \\
\dot{z}_1 &= \beta + z_1(x_1 - \mu), \\
\dot{x}_2 &= -\Omega_2 y_2 - z_2 + \gamma(x_1 - x_2), \\
\dot{y}_2 &= \Omega_2 x_2 + \alpha y_2, \\
\dot{z}_2 &= \beta + z_2(x_2 - \mu).
\end{aligned} \tag{2.88}$$

Here, the parameters $\Omega_{1,2} = \Omega_0 \pm \Delta$ determine the frequencies of the partial oscillators, $\Delta$ denotes the frequency mismatch, $\gamma$ is the strength of coupling, and the parameters $\alpha$ and $\mu$ govern the dynamics of each oscillator. For this system three of the above-listed characteristics of synchronization, are calculated and plotted in Fig. 2.39a–c as functions of the coupling parameter for a fixed frequency mismatch.

Another kind of synchronization, namely, generalized synchronization, was introduced in [198–200]. Two systems are considered to be synchronized if there is some functional relation between their states. A theory of generalized synchronization is proposed in [199] for unidirectionally coupled chaotic systems which are described by the following equations:

$$\begin{aligned}
\dot{x} &= F(x), \\
\dot{y} &= G(y, v) = G(y, h(x)).
\end{aligned} \tag{2.89}$$

$x \in \mathbf{R}^N$ and $y \in \mathbf{R}^n$ are the state vectors of the first and second systems, respectively. Vector $v \in \mathbf{R}^k$ is defined by the instantaneous state of the first

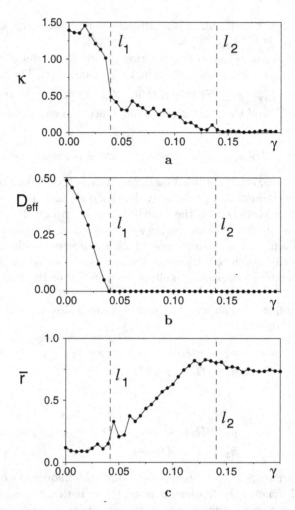

**Fig. 2.39.** Dependences of different quantitative characteristics of synchronization of interacting oscillators (2.88) versus the coupling parameter for $\alpha = 0.165$, $\beta = 0.2$, $\mu = 10$, $\Omega_0 = 0.97$, and $\Delta = 0.02$. (a) The minimum of similarity function $\kappa(\gamma)$; (b) the effective diffusion coefficient $D_{\text{eff}}(\gamma)$; and (c) the mean value of coherence coefficient $\bar{r}(\gamma)$. The dashed lines $l_1$ and $l_2$ indicate the borders of phase and lag synchronization, respectively

system, i.e., $v = h(x)$. The first partial system is called the drive, and the second one is called the response. In the regime of generalized synchronization the instantaneous states $x$ and $y$ are connected by the relation $y = Q(x)$, and all phase trajectories from some basin of attraction $B$ approach the manifold $M = \{x, Q(x)\}$. If $Q$ is an identity transformation, generalized synchronization degenerates into complete synchronization. It is shown that the generalized synchronization effect can occur in (2.89) if and only if the response system $\dot{y} = G(y, h(x))$ is asymptotically stable, i.e., for any initial conditions $y_1(0)$ and $y_2(0)$ from some open subset $B_y$ the following limit exists:

$$\lim_{t \to \infty} \|y(t, x(0), y_1(0)) - y(t, x(0), y_2(0))\| = 0. \tag{2.90}$$

If this condition is satisfied, the effect of generalized synchronization is realized even if the drive and response systems are completely different. Synchronization can easily be detected by using conditional Lyapunov exponents

$$\lambda_{\text{con}}^j = \lim_{t \to \infty} \frac{1}{t} \ln \|z^j(t, x(0), y(0))\|, \quad j = 1, 2, \dots, n, \tag{2.91}$$

where $z^j$ is the $j$th fundamental solution of the linearized system

$$\dot{z} = \left[ \frac{\partial G(y, h(x))}{\partial y} \right]_{y=Q(x)} \cdot z. \tag{2.92}$$

In the regime of synchronization all the $\lambda_{\text{con}}^j$ must be negative.

### 2.3.4 Phase Multistability in the Region of Chaos Synchronization

The presence of a multistability region of periodic and chaotic regimes inside an Arnold's tongue is a peculiarity of mutual synchronization of oscillators in which onset of chaos follows the Feigenbaum scenario [183, 192, 194, 208, 234–237]. Such multistability can be called phase multistability because it is related to mutual synchronization of oscillations whose spectrum contains subharmonics of basic frequency and which are distinct in their phases. The larger is the number of the basic frequency subharmonics contained in the spectrum, the larger the number of possible synchronous regimes which are distinguished by a phase shift between partial oscillators.

For initial (generating) periodic oscillations with period $T_0$ the phase difference $\phi_0$ between partial oscillators is given as $\phi_0 \pm 2\pi m$, $m = 1, 2, \dots$ The spectrum of double period $(2T_0)$ oscillations contains the subharmonic $\omega_0/2$, and in this case the phase differences $\phi_0$ and $\phi_0 + 2\pi$ correspond to two different limit cycles in the phase space of interacting systems. The number of possible period-$2^n T_0$ limit cycles can grow up to $2^n$. They are distinguished by a phase shift between partial oscillators, which can take the values $\phi_0 + 2\pi m$ with $m = 2^k$, where $k = 0, 1, \dots, n-1$. Phase multistability in the region of periodic oscillations can also be observed in a chaotic zone if the chaotic attractor consists of $2^n$ bands around a saddle-focus. A hierarchy of phase multistability in dissipatively coupled identical systems has been explored in detail in numerical simulation of the dynamics of coupled logistic maps [234, 236], as well as in full-scale experiments with in-phase excited nonlinear radio-technical circuits [235]. The hierarchy has certain universal features which can also manifest themselves under a dissipative interaction of continuous-time self-sustained systems [192, 194, 237].

Phase multistability in dissipatively coupled identical oscillators with the period-doubling route to chaos can be studied by using, as a base model, the system of two coupled logistic maps [183, 234, 238]:

$$x_{n+1} = r - x_n^2 + \gamma(x_n^2 - y_n^2),$$
$$y_{n+1} = r - y_n^2 + \gamma(y_n^2 - x_n^2). \tag{2.93}$$

Here, $r$ is the control parameter and $\gamma$ is the strength of coupling. For this discrete-time system the phase shift is understood as a shift between the time realizations in the subsystems by $m$ iterations. Oscillations for which $m = 0$ are called *in phase*, and the corresponding limit sets belong to the symmetric subspace $x = y$. Otherwise, oscillations are *out of phase* and do not lie in the symmetric subspace.

With increasing parameter $r$ the phase multistability evolves as follows: The initial cycle loses its stability and becomes a saddle. As $r$ grows, the saddle cycle undergoes a period-doubling bifurcation, which is a symmetry-breaking event. The period-2 cycle being born is no longer in the symmetric subspace but has mirror symmetry with respect to the line $x = y$. This cycle is initially a saddle one and becomes stable as the parameter $r$ is further increased. Each of the in-phase cycles gives rise to its own branch of out-of-phase regimes. Out-of-phase cycles emerging from the in-phase ones always undergo a torus birth bifurcation. A resonance on the torus results in the appearance of new pairs of cycles, and so on. Multistability can also be observed in the region of chaotic regimes. Figure 2.40 illustrates phase portraits of limit cycles of system (2.93), which correspond to different $m$, and of chaotic attractors produced by these cycles.

The evolution of different types of oscillations is schematically shown in Fig. 2.41 for a small fixed value of $\gamma$ and under variation of parameter $r$. Solid lines indicate stable regimes, and dashed lines correspond to unstable ones. The bifurcational transitions are marked in Fig. 2.41 by points. Symbol $2^n C^m$ labels a period-$2^n$ cycle and corresponds to a shift by $m$ iterations between oscillations of the subsystems. A $2^n$-band chaotic attractor which emerges from a cycle with a shift $m$ is symbolized by $2^n C A^m$. Four branches, $A, B, C$ and $D$, are distinguished in the diagram. Branch $A$ indicates the evolution of in-phase regimes ($m = 0$) and $B, C$ and $D$ correspond to the evolution of out-of-phase regimes ($m \neq 0$). At the points of branching, the initial cycles lose their stability and become saddle ones. Inside the chaotic zone a sequence of band merging bifurcations of chaotic attractors $2^n C A^m$ is realized as the parameter increases. At the same time, the number of chaotic attractors decreases in reverse order to the emergence of cycles $2^n C^m$. On each of the branches the band-merging sequence results in attractors whose $n$ and $m$ numbers are the same as indices of resonance cycles on a torus. As $r$ is further increased, relevant pairs of attractors merge, namely, $4CA^1$ and

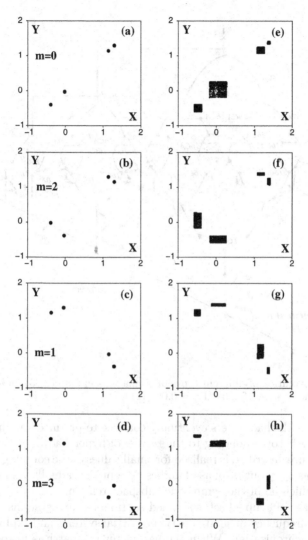

**Fig. 2.40.** Illustration of phase multistability in system (2.93) for $\gamma = 0.002$. (a)–(d) Cycles with different phase shift $m$ for $r = 1.3$; and (e)–(h) chaotic attractors generated by these cycles for $r = 1.415$

$4CA^3$ on branch $B$, $8CA^2$ and $8CA^6$ on branch $C$, and so on. The attractor merging is accompanied by the phenomenon of "chaos–chaos" intermittency and results in the appearance of chaotic attractors $2^n CA^m$, with $n$ and $m$ being equal to the indices of a cycle generating a given branch. Starting with large $n$ and $m$ the remaining chaotic sets merge at the points marked with a cross. In addition, one of the two sets becomes nonattractive before they merge. Thus, a crisis "chaotic attractor – chaotic saddle" occurs at

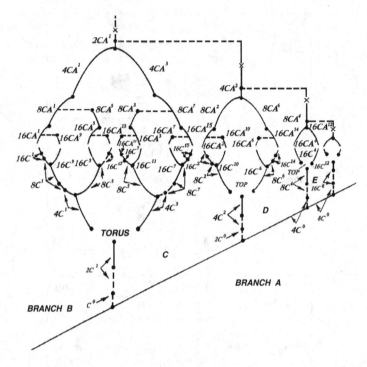

**Fig. 2.41.** Evolution of different types of oscillations in map (2.93) under variation of the parameter $r$ and for fixed $\gamma = 0.002$ [234]

these points. All chaotic sets eventually coalesce to produce a unified chaotic attractor $CA^{\Sigma}$ corresponding to the regime of hyperchaos.

The picture described is realized for small values of the coupling strength $\gamma$. With increasing $\gamma$ out-of-phase families "die out" gradually and only those limit sets which lie in the symmetric subspace remain.

In the case of coupled self-sustained continuous-time systems one can introduce the frequency mismatch between partial systems and explore a region of phase synchronization. When the parameter mismatch is introduced, the symmetry relations of limit sets are violated and certain bifurcations of limit cycles change their character [239]. The study of system (2.88) and other similar systems has shown that the presence of a small frequency mismatch under weak coupling [for (2.88) $\Delta \leq 0.001$ with $\gamma \approx 0.02$] does not considerably change the evolution scheme of different kinds of oscillations found for the discrete-time model (2.93). If the frequency mismatch is large enough, the order and the character of bifurcations of periodic and chaotic regimes can significantly change.

Depending on the choice of parameter values of (2.88), the boundary of phase synchronization may pass into the region of multistability or of merged chaos $CA^{\Sigma}$. Figure 2.42 shows a fragment of the bifurcation diagram in the

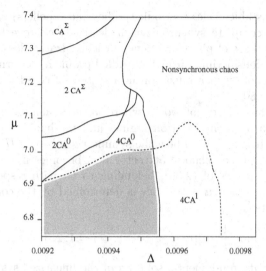

**Fig. 2.42.** A sketch of the bifurcation diagram of system (2.88) near the phase locking boundary and in the presence of chaotic multistability. The parameter values are $\alpha = 0.165, \beta = 0.2, \gamma = 0.02$ and $\Omega_0 = 1$

$\Delta - \mu$ plane near the synchronization region boundary for system (2.88). Two chaotic attractors, $4CA^0$ and $4CA^1$, exist in the shaded region. Attractor $2CA^\Sigma$ emerges when chaotic sets of two families merge (before this, $4CA^1$ becomes a chaotic saddle).

The presence of different attracting and saddle limit sets of out-of-phase families for small coupling strengths makes the behavior of interacting chaotic oscillators essentially complicated. In such systems one can observe various phenomena, including multistability, fractality of basin boundaries, riddled[7] basins of attraction and crises of chaotic attractors, the latter being accompanied by intermittency and a transition to hyperchaos. With increasing strength of the coupling "nonsymmetric" limit sets are degenerated and the transition to the region of lag synchronization or complete synchronization (for $\Delta = 0$) takes place. In this region there is only one chaotic attractor, $2^n CA^0$, which is topologically equivalent to the chaotic attractor in a partial oscillator.

### 2.3.5 Bifurcation Mechanisms of Partial and Complete Chaos Synchronization Loss

The transition from synchronous chaotic regimes to nonsynchronous ones has a certain peculiarity. Such transitions result from a sequence of hidden (at first glance) internal bifurcations of a chaotic attractor due to the presence

---

[7] A detailed explanation of the notion "riddling" will be given in Sect. 2.3.5.

of embedded unstable cycles (periodic saddles and repellers) in it. Moreover, mechanisms of complete synchronization destruction prove to be in many ways similar to those of phase–frequency synchronization loss.

We start by considering a recently studied problem concerning the loss of the complete chaotic synchronization in a system of coupled identical oscillators of type (2.80).

There is a large body of works where authors analyze the robustness of complete chaos synchronization regime and mechanisms of its destruction [239–250]. It is examined how the symmetric subspace $U$ of the system, in which an "in phase" chaotic attractor lies, becomes nonattracting. The stability of trajectories of (2.80), belonging to $U$, with respect to a small transverse perturbation $\boldsymbol{u} = \boldsymbol{x}_2 - \boldsymbol{x}_1$ is determined by the conditional Lyapunov exponents [186]:

$$\lambda_{\text{con}}^{j} = \lim_{t \to \infty} \frac{1}{t} \ln \|\boldsymbol{u}^{j}(t, \boldsymbol{x}_1(0))\|, \quad j = 1, 2, 3, \dots, N, \qquad (2.94)$$

where $\boldsymbol{u}^{j}$ is the $j$th fundamental solution of the linearized system

$$\dot{\boldsymbol{u}} = \frac{\partial}{\partial \boldsymbol{x}_2} \Big[ \boldsymbol{F}(\boldsymbol{x}_2, \alpha) + \gamma_2 \boldsymbol{g}(\boldsymbol{x}_1, \boldsymbol{x}_2) - \gamma_1 \boldsymbol{g}(\boldsymbol{x}_2, \boldsymbol{x}_1) \Big] \Big|_{\boldsymbol{x}_2 = \boldsymbol{x}_1} \cdot \boldsymbol{u}. \qquad (2.95)$$

If all the $\lambda_{\text{con}}^{j}$, $j = 1, 2, \dots, N$, are negative, the regime of complete chaotic synchronization is asymptotically stable. When at least one of the $\lambda_{\text{con}}^{j}$ becomes positive, the symmetric subspace $U$ is no longer stable. As a result, the complete chaos synchronization regime is destroyed in a so-called blowout bifurcation [244]. This process is usually accompanied by the phenomenon of transient (on finite time intervals) or "true" intermittency (Yamada–Fujisaka intermittency or "on–off" intermittency) [241–244, 251].

However, the conditional Lyapunov exponents are averaged over attractor characteristics which cannot reflect all the local changes in the structure of the limit set. In [240–244] it was shown that before the transverse direction becomes unstable, a set of points of zero measure appears in the symmetric subspace $U$, at which the transverse instability occurs. These points belong to some of the unstable cycles lying in $U$. Having fallen into the neighborhood of such a cycle, the phase point, if it does not strictly lie in $U$, moves away from the symmetric subspace. If for these parameter values the system does not possess another attractor, except the "in phase" one, i.e., lying in $U$, in some time the trajectory will return to its neighborhood and then reach the symmetric subspace. At the same time, a long transient process of "on–off" intermittency can be observed. The effect of weak noise on "in phase" chaos causes the intermittency process to be constantly renewed. Due to the noise influence an experimentally observed chaotic attractor no longer lies in the symmetric subspace and one can observe dynamics with temporal bursting. This phenomenon is referred to as *attractor bubbling* [241–244].

The presence of unstable cycles embedded in the "in phase" chaos is responsible for the formation of certain regions which have a tongue-like

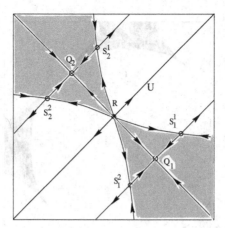

**Fig. 2.43.** Repelling tongue in the basin of an "out of phase" period-2 cycle (points $Q_{1,2}$), which originates from the repeller $R$ in the symmetric subspace $U$ (bisecting line). $S^1_{1,2}$ and $S^2_{1,2}$ are points of saddle period-2 cycles. The tongue is bounded by the stable manifolds of the "out of phase" saddle cycles $S^{1,2}$

shape in their sections and originate from the points of the unstable cycles [241, 242, 248]. The trajectory from such a tongue is repelled from the "in phase" attractor. If for the same parameter values the system has some "out of phase" attractor, the trajectory falls on it. Figure 2.43 shows a typical tongue on the phase plane. The emergence of a countable set of tongues in the phase space leads to the "riddling" of the local neighborhood of the chaotic attractor lying in $U$. This phenomenon is thus called *riddling* [241–246]. The chaotic attractor in $U$ is no longer attracting in the usual sense and is called a *Milnor attractor* [252]. Figure 2.44 demonstrates the riddled neighborhood of the chaotic attractor in $U$ (on the bisecting line) for the system of coupled logistic maps. For the given parameter values the transverse Lyapunov exponent is still negative, i.e., a blowout bifurcation does not yet occur. When the transverse direction becomes unstable, on average, over the attractor, the chaotic limit set in the symmetric subspace is no longer attractive even in the Milnor sense. This situation corresponds to the blowout bifurcation.

Bifurcations of saddle cycles embedded in a synchronous attractor play an important role in the appearance and destruction of phase–frequency synchronization. In [192] it was found that on the control parameter plane the lines corresponding to tangent bifurcations of saddle cycles of different periodicity are accumulated on the boundary of chaos synchronization. The boundary itself can be treated as a critical line which the tangent bifurcation points of cycles with increasing periods approach. The role of saddle cycles embedded in a chaotic attractor has been considered in detail in [196] using a noninvertible two-dimensional map, which models synchronization of a chaotic oscillator by an external periodic force. The map has the following form:

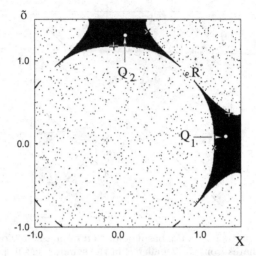

**Fig. 2.44.** Bistability and riddling in the system of coupled logistic maps (2.93). There exist the chaotic "in phase" attractor, located on the bisectrix $x = y$, and the "out of phase" stable period-2 cycle (points $Q_{1,2}$). The basin of attraction of the "out of phase" attractor is marked by black points

$$x_{n+1} = f(x_n, \phi_n),$$
$$\phi_{n+1} = \phi_n + \Omega + \varepsilon \cos\left(2\pi\phi_n + g(x_n)\right), \mod 1, \qquad (2.96)$$

where $\phi$ describes the phase difference of the oscillator and the external force, $\Omega$ is the frequency mismatch, and $\varepsilon$ denotes the amplitude of the external force. Function $g(x)$, which defines chaotic modulation of the phase, is taken to be $\delta \cdot x$, and $f(x, \phi)$ describing the amplitude dynamics of the chaotic oscillator is set to be $f(x, \phi) = 1 - a|x| + \varepsilon\rho\sin(2\pi\phi)$. A similar map has been also studied in [197]. Saddle cycles forming a skeleton of synchronous chaos undergo tangent bifurcations together with relevant periodic repellers. The latter create a skeleton of a chaotic repeller that touches the chaotic attractor at separate points, namely, at the points of saddle and unstable cycles at the moments of their merging. Each pair of the skeleton cycles belongs to an unstable invariant curve corresponding to a saddle torus in a continuous-time system. As a result of tangent bifurcation, the motion on the invariant curve becomes ergodic. This means that there appears a direction along which the phase point is repelled from the synchronous attractor and after a revolution along the invariant curve returns again. With this, the phase difference changes by $2\pi$.

It may take a long time to detect numerically the disappearance of certain pairs of cycles and the occurrence of instability directions. Nevertheless, the accumulation of local changes in the synchronous chaos structure ultimately leads to the complete destruction of phase synchronization. The corresponding bifurcation is similar to the blowout bifurcation. It can be

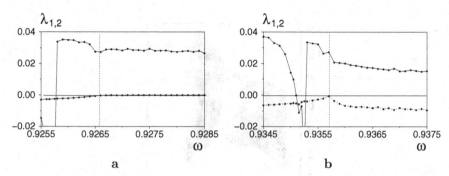

**Fig. 2.45.** Dependences of the two largest Lyapunov exponents on the frequency mismatch $\omega$ for the nonautonomous Rössler oscillator (2.79) with $\alpha = \beta = 0.2$, $\gamma = 0.02$, and (**a**) $\mu = 3.71$ and (**b**) $\mu = 3.8$

clearly identified from the behavior of Lyapunov exponents. One of the negative Lyapunov exponents of synchronous chaos becomes equal to zero at the boundary of the phase-locking region [195, 253]. Figure 2.45 illustrates dependences of the two largest Lyapunov exponents on the detuning calculated for the nonautonomous Rössler oscillator (2.79) when one crosses the boundary of the phase-locking region. One of the exponents, which corresponds to periodic perturbations, is identically equal to zero. Outside the main synchronization region either only one exponent (Fig. 2.45a) or two exponents are equal to zero (Fig. 2.45b). In any case there are two zero Lyapunov exponents at the synchronization boundary.

Numerical experiments carried out with the nonautonomous Rössler system and described in [254] verify the assumed bifurcation mechanism of phase locking of chaotic oscillations. Synchronous oscillations being shifted in phase relative to the periodic external force by $\theta(t)$ and $\theta(t) + 2\pi$ are considered as belonging to two co-existing chaotic attractors $R$ and $L$. The basin boundary of these attractors gives an idea of where a chaotic saddle may be located. The chaotic saddle collides with a chaotic attractor and phase synchronization is destroyed. Figure 2.46, taken from [254], shows projections of attractor $R$ and a part of the attractor $L$ basin of attraction on the plane of instantaneous amplitude $r$ and phase difference $\theta$. It is seen that the tongues of the basin of attraction of $L$ approach $R$. Unstable cycles (saddle cycles in a continuous-time system) are arranged at the ends of these tongues. If the initial point falls into one of the tongues, the trajectory is repelled from $R$ and a phase slip occurs.

## 2.3.6 Summary

Synchronization is one of the fundamental phenomena in nature and is observed in a variety of chaotic systems. A unified concept of chaotic synchronization is practically complete and is based on classical methods and

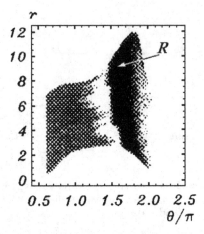

**Fig. 2.46.** Crisis of an attractor and a chaotic saddle in the nonautonomous Rössler oscillator. The chaotic saddle is located at the boundary of the grey zone [254]

approaches and includes knowledge of complete (in-phase) and partial chaos synchronization. However, a large number of problems is still insufficiently studied. They include synchronization of certain kinds of chaos (for example, torus chaos), bifurcation mechanisms of partial and complete synchronization destruction and numerous synchronization-related phenomena in ensembles of oscillators and in distributed systems.

The comprehensive study of the synchronization effect of complex DS and ensembles has principal importance not only in the framework of nonlinear dynamics but also for the whole range of natural sciences. This research direction allows us to gain deep insight into the mechanisms of self-organization of interacting systems.

## 2.4 Effects of Synchronization in Extended Self-Sustained Oscillatory Systems

### 2.4.1 Introduction

Studying dynamics of nonlinear extended systems is one of the basic directions in development of the theory of oscillations and waves. Spatially extended systems can be represented by continuous media usually described by differential equations in partial derivatives as well as by ensembles of interacting elements (oscillators, self-sustained oscillators, maps, etc.) locally coupled in chains and arrays. In the latter case spatial coordinates are discrete, i.e., they take a countable set of values corresponding to numbering of ensemble elements. Chains and arrays consisted of a large number of nonlinear elements with regular, chaotic, or stochastic behavior are widely applied in mathematical modeling of physical, optical, and radio-electronic extended systems [179,204,205,255–265], as well as of chemical and biological processes [217–219, 221, 222, 255, 266–271]. Nonlinear phenomena in models of self-sustained oscillatory media with continuous spatial coordinates have also been studied extensively. Especially a large number of works have been devoted to such basic models as the Ginzburg–Landau equations [272–275] and the Kuramoto-Sivashinsky equations [267, 276, 277].

Synchronization of elements of a medium or an ensemble is the major factor in dynamics of extended self-sustained oscillatory systems that leads to an ordered spatiotemporal behavior. Synchronization effects in ensembles of self-sustained oscillators and phase oscillators with local coupling have been studied in [172,217,218,278–283] and in many other papers and monographs. Partial frequency–phase synchronization in chains and arrays of quasi-harmonic self-sustained oscillators and phase oscillators manifests itself in the formation of phase and frequency clusters in the presence of basic frequency mismatch [172, 173, 218, 284–287]. Similarly, the partial phase synchronization can also lead to cluster formation in a chain of spiral chaos oscillators [213]. A large number of publications are devoted to the study of global and partial synchronization, to the formation of clusters of synchronous states and of ordered spatial structures in chains and arrays of identical chaotic self-sustained oscillators and in model chaotic maps [207, 282, 288–293].

Synchronization in extended self-sustained oscillatory systems serves as a reason that restricts the growth of attractor dimension [205, 294–296]. The possibility of realizing synchronous regimes with different phase shits is closely related with a phenomenon of multistability that implies the coexistence of a set of regular and chaotic attractors in phase space [208]. Multistability in turn can lead to crises of attractors, to fractalization of basins of attraction, and to other nontrivial effects.

Forced synchronization of a continuous self-sustained oscillatory medium has been studied in [297, 298]. Mutual synchronization of spatiotemporal regimes in interacting extended systems has been considered in [215,299–302].

Despite a large number of works dedicated to effects of synchronization in extended systems, this research area still contains a lot of insufficiently studied problems. They are related to investigation of inhomogeneous media, to study of effects induced by noise and local external forcings, to revealing the role of anharmonicity, to analysis of multistability of synchronous spatiotemporal regimes, etc. This section is devoted to the effects of frequency cluster formation in an inhomogeneous extended system, to the influence of noise on cluster synchronization regime, and to generalization of the effective synchronization notion to the case of spatially extended systems. These effects are analyzed in models of different level, i.e., in chains of oscillators each being described by truncated amplitude and phase equations, in chains of phase oscillators and in an inhomogeneous self-sustained oscillatory medium defined by the Ginzburg–Landau equation. We also study mutual synchronization of cluster structures in two interacting inhomogeneous media and forced synchronization of a chain of chaotic oscillators by external harmonic force. Effects of synchronization and multistability are described in a ring of identical quasiharmonic and relaxation oscillators as well as of chaotic oscillators with Feigenbaum's scenario of chaos development.

## 2.4.2 Cluster Synchronization in an Inhomogeneous Chain of Quasiharmonic Oscillators

Consider an inhomogeneous chain of diffusively coupled quasiharmonic oscillators, being an analog of the chain studied in [173]. The model can be defined by the following system of equations with respect to complex amplitudes:

$$\dot{a}_j = r(1 - |a_j|^2)a_j + i\omega_j a_j + g(a_{j+1} + a_{j-1} - 2a_j), \qquad i = \sqrt{-1}. \quad (2.97)$$

This system can also be rewritten in terms of real amplitudes $\rho_j$ and phases $\Phi_j$ as follows:

$$\dot{\rho}_j = r(1 - \rho_j^2)\rho_j + g(\rho_{j-1}\cos(\Phi_j - \Phi_{j-1}) + \rho_{j+1}\cos(\Phi_{j+1} - \Phi_j) - 2\rho_j),$$
$$\dot{\Phi}_j = \omega_j + g\left(\frac{\rho_{j+1}}{\rho_j}\sin(\Phi_{j+1} - \Phi_j) - \frac{\rho_{j-1}}{\rho_j}\sin(\Phi_j - \Phi_{j-1})\right), \qquad (2.98)$$

where $\rho_j = |a_j| = \sqrt{(\text{Re } a_j)^2 + (\text{Im } a_j)^2}$ and $\Phi_j = \arctan(\text{Im } a_j/\text{Re } a_j)\pm\pi k$, $k = 0, 1, 2, \ldots$. The quantity $\pm\pi k$ is added as the phase changes continuously in time. In Eqs. (2.97) and (2.98), $j = 1, 2, 3, \ldots, m$ denotes the number of an oscillator (the discrete spatial variable), $r$ is the parameter of excitation being the same for all oscillators, and $\omega_j$ is the frequency of self-sustained oscillations of the $j$th oscillator without interaction (the unperturbed basic frequency). The values of basic frequencies are distributed linearly along the chain:

$$\omega_j = \omega_1 + (j - 1) \cdot \Delta, \qquad (2.99)$$

where $\Delta = \omega_{j+1} - \omega_j$ is the frequency mismatch of neighboring oscillators. $\omega_1$ is the frequency of the first oscillator and can take any value since the variables and the time here are dimensionless. In this case distributions of both unperturbed frequencies and average frequencies of oscillations in the presence of coupling are simply shifted on the corresponding quantity. In what follows, we put $\omega_1 = 0$. The boundary conditions are defined in the form

$$\rho_0 = \rho_1, \qquad \Phi_0 = \Phi_1, \qquad \rho_{m+1} = \rho_m, \qquad \Phi_{m+1} = \Phi_m. \quad (2.100)$$

The length of the chain is fixed as $m = 100$ for the parameter of excitation $r = 0.5$.

When oscillators interact at $g \neq 0$, the frequencies of self-sustained oscillations change. Since oscillations of partial oscillators are, in general, nonperiodic, average frequencies can be introduced as follows:

$$\Omega_j = |\langle \dot{\Phi}_j(t) \rangle| = \lim_{T \to \infty} \frac{|\Phi_j(t_0 + T) - \Phi_j(t_0)|}{T}. \tag{2.101}$$

When the basic frequencies of oscillators $\omega_j$ are linearly detuned along the chain, regimes of partial synchronization can be observed in a certain parameter region $(\Delta, g)$. The partial synchronization manifests itself in the formation of frequency clusters which represent groups of oscillators with strictly equal average frequencies $\Omega_j = \Omega^{(k)}$, where $j \in [j_{\min}, j_{\max}]$ is the number of an oscillator in the chain, and $k = 1, 2, \dots, k_{\max}$ is the cluster number (Fig. 2.47a). Such structures are called *perfect clusters*[8]. In the regime of perfect clusters the oscillators behave regularly in time [173]. Amplitudes and phases fluctuate periodically with respect to their average values. Spectra of amplitude and phase fluctuations of any oscillator exhibit only lines at the intercluster frequency $\Delta\Omega = \Omega^{k+1} - \Omega^k$, being the same for any $k$, as well as at its harmonics. Figure 2.47b exemplifies the power spectrum of amplitude fluctuations $\tilde{\rho}_j(t)$ in the 50th oscillator that corresponds to the perfect clusters shown in Fig. 2.47a. For a more illustrative presentation of oscillations, a projection of trajectories on the plane $(\tilde{\rho}_j(t), H\tilde{\rho}_j(t))$ is shown in Fig. 2.47c for $j = 50$. $H\tilde{\rho}_j(t)$ is the Hilbert transform:

$$H\tilde{\rho}_j(t) = \frac{1}{\pi} \int_{\infty}^{\infty} \frac{\tilde{\rho}_j(\tau)}{t - \tau} d\tau. \tag{2.102}$$

When the parameters are varied, perfect structures with different numbers of clusters alternate with *imperfect* (intermediate) ones for which the average frequencies $\Omega_j$ of certain groups of oscillators are quite close but do not completely coincide (see Fig. 2.47d). Such groups of the oscillators can be considered as frequency clusters if the cluster frequency $\Omega^{(k)}$ is equal to the average frequency value at the center of the $k$th group. A group of oscillators is located in the intercluster area, and the boundaries of clusters and of intercluster regions are defined arbitrarily. In the case of imperfect clusters oscillations are irregular [173]. The spectra of amplitude and phase fluctuations become continuous and the basic spectral frequency depends on how close the considered oscillator is located to the cluster center. The power spectrum of amplitude fluctuations $\tilde{\rho}_j(t)$ is presented in Fig. 2.47e for the 50th oscillator and corresponds to the imperfect cluster structure depicted in Fig. 2.47d. A projection of oscillations on the plane $(\tilde{\rho}_j(t), H\tilde{\rho}_j(t))$ for $j = 50$ is shown in Fig. 2.47f.

The results presented in Figs. 2.47e and f enable one to assume that self-sustained oscillations in the inhomogeneous chain of oscillators are chaotic in the regime of imperfect clusters. The dynamical chaos mode can be strongly defined from a linear analysis of stability of oscillations in the chain. For this

---

[8] It is worth taking into account that it is the average values of oscillation frequencies that are used here. The spectrum of oscillations in each spatial point will contain the same set of frequency components but the spectral powers of these components will be redistributed according to which cluster is considered.

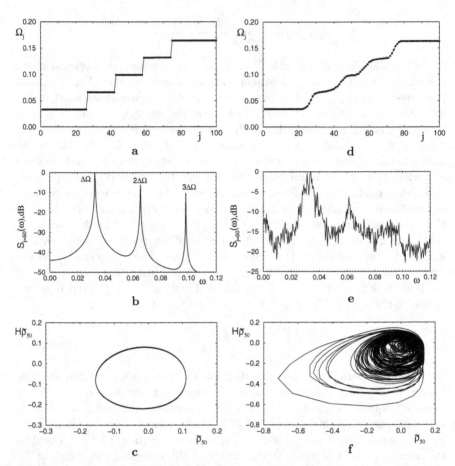

**Fig. 2.47.** Distribution of average frequencies $\Omega_j$, the normalized power spectrum of oscillations $\tilde{\rho}_j(t)$ for $j = 50$ and projection of phase trajectory on the plane $(\tilde{\rho}_j(t), H\tilde{\rho}_j(t))$ for $j = 50$ in the regime of perfect clusters at $\Delta = 0.002$, $g = 1.2$ **(a)**, **(b)**, **(c)** and in the regime of imperfect clusters at $\Delta = 0.002$, $g = 1.45$ **(d)**, **(e)**, **(f)**

purpose, the equations of chain (2.97) are integrated simultaneously with a system of linearized equations that describe the evolution of perturbations of complex amplitudes $u_j(t) = a'_j(t) - a_j(t)$ along a given solution. Equations for the perturbations read

$$\dot{u}_j = r(1 - 2|a_j|^2)u_j + i\omega_j u_j - ra_j^2 u_j^* + g(a_{j+1} + a_{j-1} - 2a_j), \qquad (2.103)$$

where the mark * denotes the complex-conjugate quantity. The boundary conditions for (2.103) have the form $u_0 = u_1$, $u_{m+1} = u_m$.

Our calculations indicate that the perturbation norm

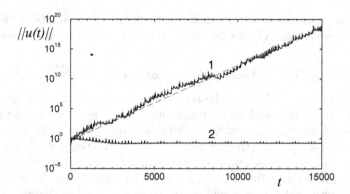

**Fig. 2.48.** Behavior of the perturbation norm $||u(t)||$ in system (2.97) in the regime of imperfect clusters for $\Delta = 0.002$, $g = 1.45$ (curve 1) and of perfect clusters for $\Delta = 0.002$, $g = 1.2$ (curve 2). The dashed line shows the exponential approximation of the perturbation norm growth $||u(t)|| = 0.5\exp(\lambda_1 t)$ with increment $\lambda_1 \approx 0.0029$

$$||u(t)|| = \left( \sum_{j=1}^{m} (Re\ u_j(t))^2 + (Im\ u_j(t))^2 \right)^{1/2} \qquad (2.104)$$

grows, on the average, exponentially in the regime of imperfect clusters, whereas it remains bounded in the regime of perfect clusters. The average increment of growth of $||u(t)||$ is the maximal Lyapunov exponent of the studied solution $a_j(t)$, $j = 1, 2, \ldots, m$. The temporal dependence of the perturbation norm is shown in Fig. 2.48 for the above considered regimes of perfect and imperfect clusters. The maximal Lyapunov exponent $\lambda_1$ in the regime of imperfect clusters (for $\Delta = 0.002$ and $g = 1,45$) is positive and takes the value of $\lambda_1 \approx 0.0029$.

The number of clusters decreases with increasing coupling strength $g$. If the following condition is satisfied

$$\left| \frac{\Delta m^2}{8g} \right| < 1, \qquad (2.105)$$

the global synchronization takes place. In this case, all the oscillators are synchronized at the same frequency.

### 2.4.3 Effect of Noise on Cluster Synchronization in a Chain of Quasiharmonic Oscillators

In this part we present numerical results concerning the influence of noise on regimes of cluster synchronization and generalize the notion of effective synchronization, i.e., synchronization in the presence of fluctuations [148,165], to spatially extended systems [303].

Taking into account external fluctuations, the equations for the inhomogeneous chain of quasiharmonic oscillators (2.97) can be written as follows:

$$\dot{a}_j = r(1 - |a_j|^2)a_j + i\omega_j a_j + g(a_{j+1} + a_{j-1} - 2a_j) + \sqrt{2D}\xi_j(t), \quad (2.106)$$

where $\xi_j(t) = \nu_j(t) + i\eta_j(t)$ is the normalized random force affecting the $j$th oscillator. Imaginary and real components of random forces are defined by uncorrelated sources of normalized white Gaussian noise $\langle \nu_j(t) \rangle \equiv \langle \eta_j(t) \rangle \equiv 0$; $\langle \nu_j(t)\nu_k(t+\tau) \rangle \equiv \langle \eta_j(t)\eta_k(t+\tau) \rangle \equiv \delta_{jk}\delta(\tau)$; $\langle \nu_j(t)\eta_k(t+\tau) \rangle \equiv 0$, where the brackets $\langle \ldots \rangle$ denote statistical averaging, $\delta_{jk}$ is the Kronecker symbol, and $\delta(\tau)$ is the Dirac function. Parameter $D$ characterizes the intensity of random forces, which is assumed to be the same for all oscillators. The numerical calculations are carried out for $r = 0.5$ in the chain with boundary conditions (2.100). The initial conditions for oscillators are chosen to be homogeneous and randomly distributed within the range $[0.9, 1.1]$.

Consider the effect of noise on the structure of frequency clusters in the chain of oscillators whose basic frequencies are linearly distributed along the spatial coordinate according to relation (2.99). The numeric results are presented in Fig. 2.49 for two different values of the coupling strength $g$ and for the mismatch $\Delta = 0.002$.

It is clearly seen in the graphs that when noise intensity synchronization increases, clusters are destroyed for both values of the coupling strength. Weak noise (parts (II)) leads to the flattening of the boundaries of cluster steps. As the noise intensity $D$ grows, the flattening and gradual destruction of clusters are started with the middle of the chain. If the noise intensity is strong enough (parts (III)), all the middle clusters are completely destroyed. Numerous numerical calculations performed for different values of the parameters $\Delta$ and $g$ have shown that two boundary clusters are highly stable toward the influence of noise. They can be destroyed only by applying a very strong noise.

The presence of fluctuations destroys the sharp boundary of phase synchronization. Numerical experiments demonstrate that without noise the phase difference of oscillators belonging to different clusters grows, on the average, linearly in time. However, the phase difference of oscillators from the same cluster remains constant excluding oscillations within the range $[-\pi, \pi]$ with respect to the average value. The influence of noise causes the phase difference of any neighboring oscillators to increase indefinitely in time but this growth is not linear for any $j$. The mean rate of the phase growth is different for various $j$. However, one can distinguish certain parts of the chain for which this rate is small. As a result, it becomes possible to define clusters of effective synchronization in the presence of noise [165].

The boundaries of effective synchronization clusters in the presence of noise can be estimated by using the effective diffusion coefficient $D_{\text{eff}}$ of the phase difference of neighboring oscillators [148]. Its mean value for the $j$th and $(j + 1)$th elements of the chain reads

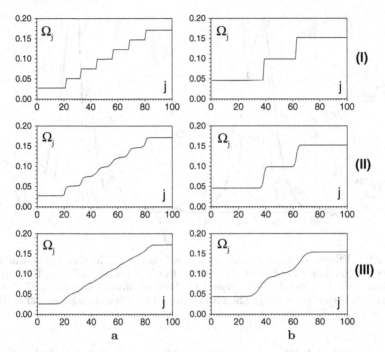

**Fig. 2.49.** Distribution of average frequencies of oscillators $\Omega_j$ in chain (2.106) for $\Delta = 0.002$ and for two values of the diffusive coupling: (a) $g = 0.55$ and (b) $g = 3.8$. Parts (I), (II), and (III) correspond to the noise intensity $D = 0$, $D = 0.00001$, and $D = 0.001$, respectively

$$D_{\mathrm{eff}}(j) = \lim_{t \to \infty} \frac{1}{2} \left( \frac{\sigma^2_{\theta_j}(t)}{t} \right), \qquad (2.107)$$

where $\sigma^2_{\theta_j}(t) = \langle \theta^2_j(t) \rangle - \langle \theta_j(t) \rangle^2$.

Dependences of the effective diffusion coefficient on the spatial coordinate within a single cluster ($39 \le j \le 62$) are shown in Fig. 2.50 for different values of the noise intensity $D$. The graphs testify that the cluster boundaries are gradually destroyed when the noise intensity increases. It can be mentioned that the dependence $D_{\mathrm{eff}}$ on $j$ is similar (taking into account that the variable $j$ is discrete) to the known dependence of the effective diffusion coefficient of the phase difference between a self-sustained oscillatory system and an external forcing on the mismatch. The boundaries of the effective synchronization cluster can be defined if a certain maximally admissible diffusion level $D_{\mathrm{eff}}^{\max}$ is given. In this case one can conclude that the oscillators for which $D_{\mathrm{eff}} \le D_{\mathrm{eff}}^{\max}$ belong to the same cluster. Such a definition of cluster boundaries is sufficiently arbitrary since the value of $D_{\mathrm{eff}}^{\max}$ can be set in different ways depending on the problem under consideration. However, in any case the cluster length will decrease as the noise intensity

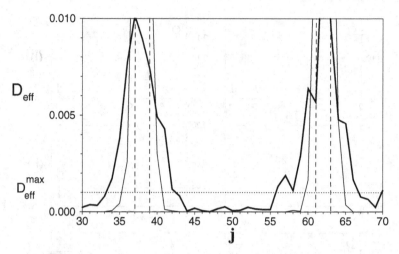

**Fig. 2.50.** Effective diffusion coefficient as a function of the spatial coordinate $j$ for $D = 10^{-8}$ (thin dashed line), $D = 10^{-5}$ (thin solid line), and $D = 10^{-3}$ (thick solid line). The horizontal dot line marks the level of diffusion coefficient $D_{eff}^{max}$ that defines the cluster boundaries. The mismatch and coupling strength are $\Delta = 0.002$ and $g = 3.8$

grows. For example, if we set $D_{eff}^{max} = 0.001$, the boundaries of the cluster shown in Fig. 2.50 for $D = 0.001$ correspond to the chain elements with numbers $j = 43$ and $j = 56$.

Thus, perfect frequency clusters cannot exist in the presence of random forces and are transformed into imperfect ones. The following question arises: What is the character of self-sustained oscillations that corresponds to noise-induced imperfect cluster structures? It is difficult to claim that self-sustained oscillations in the noisy inhomogeneous chain will be always chaotic. However, for the chosen parameter values of $\Delta = 0.003$ and $g = 3.8$, and of the noise intensity $D = 0.001$ the solution of linearized system (2.103) confirms that the norm $||u(t)||$ grows exponentially. Evaluating the mean rate of the exponential growth of $||u(t)||$ enables one to define the maximal Lyapunov exponent $\lambda_1 \approx 0.0005$. In other words, in this case the destruction of perfect clusters in the presence of noise is accompanied by the appearance of noise-induced weak chaos. Figure 2.51 illustrates the corresponding power spectrum of amplitude fluctuations $\tilde{\rho}_j(t)$ in the 50th oscillator and a projection of oscillations on the plane $(\tilde{\rho}_j(t), H\tilde{\rho}_j(t))$ for $j = 50$.

Studying dynamics of ensembles of oscillators is often restricted to phase equations only. With this, amplitudes of oscillators are considered to be equal and constant in time. Such an approach enables one to qualify phenomena related with frequency-phase locking and in some cases to solve the task analytically [172, 267, 284, 285, 304]. However, neglecting amplitude relations can lead to the loss of certain effects, such as, for example, the effect of "oscillation death" [173, 213, 305–307]. The influence of amplitude relations on

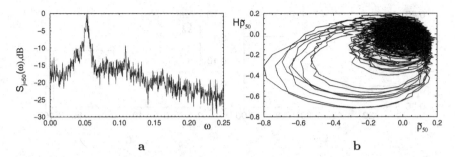

**Fig. 2.51.** Characteristics of oscillations in the chain of self-sustained oscillators in the regime of noise-induced chaos for $\Delta = 0.002$, $g = 3.8$, and $D = 0.001$. (a) The normalized power spectrum for $\tilde{\rho}(t)$ for $j = 50$, (b) projection of the phase trajectory on the plane $(\tilde{\rho}_j(t), H\tilde{\rho}_j(t))$, $j = 50$

the formation of cluster structures was mentioned in [173]. In the course of our studies, we compare the peculiarities of cluster synchronization in the chain of oscillators that is described by the total system of truncated equations for amplitudes and phases (2.106) and by the corresponding system of phase equations. The latter system can be easily derived from (2.106) by setting $\tilde{\rho}_j = 1$ for any $j$, i.e., by fixing amplitudes of all oscillators to be equal to their unperturbed value. The system of phase equations reads

$$\dot{\Phi}_j = \omega_1 + (j-1)\Delta + g(\sin(\Phi_{j+1} - \Phi_j) -$$
$$- \sin(\Phi_j - \Phi_{j-1})) + \sqrt{2D}\eta_j(t), \qquad (2.108)$$
$$j = 1, 2, \ldots, m.$$

We use the following boundary conditions: $\Phi_0 = \Phi_1$, $\Phi_{m+1} = \Phi_m$. The calculation results for the frequency distribution along the chain described by system (2.108) are presented in Fig. 2.52 for $\Delta = 0.002$ and for different values of the coupling strength $g$.

Figure 2.52a illustrates the frequency distribution for $g = 0.55$ and without noise, $D = 0$. In contrast to the result obtained in a similar case for system (2.106) (see Fig. 2.49a), here one can see only two clusters adjoining the chain boundaries. The middle clusters appear for a stronger coupling (see Fig. 2.52b,c). However, the formed cluster structures slightly differ from those that were observed in the chain described by the total system of truncated equations for amplitudes and phases (2.106). In the case being considered the extreme clusters are lengthened but the middle ones become shorter. The height of cluster steps (the frequency difference in neighboring clusters) is less here than for system (2.106) and decreases rapidly as the coupling grows. Thus, when only the phase dynamics is taken into account, the region where the cluster synchronization exists is significantly contracted. Besides, the cluster structure is appeared to be less stable toward the influence

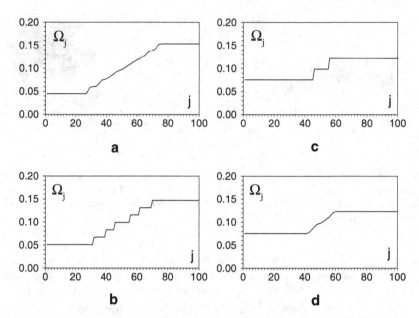

**Fig. 2.52.** Distribution of average frequencies in the chain of oscillators described by equations (2.108) for $\Delta = 0.002$ and different values of the coupling strength: (**a**) $g = 0.55$, (**b**) $g = 0.7$, and (**c**) $g = 1.5$. Graphs (**a**)–(**c**) correspond to the absence of noise, $D = 0$. Graph (**d**) is obtained for $g = 1.5$ and $D = 10^{-5}$

of fluctuations. Introducing weak noise $D = 10^{-5}$ immediately leads to the destruction of the middle clusters (see Fig. 2.52d).

### 2.4.4 Cluster Synchronization in an Inhomogeneous Self-Sustained Oscillatory Medium

Consider an inhomogeneous self-sustained oscillatory medium being analogous to the discrete model (2.97). An equation for the medium can be derived by passing to the limit from the chain of self-sustained oscillators to a continuous spatial coordinate. This equation represents the Ginzburg–Landau equation with real parameters and the oscillation frequency linearly depending on the spatial coordinate:

$$a_t = i\omega(x)a + \frac{1}{2}(1 - |a|^2)a + ga_{xx} + \sqrt{2D}\xi(x,t). \qquad (2.109)$$

Here, $a(x,t)$ is the complex amplitude of oscillations, independent variables $t$ and $x \in [0, l]$ are the time and the normalized spatial coordinate, respectively. $a_t$ is the first temporal derivative, $a_{xx}$ is the second derivative with respect to the spatial coordinate, and $\xi(x,t) = \nu(x,t) + i\eta(x,t)$ is the normalized random force acting at each point of the medium. The imaginary and real components of the random force $\nu(x,t)$ and $\eta(x,t)$ are assumed to be statistically independent Gaussian forcings being temporarily and spatially uncorrelated: $\langle \nu(x,t) \rangle \equiv \langle \eta(x,t) \rangle \equiv 0$; $\langle \nu(x_1,t_1)\nu(x_2,t_2) \rangle \equiv \langle \eta(x_1,t_1)\eta(x_2,t_2) \rangle \equiv \delta(x_2 - x_1)\delta(t_2 - t_1)$; $\langle \nu(x_1,t_1)\eta(x_2,t_2) \rangle \equiv 0$, where the brackets $\langle \ldots \rangle$ denote statistical averaging, and $\delta(\ldots)$ is the Dirac function. Parameter $D$ characterizes the random force intensity that is considered to be constant in time and space. In numerical experiments the medium length is fixed as $l = 50$. The diffusion coefficient $g$ is assumed to be the same in every point of the medium. For $g \to 0$, oscillations in various points of the medium possess different frequencies which are given by the function $\omega(x) = x\frac{\Delta}{l}$, where $\Delta$ is the maximal mismatch (the mismatch between boundary points of the medium). The boundary conditions are set in the form $a_x(x,t)|_{x=0,l} \equiv 0$. The initial condition of the medium is chosen randomly near some homogeneous distribution $a_0 = \text{const}$. An analogous model of the medium was studied in [174] in the deterministic case ($D = 0$).

Passing to spatio-continuous model (2.109) and to a different scheme of numerical integration enables one to answer the following question: How much a chain composed from a finite (and, strictly speaking, not very large) number of self-sustained oscillators can adequately reflect the phenomena being characteristic for an inhomogeneous self-sustained oscillatory medium and can designate more validly a class of such effects.

Equation (2.109) is integrated numerically by means of the finite difference method with regard to the influence of random forces [308] and according to an implicit scheme of forward and backward sweeps. The real amplitude $\rho(x,t)$ and the phase $\Phi(x,t)$ of oscillations are calculated as follows:

$$\rho(x,t) = \mid a(x,t) \mid = \sqrt{(\text{Re } a)^2 + (\text{Im } a)^2},$$

$$\Phi(x,t) = \arg a(x,t) = \text{arctg} \, \frac{\text{Im } a}{\text{Re } a} \pm \pi k, \qquad k = 0,1,2,\ldots \, .$$

The average frequency of oscillations in a medium point with coordinate $x$ is estimated in a similar way with (2.101):

$$\Omega(x) = \langle \Phi_t(x,t) \rangle = \lim_{T \to \infty} \frac{\Phi(x,t_0 + T) - \Phi(x,t_0)}{T}. \qquad (2.110)$$

The temporal dynamics in fixed points of the medium is analyzed by calculating the power spectra of fluctuations $\tilde{\rho}(x,t)$. For a more illustrative presentation of oscillations projections of a phase trajectory are drawn on the plane of variables $(\tilde{\rho}, H\tilde{\rho})$, where $H\tilde{\rho}$ is the Hilbert conjugate process. Besides, the maximal Lyapunov exponent $\lambda_1$ is calculated for different oscillatory regimes of the medium. For this purpose, Eq. (2.109) is integrated together with a linearized equation for a small perturbation $u(x,t)$ of the complex amplitude $a(x,y)$ that has the following form:

$$u_t = \mathrm{i}\omega(x)u + \frac{1}{2}(1 - 2|a|^2)u - \frac{1}{2}a^2 u^* + gu_{xx}. \qquad (2.111)$$

The maximal Lyapunov exponent is introduced as an average over time increment of exponential growth of the Euclidean perturbation norm $||u(x,t)||$. Taking into account that the spatial coordinate is discrete, $||u(x,t)||$ is reduced to the sum of a finite number of terms:

$$||u(x,t)|| = \left( \int_0^l ((\text{Re } u(x,t))^2 + (\text{Im } u(x,t))^2)dx \right)^{1/2} \approx$$

$$\approx \left( \sum_{k=1}^m (\text{Re } u(x_k,t))^2 + (\text{Im } u(x_k,t))^2 \right)^{1/2}, \qquad (2.112)$$

where $m$ is the number of integration steps $h_x$ all along the medium length.

Analyze the basic characteristics of oscillatory regimes of the medium (2.109) in a purely deterministic case $(D = 0)$. In contrast to a more general case of the Ginzburg–Landau model with complex parameters, a turbulent regime cannot arise in the homogeneous medium described by Eq. (2.109). Without frequency mismatch, medium (2.109) can demonstrate the homogeneous regime, $a(x,t) \equiv 1$, only. The introduction of frequency mismatch causes the mean frequency of oscillations to change along the spatial coordinate $x$. In some ranges of parameter $\Delta$ and $g$ values, regimes of partial synchronization can be realized which are accompanied by the formation of perfect frequency clusters. They are separated by regions where intermediate (imperfect) cluster structures exist. Qualitatively, cluster structures in the continuous medium (2.109) are completely similar to the analogous structures in the chain of self-sustained oscillators (2.97). The behavior of medium (2.109), as well as of chain (2.97), depends on the character of cluster structure. In the regime of perfect clusters, oscillations are regular (periodic or quasiperiodic). The destruction of perfect clusters leads to a chaotic temporal dynamics. Figure 2.53 illustrates the basic characteristics of oscillations in the fixed spatial point $x = 25$ (the center of the middle cluster) for regimes of perfect and imperfect clusters. Distributions of mean frequencies $\Omega(x)$ along the medium are shown in Fig. 2.53a,d and indicate the character of the cluster structure. Normalized power spectra of amplitude fluctuations $\tilde{\rho}(t)$ testify to a periodic behavior of the amplitude in the regime of perfect clusters (Fig. 2.53b) and to an irregular process in the regime of imperfect clusters (Fig. 2.53e). The basic frequency of the fluctuation $\tilde{\rho}(t)$ spectrum is equal to the intercluster frequency $\Delta\Omega = \Omega_{j+1} - \Omega_j$ that is the same for all neighboring clusters. On the plane $(\tilde{\rho}(t), H\tilde{\rho}(t))$, a limit cycle corresponds to perfect clusters (Fig. 2.53c), while in the regime of imperfect clusters the phase trajectories obtained are similar to a chaotic attractor of the saddle-focus type (Fig. 2.53f). If to compare the results presented in Figs. 2.53 and 2.47, one can conclude that the behavior of medium (2.109) and of chain (2.97) are completely similar in the regimes of perfect and imperfect clusters.

There is no point in speaking about spatial order or disorder in the system being studied as the system length includes only two or three spatial

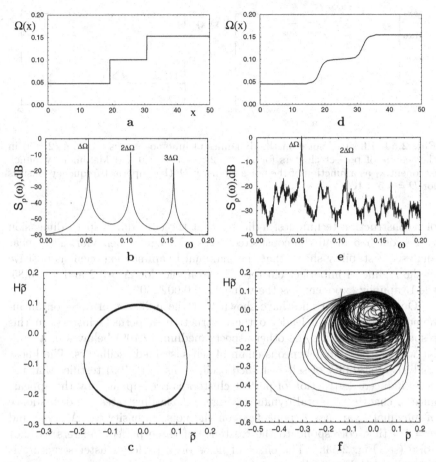

**Fig. 2.53.** Distribution of average frequencies, normalized power spectrum of oscillations $\tilde{\rho}(t)$ at point $x = 25$ and projection of the phase trajectory on the plane $(\tilde{\rho}(t), H\tilde{\rho}(t))$ at point $x = 25$ in the regime of perfect clusters for $\Delta = 0.02$ and $g = 1.0$ (**a**) and in the regime of imperfect clusters for $\Delta = 0.02$ and $g = 0.85$ (**b**). Discretization steps are $h_t = 0.01$ and $h_x = 0.001$

oscillations for the chosen values of the parameters $g$ and $\Omega$. We cannot respectively judge about turbulence since the latter notion implies a nonregular behavior of the medium both in time and space. However, as the system length $l$ increases or the parameter $g$ decreases, the medium with imperfect clusters can also exhibit a spatial disorder.

Analyzing such characteristics as power spectra, projections of phase trajectories as well as autocorrelation functions of oscillations in the regime of imperfect clusters, one can conclude a chaotic temporal behavior of the system. However, the dimension of this limit set is not large. The time series analysis of the process $\rho(x,t)$ in fixed spatial points, performed by means

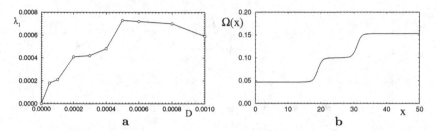

**Fig. 2.54.** Effect of noise on the dynamics of inhomogeneous medium (2.109) in the regime of perfect clusters for $\Delta = 0.2$ and $g = 1.0$. (a) Maximal Lyapunov exponent $\lambda_1$ as a function of the noise intensity $D$; (b) imperfect frequency clusters for $D = 0.5 \times 10^{-4}$

of reconstruction techniques, indicates that the embedding space dimension is of order 3–5 in all the considered cases of irregular behavior. The linear analysis of stability shows that the maximal Lyapunov exponent is positive in the regime of imperfect clusters. For example, for $\Delta = 0.2$ and $g = 0.85$, the Lyapunov exponent has the value $\lambda_1 \approx 0.002$ [309].

Our numerical studies have shown that the influence of noise on an inhomogeneous medium leads to the destruction of perfect clusters. In this respect as well as in some other aspects medium (2.109) behaves in a similar way as the inhomogeneous chain of self-sustained oscillators. The linear analysis of stability of self-sustained oscillations in (2.109) testifies that the nosie-induced destruction of perfect clusters is accompanied by the appearance of chaotic temporal dynamics. Figure 2.54a illustrates the dependence of maximal Lyapunov exponent $\lambda_1$ on the noise intensity for $\Delta = 0.2$ and $g = 1.0$ that correspond to the existence of three perfect clusters without noise (see Fig. 2.53a). The effect of noise on a perfect cluster structure is exemplified in Fig. 2.54b.

Analyze in more detail the transition of the medium to a chaotic behavior in the presence of noise. Figure 2.55 illustrates the spectrum of amplitude fluctuations at point $x = 25$ and the corresponding phase projection for $\Delta = 0.2$, $g = 1.0$, and $D = 0.5 \times 10^{-4}$. This regime represents a weak noise-induced chaos with Lyapunov exponent $\lambda = 0.00018 \pm 10^{-5}$. The trajectory on the plane $(\tilde{\rho}(t), H\tilde{\rho}(t))$ is mainly rotated in the vicinity of a limit cycle corresponding to $D = 0$ but can go sufficiently far away from this area (see Fig. 2.55b). Such behavior indicates that in the system without noise there exists some unattracting chaotic set in the vicinity of the regular solution. Weak noise excites the system state and demands trajectories to move along this set. An analogous mechanism of noise-induced chaos is well known for major finite-dimensional systems (see, for example, [310]).

Thus, the noise "activates" the same chaotic set that becomes attracting as a result of the hard bifurcation taking place in the deterministic case.

The studies performed have shown that the discrete model of inhomogeneous self-sustained oscillatory medium (2.97), even for a not large number

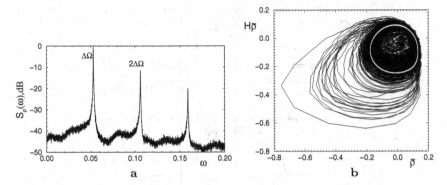

**Fig. 2.55.** Characteristics of noise-induced chaotic oscillations of medium (2.109) for $\Delta = 0.2$, $g = 1.0$, and $D = 0.5 \times 10^{-4}$ in fixed spatial point $x = 25$. (a) Normalized power spectrum of oscillations $\tilde{\rho}(t)$, (b) projection of phase trajectory on the plane $(\tilde{\rho}(t), H\tilde{\rho}(t))$ (the white curve corresponds to a limit cycle for $D = 0$)

of elements ($m = 100$), can demonstrate all the effects of partial synchronization that are observed in the continuous model (2.109). The addition of noise plays a significant role for both models because it destroys perfect cluster structures and leads to chaotization of regular oscillations. Thus, the behavior of both models is completely analogous in the framework of qualitative description.

### 2.4.5 Cluster Synchronization in Interacting Inhomogeneous Media

Two interacting extended systems can be mutually synchronized. Effects of mutual synchronization of spatiotemporal dynamics of interacting extended systems are described in a series of publications (for example, [301,311–313]). However, the majority of works consider mutual synchronization of homogeneous systems and media. We have studied numerically effects of synchronization of interacting active media in regimes of frequency clusters. The results obtained were first published in [314] and are presented below.

For our research we choose a system of two coupled Ginzburg–Landau equations (2.109) without noise sources:

$$a_t = i\omega_1(x)a + \frac{1}{2}(1 - |a|^2)a + g_1 a_{xx} + \varepsilon(b - a), \qquad (2.113)$$

$$b_t = i\omega_2(x)b + \frac{1}{2}(1 - |b|^2)b + g_2 b_{xx} + \varepsilon(a - b).$$

Here, $a(x,t)$ and $b(x,t)$ are complex amplitudes of the first and second systems, and $\varepsilon$ is the parameter of coupling that characterizes the degree of interaction at each spatial point. The frequency mismatch at different points of each separate medium is defined by functions $\omega_1(x)$ and $\omega_2(x)$, respectively,

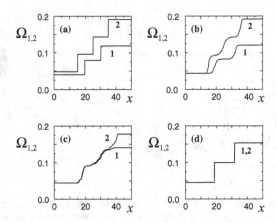

**Fig. 2.56.** Average frequencies $\Omega_{1,2}$ as a function of spatial coordinate $x$ for fixed $\delta = -0.08$ and for different values of the coupling strength: (**a**) $\varepsilon = 0$, (**b**) $\varepsilon = 0.005$, (**c**) $\varepsilon = 0.03$, and (**d**) $\varepsilon = 0.045$. Curves 1 and 2 correspond to the first and second medium

and $\Delta_{1,2}$ is the maximal mismatch (i.e., the detuning of the boundary points of the medium). The length $l$ is set to be constant $l = 50$ and the same for both systems. We integrate numerically Eqs. (2.113) and find how the mean frequency of oscillations for each partial system ($\Omega_1$ and $\Omega_2$, respectively) depends on the spatial coordinate $x$.

Consider two media (2.113) in the regime of frequency clusters. We fix $g_1 = g_2 = 0.9$, $\Delta = 0.16$ and vary $\Delta_2$ and $\varepsilon$. We analyze spatial distributions of average frequencies $\Omega_{1,2}(x)$ depending on the mismatch between corresponding elements of the two media, $\delta = \Delta_1 - \Delta_2$, and of the coupling strength $\varepsilon$. When $\varepsilon = 0$ and $\Delta_2 = \Delta_1$, both systems demonstrate spatial structures with three perfect frequency clusters. However, if $\Delta_1 \neq \Delta_2$, the average frequencies of oscillations in the corresponding points of the two media do not coincide. For large values of the mismatch $\delta$, the second medium contains $N = 2$ or $N = 4$ clusters depending on the sign of $\delta$.

Choose the value of mismatch $\delta$ in such a way that the uncoupled partial media can demonstrate regimes with different numbers of perfect clusters as shown in Fig. 2.56a.

For $\varepsilon \neq 0$ the elements of both media tend to synchronize their oscillations at equal frequencies. With increasing $\varepsilon$ the transition to synchronization of the two media can be observed. With this, the synchronization begins from the left boundary (for small $x$) since the frequency mismatch $\omega_1 - \omega_2$ is minimal for the initial segments of the media. When oscillations of the initial segments become synchronized, perfect clusters are destroyed and intermediate structures are formed. Then, more and more elements of the two media are involved in the synchronization process. While coupling is small enough, the number of imperfect clusters remains the same in both media

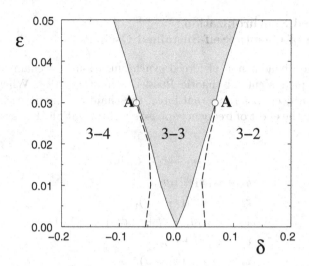

**Fig. 2.57.** Region of cluster synchronization on the parameter plane $(\delta, \varepsilon)$ for $\Delta_1 = 0.16$ and $g_{1,2} = 0.9$. The gray region corresponds to synchronization in each spatial point of the media. Inside the region bounded by dashed lines both media have an equal number of clusters. Figures denote the numbers of clusters in the first and second media

(see Fig. 2.56b,c). Starting with a certain value of the coupling strength, dependences of average frequencies on the spatial coordinate $\Omega_{1,2}(x)$ become completely identical (Fig. 2.56d). The resulting synchronized cluster structures are similar to those observed at $\varepsilon = 0$ and $\Delta = 0$.

Figure 2.57 illustrates a portion of the diagram of regimes for two interacting media (2.113) on the $(\delta, \varepsilon)$ plane. Two main regions are marked in the Figure. In the first region shown in gray, cluster structures are strongly synchronized. This means that the oscillation frequencies of the subsystems are strongly equal to each other at each spatial point, $\Omega_1(x) \equiv \Omega_2(x)$. In this case the synchronization of spatial structures are directly related with the classical phenomenon of frequency locking that can be observed at each point of self-sustained oscillatory media. The region of synchronization has a tongue-like form that is typical for regions of classical synchronization [315].

Besides, inside the area bounded by dashed lines both media are characterized by the same number of clusters but their frequencies do not coincide. Hence, the observed clusters are imperfect (Fig. 2.48b,c). The synchronization region is expanded as the coupling strength $\varepsilon$ increases. For a sufficiently strong coupling (the boundary over points $A$) there exists only the region of strong synchronization of the two media. In this case, with decreasing mismatch $\delta$, the transition into the synchronization region leads to a sharp rebuilding of cluster structures from "3-4" to "3-3" clusters. This change is accompanied by strong frequency locking of the partial systems at each spatial point.

## 2.4.6 Forced Synchronization
## of a Chain of Chaotic Self-Sustained Oscillators

Consider the phenomenon of forced synchronization in a chain of unidirectionally coupled identical chaotic Rössler oscillators [316]. When the first element is subject to an external force, the chain of chaotic oscillators can demonstrate the effect of frequency–phase synchronization. The system under study reads

$$
\begin{aligned}
\dot{x}_1 &= -\omega_1 y_1 - z_1 + C \sin \omega_{\text{ext}} t, \\
\dot{y}_1 &= \omega_1 x_1 + \alpha y_1, \\
\dot{z}_1 &= \beta + z_1(x_1 - \mu), \\
\dot{x}_j &= -\omega_j y_j - z_j + \gamma(x_{j-1} - x_j), \\
\dot{y}_j &= \omega_j x_j + \alpha y_j, \\
\dot{z}_j &= \beta + z_j(x_j - \mu), \\
j &= 2, 3, \ldots, m,
\end{aligned}
\tag{2.114}
$$

where $j$ is the oscillator number (discrete spatial coordinate), $m$ is the length of the chain, and $\gamma$ is the coupling coefficient. Parameters $\alpha$, $\beta$, and $\mu$ define the dynamics of partial systems, $\omega_j$ are parameters controlling the frequencies of oscillators, and $C$ and $\omega_{\text{ex}}$ denote the amplitude and the frequency of the external harmonic force affecting the first element of the chain. The system consists of 50 oscillators possessing equal basic frequencies, $\omega_j = \omega_1 = 0.924$. The fixed values of the parameters $\alpha = 0.2$, $\beta = 0.2$, and $\mu = 4$ correspond to the regime of spiral chaos [8, 162] in the Rössler system. The initial conditions of the oscillators were chosen with small random deviations from a homogeneous state.

To introduce instantaneous amplitudes $\rho_j$ and phases $\Phi_j$ of chaotic oscillations, the following change of variables is used:

$$
x_j = \rho_j \cos \Phi_j, \qquad y_j = \rho_j \sin \Phi_j,
\tag{2.115}
$$
$$
j = 1, 2, 3, \ldots, m.
$$

Mean frequencies $\Omega_j$ of oscillations were calculated according to relation (2.101). Without external signal, the mean frequencies of oscillations are equal in all chaotic oscillators, $\Omega_j = \omega_0 \approx 0.9981$, $j = 1, 2, \ldots, m$. The frequency of basic maximum in the power spectrum coincides with the mean frequency $\omega_0$ within the accuracy of numerical calculations. When the external signal frequency $\omega_{\text{ex}}$ is close to $\omega_0$, the phenomenon of forced frequency–phase synchronization can be observed in the chain of chaotic oscillators. The instantaneous phase of chaotic oscillations of the $j$th oscillator is considered to be locked if the following condition is fulfilled:

$$
\lim_{t \to \infty} |\Phi_j(t) - \Phi_{\text{ex}}(t)| < \infty.
\tag{2.116}
$$

This relation means that the phase difference is bounded. Here $\Phi_{ex} = \omega_{ex}t$ denote the phase of external signal. The mean frequency of oscillations of the $j$th oscillator is locked simultaneously with phase locking, $\Omega_j = \omega_{ex}$. The frequency ratio defines the rotation number for each oscillator of the chain $W_j = \Omega_j/\omega_{ex}$, $j = 1, 2, \ldots, m$. In numerical simulation, the $j$th element of the chain is assumed to be synchronized if the condition $|W_j - 1| \leq \varepsilon$ holds, where $\varepsilon = 10^{-4}$.

Consider the process of synchronization of chain elements when the amplitude $C$ and external signal frequency $\omega_{ex}$ are varied and the coupling coefficient is fixed, $\gamma = 0.01$. On the parameter plane $(C, \omega_{ex})$ synchronization regions of each element of the chain with $j \geq 2$ have the same form. The region boundaries basically coincide and cannot be distinguished within the accuracy of numerical experiments. The region $S$ of global synchronization (i.e., synchronization of all the elements) on the plane $(C, \omega_{ex})$ is shown in Fig. 2.58 for the chain with 50 oscillators. It is worth noting that we have not studied the problem on the existence of synchronization threshold [197]. Even if such a threshold exists, it approaches zero for the chosen parameter values, and the region $S$ in Fig. 2.58 rests virtually on the abscissa axis.

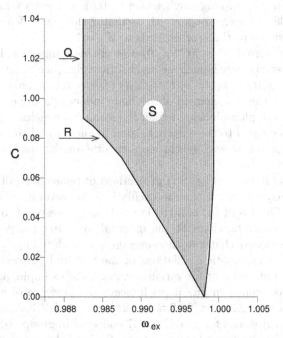

**Fig. 2.58.** Region of global synchronization $S$ for chain (2.115) with $m = 50$ on the parameter plane $(C, \omega_{ex})$. The other parameters are $\alpha = \beta = 0.2$, $\mu = 4$, $\omega_j = 0.924$, $j = 1, 2, \ldots, 50$, $\gamma = 0.01$. The marked directions $R$ and $Q$ correspond to different mechanisms of global synchronization

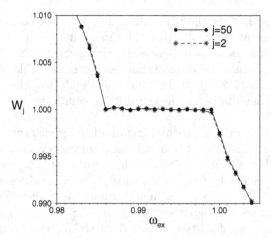

**Fig. 2.59.** Rotation number $W_j$ as a function of external signal frequency $\omega_{ex}$ for the $j = 2$nd and $j = 50$th elements of the chain at $\gamma = 0.01$ and $C = 0.08$

The rotation number $W_j$ as a function of frequency $\omega_{ex}$ is presented in Fig. 2.59 for fixed coupling coefficient $\gamma = 0.01$ and external signal amplitude $C = 0.08$. As seen, for any $j$th oscillator this dependence possesses a horizontal part (step) that corresponds to $W_j = 1$.

Without external force ($C = 0$) the chain being considered demonstrates a robust spatially homogeneous chaotic regime of oscillations for which $x_j(t) = x_k(t)$, $y_j(t) = y_k(t)$, $z_j(t) = z_k(t)$ ($j, k = 1, 2, \ldots, m$). The external force applied to the first element of the chain induces the appearance of inhomogeneity. In the phase locking region the mean frequencies of all oscillators of the chain are equal to the external signal frequency and the instantaneous phases are locked. However, synchronous chaotic oscillations are not spatially homogeneous.

Figure 2.60 illustrates $(x_1, x_j)$ projections of phase portraits of synchronous and nonsynchronous chaotic oscillations for several values of spatial coordinate $j$. Groups of (a) and (b) in the figure correspond to global phase synchronization of the chain at the external force frequency $\omega_{ex} = 0.992$. It can be mentioned that the time-average phase shift $\langle \Phi_1(t) - \Phi_j(t) \rangle$ between synchronous chaotic oscillations of the first and $j$th oscillators of the chain is finite but different for various $j$ values. For example, projections for $j = 9, 11, 13$ combined in group (a) indicate the presence of average phase shift that is close to $\pm 2\pi k$, $\pi/2 \pm 2\pi k$, $\pi \pm 2\pi k$, ($k = 0, 1, 2, 3, \ldots$), respectively. The form of projections for $j = 42, 44, 47$ collected in group (b) virtually repeats the one for the group (a) projections. Such a behavior can be explained by the fact that the average phase shift $\langle \Phi_1(t) - \Phi_j(t) \rangle$ either monotonically grows or decreases (depending on the value of $\omega_{ex}$) along the chain. Thus, one can claim that a running phase wave is propagated along the chain of chaotic oscillators. Projections of phase portraits of nonsynchronous oscilla-

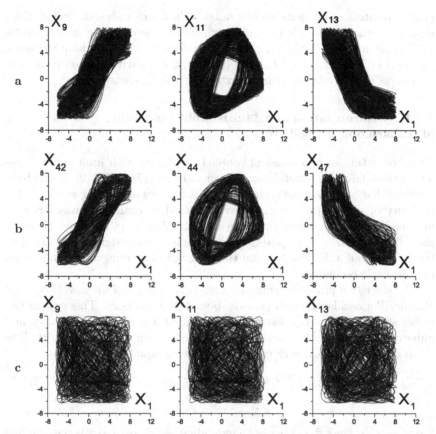

**Fig. 2.60.** Projections of synchronous ((a) and (b) groups) and nonsynchronous (group (c)) oscillations for different values of spatial coordinate $j$ and for $C = 0.1$ and $\gamma = 0.006$

tions obtained for $\omega_{\mathrm{ex}} = 1.025$ are shown in Fig. 2.60c. They differ noticeably from the projections corresponding to synchronous oscillations.

Chaotic dynamics of the chain elements was identified from the form of phase projections and by Lyapunov characteristic exponents. When the elements are unidirectionally coupled in the chain, each successive element can be individually considered as an nonautonomous oscillator subjected to a complex signal. Thus, the Lyapunov characteristic exponent (LCE) spectrum is divided into partial exponent spectra produced by individual elements of the chain. The presence of a positive exponent in the LCE spectrum of the first oscillator already indicates a chaotic behavior of all the subsequent oscillators. They, in turns, may add their positive exponents (regime of hyperchaos) into the total LCE spectrum or not (in the case of complete or general synchronization [189,190,198]). For the parameter values chosen, the chain can exhibit a hyperchaotic regime both in the case of frequency–phase synchronization of all elements and outside the synchronization region. Each element of the chain adds its positive characteristic exponent into the total LCE spectrum. Inside the synchronization region of all the chain elements, the total LCE spectrum contains $m$ positive, $2m$ negative, and only one zero exponents that is related with the external periodic signal.

### 2.4.7 Synchronization and Multistability in a Ring of Anharmonical Oscillators

Synchronization in ensembles of coupled oscillators with limit cycles is referred to as one of the well-known classical tasks [278, 317–319]. It has been revealed that such systems typically exhibit regimes of running waves. In the majority of papers these regimes were studied in chains of quasiharmonic or phase oscillators, their existence and stability were analyzed theoretically [283, 320–323]. It appears to be important and interesting to explore the influence of anharmonicity on the behavior of running waves in a chain of limit cycle oscillators.

Consider this problem using as an example a chain of diffusively coupled Van der Pol oscillators with periodic boundary conditions. This system can demonstrate both nearly harmonic oscillations for small nonlinearity and anharmonical and relaxational ones for larger values of nonlinearity. The chain of oscillators is described by the following equation:

$$\ddot{x}_i - (\varepsilon - x_i^2)\dot{x}_i + x_i - \gamma(\dot{x}_{i+1} - 2\dot{x}_i + \dot{x}_{i-1}) = 0, \qquad (2.117)$$

$$i = 1, 2, \ldots, N ; \qquad x_1 = x_{N+1}, \ \dot{x}_1 = \dot{x}_{N+1}.$$

Parameter $\varepsilon$ defines dynamics of an individual oscillator, and $\gamma$ is the coupling coefficient. For zero coupling, the chain of oscillators demonstrates periodic oscillations with equal amplitudes and arbitrary initial phases. As the chain is closed in a ring, the total phase shift along the chain must be proportional to $2\pi$:

$$\sum_{i=1}^{N}(\varphi_{i+1} - \varphi_i) = 2\pi k, \qquad k = 0, 1, 2, 3, \ldots.$$

We deal with regimes having an equal phase difference between oscillations of neighboring oscillators, i.e., running waves propagating directly with a constant phase velocity along the ring ($\Delta\varphi \geq 0$). Since the extended system is discrete, the wavelength must be integer and is defined as $\Lambda_k = 2\pi/\Delta\varphi_k = N/k$. The values of $k$ can change from 0 to $N/2$. For definiteness, set $N = 30$. In this case we have $k = 0, 1, 2, 3, 5, 6, 10, 15$. The value of $k = 0$ corresponds to a spatially homogeneous regime while the other values of $k$ associate with regimes of running waves with wavelength $\Lambda_k = 30, 15, 10, 6, 5, 3, 2$.

Consider the regions where the aforementioned wave modes exist and are stable, provided each oscillator exhibits quasiharmonic oscillations. This task can be solved analytically by using standard methods (see, for instance,

[320,322]). Our case is conveniently studied by passing to truncated equations for amplitudes and phases:

$$\dot{\rho}_i = \frac{\varepsilon}{2}\rho_i - \frac{\rho_i^3}{8} + \frac{\gamma}{2}(\rho_{i+1}\cos(\varphi_{i+1} - \varphi_i) + \rho_i\cos(\varphi_i - \varphi_{i-1}) - 2\rho_i),$$

$$\dot{\varphi}_i = \frac{\gamma}{2}\left(\frac{\rho_{i+1}}{\rho_i}\sin(\varphi_{i+1} - \varphi_i) - \frac{\rho_{i-1}}{\rho_i}\sin(\varphi_i - \varphi_{i-1})\right), \qquad (2.118)$$

$$\rho_1 = \rho_{N+1}, \qquad \varphi_1 = \varphi_{N+1}.$$

As all elements of the ring are identical, in Eqs. (2.118) one can assume that all amplitudes are equal to $\rho$ and phase differences between any neighboring oscillators are equal to $\Delta\varphi_k$. Then $\rho$ and $\varphi_i$ obey the following equations:

$$\dot{\rho}^{(k)} = \frac{\acute{\varepsilon}_k}{2}\rho^{(k)} - \frac{(\rho^{(k)})^3}{8},$$

$$\varphi_{i+1} - \varphi_i = \Delta\varphi_k = 2\pi\frac{k}{N} = \text{const.}, \qquad (2.119)$$

where $\acute{\varepsilon}_k = \varepsilon - 2\gamma(1 - \cos\Delta\varphi_k)$, index $k$ denotes that the solution represents a wave with index $k$. The amplitude equation (2.119) is identical in form with a truncated equation of the single Van der Pol oscillator. Correspondingly, as in case of a single oscillator, when the parameter $\acute{\varepsilon}_k$ passes through zero, the solution $\rho^{(k)} = 0$ loses its stability that conforms to the Andronov–Hopf bifurcation for an equilibrium at the origin in the initial system (2.117). The resulting new solution $\rho^{(k)} = 2\sqrt{\acute{\varepsilon}_k}$ corresponds to a running wave with a given value of index $k$. Thus, lines $\varepsilon = 2\gamma(1 - \cos\Delta\varphi_k)$ are bifurcational. With increasing parameter $\varepsilon$, the bifurcational condition $\varepsilon = 0$ is first satisfied that results in a spatially homogeneous regime, $\Delta\varphi = 0$. Then, at $\varepsilon = 2\gamma(1 - \cos(2\pi/N))$ the bifurcational condition is fulfilled for regimes with $k = \pm 1$. They are associated with waves running in opposite directions and having the length $\Lambda_1 = N$. Similar bifurcations occur for the overall sequence of possible indexes $k$. With this, new regimes that appear are unstable (except the homogeneous state). Finally, a pair of unstable running waves with $\Lambda_{15} = 2$ arises at the line $\varepsilon = 4\gamma$. Hence, a stable fixed point is born on the line $b_0$ (see Fig. 2.61) and corresponds to a spatially homogeneous regime $x_i(t) = 2\sqrt{\varepsilon}\cos(t)$. Saddle fixed points appear on the lines $b_1$–$b_{15}$ and are associated with running along the ring self-sustained waves $x_i(t) = 2\sqrt{\acute{\varepsilon}_k}\cos\left(t - \frac{2\pi k}{N}i\right)$. These spatially periodic regimes can become stable for certain values of the controlling parameters. The stability conditions can be derived by means of a standart method from the first approach. The application of this method is described in [322] for a ring of harmonic oscillators. Analyzing the system (2.118) stability one can obtain the eigenvalues for a linearization matrix that are as follows:

$$\lambda_{0,1}^{(k)} = 0, \qquad (2.120)$$

$$\lambda_{0,2}^{(k)} = -\acute{\varepsilon}_k, \qquad (2.121)$$

$$\lambda_{i,1}^{(k)} = -\frac{\acute{\varepsilon}_k}{2} + \sqrt{\frac{\acute{\varepsilon}_k^2}{4} + \gamma^2 \sin^2(\Delta\varphi_k)\sin^2\left(\frac{2\pi i}{N}\right)}$$
$$- \gamma\cos(\Delta\varphi_k)\left(1 - \cos\left(\frac{2\pi i}{N}\right)\right),\tag{2.122}$$

$$\lambda_{i,2}^{(k)} = -\frac{\acute{\varepsilon}_k}{2} - \sqrt{\frac{\acute{\varepsilon}_k^2}{4} + \gamma^2 \sin^2(\Delta\varphi_k)\sin^2\left(\frac{2\pi i}{N}\right)}$$
$$- \gamma\cos(\Delta\varphi_k)\left(1 - \cos\left(\frac{2\pi i}{N}\right)\right),\tag{2.123}$$

$$i = \frac{1}{N-1}.$$

The zeroth eigenvalue $\lambda_{0,1}^{(k)}$ results from the fact that system (2.118) solutions are invariant with respect to the choice of the initial phase. The transition of eigenvalues $\lambda_{0,2}^{(k)}$ through zero corresponds to the fixed point birth from the bifurcation of the equilibrium at the origin.

Analysis of expressions (2.122) and (2.123) yields the following conditions for regions of stability of running waves in the chain of harmonic oscillators:

$$\varepsilon > \gamma\left[2 - 4\cos(\Delta\varphi_k) + \frac{1 + \cos(\frac{2\pi}{N})}{\cos(\Delta\varphi_k)}\right], \qquad \cos(\Delta\varphi_k) > 0.\tag{2.124}$$

The associated birth and stability lines for the spatially periodic regimes under consideration are constructed in Fig. 2.61. The line $\varepsilon = 0$ separates regions where stable long-wave $C^0 - C^6$ ($\gamma > 0$, $\cos(\Delta\varphi_k > 0)$) and short-wave $C^{10}$, $C^{15}$ ($\gamma < 0$, $\cos(\Delta\varphi_k < 0)$) regimes exist.

First consider the region where the coupling parameter takes positive values. The solid line $b_0$ corresponds to the birth of a stable regime of spatially homogeneous oscillations. Unstable oscillatory regimes with different spatial periods are born on lines shown dashed in Fig. 2.61 and marked by letters $b_k$, where $k = 1, 2, 3, \ldots$ is the index of an appropriate running wave. On lines $s_1$, $s_2$, $s_3$, $s_5$, and $s_6$ regimes with relevant values of $k$ become stable. Thus, inside the cone enclosed by the lines $\gamma = 0$ and $s_6$ and shown in the figure in dark, there exist stable regimes of running waves $C^0$, $C,^1$ $C^2$, $C^3$, $C^5$, and $C^6$ with wavelengths $\Lambda = \infty$, 30, 15, 10, 6, and 5, respectively. In the region between the lines $s_6$ and $s_5$ regime $C^6$ is no longer stable that results in the occurrence of five coexisting waves $C^0$, $C,^1$ $C^2$, $C^3$, $C^5$. Furthermore, after line $s_5$ four regimes $C^0$, $C,^1$ $C^2$, $C^3$ remain stable, and so on. In the parameter space an embedding structure is formed that consists of the regions where running waves with different wavelengths are stable. The shorter the wavelength, the closer the stability region of appropriate oscillatory regime is located to the line $\gamma = 0$. For strong coupling, i.e., below the line $s_1$, the system can demonstrate only a spatially homogeneous oscillatory regime.

When the coupling is negative ($\gamma < 0$), all the above considered regimes become unstable. On the line $s_{15}$ a stable regime with wavelength $\Lambda = 2$ is

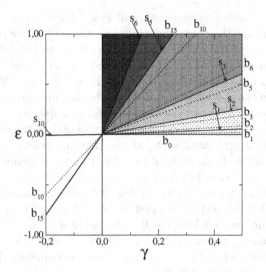

**Fig. 2.61.** Bifurcation diagram of the regimes in system (2.118). Solid line $b_0$ corresponds to the birth of a spatially homogeneous regime of oscillations; dashed lines $b_k$ mark the region boundaries where regimes of running waves with wave number $k$ exist; solid lines $s_k$ indicate the boundaries of regions inside which the corresponding wave regimes are stable. The regions of stable regimes of running waves are shown in gray: the darker background corresponds to the area where a shorter-wave regime is stable

born from the equilibrium at the origin when neighboring oscillators oscillate in antiphase, $x_i = -x_{i+1} = -x_{i-1}$, $\dot{x}_i = -\dot{x}_{i+1} = -\dot{x}_{i-1}$. This kind of wave is observed for any values of the parameters $\varepsilon > -4\gamma$, $\gamma < 0$. It can be easily seen that antiphase oscillations emerge for negative values of the parameter of excitation $\varepsilon$ when a single oscillator does not oscillate yet. Then, on the line $b_{10}$ ($\gamma < 0$) and in the vicinity of the already unstable equilibrium state a regime $C^{10}$ with wavelength $\Lambda = 3$ appears at which phase differences between neighboring oscillators are equal to $2\pi/3$. The given regime becomes stable above the line $s_{10}$.

The two-parameter analysis carried out for stability regions of harmonic running waves has shown that the considered system is multistable. The parameter space consists of overlapping leaves each of which is unbounded on the parameter $\varepsilon$ but is bounded on the coupling $\gamma$. When $\varepsilon$ is fixed and the coupling increases, one can detect typical transitions from shorter-wave regimes to longer-wave ones, which are described in [323]. The absence of restrictions for stability regions on the parameter of excitation leads to the fact that for any $\varepsilon > 0$ a sequence of these transitions appears to be the same, i.e., one-parameter analysis of stability is sufficient for harmonic waves. However, this conclusion is not valid for anharmonic waves.

We now turn to the analysis of the original system (2.117) behavior. With increasing $\varepsilon$, the oscillation amplitudes grow and the form of oscillations also changes that is resulted in generation of new harmonics in the oscillation spectrum. With this, the truncated equations are inadequate properly to describe the system dynamics. The regions of existence and stability of running wave regimes in system (2.117) were defined by numerical methods. The Poincaré section is best suited for our study. Dynamical variable values of all the oscillators in the ensemble $\{x_i\}$ are fixed in the time moments when time derivative of one of the dynamical variables, for example, $\dot{x}_1$, is equal to zero. By this means one can get instantaneous "snapshots" of the oscillator states , i.e., the profile of a running wave, at the time moments corresponding to a certain phase of the basic element ($x_1$ in our case). Regimes with different wavelengths are presented in Fig. 2.62.

The oscillatory regimes shown in Fig. 2.62 do not exhaust all possible stable regimes that can be found in system (2.117). In particular, it is possible to detect running waves whose oscillation amplitudes and phase difference between neighboring oscillators are not constant along the chain. This is typical for the cases when the chain length is not divisible by the wavelength of the regime under consideration.

The numerical results for stability regions of running waves are shown in Fig. 2.63 for positive values of the coupling coefficient. The wave regimes presented in Fig. 2.62 are born strictly at the same lines $b_1$–$b_{15}$ as their associated stationary states in truncated equations (2.118) (see Fig. 2.61). This is so indeed since the oscillations are harmonic at the moment of their birth and, consequently, are described by truncated equations. However, with increasing $\varepsilon$ the form of oscillations becomes anharmonic, and the oscillation spectrum exhibits harmonics at multiple frequencies. Thus, as we recede from the generation threshold, the initial system of Van der Pol oscillators (2.117) behaves differently than the simplified model (2.118). Correspondingly, the stability regions for running waves, presented in Fig. 2.63, qualitatively differ from the corresponding regions depicted in Fig. 2.61. For comparison, in Fig. 2.63 the boundaries of stability regions $s_1$–$s_6$ are duplicated from Fig. 2.61 as dashed. If to compare Figs. 2.61 and 2.63, one can see that unlike harmonic oscillators, each wave regime, except the spatially homogeneous one, in the chain of full equations possesses a finite stability region. For weak coupling, the lower boundaries of these regions (lines $l_1$–$l_6$) nearly coincide with relevant boundaries ($s_1$–$s_6$) for system (2.118). A difference begins for large values of $\gamma$ and is especially well marked in the case of short-wave regimes. Consider, for example, the stability region boundaries (lines $l_5$ and $l_6$) for a running wave with length $\Lambda = 6$. Before coupling $\gamma \simeq 0.5$, the lower boundary of stability of the considered regime in (2.117) nearly matches the dashed line $s_5$. However, for a strong coupling, it deviates significantly from the straight line. Besides, for each of the studied regimes, except the spatially homogeneous one, there is a borderline bounding the stability region above. The upper and lower boundaries of the stability region, together with the bifurcational line $\gamma = 0$, close the stability region for the running wave under question. Figure 2.63 shows a

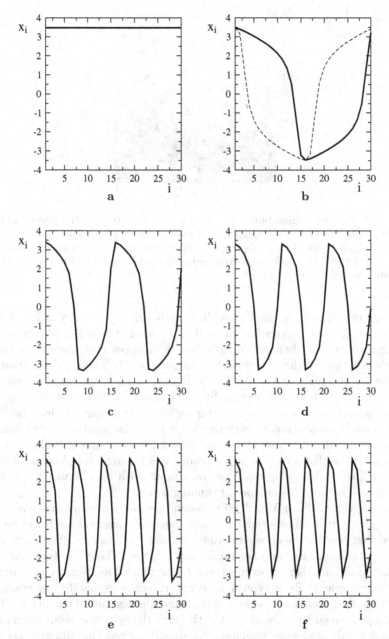

**Fig. 2.62.** Oscillatory regimes in system (2.117). (**a**) Spatially homogeneous oscillations and regimes of running waves propagating directly with wavelength $\Lambda = 30$ (**b**), $\Lambda = 15$ (**c**), $\Lambda = 10$ (**d**), $\Lambda = 6$ (**e**), and $\Lambda = 5$ (**f**). The order numbers of oscillators in the chain (discrete spatial coordinate) are laid off as abscissa, and the value of a dynamical variable of the given oscillator in the section $\dot{x}_1 = 0$ is laid off as ordinate. The dashed line in (**b**) shows the wave profile propagating in an opposite direction

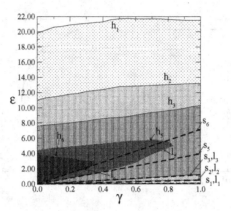

**Fig. 2.63.** Stability region boundaries for running wave regimes in system (2.117). $l_k$ and $h_k$ are the lower and upper boundaries of stability region for the wave $C^k$. Stability regions are shown in gray; a darker color corresponds to shorter-wave regimes. Dashed lines mark the corresponding boundaries of stability regions $s_1$–$s_6$ for harmonic oscillators (2.118)

fragment of stability regions for coupling coefficient $0 \le \gamma \le 1$, which includes the entire stability regions for waves $\Lambda = 5$ and $\Lambda = 6$. Stability regions for longer wave regimes behave similarly but their upper and lower boundaries close when the coupling coefficient is larger than 1. Thus, the diagram of regimes for the chain of Van der Pol oscillators (2.117) represents an embedding structure of bounded regions. Stability regions of shorter waves regimes are arranged inside stability regions for regimes with larger wavelengths. The lower boundaries of stability regions emerge from the point with coordinates $\gamma = 0$ and $\varepsilon = 0$ and represent straight lines for weak coupling (as in the case of harmonic oscillators). The upper boundaries for weak coupling are nearly straight lines each issuing out of its own point with coordinates $\gamma = 0$ and $\varepsilon_u^k$, where $k$ is the wave number of running wave.

Hence, the chain of Van der Pol oscillators demonstrates the developed multistability in a definite range of controlling parameter values. It can be assumed that besides the above described running wave modes, the system can exhibit simultaneously another oscillatory regimes. Therefore, the following regular equation arises: How much the above mentioned regimes of running waves are typical? By a typical aspect is meant a probability of reaching one or another oscillatory regime at randomly defined initial conditions. The studies performed have shown that in this case the system demonstrates only the above indicated wave regimes. Probabilities of reaching different spatial modes are plotted in Fig. 2.64. Clearly the parameter values were chosen so that all the above mentioned modes are stable. Probabilities of two longest wave regimes with $k = 0$ and $k = 1$ are practically equal and close to 0.5. It is precisely these modes that the system reaches in the majority of cases. With increasing $k$ the probability for other regimes decreases nearly expo-

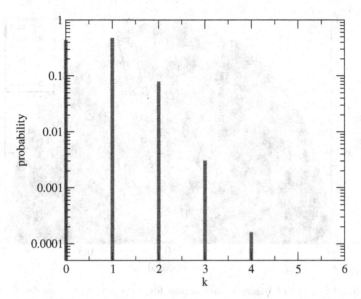

**Fig. 2.64.** Probability of reaching spatiotemporal regimes corresponding to different values of index $k$ at initial conditions $x_i$ randomly chosen within the interval $x \in [0.1]$. The other parameters are $\varepsilon = 2.0$ and $\gamma = 0.1$

nentially (for a more illustrative presentation the $y$-axis is logarithmically scaled). Regimes with $k = 5$ and $k = 6$ have not been reached in our numerical simulation. As can be assumed from Fig. 2.64, their probability is approximately one and two orders, respectively, less than the probability for the regime with $k = 4$. Unfortunately, the calculation time taken in this case is beyond the possibilities of computer experiments. A similar dependence of the probability on index $k$ are also observed for different values of parameters $\varepsilon$ and $\gamma$. For large values of coupling, the probability of reaching the spatially homogeneous regime starts to slightly prevail over the probability of reaching the regime with $k = 1$.

Of particular interest is to define the basin of attraction of an attractor associated with one or another spatiotemporal regime. Clearly that since the system is of a high dimension, it is practically impossible to construct complete basins of attraction. Regions of attraction can be certainly estimated if initial conditions are chosen with a specific phase shift with respect to the spatial coordinate:

$$x_i = \rho \sin(i\phi), \qquad y_i = \rho \cos(i\phi).$$

Varying $\rho$ and $\phi$ one can get a characteristic diagram for basins of attraction, which is shown in Figs. 2.65 and 2.66 for two different values of the coupling coefficient.

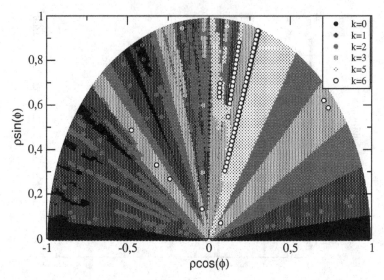

**Fig. 2.65.** Diagram of basins of attraction for different spatial modes ($\varepsilon = 2.0$, $\gamma = 0.1$)

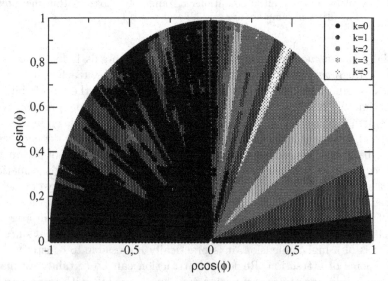

**Fig. 2.66.** Diagram of basins of attraction for different spatial modes ($\varepsilon = 2.0$, $\gamma = 0.42$)

As the coupling grows, zones of short-wave modes are narrowed or completely disappear (the mode with $k = 6$ in Fig. 2.66), while zones of long-wave regimes are correspondingly widened. It is seen from the diagram that when angle $\phi$ values are less than $90°$, the regions of attraction of different modes are structured regularly enough: they fill the sectors adjacent to the angle value that is equal to the phase difference between oscillations of neighboring oscillators for a given regime. For angle values less than 90°, stability regions behave less predictable.

If the parameter of nonlinearity is chosen sufficiently large ($\varepsilon \sim 10$), then in parallel with the already described regimes of running waves one can observe running waves with moving "defects." Such defects reflect phase and amplitude slips (i.e., a spatial point and the number of oscillators for which the phase difference differs from a characteristic value for a given mode) that move along the ring with a constant velocity. One of such examples is shown in Fig. 2.67. In the presence of one defect, the temporal dynamics represents two-frequency quasiperiodic oscillations. If there are several defects, then they move with their own velocity and a multidimensional torus is accordingly formed in the phase space. If the chain of oscillators is long enough,

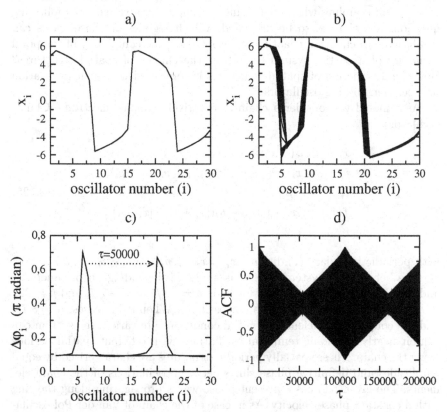

**Fig. 2.67.** (a) Initial wave without "defect" with $k = 2$, (b) wave with phase "defect," (c) dependence of the phase difference between oscillations of neighboring oscillators in two different time moments, and (d) autocorrelation function computed from time series of one of the oscillators. The parameters are $\varepsilon = 10, \gamma = 0.01$

then the resulting tori can be of a high dimension. In this case the presence of weak noise can lead to the behavior being very similar to a chaotic one.

The regions of parameter space where regimes with phase defects can be observed exist for large nonlinearity and weak coupling. To reach a regime with phase defects that is formed from wave $C^k$ it is necessary to choose parameter values when a "pure" regime is stable and to cross the upper stability boundary of the initial wave by increasing the parameter $\varepsilon$. The transition between the "pure" regime and a regime with defects can be accompanied by hysteresis, i.e., both regimes coexist in a certain region of the parameter space.

### 2.4.8 Synchronization and Multistability in a Ring of Oscillators with Period Doubling

In this part we will study the evolution of phase waves in a ring of oscillators with period doubling when controlling parameters are varied. The following questions are planned to be discussed. Which regimes of phase waves can be observed in chains of chaotic oscillators? How the complexity of temporal dynamics of a partial system can affect the change of spatial structures? How the destruction of spatial structures is connected with synchronization between neighboring oscillators?

As a model we consider a chain of resistively coupled electronic Chua's oscillators

$$
\begin{aligned}
\dot{x}_i &= \alpha(y_i - x_i - f(x_i)), \\
\dot{y}_i &= x_i - y_i + z_i + \gamma(y_{i-1} + y_{i+1} - 2y_i), \\
\dot{z}_i &= -\beta y_i, \\
f(x_i) &= b x_i + 0.5(a - b)(|x_i + 1| - |x_i - 1|), \\
i &= 1, 2, \ldots, N
\end{aligned}
\tag{2.125}
$$

with periodic boundary conditions, $x_1 = x_{N+1}, y_1 = y_{N+1}, z_1 = z_{N+1}$.

The behavior of system (2.126) is analyzed depending on parameter $\alpha$ and coupling coefficient $\gamma$ values for fixed $a = -8/7$, $b = -5/7$, and $\beta = 22$.

At $\alpha < 8.78$, the system does not oscillate. When $\alpha$ increases, $\alpha \geq 8.78$, and the coupling coefficient and initial conditions are varied, the system can exhibit nearly harmonic temporal oscillations with different spatial periods along the chain. These spatially periodic structures are characterized by equal amplitudes and definitive phase shifts between oscillations of the chain elements. The given regimes represent phase waves propagating along the ring with a constant phase velocity. As in case of the chain of Van der Pol oscillators, the Poincaré section will be used to visualize spatial structure of phase waves. For this purpose, values of dynamical variables $x_i$ of each oscillator are fixed at the time moment when the variable $y_1$ crosses a zero level from positive values to negative ones.

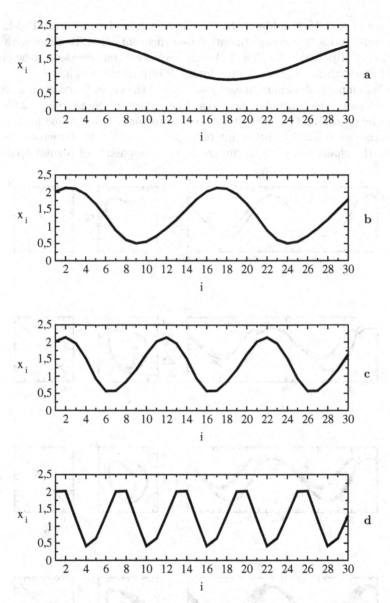

**Fig. 2.68.** Simplest spatial structures with different wavelengths: (a) $n = 1$, (b) $n = 2$, (c) $n = 3$, and (d) $n = 5$

The simplest structures with different spatial periods are shown in Fig. 2.68. These states correspond to phase shifts between oscillations of neighboring oscillators by $\pi/15$, $2\pi/15$, $\pi/5$, and $\pi/3$. In our case (the chain of 30 oscillators), among the possible structures, one can also observe regimes with

$n = 6$ and $n = 15$ in addition to the aforementioned structures. The latter case corresponds to a complete antiphase synchronization. Regimes with the given spatial periods, $\Lambda = 5$ and $\Lambda = 2$, have not been revealed in the chain under study within the considered intervals of parameter values.

The temporal dynamics in each oscillator of the chain becomes more complicated as the parameter $\alpha$ increases. The states presented in Fig. 2.68 undergo bifurcations. Period doubling and torus birth bifurcations for regular attractors as well as band-merging bifurcations for chaotic attractors can occur in the chain. Figure 2.69 illustrates the complexity of regular spatially

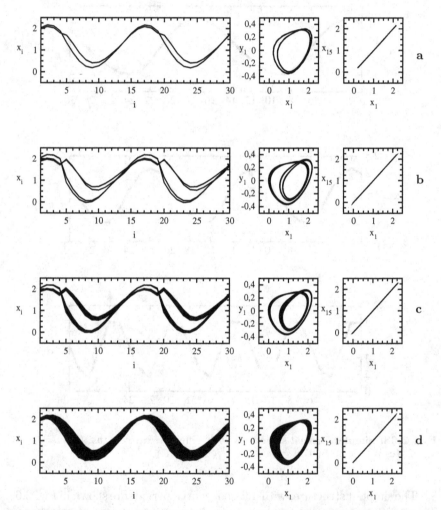

**Fig. 2.69.** Complexity of regular phase waves with increasing parameter $\alpha$: (a) $\alpha = 1.14, \gamma = 0.02$, (b) $\alpha = 11.48, \gamma = 0.02$, (c) $\alpha = 11.54, \gamma = 0.1$, and (d) $\alpha = 11.4, \gamma = 0.2$

periodic regimes with increasing $\alpha$. One can see the evolution of regimes from the initial condition with spatial period $\Lambda = N/2$ (Fig. 2.68b). When phase waves are complicated in each oscillator, the temporal dynamics also becomes more complicated. Period-two cycle $2C_{N/2}$ (Fig. 2.69,a), period-four cycle $4C_{N/2}$ (Fig. 2.69b), doubled torus $2T_{N/2}$ (Fig. 2.69,c), and torus $1T_{N/2}$ (Fig. 2.69d) can be observed. Figure 2.69 presents spatial structures and projections of phase portraits on the planes $(x_1 - y_1)$ and $(x_1 - x_{16})$. The latter projection shows that the regular regimes listed are precisely spatially periodic, $x_i = x_{i+15}$. It is seen that the increase of the temporal period does not cause the spatial period to change, and temporal quasiperiodicity does not induce spatial quasiperiodicity. In this case temporal regimes are developed from the initial spatial structure that is destroyed only after transition to a temporal chaotic behavior of the oscillators.

The evolution of chaotic regimes are depicted in Fig. 2.70 where four-band (Fig. 2.70a), two-band (Fig. 2.70b) and one-band (Fig. 2.70c) attractors can be seen. Temporal chaotic regimes are not spatially periodic that can be seen from the mutual projection $(x_1 - x_{16})$ that is no longer a thin line on the bisectrix and has a finite "thickness." Projections of phase portraits for multiband chaotic attractors are sufficiently thin and are located in the vicinity of bisectrix, i.e., these regimes remain nearly periodic in space. As the chaos develops, the projections become "thicker" and the structures are no longer spatially periodic. In the regime of a developed one-band chaotic attractor, the phase trajectory does not stay in the vicinity of bisectrix and fills the whole square (Fig. 2.70c). The mutual projections of phase portraits and spatial diagrams demonstrate that chaotic oscillations are no longer spatially periodic. However, as will be shown below, they remain spatially periodic on the average.

Stability of phase waves depends on their wavelength and the coupling strength. The increase of coupling coefficient $\gamma$ can cause the stability loss by short-wave modes and the transition to longer wave regimes.

Figure 2.71 shows stability regions for a set of regimes originating from the phase waves presented in Fig. 2.68. When the upper boundaries marked by symbols are crossed, the relevant spatial structure is destroyed as a result of the development of temporal chaotic dynamics. When passing through the right boundaries indicated by lines a given regime loses its stability and a sharp transition to a spatiotemporal regime with a larger wavelength is realized. A set of regimes with spatial period $N$ (see Fig. 2.68a) is bounded by the line 1. It goes practically horizontal to the abscissa. Only a spatially homogeneous regime can exist below this line. Families of regimes with spatial periods $N/2$, $N/3$, and $N/5$ are bounded by the lines 2, 3, and 4, respectively. These regimes are observed to the left of the indicated lines. The system under study demonstrates the phenomenon of multistability. Beyond the line 4 families of regimes with wavelengths $N/5$, $N/3$, $N/2$, and $N$ coexist with spatially homogenenous states. Regimes with wavelengths $N/3$, $N/2$, and $N$

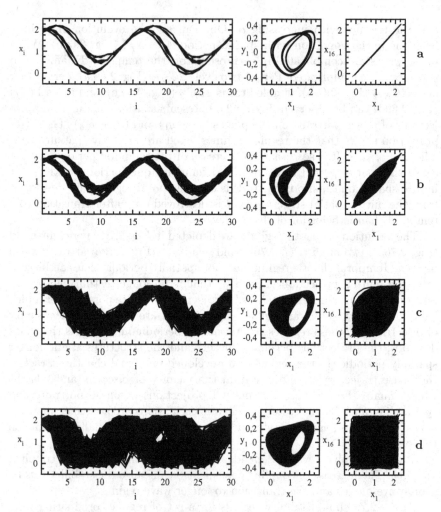

**Fig. 2.70.** Development of chaotic phase waves with increasing parameter $\alpha$ ($\gamma = 0.05$): (**a**) $\alpha = 11.57$, (**b**) $\alpha = 11.61$, (**c**) $\alpha = 11.66$, and (**d**) $\alpha = 11.79$

and spatially homogeneous regimes coexist inside the region between the lines 3 and 4. Spatially homogeneous regimes and modes with wavelengths $N/2$ and $N$ coexist between the lines 2 and 3. Spatially homogeneous regimes can be observed for any values of the coupling coefficient. It is seen that stability of phase waves depends on the coupling strength. With increasing coupling, regimes with small spatial periods lose their stability. Typical bifurcational transitions and a structure of the controlling parameter plane are described in detail in [324] for a family of spatially periodic regimes.

It has already been mentioned that the exact spatial periodicity is destroyed when passing to a temporal chaotic behavior. There is no exact pe-

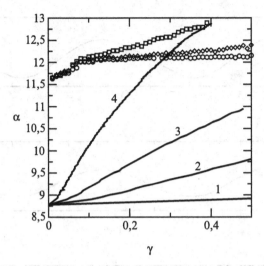

**Fig. 2.71.** Regions of stability for a family of regimes with different wavelengths

riodicity in the chaotic region but spatial structures retain their periodicity on the average in time. Averaged spatially temporal diagrams are presented in Fig. 2.72 when the coupling decreases.

It is seen that the averaged spatial periodicity of structures is gradually destroyed. The temporal behavior is associated with a one-band chaotic attractor. For strong coupling, the averaged spatial diagram looks like the initial strongly periodic regime. Then, with decreasing $\gamma$, the spatial structure has a more plane form but the initial picture with two maxima is still retained. Further for small values of coupling, the averaged spatial periodicity is destroyed only in the regime of developed temporal chaos.

From a viewpoint of mutual synchronization of oscillators, spatial periodicity of structures is broken down as follows. Before passing to a chaotic temporal behavior, complete synchronization takes place. These regimes are characterized by rigorous equalities $x_i(t) = x_{i+N/2}(t)$ and $x_i(t) = x_{i+1}(t+\tau_i)$ for each oscillator in the chain. With increasing parameter $\alpha$ when oscillations turn to be chaotic, these equalities become approximate and are then violated. It has been shown in [324] that in this case a quantitative measure of synchronization decreases from 1 (completely synchronous oscillations) to 0 (lack of synchronization). It has also been established that spatial periodicity is connected with coherence of oscillations at the basic spectral frequencies. While the oscillation spectrum contains the frequencies at which motions are completely coherent, the chain still possesses spatial structures being periodic on the average. If the coherence function decays for each frequency, the spatially periodic structure is destroyed.

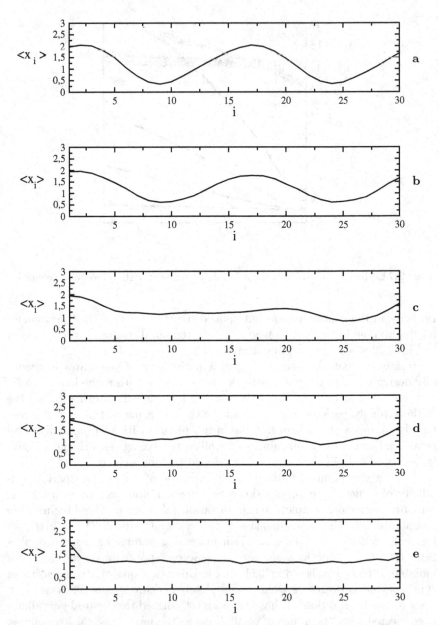

**Fig. 2.72.** Averaged spatial structures for $\alpha = 11.78$ and different values of $\gamma$: (**a**) 0.15, (**b**) 0.045, (**c**) 0.035, (**d**) 0.03, and (**e**) 0.01

## 2.4.9 Summary

In the presented section we have considered effects of synchronization in extended systems, such as chains of locally coupled self-sustained oscillators and a self-sustained oscillatory medium described by the Ginzburg–Landau equation. Particularly, formation of frequency clusters has been studied for inhomogeneous extended systems. It has been revealed that cluster structures in a chain with a finite number of elements qualitatively correspond to those in a medium with a continuous spatial coordinate. The chain and the medium demonstrate similar regimes. The influence of noise on the chain and the medium also leads to the same effects. Thus, a medium can be justly modeled by a chain of moderate length ($m = 100$) to study qualitatively the medium dynamics. It has been established that both in the chain of self-sustained oscillators and in the medium the transition to chaotic dynamics is accompanied by the destruction of perfect clusters.

We have also analyzed the influence of noise on inhomogeneous systems in the regime of cluster synchronization. Noise typically causes perfect clusters to destroy that is accompanied by transition to noise-induced chaos. The notion of effective synchronization introduced for a single Van der Pol oscillator and characterized by the effective diffusion coefficient can be generalized for the case of cluster synchronization in extended systems.

The studies performed have shown that the amplitude dynamics can play an essential role in forming cluster structures. Cluster synchronization can also be observed in a chain modeled by phase equations only. However, this effect is realized within a significantly narrow range of coupling parameter values. Besides, the appearing cluster structures are more sensitive toward the influence of noise.

The phenomenon of synchronization of cluster structures has been revealed and studied in interacting inhomogeneous self-sustained oscillatory media. This effect is based on a classical mechanism of frequency locking of relevant elements of the two media.

The effect of forced frequency–phase synchronization has been detected in a chain of unidirectionally and diffusively coupled chaotic oscillators. The region of global phase locking of the chain has been constructed on the plane of controlling parameters "external signal amplitude – external signal frequency."

The influence of anharmonicity on multistability of spatially periodic regimes has been explored in a ring of oscillators with limit cycles. It has been demonstrated that different wave modes can typically appear for random and spatially periodic initial conditions. The regularities of their disappearance have also been studied.

Dynamics of spatially periodic regimes has been analyzed in a ring of identical oscillators with period doubling. A map of typical regimes has been constructed on the plane of controlling parameters and typical bifurcational transitions have been studied.

# 2.5 Synchronization in Living Systems

## 2.5.1 Introduction

The phenomenon of synchronization considered above was studied in rather simple generic dynamical systems using methods of oscillation theory. Indeed, synchronization phenomena are ubiquitous in living systems [255,325–327,338]. Synchronization was observed and studied on various levels of biological systems, starting from processes in single cells and cell communications to collective dynamics of complex networks and population dynamics of biological organisms. Synchronization effects may have an important physiological significance. For example, synchronization is believed to play a crucial role in the processing of sensory information [328,329]. Neuronal synchrony has recently been suggested as a collective mechanism for intentional selection [330]. On the other hand, several human neurological disorders such as essential tremor and tremor caused by Parkinson's disease appear to be caused by abnormal synchronization of neuronal populations [331].

In this section we present examples of two experimental studies of synchronization of biological systems by external perturbations. Both systems are rather complex. The first system is represented by electroreceptors of paddlefish and is synchronized by external electric field. The second example is synchronization of human cardiorythms by external stimulation. Presently, no detailed mathematical models exist for these systems. Nevertheless, as we will see both systems show fundamental synchronization phenomena which can be understood in the framework of the theory of oscillations using simple generic models.

## 2.5.2 Stochastic Synchronization of Electroreceptors in the Paddlefish

The paddlefish (Fig. 2.73a), *Polyodon spathula*, named for its long flattened spatula-like appendage extending in front of the head, the "rostrum." The rostrum is covered with tens of thousands of sensory receptors [332], morphologically similar to the ampullae of Lorenzini of sharks and rays. Clusters of electroreceptors also cover the head and the gill covers. It was established that paddlefish uses electrosense to locate prey such as *Daphnia* during feeding behavior [333,334]. Electroreceptors in paddlefish form a passive sensory system, meaning that paddlefish only receive signals from external sources. An external opening (pore) in the skin, 80–210 $\mu$ diameter, leads into a short canal $\approx$200 $\mu$ long. The pores are organized into clusters of 5–30 on the rostrum (see enlarged part of the rostrum in Fig. 2.73). The internal end of each canal is covered with a sensory epithelium. The epithelium contains two types of cells: hair cells and support cells. It is the hair cells which are considered electrosensitive, named for their kinocilium projecting into the lumen of the ampulla. The number of receptor cells per epithelium is $\leq$400. On the

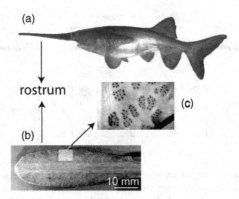

**Fig. 2.73.** (a) Photo of a small paddlefish, *Polyodon spathula*. (b) Photo of the underside of the rostrum from a 21 cm paddlefish. (c) Enlarged photo of clusters of electroreceptor pores

basal side of the epithelium, the hair cells make excitatory chemical synapses onto "primary afferent" axons. The synapse from each hair cell, together with the spiking properties of the primary afferent endings, converts the analog signal from the hair cells into spike trains (series of action potentials), coding the electrosensory information as a time series (the intervals between afferent spikes).

The term "electroreceptor" unifies a complex structure of pore + canals + epithelium + primary afferent neuron. A key feature of the spike trains recorded from the primary afferents of an electroreceptor is their spontaneous quasiperiodic firing pattern. It was shown that the electroreceptors in paddlefish possess a novel type of organization of being composed from two distinct types of oscillators [335]. One oscillator resides in a population of epithelial cells and is unidirectionally coupled with the second oscillator, located in the afferent terminal. The fundamental frequency of epithelial oscillator is 25–27 Hz at 22°C for different electroreceptors, while the mean firing rate for different afferents varies in a wide range of 30–70 Hz. The unidirectional coupling of these oscillators results in a specific biperiodic firing patterns.

*In vivo* electrophysiological experiments have been performed with paddlefish 35–40 cm in length. A detailed description of the experimental setup and procedures can be found in [335]. In brief, extracellular recordings were performed using tungsten microelectrodes. Stimulation with time-dependent external electric field was performed by placing a small dipole electrode near (5 mm) receptive field on the rostrum of the fish or by parallel plates placed at the ends of the epxerimental chamber.

Electroreceptors are tuned to respond to low frequency signals with the maximal frequency response at about 5 Hz [333]. Thus, we can expect that synchronization of higher frequency (30–70 Hz) afferent neurons by low frequency periodic stimuli should occur in the form of high order mode locking. The phase of the periodic signal is $\Psi(t) = 2\pi f_0 t$. Since recordings of neuron

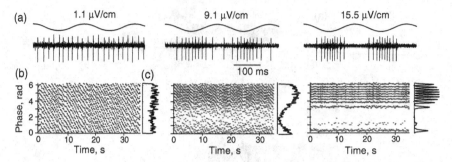

**Fig. 2.74.** Synchronization of an electroreceptor by 5 Hz electric field. Each of the three columns in the graph corresponds to the indicated values of the electric field amplitude. Each column shows: (**a**) electric field waveform and raw recordings from the afferent (below); (**b**) synchrograms, $\phi_k(t_k)$, and (**c**) distributions of the cyclic phase

activity are represented by stochastic point processes, it is natural to present the phase difference as a Poincaré "stroboscopic" map [344]: we calculate the phase of the stimulus at the moments of time, $t_k$, when the neuron fires and then define the result on a circle $[0 : 2\pi]$:

$$\phi_k = [\Psi(t_k) \bmod 2\pi] . \tag{2.126}$$

Plotting the "cyclic" phase, $\phi_k$ against the firing times $t_k$ results in so-called synchrograms plots [344].

A representative example of synchronization of an electroreceptor is shown in Fig. 2.74. For small amplitude electric field (1.1 $\mu$V/cm) the cyclic phase is distributed uniformly. With the increase of the amplitude (9.1 $\mu$V/cm) the firing frequency of the neuron became modulated. However, no synchronization occurs as position of individual spikes with respect to stimulus changes continuously from cycle to cycle. Situation changes with further stimulus increase (15.5 $\mu$V/cm) when individual spikes are locked to specific positions with respect to the stimulus period, demonstrating synchronization pattern of 1:10, that is there are exactly 10 spikes per period of stimulation. Indeed, synchronization is still stochastic due to noise in the system, which is indicated by finite width of the peaks in the phase distribution.

For afferent neurons with a low variability a different synchronization pattern was observed [336], when synchronization occurs without significant modulation of the firing rate.

Another interesting type of synchronization occurs when a population of electroreceptors are exposed to a common electric field. Individual electroreceptors are not coupled [332] and are characterized by a distribution of natural frequencies of their afferent neurons, as we noted above. Nevertheless, apparently electroreceptors shares a common slow time scale which is revealed by broad-band stimuli [337]. Noise electric field induces qualitative change in the firing pattern of afferent neurons, which changes from quasiperiodic to

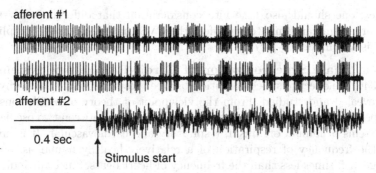

**Fig. 2.75.** Example of simultaneous recording from a pair of electroreceptor afferents stimulated with a computer generated exponentially correlated noise with the correlation time of 2 ms

bursting mode. Bursting mode is characterized by two well separated time scales: short interspike intervals inside bursts and long intervals of no firing between bursts. This slow time scale is similar for different electroreceptors and thus they can be synchronized by common noise stimulus. That is, common noise synchronizes onsets of bursts, while the infrastructure of bursts is still incoherent [337]. Figure 2.75 shows an example of noise-induced burst synchronization. A common broad-band noise stimulus was introduced to the whole population of electroreceptors and simultaenous recordings from a pair of electroreceptors afferents were obtained. Noise with large enough amplitude induced transition to bursting mode in both electroreceptors with consequent synchronization of onsets of bursts.

Small paddlefish use electrosensitivity to find and feed on individual zooplankton, which generates electric fields having prominent periodic low-frequency (5 to 12 Hz) components. The synchronization effects could be a possible mechanism that the animal uses to detect and track its target prey. However, further experiments are necessary to verify connections between the feeding behavior of the fish and electrophysiological data.

### 2.5.3 Synchronization of Cardiorhythm

Cardiovascular system (CVS) of a human is one of the examples of the most complicated nonlinear oscillatory system. CVS can be obviously treated as self-sustained that is proven by the existence of nondamping oscillations. At the same time, the mechanism of oscillations of CVS has not been completely understood yet. It is known that oscillations of the heart of a human (or animals) are not strictly periodic [338–340]. This can be a consequence of the influence of different noise sources on CVS as in the case of self-sustained oscillations of noisy Van der Pol oscillator that was considered in Sect. 1.3.4. However, there are reasons to assume that aperiodicity of heart oscillations can be principally stipulated by chaotic dynamics of CVS [338, 340, 341].

Besides, one should also take into consideration that different subsystems of an organism can consistently affect the CVS that significantly complicate analysis and modeling [327].

**Mutual Phase Synchronization of Cardio-Respiratory System.** In [344–346] the effect of synchronization between respiration and cardiorhythm was revealed and studied. From the viewpoint of theory of oscillations, we can speak about the effect of mutual synchronization between two oscillatory subsystems of human organism, respiratory and cardiovascular. It is known that the frequency of respiration of a relatively healthy human is, on the average, 3–5 times less than the frequency of heart beats. Since the indicated subsystems are naturally interconnected in a whole organism, synchronization phenomena between respiration and heart beats are observed when their frequencies become integer multiple.

Leaving the details of experiments described in [344–346] the following important results should be noted. First, it was established that phase synchronization of the cardio-respiratory system of a human takes place and corresponds to complex resonances 3 : 1, 5 : 2, and 8 : 3. Second, it was shown that the effect of phase and frequency locking occurs during finite epochs exceeding the average period of respiration and, consequently, the average duration of cardio-interval.

**Phase Synchronization of Cardiorhythm by External Periodic Forcing.** To study synchronization of cardiorhythm by external stimuli the following experiment protocol was proposed [342, 343]. A human was subjected to a weak external signal represented by periodic light flashes on a computer monitor and simultaneously by a weak sound coming from the computer speaker. These stimuli were derived from a periodic sequence of rectangular pulses with the frequency of periodic $f_F$ which was close to the average frequency of heart rate of the subject $\langle f_H \rangle$. The duration of pulses $\tau$ was $\simeq 10\%$ from the average duration of one cardio-interval (Fig. 2.76). The experimental scheme is illustrated in Fig. 2.77.

Effects of synchronization were studied by processing numerically the electrocardiogram (ECG). The ECG of the subject was registered both at rest and in the presence of external stimulation. The duration of stimulation was 300–600 s. Data collected from EEG and from the stimulus sequence were then analyzed by computing the instantaneous phase difference between the stimulus and the ECG of the subject. The calculations were performed with discrete sequences of RR-intervals of the ECG and by applying the instantaneous phase definition (1.295). Forty experiments were undertaken in which 16 young people without signs of any heart disease took part.

In Fig. 2.78a a representative temporal dependence of the $2\pi$ normalized instantaneous phase difference is presented for the case of 3% detuning between the average heart rate of the subject and the external forcing. The instantaneous phase difference remains bounded and close to zero or to an

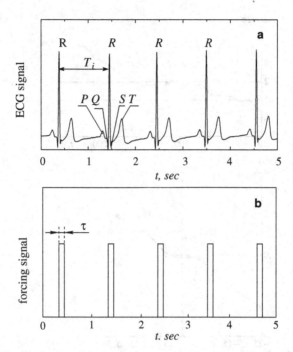

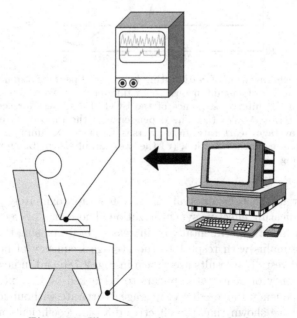

**Fig. 2.76.** (a) Typical human ECG. $T_i$ indicates a RR interval. (b) Rectangular pulses were converted to light flashes on a computer monitor and to sound clicks. $\tau$ indicates the duration of the stimulation pulse

**Fig. 2.77.** Illustration of experimental scheme

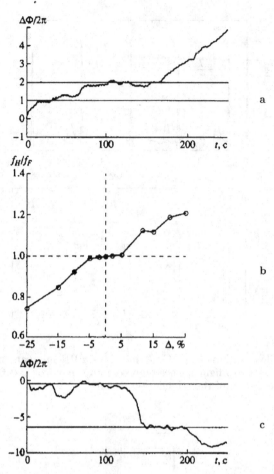

**Fig. 2.78.** Synchronization of cardiorythm by external periodic stimulus. (a) Representative example of instantaneous phase difference between external periodic stimulus and a RR interval sequence of the ECG of a subject for the case when $\Delta = |\langle f_H \rangle - f_F|/f_F = 3\%$. (b) The dependence of the ratio of external forcing frequency to average heart rate $f_H/f_F$ versus frequency detuning $\Delta$, indicating the effective phase synchronization, (c) the same graph as (a) but for the case of aperiodic forcing

integer of $2\pi$ in the time interval $\Delta t \simeq 150$ s (approximately 150 cardio-intervals), indicating effective synchronization. The region of frequency locking was determined experimentally by increasing or decreasing the frequency of external stimulus with respect to the average frequency of heart rate of the subject at rest. The results are shown in Fig. 2.78b and indicate that the average frequency of heart rate appears to be locked by the external signal within the frequency range ±5% of average heart rate without forcing. The calculations have shown that the effective diffusion coefficient of the phase difference $D_{\text{eff}}$ had a well-pronounced minimum being close to $10^{-2}$ in the synchronization region and grows abruptly when leaving the region.

**Phase Synchronization of Cardiorhythm by External Aperiodic Signal.** It is interesting to study whether the phase locking phenomenon can be observed when a subject is forced aperiodically. In our experiments an external signal applied to a subject was defined by the sequence of RR intervals of the ECG of another human subject. Moreover, the human subjects were chosen such that the average durations of RR intervals of the controlling and controlled subjects should differ more than 5%. This means that synchronization might occur not at the basic tone. Below we consider the results obtained in the case when the average frequency of heart rate of the controlling ECG was 1 Hz and the average heart rate of the controlled individual (without forcing) was 0.85 Hz.

If the average frequencies of controlling and controlled oscillatory processes are noticeably different, then the condition for phase synchronization obeys the following general relation:

$$\lim_{t \to \infty} |m\Phi_1(t) - n\Phi_2(t)| < M = \text{const.}, \tag{2.127}$$

where $m$, $n$ are integers, and $\Phi_{1,2}(t)$ are instantaneous phases of the compared oscillatory processes.

Typical experimental results are shown in Fig. 2.78c. The plot illustrates the temporal dependence of phase difference for $m = 7$ and $n = 6$. The phase difference is close to zero in the time interval $\Delta t \simeq 50$ s and does not exceed $2\pi$ in the interval $\Delta t \simeq 100$ s. Thus, we can speak about effective phase synchronization with $m : n = 7 : 6$. Synchronization by aperiodic forcing was verified in 19 of the 20 performed experiments. Note that from the viewpoint of condition (2.127), the results of the previous section is treated as effective synchronization on the basic tone when $m = 1$ and $n = 1$.

### 2.5.4 Summary

The presented results of the study of paddle-fish electroreceptors and of cardiorhythm of a human demonstrate that the nonlinear dynamics methods can be successfully applied to complicated systems that are quite difficult to be modeled or cannot be simulated at all. For example, our investigations convincingly show that effects of forced and mutual synchronization of cardiorhythms do not qualitatively differ from the classical phenomena of synchronization of van der Pol oscillator. This fact, particularly, proves that the cardiovascular system of a human organism can be treated as a self-sustained oscillatory system. This conclusion enables one to confidently use the nonlinear dynamics methods for deeper studies in experimental physiology and other sciences.

# 2.6 Controlling Chaos

## 2.6.1 Introduction

Until recently, chaos was associated with absolutely unpredictable and uncontrolled processes and the combination of words "chaos control" sounded nonsensical. But over the last decade this view has fundamentally changed. Chaotic systems are highly susceptible to controlling perturbations and include a wider range of possible regimes of oscillations as compared with systems exhibiting regular dynamics only. To significantly change the dynamics of a nonchaotic system one needs, as a rule, to change considerably the conditions of its performance. In a system which behaves chaotically, the same results can be achieved by making small specifically defined controlling perturbations. In addition, a chaotic attractor has embedded densely within it a countable set of unstable periodic orbits. This fact provides an unbounded choice of possible regimes of system function. Small perturbations enable not only the transitions between these states but also the duration of transient processes to be controlled. All the advantages listed are first conditioned by the structure and properties of chaotic attractors. Chaotic oscillations are encountered very often and can occur in various systems. In practice, however, it is often desirable for chaos to be avoided and/or the system performance to be improved or changed in some way. Thus, the ability to control chaos is of much practical importance. This problem is one of the typical ones in control theory formulated for chaotic systems.

Controlling chaos is often associated with the problem of suppression of chaotic oscillations, i.e., transformation of a system into either stable periodic motions or equilibrium states. In a wider sense, the control of chaos consists in converting the chaotic behavior of the system to a periodic or chaotic motion with different properties.

The problem of control of chaos was first stated in the papers of Hübler and Lüscher [370] and Jackson [371,372], and in what has become the classical paper of Ott, Grebogi, and Yorke (OGY) [373]. References to earlier papers on the chaos control problem may be found in the review [374]. The effective method of controlling chaos, proposed by Ott, Grebogi, and Yorke, has attracted the attention of many researchers. This method has been widely implemented in a variety of systems [375–391] including hydrodynamical [378], mechanical [388, 390], chemical [383], and biological and medical [382, 385] systems.

The idea of the OGY method is that there exists an infinite number of unstable periodic orbits embedded in a chaotic attractor and the trajectory eventually enters the neighborhood of each orbit. Once inside, the small carefully chosen control perturbation is applied to a system parameter so that the trajectory remains near the desired periodic orbit. As a result, the system will execute the periodic motion. Thus, the problem of chaos control is reduced to one of stabilizing certain (desired) periodic orbits contained in the attractor.

It is worth noting that presently proposed techniques for controlling chaos are not restricted to the OGY method, which consists in stabilizing the already existing unstable periodic orbits. Detailed surveys of ideas and methods of controlling chaos are given in monographs and papers [203, 392–398].

The issues of chaos control in interacting systems are closely related to the problem of controlled synchronization. By applying appropriately chosen perturbations, certain chaotic subsets corresponding to synchronous motions of identical oscillators can be made stable in some eigendirections while they remain unstable in others. As a result, a controlled transition can be accomplished from nonsynchronous chaotic oscillations to the regime of complete chaos synchronization [399–406].

Formulation of the problem on controlling spatio-temporal chaos is a natural extension of this research direction [407–414]. Certain advances have been made towards the solution of this very important and complicated problem. However, most of them are concerned with the simplest models of distributed systems in the form of coupled map chains. A review of the studies on chaos control in ensembles of coupled systems as well as in distributed systems is given in [415]. Approaches utilized for controlling spatio-temporal chaos are based, as a rule, on the control techniques elaborated for systems with a few degrees of freedom. Some specific algorithms for controlling chaos in distributed systems were proposed in [416, 417]. One of the approaches consists in combining identical systems which exhibit both regular and chaotic dynamics into a chain of unidirectionally coupled elements. Such a construction enables the suppression of chaotic oscillations and thus the conversion of the motion to a periodic behavior. Another specific technique assumes the introduction of asymmetric coupling between elements in an ensemble. Chaotic oscillations can be suppressed by changing the homogeneity of couplings, e.g., when the coupling strength between elements becomes inhomogeneous for different parts of the chain.

In this section we are concerned with controlled transitions in interacting chaotic systems. We study anti-phase synchronization in interacting chaotic symmetric systems by using, as an example, a system of dissipatively coupled cubic maps. The regularities of controlled transitions from developed chaos to periodic oscillations are examined for two symmetrically coupled self-sustained systems.

We also discuss the issue of synchronization of chaotic systems by means of a parametric periodic forcing on the coupling element. Specifically, we pay attention to the possibility of synchronizing a chain of identical chaotic oscillators with periodic boundary conditions.

Finally, we describe a method of controlling spatio-temporal chaos in coupled map lattices. Particularly, the form of spatio-temporal perturbations of system parameters is determined to stabilize the desired unstable spatio-temporal states in one- and two-dimensional logistic map lattices with different kinds of coupling.

## 2.6.2 Controlled Anti-Phase Synchronization of Chaos in Coupled Cubic Maps

So-called complete chaos synchronization is one of the simplest kinds of coherent motions of interacting chaotic systems [183, 189, 239, 248, 250, 418, 419]. For coupled identical oscillatory systems the situation is possible when above a threshold value of the coupling parameter the subsystems oscillate identically, i.e., $x = y$. Here, $x$ and $y$ are the vectors of dynamical variables of the first and second oscillators, respectively. Thus, the regime of complete synchronization corresponds to a chaotic attractor located in the symmetric subspace $x = y$ of the total phase space of coupled systems. However, there is a wide class of chaotic DS, e.g., coupled cubic maps, Duffing oscillators and Chua's generators, which are symmetric both with respect to the change $x \rightarrow y$ and the substitution $x \rightarrow -y$. For such systems there exist two symmetric subspaces, namely, $x = y$ and $x = -y$, and we may distinguish between two kinds of complete chaos synchronization, each corresponding to the motion inside its own symmetric subspace. Oscillations in the first subspace are referred to the regime of complete in-phase synchronization, and motions in the second one correspond to the regime of complete anti-phase synchronization. Anti-phase chaos synchronization was described in [420] for unidirectionally coupled systems in the form of discrete maps and ordinary differential equations. In this paper, following Carroll and Pecora [186], the authors consider "master–slave" synchronization of chaos. It should be noted that the effect of anti-phase chaos synchronization has practical importance. New possibilities appear for developing methods for secure communication, and this aspect is discussed in detail in [420].

It is well known that stable and robust regimes of in-phase synchronization of chaos can be realized only for certain kinds of coupling and above a particular threshold value. However, for many kinds of interactions in-phase and anti-phase synchronous chaotic motions can occur in a system within a wide range of coupling strengths. These motions are unstable to perturbations directed transversely to the symmetric subspace. In such cases nonsynchronous chaotic oscillations can be synchronized by using the chaos control methods. By this means, certain chaotic sets corresponding to in-phase and anti-phase synchronous motions can be made stable in one eigendirection while they remain unstable in others.

We consider a system of coupled maps in the form:

$$x_{n+1} = f(x_n) + \gamma\big(f(y_n) - f(x_n)\big),$$
$$y_{n+1} = f(y_n) + \gamma\big(f(x_n) - f(y_n)\big), \tag{2.128}$$

where $x_n$ and $y_n$ are the dynamical variables of the first and second subsystems and $\gamma$ is the strength of coupling. The dynamics of a partial system is described by the map

$$x_{n+1} = f(x_n) = (a-1)x_n - ax_n^3, \tag{2.129}$$

where $a$ is the control parameter of the partial system.

Without coupling ($\gamma = 0$) and for $0 < a < 2$, the system (2.129) has a stable fixed point $C_0$ with coordinate $x = 0$. At $a = 2$ its eigenvalue $\mu = a - 1$ becomes equal to $+1$ and $C_0$ undergoes the symmetry-breaking bifurcation (the pitchfork bifurcation). As a result, point $C_0$ loses its stability and a pair of stable symmetric fixed points $C_1$ and $C_2$ is born in its neighborhood. Their coordinates are $x_1 = \{(a-2)/a\}^{1/2}$ and $x_2 = -\{(a-2)/a\}^{1/2}$, respectively. When $a$ is further increased, points $C_1$ and $C_2$ undergo a cascade of period-doubling bifurcations and at $a = 3.3$ two symmetric chaotic attractors arise. At $a = 3.6$ they merge, forming a single chaotic attractor.

When coupling is introduced, the dynamics of (2.128) becomes complicated. Note that this system is symmetric with respect to the changes of variables $x \leftrightarrow y$ and $x \leftrightarrow -y$. We shall call the symmetric subspace $x = -y$ the "anti-symmetric" subspace. The motions in the subspace $x = y$ of the total phase space of interacting systems correspond to complete in-phase synchronization. The motions in the subspace $x = -y$ correspond to complete anti-phase synchronization. Synchronous in-phase and anti-phase oscillations are observed in the system if the corresponding limit sets are transversely stable with respect to the symmetric or the anti-symmetric subspace. Their stability depends on the sign of the transverse Lyapunov exponent. If the exponent is negative, the synchronous oscillations are transversely stable. Otherwise, they are transversely unstable and cannot be observed in experiments. The studies show that in-phase synchronous chaotic oscillations can occur in the system of coupled maps in a certain coupling parameter range. At the same time, anti-phase synchronous chaotic oscillations are not observed.

We try to explore in detail how anti-phase chaos synchronization can be realized in the system. In this case it is convenient to use new variables

$$u = \frac{x+y}{2}, \quad v = \frac{x-y}{2}. \tag{2.130}$$

Adding and subtracting the equations in (2.128) we derive

$$\frac{x_{n+1} + y_{n+1}}{2} = \frac{1}{2}[f(x_n) + f(y_n)], \tag{2.131a}$$

$$\frac{x_{n+1} - y_{n+1}}{2} = \frac{1}{2}[(1 - 2\gamma)(f(x_n) - f(y_n))]. \tag{2.131b}$$

Suppose that the phase point is in the neighborhood of the anti-symmetric subspace:

$$x = \frac{x-y}{2} + \Delta x, \quad y = \frac{y-x}{2} + \Delta y,$$

where $\Delta x$ and $\Delta y$ are small. In this case, we can expand $f(x)$ and $f(y)$ into a Taylor series in the neighborhood of $(x-y)/2$ and restrict ourselves by the linear expansion terms

$$f(x) = f\left(\frac{x-y}{2}\right) + f'\left(\frac{x-y}{2}\right)\left(x - \frac{x-y}{2}\right),$$

$$f(y) = -f(-y) = -f\left(\frac{x-y}{2}\right) - f'\left(\frac{x-y}{2}\right)\left(-y - \frac{x-y}{2}\right).$$

Substituting these relations into (2.131a) and (2.131b), we derive

$$u_{n+1} = f'(v_n)u_n, \tag{2.132a}$$

$$v_{n+1} = (1 - 2\gamma)f(v_n). \tag{2.132b}$$

Equation (2.132b) describes the motion inside the anti-symmetric subspace. Unlike the in-phase synchronization, the form of synchronous oscillations depends on the coupling $\gamma$. Stability of oscillations inside the anti-symmetric subspace (for tangent perturbations) is determined by the tangent Lyapunov exponent

$$\Lambda_\parallel = \lim_{N\to\infty} \frac{1}{N} \sum_{n=1}^{N} \ln|(1 - 2\gamma)f'(v_n)|. \tag{2.133}$$

Equation (2.132a) governs the dynamics in the direction normal to the anti-symmetric subspace. It implicitly depends on the coupling parameter through the variable $v$. The solution $u = 0$ of (2.132a) corresponds to anti-phase oscillations. Its stability determines the transverse stability of anti-phase oscillations and is defined by the transverse Lyapunov exponent

$$\Lambda_\perp = \lim_{N\to\infty} \frac{1}{N} \sum_{n=1}^{N} \ln|f'(v_n)|. \tag{2.134}$$

From (2.133) and (2.134) it follows that both Lyapunov exponents are related to each other:

$$\Lambda_\perp = \Lambda_\parallel - \ln|1 - 2\gamma|. \tag{2.135}$$

In our computation, the coupling strength is taken to be positive and small, $0 \le \gamma \le 0.5$. With this, $\ln|1 - 2\gamma| \le 0$ and, consequently,

$$\Lambda_\perp \ge \Lambda_\parallel.$$

This relation shows that all oscillatory regimes whose limit sets are arranged in the anti-symmetric subspace first become transversally unstable and only then lose their stability in the tangent direction. Above the chaos critical line, a chaotic limit set is formed in the anti-symmetric subspace, and its tangent Lyapunov exponent is always larger than zero. With this, according to (2.135) the normal Lyapunov exponent is also positive. Hence, the anti-phase chaotic oscillations in the system of diffusively coupled maps cannot be transversally stable, and thus the anti-phase self-synchronization cannot be realized in such a system.

In order to stabilize previously unstable anti-phase chaotic oscillations we use an additional feedback controlling influence. Adding a feedback term $r(f(x_n)+f(y_n))$ to the right-hand side of the first equation of system (2.128), we derive

$$x_{n+1} = f(x_n) + \gamma(f(y_n) - f(x_n)) + r(f(x_n) + f(y_n)),$$
$$y_{n+1} = f(y_n) + \gamma(f(x_n) - f(y_n)). \qquad (2.136)$$

The controlling perturbation is chosen so that the additional term vanishes when control, i.e., the anti-phase synchronization, is achieved. Consequently, the controlling function does not change the form of anti-phase oscillations but affects their stability in the transverse direction.

Having passed to the new variables $u$ and $v$, we write the linearized equations:

$$u_{n+1} = (1+r)f'(v_n)u_n, \qquad (2.137a)$$
$$v_{n+1} = (1-2\gamma)f(v_n) + rf'(v_n)u_n. \qquad (2.137b)$$

In the case of anti-phase oscillations we have $u_n = 0$ and (2.137b) transforms to (2.132b). Using (2.137a) we can determine the normal Lyapunov exponent for the system with control:

$$\Lambda_{\perp\text{contr}}^a = \lim_{N\to\infty} \frac{1}{N} \sum_{n=1}^{N} \ln|(1+r)f'(v_n)| \qquad (2.138)$$

or

$$\Lambda_{\perp\text{contr}}^a = \Lambda_\perp + \ln|1+r|. \qquad (2.139)$$

Setting $r$ to be close to $-1$, we can achieve an arbitrary small value for the normal Lyapunov exponent and thus stabilize the anti-phase oscillations.

The normal Lyapunov exponent is plotted in Fig. 2.79a as a function of the control parameter $r$. The other parameter values correspond to the regime of a nonsynchronous united chaotic attractor (for $r = 0$). The transition to anti-phase synchronization can be accomplished as follows: We fix $a = 3.8$ and $\gamma = 0.04$ for which the nonsynchronous chaos regime is realized (see Fig. 2.80a) and choose initial conditions from the basin of attraction of the chaotic attractor. At every time moment we estimate the distance between the phase point and the anti-symmetric subspace, $\rho = |x + y|$, and compare it with a given threshold value $\delta$. If $\rho > \delta$, the phase point is far from the anti-symmetric subspace and control is not activated. When the phase point enters the small neighborhood of the anti-symmetric subspace (for $\rho \le \delta$), the controlling perturbation begins to act on the system. With this, if the chosen value of $r$ is located within the interval where $\Lambda_{\perp\text{contr}} < 0$, the chaotic set inside the anti-symmetric subspace becomes transversally stable and the phase

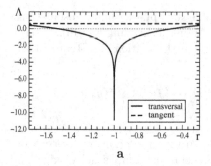
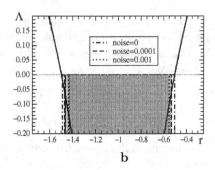

a

b

**Fig. 2.79.** (a) Normal Lyapunov exponent (*solid line*) as a function of the control parameter $r$ for $a = 3.8$ and $\gamma = 0.04$. *Dashed line* denotes the values of the tangent Lyapunov exponent. (b) The same dependence on a larger scale with regions of controlled synchronization

trajectory is attracted to the subspace. In this case the controlling perturbation tends to zero. In our computation, we take $\delta = 0.01$. Figure 2.80 illustrates system phase portraits and the corresponding time series $u$ of the system without control and in the presence of controlling perturbations. Without control, the trajectory uniformly covers a square-like region (Fig. 2.80a). When the controlling perturbation is applied, the diagonal $x = -y$ becomes visible in this domain (Fig. 2.80c), and one can observe intervals of synchronous behavior (Fig. 2.80d). Finally, if the parameter $r$ is chosen such that the normal Lyapunov exponent is negative, the system demonstrates the regime of completely synchronous chaos (Fig. 2.80e,f).

Intervals of the parameter $r$ where the controlled synchronization is achieved are shown in Fig. 2.79b for different noise levels. The dark region corresponds to larger noise intensities. Without noise the controlled synchronization region completely coincides with the parameter $r$ range where the normal Lyapunov exponent is negative. The addition of noise results in shrinking of the controlled synchronization domain. In several works [239, 248, 250, 418] where the mechanisms of loss of complete in-phase chaos synchronization are studied for symmetrically coupled systems with period-doubling, it was shown that the synchronization loss occurs according to a certain scenario. The first step in the desynchronization of chaos is attractor bubbling, when in the presence of arbitrary small noise the phase trajectory leaves from time to time the small vicinity of the symmetric subspace (turbulent bursts) [242]. The attractor bubbling can be followed by the riddled basin of its attraction, when tongues of other basins of attraction appear near the chaotic attractor [246].

It is reasonable to elucidate whether the loss of controlled anti-phase chaos synchronization is accompanied by the bubbling phenomenon and riddled basins of attraction. To answer this question we consider the evolution of a chaotic attractor as the controlling parameter $r$ is varied. In numeric simu-

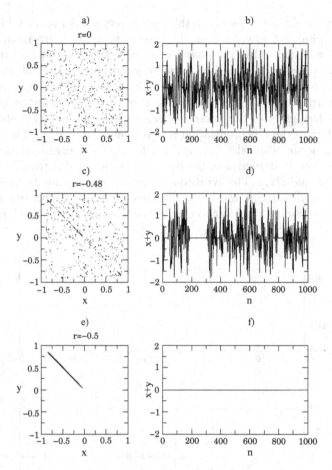

**Fig. 2.80.** Phase portraits and time series of oscillations (**a,b**) without control, (**c,d**) with partial control and (**e,f**) in the regime of contr olled synchronization

lation the initial conditions are chosen near the anti-symmetric subspace and the controlling perturbation is always activated. When $-1 < r < -0.525$, the chaotic attractor located inside the anti-symmetric subspace is transversally stable. The initially chosen perturbation decreases in time, and the phase trajectory is attracted to the subspace. Adding weak noise does not cause a significant change of the systems behavior. With increasing $r$ ($r > -0.525$) a bubbling attractor is observed in the system. The chaotic attractor in the anti-symmetric subspace remains transversally stable but the duration of transient process becomes extremely large and is very sensitive to the initial conditions. When a small amount of noise is added to the system, the phase portrait of the attractor is significantly modified. The attractor achieves a finite transversal size. The phase trajectory begins to switch between the chaotic sets, and the time series in this case is very similar to on–off intermittency [421]. Figure 2.81a–d shows phase portraits and the corresponding time series $u(t)$ without and in the presence of noise. In the latter case the phase point first evolves for a long time in the neighborhood of the anti-symmetric subspace. Then there is a short burst away from it, after which the phase trajectory returns again to the neighborhood of the anti-symmetric subspace. The mean frequency of the bursts increases as the parameter $r$ approaches the boundary of the synchronization region. Finally, at $r = -0.406$ a blowout bifurcation occurs when the chaotic attractor inside the anti-symmetric subspace is no longer attractive in the normal direction and is transformed into a chaotic saddle [241]. The synchronous oscillations cannot be observed in the system both without and in the presence of noise. The phase portrait of the chaotic attractor is similar to that of the bubbling attractor (Fig. 2.81e).

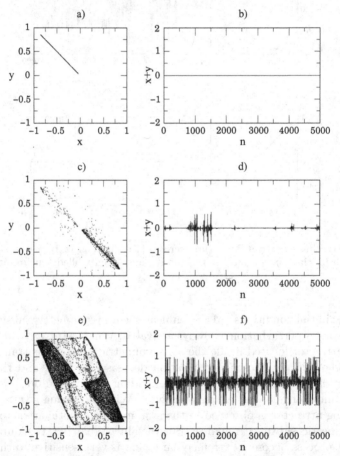

**Fig. 2.81.** Phase portraits and times series of anti-phase oscillations (**a,b**) without noise, (**c,d**) in the presence of noise with intensity 0.00001, and (**e,f**) after the blowout bifurcation

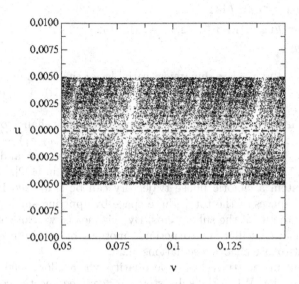

**Fig. 2.82.** A portion ($-0.005 \le u \le 0.005$) of the basin of attraction (*white color*) of the chaotic attractor inside the anti-symmetric subspace for $a = 3.8$, $\gamma = 0.04$ and $r = -1.485$. Trajectories starting from black points approach the attractor at infinity

When the parameter $r$ decreases, i.e., $r < -1.472$, the basin of attraction of the attractor in the anti-symmetric subspace is riddled by the holes from the basin of attraction of the attractor at infinity. Figure 2.82 shows a portion of the basin of attraction of the chaotic attractor located in the anti-symmetric subspace (white color). Black points indicate the basin of attraction of the attractor at infinity. The anti-symmetric subspace ($u = 0$) is marked by the dashed line.

### 2.6.3 Control and Synchronization of Chaos in a System of Mutually Coupled Oscillators

There are a lot of methods of controlling chaos whose main idea is to convert the chaotic behavior to a periodic time dependence by making small perturbations to the system. Techniques allowing the stabilization of saddle cycles embedded in a chaotic attractor are often complicated enough because they are related to the search for saddle cycles and their manifolds and to the determination of an appropriate form of controlling perturbations. For a system of two interacting identical oscillators, which possesses several kinds of symmetry, a procedure for stabilizing a certain subset of saddle cycles can be significantly simplified for certain types of coupling. As an example we consider the system of two coupled via capacity identical Chua's circuits [422]:

$$\dot{x}_{1,2} = \alpha[y_{1,2} - x_{1,2} - h(x_{1,2})],$$
$$\dot{y}_{1,2} = x_{1,2} - y_{1,2} + z_{1,2} + \gamma[(x_{2,1} - x_{1,2}) - (y_{2,1} - y_{1,2}) + (z_{2,1} - z_{1,2})],$$
$$\dot{z}_{1,2} = -\beta y_{1,2}, \qquad\qquad\qquad\qquad\qquad (2.140)$$

where $x_{1,2}, y_{1,2}$ and $z_{1,2}$ are dynamical variables of the first and second subsystems; $h(x_{1,2}) = bx_{1,2} + 0.5(a - b)(|x_{1,2} + 1| - |x_{1,2} - 1|)$; $\alpha, \beta, a$ and $b$ are the parameters; and $\gamma$ denotes the coupling coefficient.

The system (2.140) is symmetric with respect to the change of variables $\boldsymbol{X}_1 = (x_1, y_1, z_1) \longleftrightarrow \boldsymbol{X}_2 = (x_2, y_2, z_2)$ and $\boldsymbol{X}_1 = (x_1, y_1, z_1) \longleftrightarrow -\boldsymbol{X}_2 = (-x_2, -y_2, -z_2)$. Its behavior depending on the parameters $\alpha$ and $\gamma$ and for fixed $a = -8/7$, $b = -5/7$, $\beta = 22$ is described in detail in [422].

In this system a chaotic phase trajectory can be stabilized in different symmetric subspaces of the total phase space by applying small controlling perturbations to one of the subsystems. By this means, we can convert non-symmetric chaotic oscillations to periodic motions or synchronize both in-phase and anti-phase chaotic oscillations [423].

The system under study has a peculiarity which allows one to control its chaotic behavior. When the subsystems are coupled and the coupling between them is finite, chaotic dynamics arises for essentially smaller values of $\alpha$ of a partial oscillator than in the uncoupled case. The system of interacting oscillators can already exhibit developed chaos, whereas the individual system still oscillates periodically, i.e., it contains stable cycles. This means that these cycles also exist in the system of mutually coupled oscillators but as unstable ones. They are located in the symmetric subspace $\boldsymbol{X}_1 = \boldsymbol{X}_2$ of the total phase space and are stable to symmetric perturbations and unstable to transverse perturbations. If for finite coupling a chaotic attractor is created in the system and contains the saddle symmetric cycles, the phase trajectory can easily be stabilized in their neighborhood. In this case the problem reduces to the stabilization of the phase trajectory in the symmetric subspace $\boldsymbol{X}_1 = \boldsymbol{X}_2$ by applying small perturbations to one of the oscillators. Since the saddle cycle of interest is stable to symmetric perturbations, in some time the phase trajectory will eventually come close to it and then will evolve on the cycle as long as the control is activated. The perturbations for stabilizing the phase trajectory in the symmetric subspace can be determined in a simpler way than those used to stabilize it in the neighborhood of the saddle cycle. If without coupling each of the partial systems demonstrates chaotic oscillations, stabilization of the phase trajectory in the given symmetric subspace leads to the regime of chaos synchronization.

It should be pointed out that in systems possessing several kinds of symmetry the chaotic trajectory can be stabilized in different symmetric subspaces, e.g., not only in $\boldsymbol{X}_1 = \boldsymbol{X}_2$ but also in $\boldsymbol{X}_1 = -\boldsymbol{X}_2$. Which kind of motion with the indicated symmetry properties will be stabilized depends on whether the given subspace contains saddle cycles stable to symmetric perturbations or not.

In order to determine conditions for stabilizing symmetric motions we proceed as follows. First we use the new variables $u = (x_1 - x_2)/2, v = (y_1 - y_2)/2, w = (z_1 - z_2)/2, u_1 = (x_1 + x_2)/2, v_1 = (y_1 + y_2)/2 and w_1 = (z_1 + z_2)/2$, and we re-write the system (2.140) as follows:

$$\dot{u} = \alpha[v - u - c(u)], \tag{2.141a}$$

$$\dot{v} = u - v + w - 2\gamma[u - v + w], \tag{2.141b}$$

$$\dot{w} = -\beta v, \tag{2.141c}$$

$$\dot{u}_1 = \alpha[v_1 - u_1 - c(u_1)], \tag{2.141d}$$

$$\dot{v}_1 = u_1 - v_1 + w_1, \tag{2.141e}$$

$$\dot{w}_1 = -\beta v_1, \tag{2.141f}$$

where $c(u) = [h(x_1) - h(x_2)]/2, c(u_1) = [h(x_1) + h(x_2)]/2$.

In this system the evolution of symmetric motions $X_1 = X_2$ ($X_1 = -X_2$) is described by (2.141d–f) [(2.141a–c)] and their stability is determined by (2.141a–c) [(2.141d–f)].

We intend to stabilize symmetric regimes $X_1 = X_2$. We add controlling perturbation $F(u) = ru$ to (2.141a) and use the Rauth–Hurwitz criterium to define the values of $r$ at which the equilibrium $u = v = w = 0$ is stable. Thus, we can determine the stabilization conditions for the symmetric regimes of system (2.140).

From the stability analysis it follows that the chaotic trajectory can be stabilized in the subspace $X_1 = X_2$ when parameter $r$ satisfies the inequality

$$r < \alpha(c - 0.5) + 0.5\xi - \sqrt{0.25(\xi - \alpha)^2 - \xi(\beta - \alpha)}, \tag{2.142}$$

$$\xi = 1 - 2\gamma, \quad c = \begin{cases} 1 + a, & |x| \leq 1, \\ 1 + b, & |x| > 1. \end{cases}$$

The character of symmetric oscillations (regular or chaotic) depends on the dynamics of the partial system in the uncoupled case.

Adding controlling perturbation $F_1 = r_1 u_1$ to (2.141d) we can determine the stability conditions for the equilibrium $u_1 = v_1 = w_1 = 0$. In this case anti-phase oscillations $X_1 = -X_2$ will be stabilized.

As already noted, the system of coupled self-sustained oscillators can possess chaotic attractors in a certain parameter range. The phase point evolving on them enters from time to time a small neighborhood of one of the symmetric subspaces. If at this moment the corresponding perturbation, $F$ or $F_1$, is applied to one of the subsystems, the symmetric motions will be settled in the system.

To stabilize synchronous motions $X_1 = X_2$, controlling perturbation $F = r(x_1 - x_2)$ is added to the first equation of the first oscillator (2.140). Control is initiated when the phase point falls within a given small neighborhood of the symmetric subspace. This means that magnitude of the controlling perturbation is always kept small.

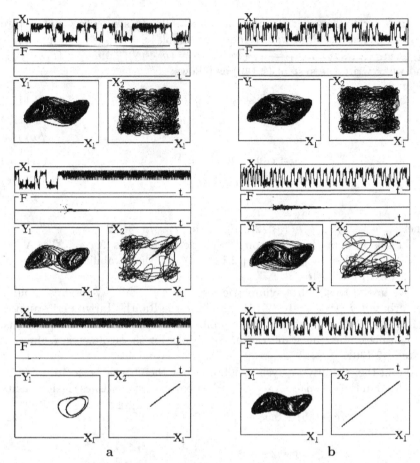

**Fig. 2.83.** Time series $X_1(t)$, controlling perturbation $F(t)$ and projections of phase portraitsunder controlled transitions from the chaotic behavior (*upper panels*) (**a**) to a symmetric period-2 cycle for $\alpha = 11.4$ and $\gamma = 0.2$ and (**b**) to synchronous chaotic motions for $\alpha = 12.7$ and $\gamma = 0.2$ (*lower panels*). The *middle panels* illustrate the intermediate stages in the transitions

Figure 2.83 illustrates controlled transitions from the chaotic behavior to regular and in-phase chaotic symmetric motions for different values of $\alpha$ and for fixed $\gamma = 0.2$. Other parameters are: $a = -8/7$, $b = -5/7$, $\beta = 22$.

From the figure it can be seen that the controlling perturbations begin to act on the subsystem when the phase trajectory enters the given $\varepsilon$-neighborhood of the symmetric subspace. As a result, the phase trajectory is attracted to the symmetric subspace and remains there. Two situations can be realized depending on the value of $\alpha$, namely, the symmetric period-two cycle (Fig. 2.83a) and the synchronous chaotic motions corresponding to a "double scroll" attractor (Fig. 2.83b). The chaos control procedure used

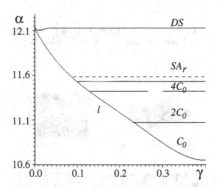

**Fig. 2.84.** Regions of symmetric regimes emerging in system (2.140) after control is achieved

enables a chaotic trajectory to stabilize in the symmetric subspace and does not affect the dynamics of the system when it evolves in this subspace. Thus, after control is achieved, the system will demonstrate the symmetric motion which is attractive in the symmetric subspace. Depending on the values of $\alpha$, one can stabilize either cycles of different periodicity or chaotic attractors. Both cases are illustrated in Fig. 2.83.

Results of the detailed study of controlled transitions are summarized in Fig. 2.84. On the $(\alpha - \gamma)$ plane, the region is schematically drawn (above the line $l$) where the system of coupled oscillators behaves chaotically in the absence of controlling perturbations. When control is activated, different regimes can be stabilized, namely, symmetric cycles with period 1, 2 and 4 ($C_0$, $2C_0$, and $4C_0$), a symmetric chaotic attractor of the Rössler type ($SA_r$), and a chaotic "double scroll" attractor ($DS$) located in the symmetric subspace. Controlled transitions to symmetric cycles of high periodicity (8, 16) have also been observed but within very narrow parameter ranges. These regions are not shown in the figure. The dashed line marks the boundary above which the synchronous chaotic motions occur in the system.

Similar controlled transitions can be realized in the symmetric subspace $X_1 = -X_2$. To stabilize the chaotic trajectory in this subspace we introduce controlling perturbation $F_1 = r_1(x_1 + x_2)$ into the first equation of the first generator (2.140). Figure 2.85 exemplifies the controlled transition from developed chaos to the regime of symmetric anti-phase chaotic oscillations. However, in contrast to the previous case, varying parameters $\alpha$ and $\gamma$ we are not able to stabilize regular motions. The application of controlling perturbations always results in the appearance of symmetric anti-phase chaotic motions.

This peculiarity can be explained as follows: For the values of $\alpha$ and $\gamma$, which correspond to chaotic regimes in coupled oscillators, the subspace $X_1 = -X_2$ does not contain saddle cycles that are stable to symmetric anti-phase perturbations. A chaotic set is already created which attracts phase

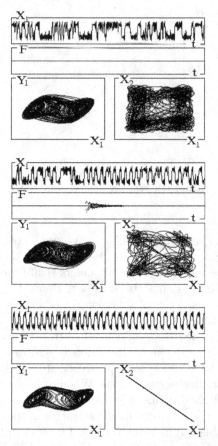

**Fig. 2.85.** Controlled transition from the developed chaos $(DS)$ to the regime of anti-phase chaotic oscillations for $\alpha = 12.7$ and $\gamma = 0.2$

trajectories from this subspace, and the system achieves this state when the control is applied.

Bifurcational transitions of unstable motions have been studied as follows: We choose initial conditions in the neighborhood of the subspace under study and apply the controlling perturbation $F_1$. By varying $\alpha$ and $\gamma$ only those oscillatory regimes which are stable to anti-phase perturbations can be observed. This result shows that the methods for stabilizing motions in certain subspaces of the total system phase space can be reliably used to perform the bifurcational analysis of a certain kind of saddle motions. Such an approach can significantly expand the possibilities of full-scale experiments.

Note that the effect of small additive noise on system (2.140) has not qualitatively changed the obtained results.

## 2.6.4 Controlled Chaos Synchronization
## by Means of Periodic Parametric Perturbations

The idea of using parametric perturbations to synchronize coupled chaotic systems is based on a well-known classical problem of the pendulum with oscillating suspension [38, 424, 425]. In such a system, beginning from certain threshold values of amplitude and frequency, the parametric perturbations change the unstable equilibrium state into a stable one.

Consider two coupled identical nonautonomous oscillators [291, 426]:

$$\ddot{x}_{1,2} + \alpha\dot{x}_{1,2} + f(x_{1,2}) - \gamma(x_{2,1} - x_{1,2}) = B\sin(\omega t), \qquad (2.143)$$

where $f(x) = (b-1)x + 0.5(b-a)(|x-1| - |x+1|)$, $\alpha$ is the dissipation parameter, $\gamma$ denotes the coupling coefficient, and $B$ and $\omega$ are the amplitude and the frequency of the external force, respectively.

When $\gamma = 0$, we have the equation of a partial system. Without external force ($B = 0$), system (2.143) describes a nonlinear damping oscillator with three equilibria: $P_1(x_1 = 0, y_1 = 0)$, $P_{2,3}(x_{2,3} = \pm(b-a)/(b-1), y_{2,3} = 0)$. If the following inequalities are satisfied, namely, $0 < a < 1$, $b > (1+\alpha^2/4)$, $\alpha > 0$, then point $P_1$ is a saddle and points $P_{2,3}$ are stable foci. When the external force is introduced, the nonlinear oscillator demonstrates chaotic oscillations in a certain parameter range. The coupled oscillators ($\gamma > 0$) show different forms of regular and chaotic motions, including the regime of nonsynchronous chaotic oscillations.

In system (2.143) synchronization of chaotic oscillations can be achieved by applying a periodic parametric perturbation to the coupling element. Suppose that the coupling coefficient $\gamma$ can be varied periodically in time about a constant level $\gamma_0$, i.e.,

$$\gamma = \gamma_0 + F(t), \qquad (2.144)$$

where $F(t)$ is a periodic function with period $T = 2\pi/\Omega$ ($\Omega$ is the frequency of parametric perturbations). In order to demonstrate the mechanism of stabilization of symmetric motions, we take the parametric perturbations in the following form:

$$F(t) = \varepsilon\Omega^2\mathrm{sgn}\big(\sin(\Omega t)\big), \qquad (2.145)$$

where $\varepsilon$ is the amplitude of the parametric perturbations.

Taking into account (2.144) and (2.145) we can re-write the equations for coupled oscillators (2.143) as follows:

$$
\begin{aligned}
\dot{x}_{1,2} &= y_{1,2}, \\
\dot{y}_{1,2} &= -\alpha y_{1,2} - f(x_{1,2}) + \big[\gamma_0 + \varepsilon\Omega^2\mathrm{sgn}\big(\sin(\Omega t)\big)\big] \\
&\quad \times (x_{2,1} - x_{1,2}) + B\sin(\omega t).
\end{aligned} \qquad (2.146)
$$

Using the variable changes $u = (x_1 - x_2)/2$, $u_1 = (x_1 + x_2)/2$, $v = (y_1 - y_2)/2$, $v_1 = (y_1 + y_2)/2$ we arrive at the following system of equations:

$$\dot{u} = v, \tag{2.147a}$$
$$\dot{v} = -\alpha v - \omega_0^2(t)u, \tag{2.147b}$$
$$\dot{u}_1 = v_1, \tag{2.147c}$$
$$\dot{v}_1 = -\alpha v_1 - f(u_1) + B\sin(\omega t), \tag{2.147d}$$

where

$$\omega_0^2(t) = A + 2\gamma_0 + 2\varepsilon\Omega_0^2\mathrm{sgn}\big(\sin(\Omega t)\big), \quad f(u_1) = \big(f(x_1) + f(x_2)\big)/2,$$

$$A = \begin{cases} a - 1, & |x| \le 1, \\ b - 1, & |x| > 1. \end{cases}$$

Equations (2.147c,d) describe synchronous motions ($u = 0$, $v = 0$) of the coupled oscillators, while (2.147a,b) determine the evolution of small deviations from the synchronous motions or the equilibrium ($u = 0$, $v = 0$). In order to define the stabilization conditions of symmetric motions of chaotic oscillators, we explore how small perturbations evolve in the neighborhood of the equilibrium ($u = 0$, $v = 0$) in the external force period $T$. This can be done by using the well-known algorithms (see, e.g., [38]).

We split the period of the parametric perturbation into two intervals, namely, $[0; T/2]$ and $[T/2; T]$. On each interval, (2.147a,b) are linear and can be solved analytically. The perturbations at the time moments 0 and $T/2$ and at $T/2$ and $T$ are connected through the matrices

$$\begin{bmatrix} u(T/2) \\ v(T/2) \end{bmatrix} = [M_1] \begin{bmatrix} u(0) \\ v(0) \end{bmatrix},$$

$$\begin{bmatrix} u(T) \\ v(T) \end{bmatrix} = [M_2] \begin{bmatrix} u(T/2) \\ v(T/2) \end{bmatrix}.$$

The stability of solution ($u = 0$, $v = 0$) is defined by the eigenvalues of the matrix $[M] = [M_2][M_1]$, which read

$$\mu_{1,2} = \exp(\alpha\pi/\Omega)(0.5S \pm \sqrt{0.25S^2 - 1}), \tag{2.148}$$

where

$$S = 2(\cosh\beta_1\cosh\beta_2 + K\sinh\beta_1\sinh\beta_2),$$
$$\beta_1 = (\pi/\Omega)\sqrt{0.25\alpha^2 - A - 2\gamma_0 - 2\varepsilon\Omega^2},$$
$$\beta_2 = (\pi/\Omega)\sqrt{0.25\alpha^2 - A - 2\gamma_0 + 2\varepsilon\Omega^2},$$
$$K = \frac{0.25\alpha^2 - A - 2\gamma_0}{\sqrt{(0.25\alpha^2 - A - 2\gamma_0)^2 - 4\varepsilon^2\Omega^4}}.$$

The symmetric motions are stable if $|\mu_{1,2}| < 1$. In addition, this condition must be satisfied for all values of the dynamical variables in the neighborhood of the symmetric subspace, i.e., both for $A = a - 1$ and for $A = b - 1$.

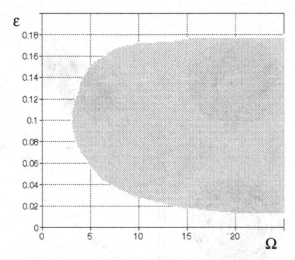

**Fig. 2.86.** Stability region of synchronous regimes on the parameter plane "amplitude – frequency" of the parametric perturbations

The stability condition for symmetric motions is fulfilled inside the shaded region on the $(\varepsilon - \Omega)$ plane, as shown in Fig. 2.86. The values of other system parameters $((B = 1.5,\ \alpha = 0.1,\ \omega = 1,\ \gamma = = 0.1,\ a = 0.5, b = 2)$ correspond to the case when nonsynchronous chaotic oscillations occur in the unperturbed system $(\varepsilon = 0)$.

Numeric experiments with the system (2.146) have verified that synchronization of chaotic oscillators can be achieved through periodic perturbations on the coupling element. Chaotic oscillations are synchronized at the same amplitude and frequency values of the parametric perturbations as predicted theoretically. Numeric simulation has been carried out as follows. For the above parameter values, the initial conditions are chosen within the small vicinity of the symmetric subspace: $x_1(0) = x_2(0) + \Delta x$, $y_1(0) = y_2(0) + \Delta y$. Usually, we set $\Delta x = 0.02$ and $\Delta y = 0.02$. Then projections of phase portraits and temporal dependences of oscillation regimes are plotted for different values of amplitude $\varepsilon$ and frequency $\Omega$. At $\varepsilon = 0$ the phase trajectory leaves the neighborhood of the symmetric subspace and nonsynchronous chaotic oscillations are observed (Fig. 2.87a). In the presence of parametric perturbation the initial deviations $\Delta x$ and $\Delta y$ decrease and the phase trajectory enters into the symmetric subspace and evolves therein (Fig. 2.87b) if the values of $\varepsilon$ and $\Omega$ are in the shaded region of Fig. 2.86. If the values are taken outside this region, the initial deviations increase in time and synchronization is not achieved.

As can be seen from Fig. 2.86, the stabilization effect of symmetric motions has a threshold character. The frequency of the parametric perturbations $\Omega$ must be several times as large as the characteristic frequency of the unperturbed motion (in the considered case it is the frequency $\omega = 1$). The

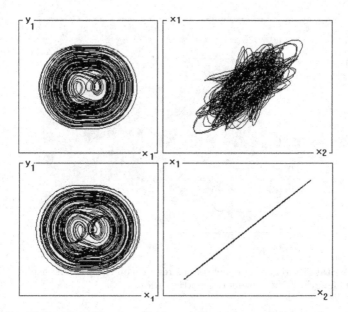

**Fig. 2.87.** Phase portraits of oscillations (**a**) without and (**b**) in the presence of parametric perturbations

smaller the amplitude of the parametric perturbation, the larger its frequency $\Omega$. In order to achieve the synchronous regime in the system under study, it is sufficient for the coupling coefficient to be slightly perturbed ($\varepsilon$ accounts for 20% of $\gamma_0 = 0.1$) at frequency $\Omega$, the latter exceeding the frequency $\omega = 1$ by $15 - 20$ times (Fig. 2.86).

The effect of chaos synchronization can also occur when the phase point is placed far from the symmetric subspace at the moment when the parametric perturbation is applied. However, in this case the duration of a transient process to the synchronous chaotic regimes substantially increases. In the perturbed system the time needed to reach the neighborhood of the symmetric subspace, where the stabilization mechanism starts to work, is significantly larger than that in the unperturbed system.

### 2.6.5 Stabilization of Spatio-Homogeneous Motions by Parametric Perturbations

This section addresses the possibility of synchronizing a chain of chaotic oscillators by applying periodic parametric perturbations to the coupling elements.

Consider the chain of nonlinear identical nonautonomous oscillators [427]:

$$\dot{x}_m = y_m, \qquad\qquad (2.149)$$
$$\dot{y}_m = -\alpha y_m - f(x_m) + (\gamma/2)(x_{m+1} - 2x_m + x_{m-1}) + B\sin(\omega t),$$

with periodic boundary conditions $x_0 = x_N$, $y_0 = y_N$, $x_{N+1} = x_1$, $y_{N+1} = y_1$, where $m = 1, \ldots, N$, and $N$ is the number of elements in the chain. An individual element of the chain is represented by the same nonlinear nonautonomous oscillator as in the previous section. The parametric perturbations are defined by (2.144) and (2.145). Thus, for $N = 2$, the system (2.149) is reduced to (2.146).

We want to find the stability conditions of synchronous motions depending on the amplitude and the frequency of parametric perturbations. We first linearize (2.149) in the neighborhood of spatio-homogeneous state $x_m = \bar{x}$, $y_m = \bar{y}$. As a result, we derive the equations for perturbations with respect to $\bar{x}$ and $\bar{y}$:

$$\dot{\eta}_m^{(1)} = \eta_m^{(2)}, \tag{2.150a}$$

$$\dot{\eta}_m^{(2)} = -\alpha\eta_m^{(2)} - [A + 2\Gamma(t)]\eta_m^{(1)} + \Gamma(t)\left[\eta_{m+1}^{(1)} + \eta_{m-1}^{(1)}\right], \tag{2.150b}$$

where $\eta_m^{(1)} = x_m - \bar{x}$, $\eta_m^{(2)} = y_m - \bar{y}$, $\Gamma(t) = (1/2)\left[\gamma_0 + \varepsilon\Omega^2\text{sgn}\left(\sin(\omega t)\right)\right]$ and

$$A = f'(\bar{x}) = \begin{cases} a - 1, & |x| \leq 1, \\ b - 1, & |x| > 1. \end{cases}$$

Using the change of variables

$$\xi_j^{(1)} = \frac{1}{N}\sum_{m=1}^{N} e^{-km}\eta_m^{(1)}, \quad \xi_j^{(2)} = \frac{1}{N}\sum_{m=1}^{N} e^{-km}\eta_m^{(2)},$$

where $k = \text{i}(2\pi j/N)$, $\text{i}^2 = -1$, $j = 1, \ldots, N$, we can re-write (2.150a,b) as follows:

$$\dot{\xi}_j^{(1)} = \xi_j^{(2)},$$

$$\dot{\xi}_j^{(2)} = -\alpha\xi_j^{(2)} - \omega_0^2(t, j)\xi_j^{(1)}, \tag{2.151}$$

where $\omega_0^2(t, j) = A + \left[\gamma_0 + \varepsilon\Omega^2\text{sgn}\left(\sin(\Omega t)\right)\right]\left[1 - \cos(2\pi j/N)\right]$.

Equations (2.150a,b) describe the dynamics of spatial perturbations in the neighborhood of synchronous state $x_m = \bar{x}$, $y_m = \bar{y}$. Spatio-homogeneous perturbations correspond to $j = N$, and spatio-inhomogeneous ones to $j = 1, \ldots, N - 1$. The synchronous state is stable if all possible out-of-phase perturbations decay in the chain.

Note that (2.150a,b) have the same form as (2.147a,b), which describe the dynamics of transverse perturbations in two coupled oscillators and fully coincide with them for $N = 2$ and $j = 1$. Thus, the results obtained in the previous section can be used to determine the eigenvalues $\mu_{1,2}$, which characterize the evolution of the $j$th transverse perturbation in a period of parametric perturbation $T = 2\pi/\Omega$. The eigenvalues are expressed as follows:

$$\mu_{1,2}(j) = e^{-\alpha\pi/\Omega}\left[0.5S(j) \pm \sqrt{0.25S^2(j) - 1}\right], \tag{2.152}$$

where

$$S(j) = 2\cosh[\beta_1(j)]\cosh[\beta_2(j)] + K(j)\sinh[\beta_1(j)]\sinh[\beta_2(j)],$$
$$\beta_{1,2}(j) = (\pi/\Omega)\sqrt{C_1(j) \mp C_2(j)},$$
$$K(j) = \frac{2C_1(j)}{\sqrt{C_1^2(j) - C_2^2(j)}},$$
$$C_1(j) = \frac{\alpha^2}{4} - A - \gamma_0[1 - \cos(2\pi j/N)],$$
$$C_2(j) = \varepsilon\Omega^2[1 - \cos(2\pi j/N)].$$

Synchronous motions in the chain of coupled chaotic oscillators are stable if for all $j = 1, \ldots, N - 1$ the condition $|\mu_{1,2}(j)| < 1$ is satisfied both for $A = a - 1$ and for $A = b - 1$.

In Fig. 2.88 the straight vertical lines indicate the regions of amplitude $\varepsilon$ values, where the stability conditions of synchronous motions are fulfilled in chains with different numbers of elements $N$ and for fixed $\Omega = 25$. The values of other system parameters ($B = 1.5$, $\omega = 1$, $\alpha = 0.1$, $\gamma_0 = 0.1$, $a = 0.5$, $b = 2$) correspond to a case when regimes of spatio-temporal chaos are observed in the absence of parametric perturbations ($\varepsilon = 0$). It is seen that as the number of elements $N$ increases, the synchronization region decreases gradually and its lower boundary is shifted up in amplitude. When the chain contains more than 10 oscillators, the parametric perturbation with $\Omega = 25$ cannot suppress all transverse perturbations. The synchronous state can be stabilized with respect to the transverse perturbation only in a certain finite interval of amplitude $\varepsilon$ values. The boundaries of these intervals do not coincide. Synchronization in the chain can appear only in a case when there exists a common (belonging to each of the intervals) domain of values of $\varepsilon$. When $N > 10$, a portion of the intervals is not overlapped.

These theoretical results show good agreement with numerical ones. Almost all the amplitude values at which synchronization is observed in the chain lie in the stability intervals plotted from theoretical data. Several points marked by crosses in Fig. 2.88 are arranged outside these intervals. This fact can be explained as follows: When analyzing the stability of synchronous motions we require the modulus of eigenvalues to be less than unity both for $A = a - 1$ and for $A = b - 1$, i.e., at every point of the phase trajectory. In fact, this requirement is overestimated. For the inhomogeneous state to be stable, it is sufficient that the averaged magnitudes of the eigenvalues be less than unity in modulo. Therefore, numerical calculations yield wider regions of synchronization than those found theoretically.

Modulating periodically the coupling coefficient we can stabilize spatio-homogeneous motions only in chains of a finite length. The maximal number of oscillators for which synchronization can be observed depends on the frequency of the parametric perturbations. For example, for $\Omega = 15, 25, 50$ and 100 this number does not exceed $8, 10, 15$ and 21, respectively. In cases when

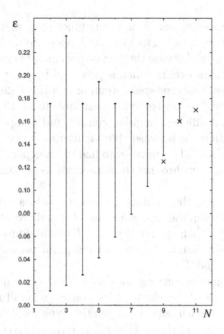

**Fig. 2.88.** Stability intervals for regimes of spatio-homogeneous chaotic oscillations in different length chains

the length of a chain is larger, the parametric perturbations provide only a partial stabilization. The spatio-homogeneous state to be stabilized becomes stable with respect to only some transverse perturbations.

### 2.6.6 Controlling Chaos in Coupled Map Lattices

Coupled map lattices are the simplest models of spatio-distributed systems and can demonstrate numerous typical spatio-temporal phenomena which are observed in a variety of systems.

Spatio-temporal chaos in chains and lattices can be effectively controlled through small spatio-distributed perturbations of a system parameter. Such local control of chain elements, which is called "pinning control" in the literature, enables to be stabilized preliminary chosen unstable spatio-temporal structures.

Consider the chains of coupled logistic maps with two kinds of coupling [428, 429]:

$$x_{n+1}(i) = \alpha - x_n^2(i) + \gamma\big(2x_n^2(i) - x_n^2(i-1) - x_n^2(i+1)\big), \qquad (2.153)$$

$$x_{n+1}(i) = \alpha - x_n^2(i) + \gamma\big(2x_n(i) - x_n(i-1) - x_n(i+1)\big), \qquad (2.154)$$

where $\alpha$ is the nonlinearity parameter, $\gamma$ is the coupling coefficient, $x_n(i)$ is a value of the dynamical variable in the point with discrete coordinate $i$ and at discrete time moment $n$, and $N$ is the number of elements in the chain ($i = 1, 2, \ldots, N$; $n = 0, 1, 2, \ldots$). The systems have periodic boundary conditions $x_n(1) = x_n(N+1)$. In this case there exist parameter regions where developed spatio-temporal chaos occurs which is preceded by complicated bifurcational transitions between different spatio-temporal regimes [207, 430–432].

In order to realize controlled transitions from the regime of spatio-temporal chaos to different regular spatial and temporal structures, controlling perturbations in an ensemble of interacting oscillators must satisfy certain requirements. They must stabilize certain periodic motions in each oscillator as well as synchronize these motions in certain phases throughout the ensemble.

Suppose that for the control parameter values corresponding to the developed spatio-temporal chaos in the chain there exist instable spatio-homogeneous states with temporal period $T = s$ and unstable spatio-periodic structures with wavelength $\lambda = 2$ and temporal period $T = 2$, $\lambda = 3$ and $T = 2$, and $\lambda = 5$ and $T = 4$.

The spatio-homogeneous regimes with $T = s$ are most simple. Assume that the $i$th element of the chain (2.153) enters a small neighborhood of the point $\bar{x}_k$ of the orbit with period $s$. Then the dynamical variable can be represented in the form $x_{n+k-1}(i) = \bar{x}_k + \tilde{x}_{n+k-1}(i)$, $x_{n+s}(i) = \bar{x}_1 + \tilde{x}_{n+s}(i)$, $k = 1, \ldots, s$, where $\tilde{x}_{n+k-1}(i)$ is a small perturbation of the dynamical variable near the point $\bar{x}_k$. Suppose also that the nonlinearity parameter $\alpha$ is able to be controlled. Write it in the form

$$\alpha = \alpha_{n+k-1}(i) = \alpha_0 + \tilde{\alpha}_{n+k-1}(i), \quad k = 1, \ldots, s, \tag{2.155}$$

where $\tilde{\alpha}_{n+k-1}(i)$ is a small perturbation of the parameter near its nominal value $\alpha_0$.

Taking into account the suggestions made, re-write the system (2.153) as follows:

$$\bar{x}_{k+1} + \tilde{x}_{n+k}(i) = \alpha_0 + \tilde{\alpha}_{n+k-1}(i) - \left(\bar{x}_k + \tilde{x}_{n+k-1}(i)\right)^2$$
$$+ \gamma\left(2\left(\bar{x}_k + \tilde{x}_{n+k-1}(i)\right)^2 - 2\bar{x}_k^2\right) - \gamma\left(\left(x_{n+k-1}^2(i+1) - \bar{x}_k^2\right)\right.$$
$$\left. + \left(x_{n+k-1}^2(i-1) - \bar{x}_k^2\right)\right), \tag{2.156}$$

where $k = 1, \ldots, s$.

Suppose, the values of the neighboring ($i - 1$)th and ($i + 1$)th elements differ from the value of fixed point $\bar{x}_k$ by such a magnitude that the last term in (2.156) is comparable with $\tilde{x}_{n+k-1}(i)$. Then, we write the equations for the fixed point and for the evolution of the dynamical variable perturbation of the $i$th element:

$$\bar{x}_{k+1} = \alpha_0 - \bar{x}_k^2, \quad \bar{x}_{s+1} = \bar{x}_1, \tag{2.157}$$

$$\tilde{x}_{n+k}(i) = \tilde{\alpha}_{n+k-1}(i) - 2\bar{x}_k(1 - 2\gamma)\tilde{x}_{n+k-1}(i)$$
$$-\gamma\Big(\big(x_{n+k-1}^2(i+1) - \bar{x}_k^2\big) + \big(x_{n+k-1}^2(i-1) - \bar{x}_k^2\big)\Big), \quad (2.158)$$

where $k = 1, \ldots, s$.

Let the parameter perturbations of the $i$th element depend on its variable and on the variables of the two neighboring elements as follows:

$$\tilde{\alpha}_{n+k-1}(i) = a_1\tilde{x}_{n+k-1}(i) + a_2\Big(\big(x_{n+k-1}^2(i+1) - \bar{x}_k^2\big) + \big(x_{n+k-1}^2(i-1) - \bar{x}_k^2\big)\Big).$$
$$(2.159)$$

Substituting (2.159) into (2.158) we can obtain coefficients $a_1$ and $a_2$ at which the perturbations of the dynamical variable of the $i$th element decrease in time. The form of the parameter perturbation of the system (2.153) to convert it from the regime of spatiotemporal chaos to the spatio-homogeneous regime with temporal period $T = s$ is as follows:

$$\tilde{\alpha}_{n+k-1}(i) = 2(1 - 2\gamma)\bar{x}_k\big(x_{n+k-1}(i) - \bar{x}_k\big)$$
$$+\gamma\Big(\big(x_{n+k-1}^2(i+1) - \bar{x}_k^2\big) + \big(x_{n+k-1}^2(i-1) - \bar{x}_k^2\big)\Big), \quad (2.160)$$

where $k = 1, \ldots, s$.

To test the obtained results, numeric simulation has been carried out with the coupled logistic map chain (2.153). Let us convert the chaotic behavior to spatio-homogeneous with $T = 1$. At first, we determine from (2.157) the value of the fixed point $\bar{x}_1$. Then, having set the values of the parameters $\alpha = 1.6$ and $\gamma = 0.01$, which correspond to the regime of spatiotemporal chaos, we wait for the moment in the evolution of the system when the value of one of the elements falls within the $\varepsilon$-neighborhood of the $\bar{x}_1$. After that, the magnitude $\gamma\Big(\big(x_n^2(i+1) - \bar{x}_1^2\big) + \big(x_n^2(i-1) - \bar{x}_1^2\big)\Big)$ is evaluated. If it is less than the given $\varepsilon$, the parameter perturbation in the form (2.160) is introduced to influence this element. When other elements enter the given neighborhood, the perturbations begin to act on their parameters, if the above-mentioned condition is satisfied. Gradual transition of the elements to the chosen state is observed. At first, there appear small randomly distributed spatio-homogeneous patterns. Then, they gradually expand and, as a result, occupy the whole chain, and a stationary spatio-homogeneous regime is observed in the system. The process of control is consistently illustrated in Fig. 2.89a. The parameter perturbations having been turned off, the system naturally returns to the regime of spatiotemporal chaos.

Chaotic behavior can be converted to a spatio-homogeneous regime with larger temporal period in a similar way. In this case the form of the parameter perturbation becomes complicated as seen from (2.160). To convert the system to the regime with temporal period-2 we need to use two kinds of perturbations. One of them acts when the system elements enter the neighborhood of the point $\bar{x}_1$, while the other perturbation acts at the next iteration when

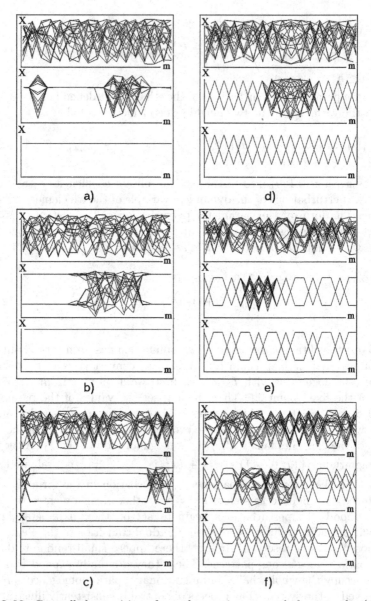

**Fig. 2.89.** Controlled transitions from the spatio-temporal chaos to the (a)–(c) spatio-homogeneous and (d)–(f) spatio-periodic regimes

they enter the neighborhood of the point $\bar{x}_2$. For the regime with temporal period-4, there are four similar expressions of the perturbations, every one of which influences the system parameter at the determined time moment and so on. The transition of the system from the spatiotemporal chaos to the spatio-homogeneous regime with temporal periods $T = 2$ and $T = 4$ is illustrated in Fig. 2.89b,c.

It should be mentioned that small patterns of inhomogeneous behavior remain uncontrollable when the system is converted to spatio-homogeneous regimes with temporal periods larger than one. To eliminate these dislocations we apply weak noise to them. As a result, the whole chain is converted to the regimes of spatio-homogeneous motions.

The more complicated situations are controlled transitions of the chain to spatio-periodic structures with $\lambda = 2$ and $T = 2$, $\lambda = 3$ and $T = 2$, and $\lambda = 5$ and $T = 4$. In this case our approach is as follows: We decompose all elements of the lattice in $\lambda$ groups. The elements of every such group, located at the distance $p\lambda$ $(p = 1, 2, \ldots)$ from each other, have equal values of the dynamical variable $x_n(i) = x_n(i + p\lambda)$. Having written the equation (2.153) for every group of the elements we obtain the system of equations:

$$x_{n+1}(\lambda m + 1) = \alpha - x_n^2(\lambda m + 1) + \gamma\big(2x_n^2(\lambda m + 1) - x_n^2(\lambda m + 2)$$
$$-x_n^2(\lambda m)\big),$$
$$x_{n+1}(\lambda m + 2) = \alpha - x_n^2(\lambda m + 2) + \gamma\big(2x_n^2(\lambda m + 2) - x_n^2(\lambda m + 3)$$
$$-x_n^2(\lambda m + 1)\big),$$
$$x_{n+1}(\lambda m + \lambda) = \alpha - x_n^2(\lambda m + \lambda) + \gamma\big(2x_n^2(\lambda m + \lambda) - x_n^2(\lambda m + \lambda + 1)$$
$$-x_n^2(\lambda m + \lambda - 1)\big), \tag{2.161}$$

$\big[m = 0, 1, \ldots, (N/\lambda) - 1\big]$. Then, using (2.161) and applying the approach considered for spatio-homogeneous regimes to every such group of the elements, we derive expressions for the parameter perturbations to convert the coupled map chain (2.153) from the regime of spatio-temporal chaos to the spatio-periodic structures mentioned above.

For example, for the controlled transition to the state with $\lambda = 2$ and $T = 2$, the following form of perturbations can be utilized:

$$\tilde{\alpha}_n(i) = 2p_1(i)(1 - 2\gamma)\big(x_n(i) - p_1(i)\big) + \gamma\Big(\big(x_n^2(i+1)$$
$$-p_2^2(i)\big) + \big(x_n^2(i-1) - p_2^2(i)\big)\Big),$$

$$\tilde{\alpha}_{n+1}(i) = 2p_{11}(i)(1 - 2\gamma)\big(x_{n+1}(i) - p_{11}(i)\big) + \gamma\Big(\big(x_{n+1}^2(i+1)$$
$$-p_{21}^2(i)\big) + \big(x_{n+1}^2(i-1) - p_{21}^2(i)\big)\Big),$$

$$p_1(i) = \begin{cases} \bar{x}_1, \\ \bar{x}_2, \end{cases} \quad p_2(i) = \begin{cases} \bar{x}_2, & \text{if } i = 2m + 1, \\ \bar{x}_1, & \text{if } i = 2m + 2, \end{cases}$$

$$p_{11}(i) = \begin{cases} \bar{x}_{11}, \\ \bar{x}_{21}, \end{cases} \quad p_{21}(i) = \begin{cases} \bar{x}_{21}, & \text{if } i = 2m + 1, \\ \bar{x}_{11}, & \text{if } i = 2m + 2, \end{cases}$$

where $\bar{x}_1, \bar{x}_2, \bar{x}_{11}, \bar{x}_{21}$ are values of the fixed points, $m = 0, 1, 2, \ldots, (N/2) - 1$.

Forms of the parameter perturbations to convert the system to the spatio-periodic structure with $\lambda = 3$ and $T = 2$ and with $\lambda = 5$ and $T = 4$ can be obtained in a similar way. Figure 2.89d–f demonstrates the results of numerical simulation of the controlled transition from spatio-temporal chaos to the spatio-periodic structure with $\lambda = 2$ and $T = 2$ (Fig. 2.89d), $\lambda = 3$ and $T = 2$ (Fig. 2.89e), and $\lambda = 5$ and $T = 4$ (Fig. 2.89f). In principle, the procedure of control is the same as in the case of spatio-homogeneous states. Only the form of the parameter perturbations becomes complicated: the form of the chosen structure is more complicated and the character of the perturbation is more complicated. It depends on both a coordinate $i$ and an iteration $n$, when a value of the dynamical variable falls within the neighborhood of the fixed point.

In the situations considered, the effect of control also occurs when we use only some of the conditions on the parameter perturbations. However, in this case the duration of the transient process grows extremely.

The approach suggested for controlling spatio-temporal chaos can be applied to chains with other kinds of coupling. For example, it is easy to obtain expressions for the parameter $\alpha$ perturbations to convert system (2.154) to different spatio-homogeneous and spatio-periodic regimes.

Numeric investigations have shown that spatio-temporal chaos in the linear coupled logistic map chain can be converted to the same spatio-homogeneous and spatio-periodic states considered in system (2.153).

Efficiency of controlling chaos in the coupled map chains (2.153) and (2.154) depends on the magnitude of the nonlinearity parameter $\alpha$ and the coupling coefficient $\gamma$. For example, transformation of system (2.153) to the spatio-homogeneous regime with temporal period-1 in the interval of parameter values $1.6 < \alpha < 1.9$ and $0 < \gamma < 0.4$ during 15 000 iterations at $N = 20$ is reached only in the parameter range shown in Fig. 2.90 by shading. We could not transfer the system from the chaotic regime to the spatio-homogeneous one during 15 000 iterations if $\gamma$ was more than 0.04. However, transition of the system to spatio-periodic regimes has been feasible outside the allocated domain.

It is evident that the possibility to convert the system from the regime of spatiotemporal chaos to the desired spatio-temporal structure depends on the probability of the chain element entering the $\varepsilon$-neighbourhood of the corresponding fixed point. From Fig. 2.91 one can see that the largest magnitudes of the probability density are located in the domain where the controlled transition to the spatiotemporal state with $T = 1$ is observed (see Fig. 2.89).

The suggested approach can be used to control spatio-temporal chaos in lattices of higher dimension.

Consider the two-dimensional coupled logistic map lattice

$$
\begin{aligned}
x_{n+1}(i,j) = \ & \alpha - x_n^2(i,j) + \gamma\big(4x_n^2(i,j) - x_n^2(i-1,j) \\
& - x_n^2(i+1,j) - x_n^2(i,j-1) - x_n^2(i,j+1)\big), \quad (2.162)
\end{aligned}
$$

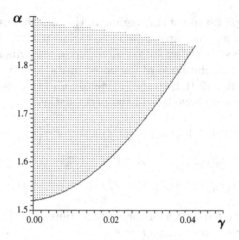

**Fig. 2.90.** Region on the $(\alpha - \gamma)$ plane where the transition to the spatio-homogeneous regime with $T = 1$ is observed

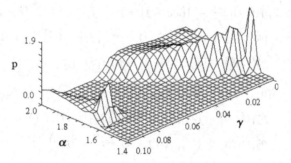

**Fig. 2.91.** Dependence of the probability density of finding the element of the chain (2.154) at the fixed point $\bar{x}_1$ on the parameters $\alpha$ and $\gamma$

with periodic boundary conditions $x_n(1, j) = x_n(N + 1, j)$, $x_n(i, 1) = x_n(i, N + 1)$, where $i, j = 1, 2, \ldots, N$, and $N \times N$ is the size of the lattice.

When the parameters $\alpha$ and $\gamma$ are varied, complex spatio-temporal dynamics is observed in the lattice. Without detailed discussion, we shall only note that there exist parameter value areas with developed spatio-temporal chaos. It is preceded by complicated bifurcational transitions between different spatio-temporal regimes [207, 432].

To convert the lattice from the regime of spatio-temporal chaos to regular states, we determine the form of the parameter perturbations in the same way as for the chains. As an example, we write the forms of perturbations for controlled transition to spatio-homogeneous regimes with temporal period $T = s$ and to the so-called "checkerboard structure". In this structure neighbouring elements of the lattice oscillate out of phase with temporal period $T = 2$. The prototype of this regime in the one-dimensional lattices is a zigzag structure which is a spatio-periodic regime with $\lambda = 2$ and $T = 2$.

From the analysis of the evolution equations of the dynamical variable perturbations in the neighborhood of the noted unstable regimes, it is seen that to convert the lattice (2.162) to the previously unstable spatio-homogeneous regimes with $T = s$ we have to add the perturbation

$$\tilde{\alpha}_{n+k-1}(i,j) = 2\bar{x}_k(1 - 4\gamma)(x_{n+k-1}(i,j) - \bar{x}_k) + \gamma\Big((x_{n+k-1}^2(i-1,j) - \bar{x}_k^2)$$
$$+ (x_{n+k-1}^2(i+1,j) - \bar{x}_k^2) + (x_{n+k-1}^2(i,j-1) - \bar{x}_k^2)$$
$$+ (x_{n+k-1}^2(i,j+1) - \bar{x}_k^2)\Big), \tag{2.163}$$

to the nonlinearity parameter of the system. Here, $k = 1, 2, \ldots, s$ and $\bar{x}_k$ is the fixed point of the orbit with $T = s$.

If the perturbation has the form

$$\tilde{\alpha}_n(i,j) = 2p_1(i,j)(1 - 4\gamma)(x_n(i,j) - p_1(i,j)) + \gamma\Big((x_n^2(i+1,j)$$
$$- p_2^2(i,j)) + (x_n^2(i-1,j) - p_2^2(i,j)) + (x_n^2(i,j-1)$$
$$- p_2^2(i,j)) + (x_n^2(i,j+1) - p_2^2(i,j))\Big),$$

$$\tilde{\alpha}_{n+1}(i,j) = 2p_{11}(i,j)(1 - 4\gamma)(x_{n+1}(i,j) - p_{11}(i,j)) + \gamma\Big((x_{n+1}^2(i+1,j)$$
$$- p_{21}^2(i,j)) + (x_{n+1}^2(i-1,j) - p_{21}^2(i,j)) + (x_{n+1}^2(i,j-1)$$
$$- p_{21}^2(i,j)) + (x_{n+1}^2(i,j+1) - p_{21}^2(i,j))\Big),$$

$$p_1(i,j) = \begin{cases} \bar{x}_1, \\ \bar{x}_2, \end{cases} \quad p_2(i,j) = \begin{cases} \bar{x}_2, \text{ if } (i+j) \text{ is even,} \\ \bar{x}_1, \text{ if } (i+j) \text{ is odd,} \end{cases}$$

$$p_{11}(i,j) = \begin{cases} \bar{x}_{11}, \\ \bar{x}_{21}, \end{cases} \quad p_{21}(i,j) = \begin{cases} \bar{x}_{21}, \text{ if } (i+j) \text{ is even,} \\ \bar{x}_{11}, \text{ if } (i+j) \text{ is odd} \end{cases}$$

(where $\bar{x}_1, \bar{x}_n, \bar{x}_{11}, \bar{x}_{12}$ are the fixed points), one can stabilize the checkerboard structure.

Figures 2.92 and 2.93 show the results of numeric simulation of spatio-temporal chaos control in the two-dimensional coupled logistic map lattice. The lattice with $20 \times 20$ elements has been used in the experiments. We have chosen $\alpha = 1.7$ and $\gamma = 0.01$ when the regime of spatio-temporal chaos is realized in the lattice (Figs. 2.92a and 2.93a). The procedure of control in the lattice is similar to that in the chain. Following the value of the variable of every element in the lattice, we wait for the moment when it enters the chosen $\varepsilon$-neighbourhood. Then, the corresponding perturbations of the system parameter are introduced if the conditions on the values of the nearest neighboring elements are satisfied. The figures consistently illustrate the

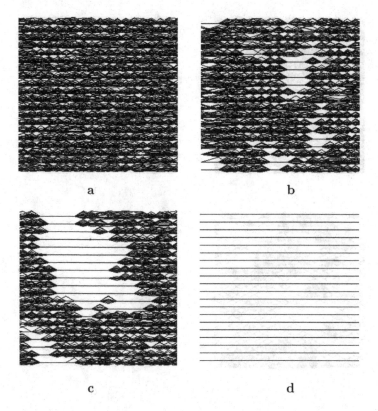

**Fig. 2.92.** Controlled transition from the regime of spatio-temporal chaos (**a**) to the spatio-homogeneous one with $T = 1$ (**d**) ((**b**) and (**c**) are intermediate stages)

controlled transitions from the regime of spatio-temporal chaos to the spatio-homogeneous one with $T = 1$ (Fig. 2.92d) and to the checkerboard structure (Fig. 2.93d). The transient process is the same as in the chains. At first, there appear small randomly distributed patterns, expanding step by step and as a result, occupying the whole lattice.

### 2.6.7 Summary

In this section we have described some problems on controlling chaos in coupled DS.

We have considered controlled anti-phase synchronization of chaos in coupled cubic maps. We have determined the stability regions of regular synchronous regimes and found the relation between the normal and the tangent Lyapunov exponents. It has been shown that diffusively coupled discrete maps cannot demonstrate anti-phase self-synchronization of chaos. Using, as an example, a system of coupled cubic maps, we have demonstrated the method

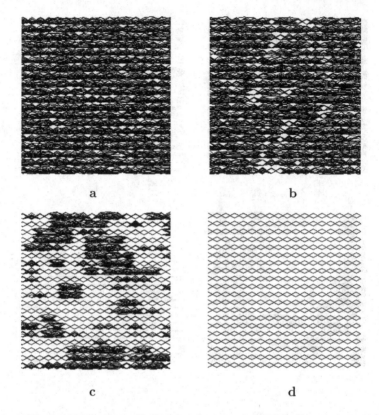

a

b

c

d

**Fig.2.93a–d.** Controlled transition to spatio-periodic structure

of controlled anti-phase chaotic synchronization for such systems. The addition of a feedback term to the system enables anti-phase chaotic oscillations within a bounded control parameter region to be stabilized. Outside this domain, the synchronization regime is destroyed. When $r > -1$, this process is accompanied by the chaotic attractor bubbling and completed by a blowout bifurcation. When $r < -1$, the loss of synchronization leads to the riddled basin of its attraction.

We have also shown that the chaotic phase trajectory of a system of two symmetrically coupled self-sustained oscillators can be stabilized by applying small controlling perturbations to one of the oscillators. By this means, we can realize (i) the controlled transitions from nonsynchronous chaos to the regimes of periodic oscillations and (ii) the regime of chaos synchronization in the form of both in-phase and anti-phase chaotic oscillations.

The results presented in this section testify that chaotic systems can be synchronized by means of parametric periodic perturbations on the coupling elements. It has been found that parametric perturbations with certain amplitude and frequency values can stabilize spatio-homogeneous motions in chains consisting of a finite number of identical chaotic oscillators.

We have also described the method of controlling spatio-temporal chaos in coupled map lattices and explored the controlled transitions from the regime of spatio-temporal chaos to the desired regular spatio-temporal structures.

## 2.7 Reconstruction of Dynamical Systems

### 2.7.1 Introduction

One of the methods for studying various processes and phenomena in real life is the construction and exploration of their mathematical models. Knowledge of the model equations for a system under consideration significantly extends the possibilities for its study, allowing one to solve the problems of prediction of the system's temporal behavior and evolution of the functioning regime of the system under variation of its control parameters. Theoretically, modeling of systems contains no problems if information about the real DS is known. For example, let the system under study be some radio-technical circuit. Usually a circuit diagram is set up, and then on the basis of physical laws the equations are formulated describing its dynamics with a certain approximation. These equations represent a model DS.

However, when exploring real processes and phenomena, the researchers may often encounter situations when all of the information about the internal arrangement and operation of a system under study cannot be obtained due to some objective reasons. Moreover, there also arise the problems of adequate determination of system state variables, the latter being a set of time-varying quantities. In general, the time evolution of the system's state may correspond to deterministic or stochastic processes, or to their superposition. This fact should be taken into account when formulating the notion of a DS.

There are two approaches for defining a DS. The first definition is mathematical, according to which a DS is an evolution operator for state variables. In the classical theory of DS a DS is defined as a set of first-order ordinary differential equations. For such a system, the state is unambiguously determined as a set of independent variables at a given time moment, while the equations themselves define the temporal evolution of the state. The number of state variables (or the order of a system of equations) defines the dimension of a DS.

The study of real evolutionary processes in the natural sciences requires extension and generalization of the classical notion of DS. A more general concept of DS should include the influence of noise and provide the possibility of using DS as evolution equations of statistical theory. From the physical point of view, we can deal with DS even in the case when one is unable to write the equations of a DS although the time evolution of the state of a physical system can be observed experimentally.

From the viewpoint of experimentalists, a DS is thought of as *a really existing system* for which one can introduce the notion of a state at every time moment and assume that there exists a continuous or discrete operator *approximately* describing its evolution (in time and/or in space). With this, noise can be understood as internal or external fluctuations or the influence of a large number of factors weakly affecting the system's behavior and, therefore, ignored when specifying a state. In this case the minimal number of independent coordinates which is sufficient to describe the system state in the given approximation is called the dimension of the system. We shall further call such systems "real" DS (RDS).

In practice, the situation when the time dependences of all coordinates of the system state can be measured is encountered very rarely. It is more likely that a limited number of the characteristic quantities can be measured. In typical situations, only one state coordinate, $a(t)$, is available for measurements. The dependence of a quantity, describing the system state, on some independent variable, very often represented by the time or a space coordinate, is named the realization (observable) of the system. It so happens that knowledge about the system under study can be obtained from its realizations only and any other information is unavailable. This circumstance has led to the notion of a "black box". Any DS is a "black box" if all available information about it is contained in the input and output signals only. Besides, the presence of the input signal is not necessary. The realization $a(t)$ (observable) sampled with some step $\Delta t$ is called the one-dimensional time series $a(i\Delta t) = a_i$, $i = 1, \ldots, N$.

At present the problem of reconstruction of DS encloses a wide class of scientific problems which are aimed at obtaining partial or possibly maximal information about the properties of a "black box" from its experimentally measured one-dimensional time series $a_i$. This problem includes reconstruction (restoration) of attractors topologically equivalent to the initial system attractor, determination of the quantitative characteristics of the dynamics (attractor dimension, Lyapunov exponents, probability measure, etc.) and, finally, global reconstruction of a DS under study. The global reconstruction problem implies restoration of the model equations of DS, which with the given accuracy reproduce the experimentally measured time series. In the present section we shall consider some of the problems stated above, describe the relevant reconstruction algorithms, and illustrate their applications with several examples.

## 2.7.2 Reconstruction of Attractors from Time Series

**Analysis of Continuous Signals.** Suppose that the dependence $a(t)$ is a one-dimensional projection of a phase trajectory of some DS defined as a set of ordinary differential equations (ODE). We assume that the phase trajectory belongs to a system attractor $A$ of dimension $d$. Until 1980 it was assumed that to describe nonlinear DS in terms of phase space one should know the time series of all the state coordinates. Packard et al. [433] outlined techniques for reconstructing the phase portrait of a DS from its scalar time series $a_i$ provided that the remaining coordinates of the state vector are specified by the same time series but taken with some delay. Mathematical results on delay coordinates for nonlinear systems were first published by Takens in 1981 [434]. He stated the following: Assume that a DS has a chaotic attractor $A$ and $a(t)$ is its one-dimensional realization. Let $A$ belong to a smooth $M$-dimensional manifold. Then, the set of vectors $\boldsymbol{x}(t)$ in $\boldsymbol{R}^n$

$$\boldsymbol{x}(t) = \Lambda_n[a(t)] = \Big(a(t), a(t+\tau), \ldots, a\big(t+(n-1)\tau\big)\Big) = (x_1, x_2, \ldots, x_n),$$
(2.164)

obtained by means of the delay method, can be viewed as an $n$-dimensional reconstruction $A_R$ of the original attractor. Besides, $n$ should satisfy the condition of the Mañé theorem [435]:

$$n \geqslant 2M + 1,$$
(2.165)

where $M$ is the nearest integer number to the attractor dimension $d$. In accordance with Takens' theorem, the map $\Lambda_n : A \to A_R$ is smooth and invertible on $A_R$ for almost any delay $\tau$ (if $N \to \infty$). The number $n$ is called the embedding dimension.

Takens' theorem was proved under conditions that a DS under study is autonomous and noise-free and that a time series was measured with a high accuracy and over an infinite time interval. Thus, this theorem cannot be formally applied to RDS (i.e., noisy systems) and real experimental conditions (any time series can be measured over a finite time interval and with a finite accuracy). In spite of this fact, since its appearance, Takens' theorem has been used by experimentalists for restoring phase portraits and estimating on their basis various characteristics of attractors of DS. Only in 1997 did Stark et al. [436] prove the theorem that extends Takens' theorem to systems driven by an external force and to noisy systems.

Takens' theorem has opened wide possibilities for solving the problems of prediction of the system's state [437–439] and calculation of the metric [440,441] and dynamical [442,443] characteristics of the original attractor. In 1987, Cremers and Hübler [444], and independently Crutchfield and Mc-Namara [445], proposed a method for restoring the equations of a DS on its one-dimensional realization (method of global reconstruction), which will be discussed in detail below.

The embedding dimension $n$ can be determined from (2.165) if the attractor dimension $d$ is known. In practice such a value of $n$ may often be overestimated [446], and the embedding space with a lower dimension may be enough. There are also methods to immediately estimate the embedding dimension, which do not require the attractor dimension and are based on the Karhunen–Loeve theorem [447]. To estimate the attractor dimension $d$ the so-called correlation dimension is often used. It is defined as follows:

$$D_{\mathrm{c}} = \lim_{\varepsilon \to 0} \lim_{N \to \infty} \frac{\ln C(\varepsilon, N)}{\ln \varepsilon} , \tag{2.166}$$

where $C(\varepsilon, N) = N^{-2} \sum_{i \neq j} v(\varepsilon - |\,\boldsymbol{x}_i - \boldsymbol{x}_j\,|)$ is the correlation integral, $\varepsilon$ is the size of a partition element of phase space, $N$ is the number of points used to estimate the dimension, $v$ is the Heaviside function, and $\boldsymbol{x}_i = \boldsymbol{x}(i\Delta t)$; $\boldsymbol{x}$ is a phase vector, reconstructed from $a_i$, in an $n$-dimensional space, and $n$ varies from the smallest possible value (e.g., 2) to the largest one, which is supposed to be an upper limit. The correlation integral is actually the number of pairs of points, separated by less than $\varepsilon$, and which are normalized by $N^2$. One can numerically estimate the correlation dimension $D_{\mathrm{c}}$ as the slope of the best fit straight line to a numerical plot of $\ln C(\varepsilon, N)$ versus $\ln \varepsilon$. In addition, the dependence of $D_{\mathrm{c}}$ on the embedding dimension can also be analyzed. If the dimension of the original attractor is finite, then as $n$ increases, the value of $D_{\mathrm{c}}$ saturates.

It is known that when calculating the correlation dimension some restrictions are imposed on the value of $\varepsilon$. If $\varepsilon$ approaches the size of the attractor $\varepsilon_{\max}$, the dependence $\ln C(\varepsilon, N)$ on $\ln \varepsilon$ undergoes saturation. This can be explained as follows: When $\varepsilon$ is larger than the attractor size, the distances between all the points on the attractor are certainly less than $\varepsilon$, and starting from $\varepsilon = \varepsilon_{\max}$ we have $C(\varepsilon, N) = 1$ and, consequently, $\ln C(\varepsilon, N) = 0$. On the other hand, with decreasing $\varepsilon$ there is a value of $\varepsilon_{\min}$ such that for $\varepsilon < \varepsilon_{\min}$ the structure of the attractor is unresolved. As a consequence, the dependence $\ln C(\varepsilon, N)$ versus $\ln \varepsilon$ becomes nonlinear again.

Despite the fact that Takens' theorem holds for any $\tau$, the choice of delay may significantly affect the value of $D_{\mathrm{c}}$. To obtain the truest information about the metric characteristics of an attractor, numerous ways of choosing an optimal value of $\tau$ have been proposed (the most complete review of them is given in [448]). If $\tau$ is too small, the $i$th and the $(i+1)$th coordinates of a phase point are practically equal to each other. In this case, the reconstructed attractor is situated near the main diagonal of the embedding space ("identity line" [448]), the latter complicating its diagnostics. When a value for $\tau$ is chosen that is too large, the coordinates become uncorrelated, and the structure of reconstructed attractor is lost [449]. The value of $\tau$ can be chosen based on the calculation of the function of mutual information [450], the first minimum of the logarithm of the generalized correlation integral [451], etc. In some cases $\tau$ is determined by a less rigorous way, namely, from the geometry of the reconstructed set (to prevent an expansion of the attractor in all directions).

Another problem in evaluating the dimension is the choice of $n$ and $\tau$. Instead of fitting individual values of $n$ and $\tau$, it is appropriate to consider the size of time window $n\tau$. At the same time, one should take into account certain fundamental limitations for estimating $D_{\mathrm{c}}$ [452], which are defined by the formula

$$D_{\max} = \frac{2 \ln N}{\ln(1/r)} , \qquad r = \frac{\varepsilon}{\varepsilon_{\max}} . \tag{2.167}$$

The latter expression means that the value of $D_c$, derived using the algorithm for dimension estimation, cannot be larger than $D_{max}$ for a given number $N$ of points. In other words, if $r = 0.1$ and $N = 1000$, then $D_{max} \leqslant 6$, and if $N = 100\,000$, then $D_{max} \leqslant 10$. As a result, certain difficulties arise when distinguishing in a system with a sufficiently large number of degrees of freedom a deterministic process from a purely stochastic process.

In agreement with Takens, a state vector can be specified by the method of delays (2.164) or the method of successive differentiation:

$$\boldsymbol{x}(t) = \left( a(t), \frac{\mathrm{d}a(t)}{\mathrm{d}t}, \ldots, \frac{\mathrm{d}^{n-1}a(t)}{\mathrm{d}t^{n-1}} \right) = (x_1, x_2, \ldots, x_n) . \tag{2.168}$$

Since the values of $a_i$ are known only at discrete time moments $i\Delta t$, the coordinates $x_j$ of vector $\boldsymbol{x}$ are defined by numerical differentiation of the initial time series using approximate mathematical expressions. Obviously, the accuracy of derivative computation depends on the sampling step $\Delta t$. A disadvantage of this method consists in its high sensitivity to noise, the latter limiting its applicability for embedding spaces of higher dimension (at least without pre-filtering).

A large number of methods have been developed (see the good survey in [453]) which allow one to identify a state vector. Among them we point out the method of integral filtration [454], assuming restoration of the phase coordinate as follows:

$$x(t) = \int_0^t a(t_1) \exp \left( \frac{t - t_1}{\beta} \right) \mathrm{d}t_1 . \tag{2.169}$$

This method provides smoothing of the initial realization and noise filtering. The same feature is characteristic for the method of moving averages [453]:

$$x_i(k) = \frac{1}{2k+1} \sum_{j=-k}^{k} a_{i+j} , \tag{2.170}$$

where $k$ is a constant parameter.

**Analysis of Interspike Intervals.** There are two ways of obtaining experimental data. The traditional way is to fix the sampling step $\Delta t$ and store the values of observables corresponding to the time moments $i\Delta t$. This method has been discussed above. In the framework of the second method, one imposes a certain condition on the value of the observable $a(t)$ and records the intervals between the time moments for which the condition is fulfilled. This method for obtaining experimental data is often used for analyzing the systems generating "events" which represent striking changes in the physical variables that are repeated more or less regularly. This situation typically arises in the biological sciences and is encountered in neurobiology (neuron firings corresponding to voltage spikes [455]), in cardiology ($R$-peaks of electrocardiograms [456]), in membrane biology (bursting oscillations of the cell membrane potential [457]), etc. Systems with this type of dynamics are often analyzed by processing time intervals between the relevant events (for example, interspike intervals (ISI) [455]).

Different models of spike generation are known. Within the framework of *integrate-and-fire* (IF) models [458]– [462], a signal $a(t)$, being a function of the variables of a low-dimensional DS, is integrated from some moment $T_0$. The times $T_i$ when spikes occur can then be defined by the equation

$$\int_{T_i}^{T_{i+1}} a(t)\mathrm{d}t = \theta , \quad I_i = T_{i+1} - T_i , \tag{2.171}$$

where $\theta$ is a firing threshold, and $I_i$ are the IF ISI. When the specified threshold $\theta$ is reached, a sharp pulse is generated (Fig. 2.94a), and the value of the integral is reset to zero.

*Threshold-crossing* (TC) models [460], on the other hand, assume the existence of a threshold level $\theta$, which defines the equation of a secant plane $a = \theta$ [$a(t)$ is now a variable of a DS], and measure time intervals between successive crossings of the given level by the signal $a(t)$ in one direction, e.g., from below and to above (TC ISI) (Fig. 2.94b). From the viewpoint of DS theory, TC ISI are the times when the phase trajectory returns to the secant plane.

The ISI analysis is important when, for whatever reason, the full signal $a(t)$ cannot be recorded, and only a sequence of firing times is available in the course of the experiment. A sensory neuron that transforms a time-varying input signal $a(t)$ into the resulting output spike trains may serve as a classical example. This transformation has previously been investigated within the framework of information theory.

A sensory neuron represents a threshold device with input and output: at the input a signal of complex structure is received; at the output a series of

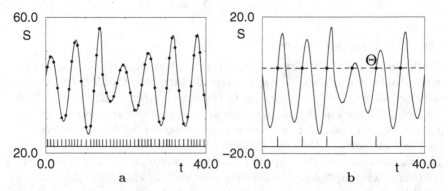

**Fig. 2.94.** Models of spike generation: (a) integrate-and-fire, and (b) threshold crossing. *Black dots* indicate the time moments when a threshold level is reached. A sequence of spikes representing the output signal is given at the bottom of each figure

pulses is measured. Since the output pulses are identical and their shape does not depend on an external force, the information about the properties of the input signal can be encoded only in time intervals between neuron firings (interspike intervals of the action potential trains or ISI). A question arises: How can a characterization of the input signal be provided when processing a spike train only? Recently this topic has become of interest from the viewpoint of reconstruction of DS [458]– [464]. When applying a reconstruction technique to a point process (a process in which the dynamical information is carried by a series of event timings) one needs to answer the following question [459]: If the point process is a manifestation of the underlying deterministic system, can the states of this system be identified from the information provided by the point process? When dealing with a sensory neuron subject to an external force, this question is as follows: Can the state of the forcing system be identified from an output ISI series?

An answer was first given by Sauer [459]. An ISI may be considered as a new state variable allowing one to characterize the low-dimensional dynamics at the input of a neuron from the observed spike train. Following Sauer [459], the attractor of a chaotic system can be reconstructed using a sequence of time intervals only, and deterministically driven IF ISI sequences can be distinguished from stochastically driven series on the basis of calculation of the prediction error. Sauer [458] also proved the embedding theorem for IF ISI. Following Hegger and Kantz [463] this theorem is valid for return times as well. From [463] it also follows that the map of return times (i.e. the set of vectors reconstructed from return times by means of the delay method (2.164)) is topologically equivalent to the Poincaré section of the initial system. A detailed study of how different properties of a chaotic forcing are reflected in an output IF ISI series was performed by Racicot and Longtin [462]. In [464] the possibility was demonstrated of evaluating the largest Lyapunov exponent of a chaotic attractor using the sequences of TC time intervals.

To reveal the interrelation between the return times and phase variables of a DS, consider the concept of the instantaneous phase. In [197] three ways of introducing the instantaneous phase of chaotic oscillations are described. The first two ways are connected with the existence of a projection of a system's attractor on the plane $(x, y)$, reminiscent of a smeared limit cycle. If this projection exists then one is able to introduce the Poincaré secant so that it will pass through an equilibrium point of the system.

According to the *first definition*, the phase is defined as follows:

$$\varphi^m(t) = 2\pi \frac{T - T_i}{T_{i+1} - T_i} + 2\pi i, \quad T_i \leqslant T < T_{i+1}, \qquad (2.172)$$

where $T_i$ are the time moments of trajectory's crossings of the secant surface.

Following the *second definition* the phase is introduced as

$$\varphi^p = \arctan\left(\frac{y}{x}\right). \qquad (2.173)$$

In the general case the two phases $\varphi^m$ and $\varphi^p$ do not coincide, and only the mean frequency defined as the average of $d\varphi^p/dt$ coincides with $2\pi/I$, where $I$ is the average return time, $I = \frac{1}{n}\sum_{i=1}^{n} I_i$.

The *third definition* of an instantaneous phase is based on the analytic signal concept and the Hilbert transform [465]. The Hilbert transform of the real signal $a(t)$ is also a real signal $a^H(t)$, such that the signal $z(t) = a(t) + i\,a^H(t) = A(t)\exp\left(i\,\varphi^H(t)\right)$ is analytic, i.e.,

$$a^H(t) = \frac{1}{\pi}\int_{-\infty}^{\infty}\frac{a(\tau)}{t-\tau}\,d\tau\,, \tag{2.174}$$

$$A(t) = (aa + a^H a^H)^{1/2}\,, \quad \varphi^H(t) = \arctan\left(\frac{a^H(t)}{a(t)}\right). \tag{2.175}$$

In (2.174) the integral is taken in the sense of the Cauchy principal value, i.e., it is supposed that the limit $\lim_{A\to-\infty, B\to\infty}\int_A^B f(x)\,dx$ may not exist, while the limit $\lim_{A\to\infty}\int_{-A}^{A} f(x)\,dx$ exists. The modulus $A(t)$ of the signal $z(t)$ represents the envelope of the original function, and argument $\varphi^H(t)$ defines the instantaneous phase.

It is known that the Hilbert transform (2.174) shifts each spectral component of the original signal by $-\pi/2$ [466]. Function $a^H(t)$ is an inertial transformation of $a(t)$ and can be, in principle, used for embedding, i.e., for reconstruction of the phase portrait in the variable space $(a, a^H)$.

In agreement with Takens' theorem, one can reconstruct the original attractor from a realization of one of the phase variables. In this case, instead of a projection on the plane $(x, y)$ of the original attractor, we can consider a projection of the attractor, reconstructed from the one-dimensional realization $a(t)$. If the Hilbert transform $a^H(t)$ is used as the embedding method, then the two latter definitions of phase coincide.

Note that consideration of instantaneous amplitude and phase (or frequency)

$$\omega^H(t) = \frac{d\varphi^H(t)}{dt} = \frac{\dot{a}^H a - \dot{a}a^H}{a^2 + (a^H)^2} \tag{2.176}$$

as new independent variables instead of $\left(a(t), a^H(t)\right)$ means substitution of coordinates which are smooth everywhere except at the origin $(a(t) = 0, a^H(t) = 0)$. If the trajectory does not cross the origin, this substitution leads to topological equivalence of the attractors in the variable space $(a, a^H)$ and $(\omega^H, A)$. Therefore, one could use either the instantaneous frequency $\omega^H(t)$ or the amplitude $A(t)$ to reconstruct the original attractor.

Now we turn to experimental data measured as time intervals between intersections by a realization $a(t)$ of some threshold level $\theta$. Suppose that the plane defined as $a(t) = \theta$ can be treated as a Poincaré secant. Then we can use the first method to introduce the instantaneous phase $\varphi^i$ and frequency $\omega^i$, i.e., we can attribute to each time moment $T_i$ the value of

$$\omega^i(T_i) = \frac{2\pi}{I_i} , \qquad (2.177)$$

where $\omega^i(T_i)$ denotes the value of instantaneous frequency $\omega^H(t)$ averaged over time $I_i$. $\omega^i(T_i)$ can be qualitatively treated as the points of the new coordinate reconstructed from $\omega^H(t)$ by means of averaging (2.170) with varying window size.

Knowing $\omega_i$ only at discrete time moments $T_i$ and trying to know for sure how the averaged instantaneous frequency behaves itself between the time moments $T_i$, one can interpolate the values of $\omega^i(T_i)$ with some smooth curve. By all means, we would never obtain exactly the true dependence, but we hope that the time series obtained in such a way will qualitatively reproduce in certain case the behavior of one of the system's coordinates and thus allow us to reconstruct approximately the view of the original attractor and its dynamical and geometrical properties.

As an example, consider the famous Rössler system,

$$\frac{dx}{dt} = -(y + z) , \quad \frac{dy}{dt} = x + ay , \quad \frac{dz}{dt} = b + z(x - c) , \qquad (2.178)$$

in the chaotic regime $a = 0.15$, $b = 0.2$, $c = 10.0$. Instantaneous frequency $\omega^H$ versus $t$ as well as its averaged value

$$\langle \omega^H \rangle(t) = \frac{1}{I_i} \int_{T_i}^{T_{i+1}} \omega^H(t)\, dt, \qquad (2.179)$$

calculated from the system's coordinate $x(t)$, are shown in Fig. 2.95a. The qualitative correspondence between $\omega^i(t)$ and $\langle \omega^H \rangle(t)$ is demonstrated in Fig. 2.95b. Here, we plot the dependence $\omega^i(T_i)$ for the values of $I_i$ measured as time intervals between zeroes crossings of the coordinate $x(t)$ and

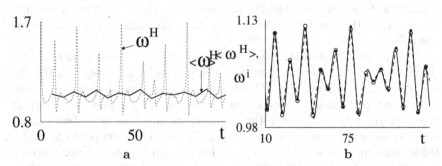

**Fig. 2.95.** (a) Time dependence of instantaneous frequency $\omega^H$ obtained from the Hilbert transform (*dashed line*) and the result of averaging $\langle \omega^H \rangle$ (*solid line*). (b) *Black dots* connected by a *dashed line* are the points of average instantaneous frequency $\langle \omega^H \rangle$; *open circles* connected by a *solid curve* represent the result of an interpolation $\omega^i(t)$

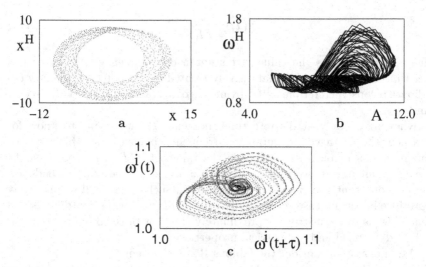

**Fig. 2.96.** Phase portraits on the planes (**a**) $(x, x^H)$, where $x$ is the first coordinate of the system (2.178) and $x^H$ is Hilbert transform of $x$; (**b**) $(A, \omega^H)$, where $A$ and $\omega^H$ are the instantaneous amplitude and frequency, respectively; and (**c**) $\left(\omega^i(t), \omega^i(t + \tau)\right)$, where $\omega^i(t)$ is a time dependence obtained by interpolation

by using interpolation technique we yield the smooth curve through all this points. Thus, a system's coordinate is reconstructed from TC ISI. The reconstructed phase portraits in coordinates $\left(a(t), a^H(t)\right)$, $\left(A(t), \omega^H(t)\right)$, and $\left(\omega^i(t), \omega^i(t + \tau)\right)$ are shown in Fig. 2.96.

The box-counting algorithm [440] was used to compute the fractal dimension of the attractors reconstructed from the $x$-coordinate of the Rössler system and from the signals $\omega^i(t)$. The plots of $\log M$ versus $\log \varepsilon$, where $M$ is the number of nonempty boxes and $\varepsilon$ is the size of a box, are given in Fig. 2.97a. It is seen that the linear segments of these two graphs are parallel to each other, and thus the dimensions of the corresponding attractors coincide.

It has been also studied how the choice of the threshold level $\theta$ (or the choice of the secant plane) influences the possibility of recovering the dynamical invariants from the TC ISI series. These investigations have been carried out in terms of an estimation of the largest Lyapunov characteristic exponent (LCE) $\lambda_1$. Consider the equations of the Rössler system (2.178) and introduce a secant plane $x = \theta$. The largest LCE computed from TC ISI at different values of the threshold level demonstrates a dependence on $\theta$ as illustrated in Fig. 2.97b. The values $\lambda_1$ are estimated as follows [464]: First, the transition from a set of time intervals $I_i = T_{i+1} - T_i$ to the points $\omega(T_i) = 2\pi/I_i$ appropriate to the values of the instantaneous (angular) frequency averaged over a return time is carried out. Here, $T_i$ are the moments of TC. Second, the series of points $\omega(T_i)$ is interpolated by a smooth function

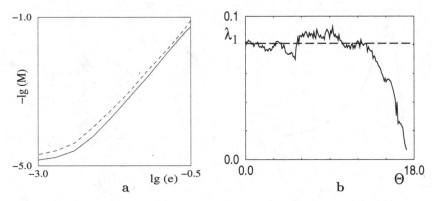

**Fig. 2.97.** (a) Plot for estimation of the Hausdorff dimension. The *solid line* is obtained from the $x$-coordinate of system (2.178), the *dashed line* is obtained from the zero-crossing ISI. (b) The largest LCE computed from TC ISI versus threshold level $\Theta$

(cubic spline) $\omega_{\mathrm{int}}(t)$ to transform it into a signal with constant time step for attractor reconstruction. Third, $\lambda_1$ is computed from $\omega_{\mathrm{int}}(t)$ using the algorithm suggested by Wolf et al. [442] with the replacement procedures after a fixed time $\overline{I} = \frac{1}{N}\sum_{i=1}^{N} I_i$. The dynamical properties of the chaotic attractor can be estimated with good accuracy if $|\theta| \leqslant 13$ (the error of estimation of $\lambda_1$ does not exceed 12% and may be less if the length of the time series is increased). The values of $\lambda_1(\theta)$ in Fig. 2.97b has been obtained for a series of about 2000 return times. We conclude that the sequence of time intervals reflects the dynamical properties of the chaotic attractor even in the case when some of the loops of the phase trajectory fail to cross the secant plane. We conclude that the value $\lambda_1$ is insensitive to the choice of a threshold level, provided that $\overline{I}$ does not exceed some temporal scale. Of course, when speaking about the sensitivity of dynamical characteristics, we understand that the largest LCE can be estimated only with some accuracy, taking into account both the finite amount of data and the dependence of the applied algorithm on the parameters of the numerical computations.

The above conclusion is not trivial. We see from Fig. 2.98 that shifting of the secant plane results in essential changes of the structure of the return times map. Even though for the correct choice of the secant (when all the phase trajectories cross it), the return times map is similar to some extent to the Poincaré map (Fig. 2.98a,b), shifting of the threshold level produces a very different phase portrait (Fig. 2.98c).

The detailed explanation of these results can be found in [464]. If the value of the average time interval $\overline{I}$ exceeds some characteristic scale (a measure of predictability [437] or the Lyapunov time), it is obviously impossible to estimate the largest LCE.

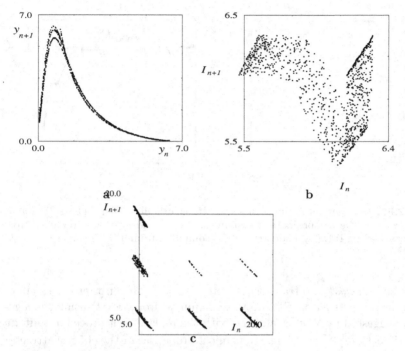

**Fig. 2.98.** (a) Poincaré map of the Rössler system; (b) and (c) return times maps appropriate to $\Theta = 0$ and 11, respectively

### 2.7.3 Global Reconstruction of DS

**History and Statement of the Problem.** The problem of recovering a DS from its one-dimensional realization refers to the problems of synthesis, or inverse problems. Unlike the problem of analysis, the task of synthesis is ambiguous, since there is an infinite number of DS of different types and complexity which can reproduce the signal under study with the given accuracy. The goal of the researcher who searches for an appropriate dynamical description for the real system under consideration consists in obtaining an approximate form of a DS producing the quantitative characteristics of the signals under study. At present only common recommendations have been developed which could help the researchers to choose one of various possible models (when the original system is not too complicated). However, there is no unified approach for solving the stated problem in the general case.

The authors of pioneering works [444, 445] proposed a method for global reconstruction of the equations of a DS from its one-dimensional realization. The suggested method that will be described in detail below is as follows: Assume that a DS is represented by a "black box" and we can measure its one-dimensional realization. First the measured time series is used to restore the phase portrait topologically equivalent to the initial system attractor [434]. In the second stage the form of equations describing the initial system is stated a priori and the unknown coefficients of the set are fitted by means of the least-squares method.

Later, the idea of global reconstruction was developed and improved in other papers. For example, for reconstruction of dynamical equations from experimental time series with a broadband continuous spectrum, the authors of the work [439] used additional information about the dynamical and statistical properties of the original system, which is contained in the time series. In particular, they took into account Lyapunov exponents and the probability distribution density calculated from the initial time series. However, the resulting evolution equations had an awkward form and were inconvenient to use. The paper of Breeden and Hübler [467] is devoted to the use of hidden variables to write model equations. A method of synchronization between the model and initial data was described in [468]. In [469], Bünner et al. proposed an algorithm for the restoration of a scalar differential equation for time-delay systems. The use of the methods described above was illustrated using simple low-dimensional model systems when the result of global reconstruction is known a priori. Nevertheless, compared with the method [444] no significant advantages were provided by any improved (and, as a rule, algorithmically very complicated) technique. The methods described are tested on well-studied low-dimensional model systems with simple enough right-hand sides to the equations, which can be easily reconstructed also by means of simpler methods [444]. Therefore, all arguments in favor of new complex algorithms appear to have little force until their workability is demonstrated with complex time series generated by real "black boxes". Unfortunately, practically no publications are available devoted to the application of the described procedures to signals from real systems whose evolution operator is unknown. Analyzing the works [470–475] one can state that the simplest originally proposed method subjected to minor modifications proves to be the most effective way for restoring DS from real experimental data. A far more complete presentation of methods of reconstruction of system dynamics can be found in [476].

**Description of the Reconstruction Algorithm.** The aim of this algorithm is to construct the mathematical model of a system under study which is assumed to be dynamical. Taking into account that the available time series is discrete, one is able to specify the sought-after DS in the form of an $n$-dimensional discrete map:

$$x_{1,i+1} = F_1(x_{1,i}, x_{2,i}, \ldots, x_{n,i})$$
$$\ldots\ldots$$
$$x_{n,i+1} = F_n(x_{1,i}, x_{2,i}, \ldots, x_{n,i}) \,,$$

(2.180)

where $x_{j,i}$ are the coordinates of a state vector at the time moments $i\Delta t$ and $F_j$ are the nonlinear functions.

If the embedding dimension is found and the phase portrait of the original system is restored, then the main problem at this stage is to fit the functions

$F_j$ into the right-hand sides of the sought-after equations. The general form of nonlinear functions $F_j$ is stated a priori. The situations are possible when the form of the evolution operator can be figured out using additional knowledge of the system, but they are encountered rarely. The functions $F_j$ are more frequently searched for by considering all available and known methods for their representation. Usually $F_j$ are presented in the form of a superposition of some basis functions $X_k$:

$$F_j(x_{1,i}, x_{2,i}, \ldots, x_{n,i}) = \sum_{k=1}^{L} C_{k,j} X_k(x_{1,i}, x_{2,i}, \ldots, x_{n,i}) . \qquad (2.181)$$

The most popular way to fit $F_j$ is to represent the sought-after functions as the polynomials of $\nu$th order:

$$F_j(\boldsymbol{x}_i) = \sum_{l_1,l_2,\ldots,l_n=0}^{\nu} C_{j,l_1,l_2,\ldots,l_n} \prod_{k=1}^{n} x_{k,i}^{l_k} , \qquad \sum_{k=1}^{n} l_k \leq \nu, \quad (2.182)$$

where $C_{j,l_1,l_2,\ldots,l_n}$ are the unknown coefficients to be found. This method is based on the Taylor theorem about the expansion of functions into a polynomial series in the vicinity of a certain point $\boldsymbol{x}^0$. The Taylor series can be restricted to a certain finite number of the expansion terms provided that the function values have a relatively small finite deviation from the given point (this condition is realized in self-sustained systems when a trajectory returns into the neighborhood of any attractor point, and the attractor itself is located in a bounded region of the phase space). Varying the number of the expansion terms left, one can change the error of function approximation in the neighborhood of the given point. The functions can be approximated by the Legendre polynomials [444] as well as by a more complicated procedure described in [472]. To define $F_j$ we shall use the expression (2.182).

The system of equations (2.180) can be written for any $i$. To find the coefficients of each function $F_j$, one needs to solve the system of $N$ ordinary differential equations,

$$x_{j,i+1} = \sum_{l_1,l_2,\ldots,l_n=0}^{\nu} C_{j,l_1,l_2,\ldots,l_n} \prod_{k=1}^{n} x_{k,i}^{l_k} , \qquad i = 1,\ldots,N, \quad (2.183)$$

with unknown $C_{j,l_1,\ldots,l_n}$, where $N$ is the number of points of a scalar time series, used for approximation of the right-hand sides and $\nu$ is the polynomial order.

For given $n$ and $\nu$, in general the number of coefficients $K$ of the polynomials (2.182) can be estimated from the formula $K = (n+\nu)!/(n!\nu!)$. As a rule, $N \gg K$. Hence to determine the evolution operator the system of equations (2.183) is solved by means of the least-squares method. The resulting mathematical model looks awkward enough, but its solution can reproduce the signal with high accuracy provided that the general form of the nonlinear functions is correctly chosen.

In a similar manner, DS can be reconstructed not only in the form of discrete maps but also as systems of first-order ODE:

$$\frac{\mathrm{d}x_1}{\mathrm{d}t} = F_1(x_1, x_2, \ldots, x_n)$$

$$\ldots\ldots \tag{2.184}$$

$$\frac{\mathrm{d}x_n}{\mathrm{d}t} = F_n(x_1, x_2, \ldots, x_n).$$

The functions on the right-hand sides have the same meaning as before. The implementation of the reconstruction algorithm at its first stage means that all $x_i$ are known. Consequently, their derivatives can be found numerically. Hence the set (2.184) really represents a system of algebraic equations, the latter being linear with respect to unknown coefficients.

If the state vector is specified by the successive differentiation method, the restored mathematical model can have a simpler form, since the interrelation between the coordinates is uniquely determined by the equalities (2.168):

$$\frac{\mathrm{d}x_1}{\mathrm{d}t} = x_2, \quad \frac{\mathrm{d}x_2}{\mathrm{d}t} = x_3, \quad \ldots, \quad \frac{\mathrm{d}x_n}{\mathrm{d}t} = f(x_1, x_2, \ldots, x_n). \tag{2.185}$$

Therefore, the algorithm described above allows one to assign the mathematical model to a scalar time series. Figures 2.99 and 2.100 exemplify the restoration of the dynamical description for two test model systems, namely, the Van der Pol oscillator,

$$\frac{\mathrm{d}x}{\mathrm{d}t} = y, \quad \frac{\mathrm{d}y}{\mathrm{d}t} = a\left(1 - bx^2\right)y - x, \tag{2.186}$$

and the Rössler system (2.178) in the chaotic regime.

The detailed study of different test models by means of the global reconstruction technique can be found, for example, in [472].

**Peculiarities and Disadvantages of the Reconstruction Algorithm.** The main problem which can arise when applying the algorithm of global reconstruction in practice is the choice of nonlinear functions on the right-hand sides of model equations. If the method of successive differentiation is used for defining the coordinates of the state vector, the mathematical model will contain only one unknown function. As has been already indicated, the simplest method for fitting the nonlinearity consists in a polynomial approximation.

With this, the legitimacy of the statement above may be questioned. As an example, consider the famous Rössler model. As was shown in [472], only in the case when the second coordinate alone is given, the Rössler system (2.178) can be reduced to the form (2.185) by substituting the variables, where $f$ is a polynomial defined by (2.182). If another coordinate is chosen as the initial one, then $f$ represents a ratio of polynomials $P/Q$.

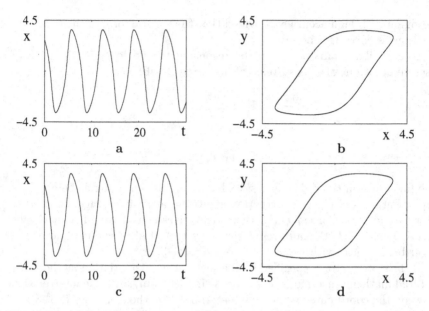

**Fig. 2.99.** (a) Initial realization and (b) projection of the phase portrait of system (2.186) for $a = 1.0$, $b = 0.3$; (c) the realization and (d) the projection of the attractor of the corresponding DS $(n = 2)$ reconstructed from the first coordinate with $\Delta t = 0.01$ by the method of delays with $\nu = 3$

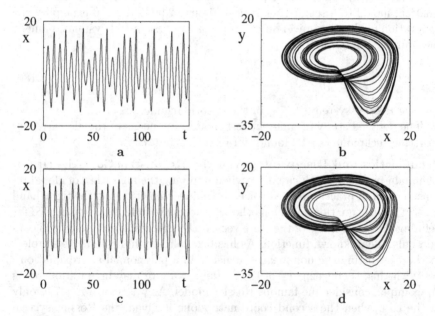

**Fig. 2.100.** (a) Initial realization and (b) projection of the phase portrait of the Rössler system in the chaotic regime; (c) the realization and (d) the projection of the attractor of the corresponding DS $(n = 4)$ reconstructed from the first coordinate with $\Delta t = 0.01$ by the method of differentiation with $\nu = 3$

In [477] it is suggested that a mathematical model be searched in the following form:

$$g_j \frac{\mathrm{d}x_j}{\mathrm{d}t} = F_j(x_1, x_2, \ldots, x_n) \,. \tag{2.187}$$

The system (2.187) is derived from (2.184) by multiplying the latter by $g_j = g_j(\boldsymbol{x})$, i.e., by one of the functions $1, x_j, x_j^2, x_j x_k, \quad j, k = 1, 2, \ldots, n$.

In some situations such an approach may appear to be very useful. However, when dealing with experimental time series generated by a DS with unknown evolution operator, the problem of fitting nonlinearities remains extremely complicated.

Another problem is associated with the necessity of operating with noisy data when processing experimental time series. On the one hand, the method of successive differentiation is more preferable for restoring a phase trajectory, since the resulting model contains in the general case $n$ times less coefficients on different nonlinearities as the one reconstructed by the method of delays. But differentiation inevitably causes a noisy component in high-order derivatives to increase. Without pre-filtering the time dependence even of the second derivative may appear to be a noisy process. Besides, the traditionally used embedding methods (2.164) and (2.168) demonstrate evident disadvantages when analyzing highly inhomogeneous realizations, i.e., the signals containing segments with quick motion followed by the segments with slow motion. Such time series are typical for systems with "pauses" and may often be encountered in biology and medicine.

Let us remark again that an arbitrary choice of nonlinearities does not always allow one to perform a good reconstruction. In particular, in [472] three typical cases were outlined. They are as follows:

(i)   Restored equations give a local description of initial signals. With this, the reconstructed model is unstable in the sense that the solution of the obtained equations reproduces the signal under study over a short time interval only.

(ii)  There exists a bad local predictability of the phase trajectory. Nevertheless, a visual similarity of the phase portraits can be observed. The solution of restored equations is stable according to Poisson. In this case the attractor of the reconstructed model has metric characteristics close to those of the initial attractor.

(iii) There exists a good local predictability of the phase trajectory from any of its points over the times exceeding a characteristic correlation time. The phase portrait of the reconstructed model is identical to the initial one, and the system itself is stable according to Poisson.

Recently, the global reconstruction technique has been used not only for obtaining mathematical models but also for classifying dynamical regimes [478–480]. The dynamical classification method supposes a transition from the phase space of the initial DS to a space of coefficients of restored equations. A feature of this approach is that the reconstruction procedure applied to experimental data does not require one to find a model, which is capable of reproducing the initial signal. Coefficients in the obtained equations represent quantitative characteristics which carry information about linear and nonlinear correlations in the initial signal. After approximating coefficients on realization segments, the researcher obtains a set of points in a coefficient space. It is shown that nonoverlapping regions in this space correspond to different classes of DS. Thus, the technique described can be effectively utilized for classification of DS.

### 2.7.4 Reconstruction from Biological Data

**Reconstruction from Inhomogeneous Realizations.** As has been mentioned in Sect. 2.4.2, there are different methods for restoration of the phase portraits of DS. The most popular are the method of delays (2.164) and the method of successive differentiation (2.168).

In this section we consider different methods for phase portrait restoration from the viewpoint of convenience of their application for further global reconstruction. The method of successive differentiation may be preferable as it leads to the simplest form (2.185) of the sought-after system. But the evident disadvantage of this method is that a noisy component of the initial realization is enhanced with each subsequent step of differentiation. One more negative peculiarity of this method is revealed when one deals with highly inhomogeneous realizations. The inhomogeneity consists in alternating segments with "quick" and "slow" motions. A typical example is the human electrocardiogram (ECG) shown in Fig. 2.101. Differentiation of a "quick" segment (between $P$ and $T$ waves) gives a quicker time dependence while differentiation of a "pause" (between $T$ and the next $P$ waves) gives a slower function of time. Thus, each subsequent phase coordinate restored by the method (2.168) becomes more and more inhomogeneous. As a consequence, the reconstructed phase portrait will be also inhomogeneous. This leads to the fact that the region with slow motion will contribute significantly to the approximation of the coefficients of a model, while the other regions of the phase space, although often being more informative, will not greatly influence the problem. Thus, to effectively apply the least-squares method the initial phase portrait should be sufficiently homogeneous.

Figure 2.102 shows phase portraits constructed from an ECG (Fig. 2.101) by using the methods of delays and differentiation. The use of the successive differentiation method leads to a smooth but sharply inhomogeneous phase portrait (Fig. 2.102a). When applying the method of delays we can choose the particular value of time lag $\tau$ for which the restored phase portrait is smooth (Fig. 2.102b), but in this case it is also inhomogeneous. To obtain a smooth attractor we have to take small values of $\tau$, namely, much less than the duration of the "pause". Resolution of the problem of inhomogeneity by means of a special choice of delay will lead to a new problem. As seen from Fig. 2.102c, the obtained phase portrait becomes nonsmooth at some points.

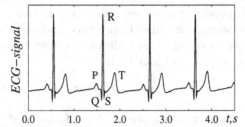

**Fig. 2.101.** Typical human ECG (after noise filtration)

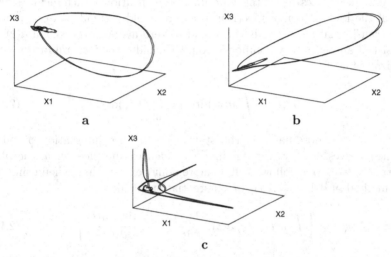

**Fig. 2.102.** Phase portraits restored on an ECG (**a**) by means of successive differentiation; (**b**) by the method of delays with $\tau$ being much less than the duration of the "pause"; and (**c**) by the method of delays with $\tau$ being close to the duration of the "pause"

The latter means that to approximate the flow at these points one needs a quickly varying function on the right-hand sides of the sought-after equations which when expanded into a Taylor series does not permit restriction to a small number of expansion terms.

In [481] a simple method was described and applied allowing one to restore the most homogeneous phase portrait possible and to derive a simple form of the reconstructed differential equations (2.185). This method is considered to be a limiting case of the integral filtration technique (2.169).

The increase in effectiveness of the method for restoring a homogeneous phase portrait is achieved by increasing the values of $\beta$. $\beta = \infty$ is the limiting case, when the reconstructed coordinate is an integral of the initial realization. Passing to the limit of $\beta$ has an important advantage: a significant decrease in the calculations necessary to restore coordinate $x$. This is explained by the fact that the exponential multiplier in the integrand of (2.169) disappears.

Consider a typical experimental realization $a(t)$ which can often be qualitatively presented as a sum of four terms:

$$a(t) = O(t) + S(t) + \xi(t) + C , \qquad (2.188)$$

where $O(t)$ is a stationary component of the self-sustained process with zero average, $S(t)$ is the "floating" of the average level (usually low frequency) which may be caused by nonstationarity, $\xi(t)$ is the additive noise, and $C$ is a constant shift of the whole realization.

After pre-processing of the experimental realization, which includes filtering of the low frequency spectrum range to avoid a trend, noise reduction, and transformation of the initial signal to zero average, the signal $a(t)$ will consist of an oscillatory component only. Consider the integral of the whole realization $a(t)$,

$$a_1(t) = \int_0^t a(t_1)dt_1 = \int_0^t O(t_1)dt_1, \qquad (2.189)$$

as one of the coordinates of the state vector. Since integration of "slow" segments gives quickly varying time dependences and vice versa, the phase portraits restored as follows will be more homogeneous than when using only the method of delays (2.164) or differentiation (2.168):

$$\mathbf{x}(t) = \left\{ \int_0^t a(t_1)dt_1, a(t), \frac{da(t)}{dt}, \dots, \frac{d^{n-2}a(t)}{dt^{n-2}} \right\} \qquad (2.190)$$

or

$$\mathbf{x}(t) = \left\{ \int_0^t a(t_1)dt_1, a(t), a(t+\tau), \dots, a(t+(n-2)\tau) \right\} . \qquad (2.191)$$

Moreover, as one uses the embedding (2.190), the reconstructed ODE have the simplest form (2.185).

**Examples of Reconstruction from Medical and Biological Data.** The time series measured from real systems of medical and biological origin are often inhomogeneous. One of the simplest examples is the time dependence of a coordinate of a point on the surface of an isolated frog's heart, which is shown in Fig. 2.103a. This realization has a rather simple shape but is inhomogeneous at the same time because of the existence of "pauses" in it.

We apply the integration embedding method (2.190) to this time series to restore the phase portrait of the system in four-dimensional phase space (Fig. 2.103b). The fitted DS of the form (2.185) modeling the given regime possesses an attractor identical to the initial one (Fig. 2.103d), whose solution is shown in Fig. 2.103c.

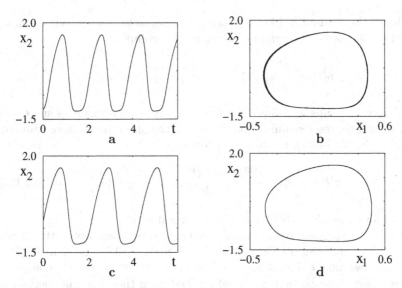

**Fig. 2.103.** (a) Initial realization of mechanical oscillations of a point on the surface of an isolated frog's heart; (b) projection of phase portrait constructed from this realization by the method (2.190); (c) the realization and (d) projection of the attractor of the corresponding reconstructed DS

Now we move to a more complicated experimental signal, namely, the ECG of a human heart. Since a typical ECG is inhomogeneous, we use the embedding method (2.190) or (2.191) instead of (2.164) or (2.168). Taking into account the fractal dimension estimates for the ECG attractor made in [482], the embedding dimension should be not less than at least 5. When restoring the phase portrait by means of the methods (2.164) or (2.191), the increase in embedding dimension leads to a more complicated form of model equations. The use of (2.168) or (2.190) even for filtered data increases the errors of computation of high-order derivatives, which leads to additional problems when fitting the right-hand sides of a model.

One of the peculiarities of an ECG of a healthy human heart is the fact that its characteristic period, i.e., the repeating sequence $P, Q, R, S, T$ peaks – "pause" (see Fig. 2.101), contains all the information about the structure of $QRS$ complex, the $P$ and $T$ waves being enough to diagnose certain pathological changes. Therefore, we can first formulate a simplified problem, namely, to model a DS whose solution will be a periodic signal reproducing with high accuracy a single beat of ECG. Consider two different electrocardiograms from which noise was pre-filtered (Figs. 2.104a and 2.105a). For each of them a single beat was arbitrarily chosen and repeated many times to obtain a periodic time series of sufficient length. The interval between two successive $R$ waves corresponds to the initial period of each realization.

Unlike the method (2.190), for these realizations we successively compute two integrals. Let $a(t)$ be the initial signal. We compute

$$b(t) = \int_0^t a(t_1)dt_1 , \qquad c(t) = \int_0^t b(t_1)dt_1 . \qquad (2.192)$$

$c(t)$ may be considered to be the initial realization. By means of successive differentiations, the remaining coordinates of the phase vector were restored, the latter finally having the following form:

$$\boldsymbol{x}(t) = \left\{ c(t) , \frac{dc(t)}{dt} , \frac{d^2c(t)}{dt^2} , \ldots , \frac{d^{n-1}c(t)}{dt^{n-1}} \right\} . \qquad (2.193)$$

It is obvious that $d^2c(t)/dt^2$ is the initial signal $a(t)$.

The phase portrait projections restored in the described way on the chosen beats of two different ECGs are shown in Figs. 2.104b and 2.105b. This method allows one to obtain more homogeneous phase portraits as compared to those ones restored by traditional methods. On this basis the method of least squares can be considered to be the most effective way for fitting the coefficients of the right-hand sides of a model. For the two cases being studied we reconstruct two models (3- and 4-dimensional) in the form of the systems of ODE (2.185), whose solutions are given in Figs. 2.104c,d and 2.105c,d.

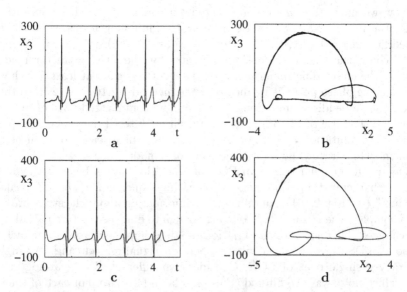

**Fig. 2.104.** Results of reconstruction of a DS from an ECG. (a) Initial periodic realization obtained by "closing" a single beat of a real ECG of the first type; (b) projection of phase portrait restored on this realization by the method (2.193); (c) the realization and (d) projection of the attractor of the reconstructed 3-dimensional model

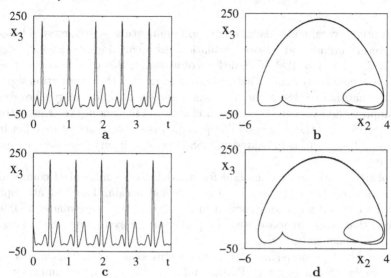

**Fig. 2.105.** Results of reconstruction of a DS from an ECG. (**a**) Initial periodic realization obtained by "closing" a single beat of a real ECG of the second type; (**b**) projection of phase portrait restored on this realization by the method (2.193); (**c**) the realization and (**d**) projection of the attractor of the reconstructed 4-dimensional model

The proposed method (the use of two integrals of the signal which is obtained by repeating a single beat of ECG) allows one to reconstruct robust mathematical models in the form of systems of discrete maps, to model different pathologies, etc. The presented results demonstrate the possibility of restoration of a sufficiently homogeneous attractor from inhomogeneous data, testifying the advantage of the used method.

## 2.7.5 Global Reconstruction in Application to Confidential Communication

Let us consider a particular case of the global reconstruction of DS consisting in the presence of a priori information about the DS producing the signal under study. The mathematical model of the DS generating the signal under observation can be written as follows:

$$\frac{\mathrm{d}\boldsymbol{x}}{\mathrm{d}t} = \boldsymbol{F}(\boldsymbol{x}, \mu), \quad \boldsymbol{x} \in \mathbf{R}^n, \ \mu \in \mathbf{R}^m, \tag{2.194}$$

where $\mu = (\mu_1, \ldots, \mu_m)$ is the control parameter vector.

Suppose that the explicit form of nonlinear vector function $\boldsymbol{F}(\boldsymbol{x}, \mu)$ is known a priori. In this case the problem of global reconstruction consists in determination of unknown values of parameters $\mu_i$. Such a supposition may be treated as a particular case of the problem under study, but it may be effectively applied in a series of practical tasks. Thus, the method of reconstruction may be successfully used for multichannel confidential communication.

Recently, several methods for secure communication were suggested based on exploiting broadband chaotic oscillations as a carrying (or masking) signal [228–231, 386, 483–488]. The authors of works on this topic used the phenomenon of chaotic systems synchronization to extract the information signal from the chaotic one [186]. The methods based on the effect of synchronization suppose the existence of identical chaotic oscillators in the transmitter and the receiver. If the corresponding parameters of both oscillators differ by more than 2%, then, as indicated in [488], the algorithms proposed become ineffective.

Consider an alternative method for multichannel confidential communication based on the technique of global reconstruction. The first attempts to use the method for reconstruction in communication were made in [489]. However, that work demonstrated the possibility of extracting only a single additive influence on the system.

The method proposed in [490] is more universal and at the same time rather simple. By analogy with Parlitz and co-workers [484, 485], information signals perform modulation of parameters of a DS. The difference consists in the method for demodulation. Instead of the auto-synchronization phenomenon, in [490] the technique of reconstruction is used.

Let us take a dynamical chaos oscillator $G_1$ as a transmitter whose mathematical model has the form (2.194). Let us restrict ourselves to the class of DS whose equations may be transformed to the form (2.185) by means of variable substitution. Many well-known models such as that of Lorenz, Rössler systems, Chua's circuit, etc. (see, e.g., [472]), satisfy this requirement. We consider relatively slow parameter $\mu_i$ modulation, the latter allowing us to transmit $m$ messages simultaneously:

$$\mu_i^*(t) = \mu_i^0 + \mu_i(t), \qquad i = 1, \ldots, m. \tag{2.195}$$

In (2.195) $\mu_i^0$ are the constant values of control parameters of system (2.185) and $\mu_i(t)$ are the information signals. The condition of slow modulation may be written as follows:

$$\left| \frac{d\mu_i}{dt} \right| \ll \left| \frac{dx_j}{dt} \right|, \tag{2.196}$$

for any $i$ and $j$.

Taking into account (2.195), (2.194) have the form

$$\frac{dx}{dt} = F\big(x, \mu^*(t)\big) . \tag{2.197}$$

To realize the message's demodulation from the carrier, which is a one-dimensional realization $x_1(t)$ of the system (2.197), the recipient of the information has to know the form of the nonlinear vector function $F(x, \mu)$. The condition (2.196) allows one to choose the temporal window $t^*$, inside which the values of $\mu_i^*$ are practically constant. The latter means that nonautonomous properties of system (2.197) may not be taken into account during the time interval $t^*$. Shifting the temporal window along the carrier signal $x_1(t)$, the recipient of the information who knows the mathematical model of oscillator $G_1$ may extract the modulation signals $\mu_i(t)$ in real time.

In practice, determination of the current values of $\mu_i(t)$ is realized as follows: The signal $x_1(t)$ of oscillator (2.197) observed during $t^*$ is differentiated $n$ times to compute the temporal dependences $x_j(t)$, $j = 2, \ldots, n$, and the left-hand sides of system (2.185). Since a scalar time series of numbers is observed $x_{1,i} = x_1(i\Delta t), i = 1, \ldots, N^*$, $N^* = t^*/\Delta t$, all the derivatives are computed at discrete time moments $i\Delta t$ using approximate formulas for numerical differentiation. As a result, one should solve

$$\frac{\mathrm{d}x_{n,i}}{\mathrm{d}t} = f(\boldsymbol{x}_i, \boldsymbol{\mu}^*) \tag{2.198}$$

to define the parameters $\mu_i^*$. This is an algebraic equation since the value of $\mathrm{d}x_{n,i}/\mathrm{d}t$ is computed by means of $n$-times differentiation of the initial signal. By writing (2.198) for $i = 1, \ldots, N^*$ and solving such a system of $N^*$ algebraic equations with $m$ unknown variables by means of least squares method $(N^* \gg m)$, the recipient of the information obtains the current values of $\mu_i^*$.

The method discussed above is also valid for discrete DS. An advantage of working with discrete maps is that one may not take into account the errors of numerical differentiation. By analogy with continuous systems one must choose a mathematical model which can be transformed into the following form:

$$x_{n+1,1} = x_{n,2}, \ x_{n+1,2} = x_{n,3}, \ \ldots, \ x_{n+1,n} = f(\boldsymbol{x}_n, \boldsymbol{\mu}). \tag{2.199}$$

Consider the equations of the well-known Hénon map

$$x_{n+1} = 1 - ax_n^2 + y_n, \ y_{n+1} = bx_n. \tag{2.200}$$

Bearing in mind that the sequence of numbers $y_i$, $i = 1, \ldots, N^*$, is available for the researcher, rewrite (2.200) as follows:

$$y_{n+1} = z_n, \ z_{n+1} = b - az_n^2/b + by_n. \tag{2.201}$$

In fact, it is enough to know the values $y_n, y_{n+1}, y_{n+2}$ and $y_{n+3}$ to determine the unknown parameters $a$ and $b$. A priori information about the general form of the mathematical model allows one to solve the global reconstruction problem for a very short scalar time series.

An example of multichannel communication on the basis of the Hénon map is presented in Fig. 2.106. Two chaotic signals [the coordinates $x(t)$ and $y(t)$ obtained by integrating the Rössler model] perform modulation of the parameters $a$ and $b$ of the system (2.200). At each iteration the values of $a$ and $b$ change according to the law shown in Fig. 2.106a,b. The restored signals are presented in Fig. 2.106d,e.

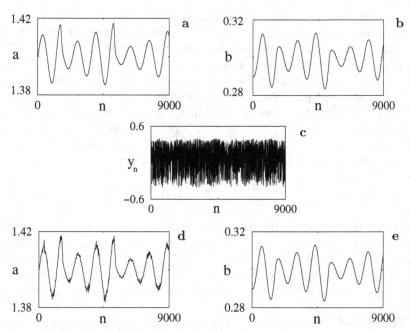

**Fig. 2.106.** Example of multichannel communication on the base of the Hénon map. (**a**) and (**b**): Information signals; (**c**) signal in the channel of communication; (**d**) and (**e**): restored signals

Let us turn to the systems with continuous time and choose the modified oscillator with inertial nonlinearity (Anishchenko–Astakhov oscillator) as a test model

$$\frac{\mathrm{d}x}{\mathrm{d}t} = m_0 x + y - xz\,, \quad \frac{\mathrm{d}y}{\mathrm{d}t} = -x\,, \quad \frac{\mathrm{d}z}{\mathrm{d}t} = -g_0 z + 0.5 g_0 (x + |x|)x\,. \quad (2.202)$$

Choose the one-dimensional realization $y(t)$ of the system (2.202) as a carrier. The equations of oscillator (2.202) may be written in the form

$$\frac{\mathrm{d}Y}{\mathrm{d}t} = Z\,, \quad \frac{\mathrm{d}Z}{\mathrm{d}t} = X\,, \quad \frac{\mathrm{d}X}{\mathrm{d}t} = f(X, Y, Z, \boldsymbol{\mu})\,, \quad \boldsymbol{\mu} = (m_0, g_0)\,, \quad (2.203)$$

$$f(X, Y, Z, \boldsymbol{\mu}) = \frac{X(X + Y)}{Z} + (m_0 g_0 - 1)Z - g_0(X + Y)$$
$$+ 0.5 g_0(|Z| - Z)Z^2\,, \quad (2.204)$$

by means of the following variable substitution:

$$Y = y\,, \quad Z = -x\,, \quad X = -m_0 x - y + xz\,. \quad (2.205)$$

Let us illustrate the possibility of simultaneous transmission of two independent messages in a single carrier. We modulate the parameter $m_0$ by a

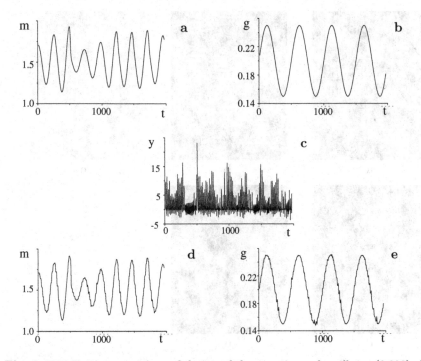

**Fig. 2.107.** Two-parametric modulation of the equations of oscillator (2.202). (a) and (b) information signals; (c) signal in the channel of communication; (d) and (e) restored signals

broadband chaotic signal from the Rössler system, while the parameter $g_0$ is modulated by an harmonic signal. To test the method used for its robustness in the presence of noise, we add a normally distributed random value with variance $10^{-5}$ to the equations of oscillator (2.202) and to its parameters.

The temporal dependences of the parameters of oscillator (2.202) are shown in Fig. 2.107a,b. The results of demodulation presented in Fig. 2.107d,e testify the reliability of the method in the case of simultaneous transmission of two independent messages in the presence of fluctuations.

The workability of the method used can clearly be demonstrated by transmitting graphic images. We choose the dependences obtained by scanning with a resolution of 200 × 250 two fragments of "Virgin of the Rocks" by Leonardo do Vinci (Fig. 2.108a,b) as information signals. The modulating signals were represented in the form of stepwise temporal dependences. The range of control parameter variation was split into 256 subranges, each corresponding to the shade of the black-and-white image. The signal in the communication channel (Fig. 2.108c) is the coordinate $y(t)$ of the Rössler system. Figure 2.108d,e illustrate the images reconstructed by the global reconstruction technique.

**Fig. 2.108.** Simultaneous transmission of two fragments by modulating parameters $b$ and $c$ of the Rössler system. (**a**) and (**b**): Initial images; (**c**) the signal in the communication channel; (**d**) and (**e**): restored images

To realize the proposed method for signal demodulation in practice, the recipient of the information must possess a specialized processor or computer and an analog-to-digital converter. We suppose that such a method for confidential communication may find effective application for the exchange of information over short-enough distances using a cable network, so that noise would not be created while broadcasting.

### 2.7.6 Summary

In the present section we have described an algorithm for global reconstruction of DS and demonstrated in principle the possibility for restoration of robust DS on signals from test models as well as on time series measured in real medical and biological experiments. Unfortunately, we were able to consider only a limited number of problems concerned with the global reconstruction of DS, which appear to be very promising.

In recent years the interest of specialists in nonlinear dynamics in the systems of biological origin has been growing. Biological systems are a wide field of research where the achievements of DS theory can find their practical application. Besides, the complexity and the presence of specific peculiarities of the signals from such systems can stimulate the development and improvement of techniques for time series analysis, including the methods for reconstruction of DS. Thus, in the section much attention has been paid to the reconstruction problem in application to highly inhomogeneous realizations, which are encountered very often among the signals of medical and biological origin, and to a human electrocardiogram, in particular.

At present, various possible applications of the global reconstruction technique are being studied yet. This can be confirmed by the fact that the authors of the majority of works on mathematical modeling use the method for global reconstruction to study test systems only. This algorithm is still rarely and not always successfully applied to experimental signals because of the complexity and ambiguity of the problem of synthesis of DS. Presently, one of the current and modern research problems is to find possible applications of the reconstruction methods. Of particular interest are applications of the global reconstruction technique for the purpose of diagnostics in medicine and biology. This problem is closely connected with construction of model systems, since a large variety of criteria, being either approximation coefficients of nonlinear functions in restored equations or any characteristics reflecting the peculiarities of a signal under study, can be used for diagnostics.

In connection with this the study of ISI data attracts great interest. Recently, the problem of attractor reconstruction from a sequence of time intervals has begun to be discussed in publications [458–463], and it offers a wide field of research for specialists in nonlinear dynamics.

We have also considered the transition from a sequence of discrete values to a smooth realization, the latter having the meaning of averaged instantaneous frequency, and possibilities of calculating the metric and dynamical characteristics for principally discrete signals. These facts allow one to believe that the nonlinear dynamics methods will complement a set of quantitative characteristics traditionally computed from ISI data for the purposes of diagnostics. In other words, the study of such signals can strengthen the role of nonlinear dynamics when solving the problem of analysis of time series of different origin.

# References

1. L.D. Landau, *Dokl. Acad. Nauk USSR* **44**, 339 (1944) (in Russian).
2. E.A. Hopf, *Commun. Pure Appl. Math.* **1**, 303 (1948).
3. D. Ruelle, F. Takens, *Commun. Math. Phys.* **20**, 167 (1971).
4. S. Newhouse, D. Ruelle, F. Takens, *Commun. Math. Phys.* **64**, 35 (1979).
5. J.P. Gollub, S.V. Benson, *J. Fluid Mech.* **100**, 449 (1980).
6. A. Lichtenberg, M.A. Lieberman, *Regular and Stochastic Motion* (Springer, New York 1983).
7. H.G. Schuster, *Deterministic Chaos* (Physik-Verlag, Weinheim 1988).

8. V.S. Anishchenko, *Complex Oscillations in Simple Systems* (Nauka, Moscow 1990).
9. P. Berge, I. Pomeau, C.H. Vidal, *Order Within Chaos* (Wiley, New York 1984).
10. R.M. May, *Nature* **261**, 459 (1976).
11. P.A. Linsay, *Phys. Rev. Lett.* **47**, 1349 (1981).
12. K.H. Simoyi, A. Wolf, H.L. Swinney, *Phys. Rev. Lett.* **49**, 245 (1982).
13. J. Testa, J. Perez, C. Jeffries, *Phys. Rev. Lett.* **48**, 714 (1982).
14. V.S. Anishchenko, V.V. Astakhov, T.E. Letchford, *Radiotech. Electron.* **27**, 1972 (1982) (in Russian).
15. V.S. Anishchenko, V.V. Astakhov, *Radiotech. Electron.* **28**, 1109 (1983) (in Russian).
16. V.S. Anishchenko, V.V. Astakhov, T.E. Letchford, M.A. Safonova, *Izv. Vuzov. Ser. Radiophys.* **26**, 169 (1983) (in Russian).
17. A. Libchaber, S. Fauve, C. Laroche, *Physica* D **7**, 73 (1983).
18. M. Henon, *Commun. Math. Phys.* **50**, 69 (1976).
19. S. Grossmann, S. Thomae, *Z. Naturforsch.* A **32**, 1353 (1977).
20. M.J. Feigenbaum, *J. Stat. Phys.* **19**, 25 (1978).
21. M.J. Feigenbaum, *J. Stat. Phys.* **21**, 669 (1979).
22. B. Hu, *Phys. Rep.* **91**, 233 (1982).
23. B. Hu, J.M. Mao, *Phys. Lett.* A **25**, 3259 (1982).
24. B. Hu, I. Satija, *Phys. Lett.* A **98**, 143 (1983).
25. J.P. Van der Welle, H.W. Cape, R. Kluiving, *Phys. Lett.* A **119**, 1 (1986).
26. D.V. Lubimov, M.A. Zaks, *Physica* D **9**, 52 (1983).
27. B.A. Huberman, J. Rudnik, *Phys. Rev. Lett.* **45**, 154 (1980).
28. V.S. Anishchenko, M.A. Safonova, *Sov. Phys. Tech. Phys.* **33**, 391 (1988).
29. A.N. Sharkovsky, *Ukr. Math. J.* **26**, 61 (1964) (in Russian).
30. T.Y. Li, J.A. Yorke, *Am. Math. Mon.* **82**, 985 (1975).
31. Y. Pomeau, P. Manneville, *Commun. Math. Phys.* **74**, 189 (1980).
32. P. Manneville, Y. Pomeau, *Physica* D **74**, 219 (1980).
33. C. Grebogi, E. Ott, J.A. Yorke, *Physica* D **7**, 181 (1983).
34. V.S. Afraimovich, Internal bifurcations and crises of attractors. In: *Nonlinear Waves. Structures and Bifurcations*, A.V. Gaponov-Grekhov and M.I. Rabinovich, eds. (Nauka, Moscow 1987) p. 189 (in Russian).
35. P. Berge, P. Dubois, P. Manneville, Y. Pomeau, *J. Phys. Lett.* **41**, L341 (1980).
36. M. Dubois, M.A. Rubio, P. Berge, *Phys. Rev. Lett.* **51**, 1446 (1983).
37. L.P. Shilnikov, Bifurcation theory and turbulence. In: *Nonlinear and Turbulent Processes*, vol. 2 (Gordon and Breach, Harward Academic Publishers 1984) p. 1627.
38. V.I. Arnold, *Mathematical Methods of Classical Mechanics* (Springer, Berlin, Heidelberg 1974).
39. V.I. Arnold, Stability loss of self-oscillations near resonances. In: *Nonlinear Waves*, A.V. Gaponov-Grekhov, ed. (Nauka, Moscow 1979) p. 116 (in Russian).
40. V.S. Afraimovich, L.P. Shilnikov, *Dokl. Acad. Nauk USSR* **24**, 739 (1974) (in Russian).
41. V.S. Afraimovich, L.P. Shilnikov, Invariant two-dimensional tori, their destruction and stochasticity. In: *Methods of the Qualitative Theory of Differential Equations* (Gorky University, Gorky 1983) p. 3 (in Russian).
42. V.S. Afraimovich, L.P. Shilnikov, Strange attractors and quasiattractors. In: *Nonlinear Dynamics and Turbulence*, G.I. Barenblatt, G. Iooss, and D.D. Joseph, eds. (Pitman, Boston 1983) p. 1–34.

43. V.S. Afraimovich, The ring concept and quasiattractors. In: *Proc. of the Int. Conf. on Nonlinear Oscillations*, vol. 2 (Naukova Dumka, Kiev 1984) p. 34 (in Russian).
44. J. Maurer, A. Libchaber, *J. Phys. Lett.* **40** L419 (1979).
45. D.G. Aranson, M.A. Chori, G.R. Hall, R.P. McGenehe, *Commun. Math. Phys.* **83**, 303 (1982).
46. V. Franceschini, *Physica* D **6**, 285 (1983).
47. V.S. Anishchenko, T.E. Letchford, M.A. Safonova, *Izv. Vuzov. Ser. Radiophys.* **28**, 1112 (1985) (in Russian).
48. J. Stavans, *Phys. Rev.* A **35**, 4314 (1987).
49. J.A. Glazier, A. Libchaber, *IEEE Trans. Circuit. Syst.* **35**, 790 (1988).
50. V.S. Anishchenko, M.A. Safonova, *Radiotech. Electron.* **32**, 1207 (1987) (in Russian).
51. V.S. Anishchenko, T.E. Letchford, D.M. Sonechkin, *Sov. Phys. Tech. Phys.* **33**, 517 (1988).
52. N.V. Butenin, Yu.I. Neimark, N.A. Fufaev, *Introduction to the Theory of Nonlinear Oscillations* (Nauka, Moscow 1987) (in Russian).
53. A.G. Mayer, *Scientific Notes of Gorky University* (Gorky University, Gorky 1939) (in Russian).
54. A.P. Kuznetsov, S.P. Kuznetsov, I.R. Sataev, *Reg. Chaotic Dyn.* **2**, 90 (1997) (in Russian).
55. S.J. Shenker, *Physica* D **5**, 405 (1982).
56. S. Ostlund, D. Rand, J. Sethna, E.D. Siggia, *Physica* D **8**, 303 (1983).
57. J. Stavans, F. Heslot, A. Libchaber, *Phys. Rev. Lett.* **55**, 596 (1985).
58. S. Martin, W. Martienssen, Transition from quasiperiodicity into chaos in the periodically driven conductivity of BSN crystals. In: *Dimensions and Entropies in Chaotic Systems*, G. Mayer-Kress ed. (Springer, Berlin, Heidelberg 1986) p. 191.
59. M.J. Feigenbaum, L.P. Kadanoff, S.J. Shenker, *Physica* D **5**, 370 (1982).
60. B.I. Shraiman, *Phys. Rev.* A **29**, 3464 (1984).
61. M.H. Jensen, P. Bak, T. Bohr, *Phys. Rev.* A **30**, 1960 (1984).
62. J.D. Farmer, F. Satiggia, *Phys. Rev.* A **31**, 3520 (1985).
63. P. Cvitanović, B. Shraiman, B. Söderberg, *Phys. Scr.* **32**, 263 (1985).
64. D.I. Biswas, R.G. Harrison, *Phys. Rev.* A **32**, 3835 (1985).
65. G.A. Held, C. Jeffries, *Phys. Rev. Lett.* **56**, 1183 (1986).
66. P.S. Linsay, A.W. Cumming, *Physica* D **40**, 196 (1989).
67. V.S. Anishchenko, T.E. Letchford, M.A. Safonova, *Sov. Tech. Phys. Lett.* **11**, 536 (1985).
68. V.S. Anishchenko, T.E. Letchford, *Sov. Phys. Tech. Phys.* **31**, 1347 (1986).
69. C. Giberti, R. Zanasi, *Physica* D **65**, 300 (1993).
70. U. Feudel, W. Jansen, J. Kurths, *Int. J. Bifurcation Chaos* **3**, 131 (1993).
71. V.S. Anishchenko, M.A. Safonova, U. Feudel, J. Kurths, *Int. J. Bifurcation Chaos* **4**, 595 (1994).
72. U. Feudel, M.A. Safonova, J. Kurths, V.S. Anishchenko, *Int. J. Bifurcation Chaos* **6**, 1319 (1995).
73. C. Baesens, J. Guckenheimer, S. Kim, R.S. MacKay, *Physica* D **49**, 387 (1991).
74. C. Grebogi, E. Ott, S. Pelikan, J.A. Yorke, *Physica* D **13**, 261 (1984).
75. C. Grebogi, E. Ott, J.A. Yorke, *Physica* D **15**, 354 (1985).
76. V.I. Arnold, *Additional Chapters of Ordinary Differential Equations Theory* (Nauka, Moscow 1978) (in Russian).
77. J.D. Farmer, E. Ott, J.A. Yorke, *Physica* D **7**, 153 (1983).
78. P.M. Battelino, C. Grebogi, E. Ott, J.A. Yorke, *Physica* D **39**, 299 (1989).
79. F.J. Romeiras, E. Ott, *Phys. Rev.* A **35**, 4404 (1987).

80. M. Ding, C. Grebogi, E. Ott, *Phys. Rev.* A **39**, 2593 (1989).
81. T. Kapitaniak, E. Ponce, J. Wojewoda, *J. Phys. A: Math. Gen.* **23**, L383 (1990).
82. J.F. Heagy, S.M. Hammel, *Physica* D **70**, 140 (1994).
83. U. Feudel, J. Kurths, A.S. Pikovsky, *Physica* D **88**, 176 (1995).
84. O. Sosnovtseva, U. Feudel, J. Kurths, A. Pikovsky, *Phys. Lett.* A **218**, 255 (1996).
85. V.S. Anishchenko, T.E. Vadivasova, O.V. Sosnovtseva, *Phys. Rev.* E **53**, 4451 (1996).
86. A.S. Pikovsky, U. Feudel, *Chaos* **5**, 253 (1995).
87. A.S. Pikovsky, U. Feudel, *J. Phys. A: Math. Gen.* **27**, 5209 (1994).
88. S.P. Kuznetsov, *Phys. Lett. A: Math. Gen.* **169**, 438 (1992).
89. D.V. Anosov, *Proc. Steklov Inst. Math.* **90**, 1 (1967).
90. S. Smale, Bull. *Am. Math. Soc.* **73**, 747 (1967).
91. J. Guckenheimer, P. Holms, *Nonlinear Oscillations, Dynamical Systems, and Bifurcations of Vector Fields* (Springer, New York 1983).
92. Ya.G. Sinai, *Russ. Math. Surv.* **25**, 137 (1970).
93. Ya.G. Sinai, In: *Nonlinear Waves*, A.V. Gaponov-Grekhov ed. (Nauka, Moscow 1979) p. 192 (in Russian).
94. L.A. Bunimovich, Ya.G. Sinai, In: *Nonlinear Waves*, A.V. Gaponov-Grekhov ed. (Nauka, Moscow 1979) p. 212 (in Russian).
95. J.-P. Eckmann, D. Ruelle, *Rev. Mod. Phys.* **57**, 617 (1985).
96. D. Ruelle, *Am. J. Math.* **98**, 619 (1976).
97. D. Ruelle, *Bol. Soc. Bras. Math.* **9**, 83 (1978).
98. L.P. Shilnikov, *Int. J. Bifurcation Chaos* **7**, 1953 (1997); *J. Circuits Syst. Comput.* **3**, 1 (1993).
99. V.S. Anishchenko, G.I. Strelkova, *Discrete Dyn. Nat. Soc.* **2**, 53 (1998).
100. R. Graham, W. Ebeling, *private communications*; R. Graham, A. Hamm, T. Tel, *Phys. Rev. Lett.* **66**, 3089 (1991).
101. E. Ott, E.D. Yorke, J.A. Yorke, *Physica* D **16**, 62 (1985).
102. C.G. Schroer, E. Ott, J.A. Yorke, *Phys. Rev. Lett.* **81**, 1397 (1998).
103. T. Sauer, C. Grebogi, J.A. Yorke, *Phys. Rev. Lett.* **79**, 59 (1997).
104. L. Jaeger, H. Kantz, *Physica* D **105**, 79 (1997).
105. Yu. Kifer, *Commun. Math. Phys.* **121**, 445 (1989).
106. S.M. Hammel, J.A. Yorke, C. Grebogi, *J. Complexity* **3**, 136 (1987); *Bull. Am. Math. Soc., New Ser.* **19**, 465 (1988).
107. V.S. Anishchenko, T.E. Vadivasova, A.S. Kopeikin et al., *Phys. Rev. Lett.* **87**, 054101 (2001).
108. V.S. Anishchenko, T.E. Vadivasova, A.S. Kopeikin et al., *Phys. Rev.* E **65**, 036206 (2002).
109. G.M. Zaslavsky, *Chaos in Dynamical Systems* (Harwood Acad. Publ., New York 1985).
110. P. Billingsley, *Ergodic Theory and Information* (Wiley, New York 1965).
111. I.P. Cornfeld, S.V. Fomin, Ya.G. Sinai, *Ergodic Theory* (Springer, New York 1982).
112. A.N. Kolmogorov, *Dokl. Akad. Nauk SSSR* **124**, 754 (1959).
113. Ya.G. Sinai, *Dokl. Akad. Nauk SSSR* **124**, 768 (1959).
114. Ya.B. Pesin, *Russ. Math. Surv.* **32**(4), 55 (1977).
115. R. Bowen, *Equilibrium States and the Ergodic Theory of Anosov Diffeomorphisms, Lecture Notes in Math.*, vol. 470 (Springer, Berlin 1975).
116. M.L. Blank, *Stability and Localization in Chaotic Dynamics* (Izd. MTsNMO, Moscow 2001) (in Russian).
117. D. Ruelle, *Commun. Math. Phys.* **125**, 239 (1989).

118. F. Christiansen, G. Paladin, H.H. Rugh, *Phys. Rev. Lett.* **65**, 2087 (1990).
119. C. Liverani, *Ann. Math.* **142**, 239 (1995).
120. G. Froyland, *Commun. Math. Phys.* **189**, 237 (1997).
121. R. Bowen, D. Ruelle, *Invent. Math.* **29**, 181 (1975).
122. R. Badii et al., *Physica* D **58**, 304 (1992).
123. V.S. Anishchenko et al., *Physica* A **325**, 199 (2003).
124. V.S. Anishchenko et al., *Fluct. Noise Lett.* **3**, L213 (2003).
125. V.S. Anishchenko et al., *J. Commun. Technol. Electron.* **48**, 750 (2003).
126. O.E. Rössler, *Phys. Lett.* A **57**, 397 (1976).
127. E.N. Lorenz, *J. Atmos. Sci.* **20**, 130 (1963).
128. V.S. Anishchenko, *Complex Oscillations in Simple Systems* (Nauka, Moscow 1990) (in Russian).
129. V.S. Anishchenko, *Dynamical Chaos – Models and Experiments* (World Scientific, Singapore 1995).
130. R.F. Williams, In: *Global Analysis, Proc. of Symp. in Pure Math.*, vol. 14, S.-S. Chern and S. Smale eds. (*Am. Math. Soc.*, Providence 1970) p. 341.
131. R.V. Plykin, *Russ. Math. Surv.* **35**, 109 (1980).
132. V.S. Afraimovich, V.V. Bykov, L.P. Shilnikov, *Sov. Phys. Dokl.* **22**, 253 (1977).
133. A.L. Shilnikov, In: *Methods of Qualitative Theory and Bifurcation Theory*, L.P. Shilnikov ed. (Izd. GGU, Gorkii 1989) p. 130 (in Russian).
134. R. Lozi, *J. Phys. Colloq. (Paris)* **39**(C5), 9 (1978).
135. V.N. Belykh, *Sbornik Math.* **186**, 311 (1995).
136. S. Banerjee, J.A. Yorke, C. Grebogi, *Phys. Rev. Lett.* **80**, 3049 (1998).
137. V.S. Anishchenko et al., *Discrete Dyn. Nat. Soc.* **2**, 249 (1998).
138. Y.-C. Lai, C. Grebogi, J. Kurths, *Phys. Rev.* E **59**, 2907 (1999).
139. Y.-C. Lai, C. Grebogi, *Phys. Rev. Lett.* **82**, 4803 (1999).
140. E. Ott, *Chaos in Dynamical Systems* (Cambridge Univ. Press, Cambridge 1993).
141. S. Dawson et al., *Phys. Rev. Lett.* **73**, 1927 (1994).
142. S. Dawson, *Phys. Rev. Lett.* **76**, 4348 (1996); P. Moresco, S.P. Dawson, *Phys. Rev.* E **55**, 5350 (1997).
143. E.J. Kostelich et al., *Physica* D **109**, 81 (1997).
144. Y.-C. Lai et al., *Nonlinearity* **6**, 779 (1993).
145. V.S. Anishchenko et al., *Phys. Lett.* A **270**, 301 (2000).
146. R.N. Madan (ed.), *Chua's Circuit: A Paradigm for Chaos* (World Scientific, Singapore 1993).
147. N.K. Gavrilov, L.P. Shilnikov, *Mathem. Sbornik* **88**, 475 (1972); *Mathem. Sbornik* **90**, 139 (1973) (in Russian).
148. R.L. Stratonovich, In: *Noise in Nonlinear Dynamical Systems, Theory of Noise Induced Processes in Special Applications*, vol. 2, F. Moss and P.V.E. McClintock eds. (Cambridge Univ. Press, Cambridge 1989), p. 16.
149. W. Horsthemke, R. Lefever, *Noise-Induced Transitions* (Springer, Berlin 1984).
150. C.W. Gardiner, *Handbook of Stochastic Methods for Physics, Chemistry and the Natural Sciences* (Springer, Berlin 1983).
151. H. Haken, *Advanced Synergetics* (Springer, Berlin 1983).
152. R. Graham, In: *Noise in Nonlinear Dynamical Systems, Theory of Continuous Fokker–Planck Systems*, vol. 1 F. Moss and P.V.E. McClintock eds. (Cambridge Univ. Press, Cambridge 1988).
153. V.S. Anishchenko, W. Ebeling, *Z. Phys.* B **81**, 445 (1990).
154. J. Guckenheimer, *Nature* **298**, 358 (1982).
155. C.S. Hsu, M.C. Kim, *J. Stat. Phys.* **38**, 735 (1985).
156. Ya.G. Sinai, *Russ. Math. Surv.* **46**, 177 (1991).

157. V.S. Anishchenko, A.B. Naiman, *Sov. Tech. Phys. Lett.* **17**, 510 (1991).
158. V.S. Anishchenko, A.B. Neiman, In: *Nonlinear Dynamics of Structures: Int. Symp. on Generation of Large-Scale Structures in Continuous Media, Perm–Moscow, USSR, 11–20 June 1990*, R.Z. Sagdeev et al. eds. (World Scientific, Singapore 1991), p. 21.
159. D. Alonso et al., *Phys. rev.* E **54**, 2472 (1996).
160. V.S. Anishchenko et al., *Nonlinear Dynamics of Chaotic and Stochastic Systems* (Springer, Berlin 2002).
161. J.D. Farmer, *Phys. Rev. Lett.* **47**, 179 (1981).
162. A. Arneodo, P. Coullet, C. Tresser, *Commun. Math. Phys.* **79**, 573 (1981).
163. M.G. Rosenblum, A.S. Pikovsky, J. Kurths, *Phys. Rev. Lett.* **76**, 1804 (1996).
164. R.L. Stratonovich, *Topics in the Theory of Random Noise*, vols. 1,2 (Gordon and Breach, New York 1963, 1967).
165. A.N. Malakhov, *Fluctuations in Self-Oscillatory Systems* (Nauka, Moscow 1968) (in Russian).
166. S.M. Rytov, Yu.A. Kravtsov, V.I. Tatarskii, *Principles of Statistical Radiophysics*, 2nd ed. (Springer, Berlin Heidelberg 1987–1989).
167. V.I. Tikhonov, M.A. Mironov, *Markovian Processes* (Sov. Radio, Moscow 1977) (in Russian).
168. E.A. Jackson, *Perspectives of Nonlinear Dynamics*, vols. 1,2 (Cambridge Univ. Press, Cambridge 1989, 1990).
169. J.P. Gollub, S.V. Benson, *J. Fluid Mech.* **100**, 449 (1980); M. Lesieur, *Turbulence in Fluids: Stochastic and Numerical Modelling* (M. Nijhoff, Dordrecht 1987); S. Sato, M. Sano, Y. Sawada, *Phys. Rev.* A **37**, 1679 (1988); S. Kida, M. Yamada, K. Ohkitani, *Physica* D **37**, 116 (1989); T. Bohr et al., *Dynamical Systems Approach to Turbulence, Cambridge Nonlinear Sci. Ser.*, vol. 8 (Cambridge Univ. Press, Cambridge 1998); I.S. Aranson, L. Kramer, *Rev. Mos. Phys.* **74**, 99 (2002).
170. Y. Kuramoto, *Chemical Oscillations, Waves, and Turbulence* (Springer, Berlin 1984); Y. Pomeau, P. Manneville, *J. Phys. Lett. (Paris)* **40**, 609 (1979); H. Chaté, P. Manneville, *Phys. Rev. Lett.* **58**, 112 (1987); P. Coullet, L. Gil, J. Lega, *Physica* D **37**, 91 (1989); H. Chaté, *Nonlinearity* **7**, 185 (1994).
171. P. Manneville, H. Chaté, *Physica* D **96**, 30 (1996); G. Grinstein, C. Jayaprakash, R. Pandit, *Physica* D **90**, 96 (1996).
172. G.B. Ermentrout, N. Kopell, *SIAM J. Math. Anal.* **15**, 215 (1984); Y. Yamaguchi, H. Shimizu, *Physica* D **11**, 212 (1984); S.H. Strogatz, R.E. Mirollo, *Physica* D **31**, 143 (1988).
173. G.V. Osipov, M.M. Sushchik, *Phys. Rev.* E **58**, 7198 (1998).
174. G.B. Ermentrout, W.C. Troy, *SIAM J. Appl. Math.* **46**, 359 (1986).
175. A.A. Akopov et al., *Tech. Phys. Lett.* **29**, 629 (2003).
176. A.A. Samarskii, A.V. Gulin, *Numerical Techniques* (Nauka, Moscow 1989) (in Russian).
177. A. Wolf et al., *Physica* D **16**, 285 (1985).
178. V.S. Anishchenko et al., *Phys. Rev.* E **69**, 036215 (2004).
179. I.I. Blekhman, *Synchronization in Nature and Technics* (Nauka, Moscow 1981) (in Russian).
180. Yu.A. Kuznetsov, P.S. Landa, A.F. Olkhovoi, S.M. Perminov, *Dokl. Acad. Nauk USSR* **281**, 1164 (1985) (in Russian).
181. G. Dykman, P. Landa, Yu. Neimark, *Chaos Soliton Fractals* **1**, 339 (1992).
182. N. Nijmeier, A.L. Fradkov, I.I. Blekhman, A.Yu. Pogromsky, Selfsynchronization and controlled synchronziation. In: *Proc. of 1st Int. Conference on Control of Oscillations and Chaos*, vol. 1 (St. Petersburg, Russia 1997) p. 36.

183. H. Fujisaka, Y. Yamada, *Prog. Theor. Phys.* **69**, 32 (1983).
184. A.S. Pikovsky, *Z. Phys.* B **55**, 149 (1984).
185. A.R. Volkovskii, N.F. Rul'kov, *Sov. Tech. Phys. Lett.* **15**, 5 (1989).
186. L. Pecora, T. Carroll, *Phys. Rev. Lett.* **64**, 821 (1990).
187. N.F. Rul'kov, A.R. Volkovskii. A. Rodriguez-Lozano, E. Del-Rio, M.G. Velarde, *Int. J. Bifurcation Chaos* **2**, 669 (1992).
188. L. Chua, M. Itoh, L. Kocarev, K. Eckert, Chaos synchronization in Chua's circuit. In: *Chua's Circuits: A Paradigma for Chaos*, R.N. Madan, ed. (World Scientific, Singapore 1993) p. 309.
189. V.S. Afraimovich, N.N. Verichev, M.I. Rabinovich, *Izv. Vuzov. Ser. Radiophys.* **29**, 1050 (1986) (in Russian).
190. V.S. Anishchenko, D.E. Postnov, *Sov. Tech. Phys. Lett.* **14**, 569 (1988).
191. V.S. Anishchenko, T.E. Vadivasova, D.E. Postnov, M.A. Safonova, *Radiotech. Electron.* **36**, 338 (1991) (in Russian).
192. V.S. Anishchenko, T.E. Vadivasova, D.E. Postnov, M.A. Safonova, *Int. J. Bifurcation Chaos* **2**, 633 (1992).
193. V.S. Anishchenko, T.E. Vadivasova, D.E. Postnov, O.V. Sosnovtseva, C.W. Wu, L.O. Chua, *Int. J. Bifurcation Chaos* **5**, 1525 (1995).
194. V.S. Anishchenko, T.E. Vadivasova, V.V. Astakhov, O.V. Sosnovtseva, C.W. Wu, L.O. Chua, *Int. J. Bifurcation and Chaos* **5**, 1677 (1995).
195. M. Rosenblum, A. Pikovsky, J. Kurths, *Phys. Rev. Lett.* **76**, 1804 (1996).
196. A. Pikovsky, G. Osipov, M. Rosenblum, M. Zaks, J. Kurths, *Phys. Rev. Lett.* **79**, 47 (1997).
197. A.S. Pikovsky, M.G. Rosenblum, G.V. Osipov, J. Kurths, *Physica* D **104**, 219 (1997).
198. N.F. Rulkov, M.M. Sushchik, L.S. Tsimring, H.D.I. Abarbanel, *Phys. Rev.* E. **51**, 980 (1995).
199. L. Kocarev, U. Parlitz, *Phys. Rev. Lett.* **76**, 1816 (1996).
200. H.D.I. Abarbanel, N.F. Rulkov, M.M. Sushchik, *Phys. Rev.* E **53**, 4528 (1996).
201. M.G. Rosenblum, A.S. Pikovsky, J. Kurths, *Phys. Rev. Lett.* **78**, 4193 (1997).
202. I.I. Blekhman, P.S. Landa, M.G. Rosenblum, *Appl. Mech. Rev.* **48**, 733 (1995).
203. V.D. Shalfeev, G.V. Osipov, A.K. Kozlov, A.R. Volkovskii, *Zarubezhnaya Radioelektronika* **10**, 27 (1997) (in Russian).
204. A.V. Gaponov-Grekhov, M.I. Rabinovich, Dynamic chaos in ensembles of structures and spatial development of turbulence in unbounded systems. In: *Selforganization by Nonlinear Irreversible Processes*, W. Ebeling and H. Ulbricht eds. (Springer, New York 1986) p. 37.
205. V.S. Anishchenko, I.S. Aranson, D.E. Postnov, M.I. Rabinovich, *Dokl. Acad. Nauk USSR* **286**, 1120 (1986) (in Russian).
206. K. Kaneko, *Physica* D **32**, 60 (1989).
207. A.P. Kuznetsov, S.P. Kuznetsov, *Izv. Vuzov. Ser. Radiophys.* **34**, 1079 (1991) (in Russian).
208. V.V. Astakhov, B.P. Bezruchko, V.I. Ponomarenko, *Izv. Vuzov. Ser. Radiophys.* **34**, 35 (1991) (in Russian).
209. J.F. Heagy, T.L. Carroll, L.M. Pecora, *Phys. Rev.* E **50**, 1874 (1994).
210. L. Brunnet, H. Chaté, P. Manneville, *Physica* D **78**, 141 (1994).
211. L. Kocarev, U. Parlitz, *Phys. Rev. Lett.* **77**, 2206 (1996).
212. A.S. Pikovsky, M.G. Rosenblum, J. Kurths, *Europhys. Lett.* **34**, 165 (1996).
213. G.V. Osipov, A.S. Pikovsky, M.G. Rosenblum, J. Kurths, *Phys. Rev.* E **55**, 2353 (1997).
214. A. Amengual, E. Hernandez-Garcia, R. Montagne, M. San Miguel, *Phys. Rev. Lett.* **78**, 4379 (1997).
215. S. Boccaletti, J. Bragard, F.T. Arecchi, *Phys. Rev.* E **59**, 6574 (1999).

216. S. Boccaletti, J. Bragard, F.T. Arecchi, H. Mancini, *Phys. Rev. Lett.* **83**, 536 (1999).
217. Yu.M. Romanovsky, N.V. Stepanova, D.S. Chernavsky, *Mathematical Biophysics* (Nauka, Moscow 1984) (in Russian).
218. A.T. Winfree, *The Geometry of Biological Time* (Springer, New York 1980).
219. J.D. Murray, *Mathematical Biology* (Springer, Berlin Heidelberg 1989).
220. G. Baier, J.S. Thomsen, E. Mosekilde, *J. Theor. Biol.* **163**, 593 (1993).
221. E. Mosekilde, O.G. Mouritsen (ed.) *Modelling the Dynamics of Biological Systems* (Springer, Berlin 1995).
222. H.D.I. Abarbanel, M.I. Rabinovich, A. Silverston, M.V. Bazhenov, R. Huerta, M.M. Sushchik, L.L. Rubchinsky, *Phys. Uspekhi* **166**, 365 (1996).
223. C. Schäfer, M.G. Rosenblum, J. Kurths, H.-H. Abel, *Nature* **392**, 239 (1998).
224. V.S. Anishchenko, N.B. Janson, A.G. Balanov, N.B. Igosheva, G.V. Bordyugov, *Int. J. Bifurcation Chaos* **10** 2339 (2000).
225. V.S. Anishchenko, T.E. Vadivasova and D.E. Postnov, Synchronization of chaos. In: *Proc. of the First International Conference on Applied Synergetics and Synergetic Engeneering* (Erlangen, Germany, 1994) p. 200.
226. V.S. Anishchenko, A.N. Silchenko, I.A. Khovanov, *Phys. Rev.* E **57**, 316 (1998).
227. V.N. Belykh, N.I. Verichev, I.V. Belykh, *Izv. Vuzov. Ser. Radiophys.* **11**, 912 (1997) (in Russian).
228. L. Kocarev, K.S. Halle, K. Eckert, L.O. Chua, U. Parlitz, *Int. J. Bifurcation Chaos* **2**, 709 (1992).
229. K.M. Cuomo, A.V. Openheim, *Phys. Rev. Lett.* **71**, 65 (1993).
230. K.M. Cuomo, A.V. Oppenheim, S.H. Strogatz, *IEEE Trans. Circuit. Syst.* **40**, 626 (1993).
231. H. Dedieu, M.P. Kennedy, M. Hasler, *IEEE Trans. Circuit. Syst.* **40**, 634 (1993).
232. L. Kocarev, U. Parlitz, T. Stojanovski, *Phys. Lett.* A **217**, 280 (1996).
233. O.V. Sosnovtseva, A.G. Balanov, T.E. Vadivasova, V.V. Astakhov, E. Mosekilde, *Physe. Rev.* E **60**, 6560 (1999).
234. V.V. Astakhov, B.P. Bezruchko, Yu.V. Gulyaev, E.P. Seleznev, *Sov. Tech. Phys. Lett.* **15**, 60 (1989) (in Russian).
235. V.V. Astakhov, B.P. Bezruchko, V.I. Ponomarenko, E.P. Seleznev, *Izv. Vuzov. Ser. Radiophys.* **31**, 627 (1989).
236. V.V. Astakhov, B.P. Bezruchko, E.N. Erastova, E.P. Seleznev, *Sov. Tech. Phys.* **60**, 19 (1990).
237. D.E. Postnov, T.E. Vadivasova, O.V. Sosnovtseva, A.G. Balanov, V.S. Anishchenko, *Chaos* **9**, 227 (1999).
238. S.P. Kuznetsov, *Izv. Vuzov. Ser. Radiophys.* **28**, 991 (1985).
239. V. Astakhov, M. Hasler, T. Kapitaniak, A. Shabunin, V. Anishchenko, *Phys. Rev.* E **58**, 5620 (1998).
240. A. Pikovsky, P. Grassberger, *J. Phys. A: Math. Gen.* **24**, 4587 (1991).
241. P. Ashwin, J. Buescu, I. Stewart, *Nonlinearity* **9**, 703 (1994).
242. P. Ashwin, J. Buescu, I. Stewart, *Phys. Lett.* A **193**, 126 (1994).
243. S.C. Venkataramani, B. Hunt, E. Ott, *Phys. Rev. Lett.* **77**, 5361 (1996).
244. E. Ott, J.C. Sommerer, *Phys. Lett.* A **188**, 39 (1994).
245. J.F. Heagy, T.L. Carroll, L.M. Pecora, *Phys. Rev. Lett.* **73**, 35 (1995).
246. Y.-Ch. Lai, C. Grebogi, J.A. Yorke, S.C. Venkataramani, *Phys. Rev. Lett.* **77**, 55 (1996).
247. M. Sushchik, N.F. Rulkov, H.D.I. Abarbanel, *IEEE Trans. Circuit. Syst.* **44**, 867 (1997).

248. V. Astakhov, A. Shabunin, T. Kapitaniak, V. Anishchenko, *Phys. Rev. Lett.* **79**, 1014 (1997).
249. M. Hasler, Yu. Maistrenko, *IEEE Trans. Circuit. Syst.* **44**, 856 (1997).
250. T. Kapitaniak, Yu. Maistrenko, *Physica* D **126**, 18 (1999).
251. H. Fujisaka, T. Yamada, *Prog. Theor. Phys.* **74**, 918 (1985).
252. J. Milnor, *Commun. Math. Phys.* **99**, 177 (1985).
253. T.E. Vadivasova, A.G. Balanov, O.V. Sosnovtseva, D.E. Postnov, E. Mosekilde, *Phys. Lett.* A **253**, 66 (1999).
254. E. Rosa, E. Ott, M.H. Hess, *Phys. Rev. Lett.* **80**, 1642 (1998).
255. A. Pikovsky, M. Rosenblum, and J. Kurths, *Synchronization: a universal concept in nonlinear science* (Cambridge University Press, Cambridge, 2003).
256. P.S. Landa, *Nonlinear Oscillations and Waves* (Nauka, Moscow 1997).
257. V.S. Afraimovich, V.I. Nekorkin, G.V. Osipov, V.D. Shalfeev, *Stability, Structures and Chaos in Nonlinear Synchronization Networks* (World Scientific, Singapore 1995).
258. P. Hadley, M.R. Beasley, K. Wiesenfeld, *Phys. Rev.* B **38**, 8712 (1988).
259. K. Wiesenfeld, P. Hadley, *Phys. Rev. Lett.* **62**, 1335 (1989).
260. V.N. Belykh, N.N. Verichev, L.J. Kocarev, L.O. Chua, *Int. J. Bifurcation Chaos* **3**(2), 579 (1993).
261. A.V. Ustinov, M. Cirillo M., B. Malomed, *Phys. Rev.* B **47**, 8357 (1993).
262. Y. Braiman, J.F. Linder, W.L. Ditto, *Nature* **378**, 465 (1995).
263. Y. Braiman, W.L. Ditto, K. Wiesenfeld, M.L. Spano, *Phys. Lett.* A **206**, 54 (1995).
264. G.V. Osipov, V.D. Shalfeev, *IEEE Trans. Circuit Syst.-I: Fundam. Theor. Appl.* **42**(10), 687 (1995).
265. G.V. Osipov, V.D. Shalfeev, *IEEE Trans. Circuit Syst.-I: Fundam. Theor. Applications* **42**(10), 693 (1995).
266. Yu.M. Romanovsky, N.V. Stepanova, D.S. Chernavsky, *Mathematical Biophysics* (Nauka, Moscow 1984).
267. Y. Kuramoto, *Progr. Theor. Phys.* **79**, 212 (1984).
268. S.K. Sarna, E.E. Daniel, J.J. Kingma, *Digestive Diseases* **17**, 299 (1972).
269. D.A. Linkens, S. Satardina, *IEEE Trans. Biomed. Eng.* **BME-24**, 362 (1977).
270. J.J. Hopfield, A.V. Herz, *Proc. Natl. Acad. Sci. USA* **92**, 6655 (1995).
271. H. Haken, *Principles of brian functioning* (Springer, Berlin, Heidelberg 1996).
272. B.I. Shraiman, A. Pumir, W. Van Saarlos, et al., *Physica* D **57**, 241 (1992).
273. M.G. Cross, P.C. Hohenberg, *Rev. Mod. Phys.* **65**(3), 851 (1993).
274. H. Chaté, *Nonlinearity* **7**, 185 (1994).
275. T. Bohr, M.H. Jensen, G. Paladin, A. Vulpiani, *Dynamical System Approach to Turbulence* (Cambridge Univ. Press, Cambridge 1998).
276. Y. Kuramoto, T. Tsuzuki, *Progr. Theor. Phys.* **55**, 356 (1976).
277. G. Sivashinsky, *Found. Phys.* **8**(9–10), 735 (1978).
278. V.M. Malafeev, M.S. Polyakova, Yu.M. Romanovsky, *Izv. VUZ. Ser. Radiophysics* **13**(6), 936 (1970) (in Russian).
279. I.K. Kostin, Yu.M. Romanovsky, *Vestnik MGU. Ser. Phys. Astron.* **13**(6), 698 (1972).
280. Y. Aizawa, *Progr. Theor. Phys.* **56**(3), 703 (1976).
281. V.S. Afraimovich, V.I. Nekorkin, *Math. Modelling* **4**(1), 83 (1992) (in Russian).
282. L.M. Pecora, *Phys. Rev.* E **58**(1), 347 (1998).
283. L. Ren, B. Ermentrout, *Physica* D **43**(1–4), 56 (2000).
284. S.H. Strogatz, R.E. Mirollo, *Physica* D **31**, 143 (1988).
285. S.H. Strogatz, R.E. Mirollo, *J. Phys.* A **21**, L699 (1988).
286. H. Sakaguchi, S. Shinomoto, Y. Kuramoto, *Progr. Theor. Phys.* **77**, 1005 (1987).

287. V.I. Nekorkin, V.A. Makarov, M.G. Velarde, *Phys. Rev.* E **58**, 5742 (1998).
288. T. Yamada, H. Fujisaka, *Prog. Theor. Phys.* **72**(5), 885 (1984).
289. K. Kaneko, *Physica* D **41**, 137 (1990).
290. V.N. Belykh, E. Mosekilde, *Phys. Rev.* E **54**(4), 3196 (1996).
291. V.V. Astakhov, V.S. Anishchenko, T. Kapitaniak, A.V. Shabunin, *Physica* D **109**(1–2), 11 (1997).
292. M. Hasler, Yu. Maistrenko, O. Popovich, *Phys. Rev.* E **58**, 6843 (1998).
293. V.N. Belykh, I.V. Belykh, E. Mosekilde, *Phys. Rev.* E **63**(3), 0362161 (2001).
294. V.S. Anishchenko, D.E. Postnov, M.A. Safonova, *Sov. Tech. Phys. Lett.* **11**, 621 (1985).
295. A. Brandstater, J. Swift, H.L. Swinny, A. Wolf, *Phys. Rev. Lett.* **5**(16), 1442 (1983).
296. B. Malreison, P. Atten, P. Bregé, M. Dubois, *J. Phys. Lett.* **44**(22), 897 (1983).
297. H. Chaté, A. Pikovsky and O. Rudzick, *Physica* D **131**, 17 (1999).
298. C. Elphik, A. Hagberg, E. Meron, *Phys. Rev.* E **59**(5), 5285 (1999).
299. G. Hu, J. Xiao, J. Yang, F. Xie, Z. Qu, *Phys. Rev.* E **56**(3), 2738 (1997).
300. Y. Jiang, P. Parmananda, *Phys. Rev.* E **57**(4), 4135 (1998).
301. L. Junge, U. Parlitz, *Phys. Rev.* E **61**(4), 3736 (2000).
302. H.-K. Park, *Phys. Rev. Lett.* **86**(6), 1130 (2001).
303. T.E. Vadivasova, G.I. Strelkova, V.S. Anishchenko, *Phys. Rev.* E **63**, 036225 (2001).
304. P.C. Matthews, S.H. Strogatz, *Phys. Rev. Lett.* **65**, 1701 (1990).
305. Y. Yamaguchi, H. Shimizu, *Physica* D **11**, 212 (1984).
306. G.B. Ermentrout, *Physica* D **41**, 219 (1990).
307. D.G. Aranson, G.B. Ermentrout, N. Kopell, *Physica* D **41**, 403 (1990).
308. J. Garcia-Ojalvo, J.M. Sancho, *Noise in Spatially Extended Systems* (Springer, New York 1999).
309. V.S. Anishchenko et al., *J. Bifurcation Chaos* **15**(11), 1 (2005).
310. V.S. Anishchenko, H. Herzel, *Z. Angew. Math. Mech.* **68**(7), 317 (1988).
311. L. Junge, U. Parlitz, *Phys. Rev.* E **62**, 438 (2000).
312. L. Kocarev, Z. Tasev, U. Parlitz, *Phys. Rev. Lett.* **51**, 79 (1997).
313. V.D. Kazantsev, V.I. Nekorkin, D.V. Artyuhin, M.G. Velarde, *Phys. Rev.* E **63**, 016212 (2001).
314. A. Akopov, V. Astakhov, T. Vadivasova, A. Shabunin, T. Kapitaniak, *Phys. Lett.* A **334**, 169 (2005).
315. V.I. Arnold, *Geometrical methods in the theory of ordinary differential equations* (Springer, New York 1983).
316. A.I. Fomin, T.E. Vadivasova, O.V. Sosnovtseva, V.S. Anishchenko, *Izv. VUZ. Appl. Nonlinear Dyn.* **8**(4), 103 (2000) (in Russian).
317. D.K. Mynbaev, M.I. Shilenkov, *J. Commun. Electron.* **2**, 361 (1981).
318. A.A. Maltsev, A.M. Silaev, *Izv. VUZ. Ser. Radiophys.* **22**(7), 826 (1979) (in Russian).
319. A.A. Dvornikov, G.M. Utkin, A.M. Chukov, *Izv. VUZ. Ser. Radiophys* **27**(11), 1388 (1984) (in Russian).
320. G.B. Ermentrout, *J. Math. Biol.* **23**(1), 55 (1985).
321. G.B. Ermentrout, *SIAM J. of Appl. Math.* **52**, 1664 (1992).
322. A.S. Gurtovnik, Yu.I. Neimark, Synchronisms in a System of Cyclely Weak-Coupled Oscillators. In: *Dynamical Systems, Collection of Scientific Works* (N. Novgorod Univ. Publisher, N. Novgorod 1991). p. 84 (in Russian).
323. V.I. Nekorkin, V.A. Makarov, M.G. Velarde, *Int. J. Bifurcation Chaos* **6**(10), 1845 (1996).
324. A. Shabunin, V. Astakhov, V. Anishchenko, *Int. J. Bifurcation Chaos* **12**(8), 1895 (2002).

325. A.T. Winfree, *J. Theor. Biol.* **16**, 15 (1967); A.T. Winfree, *The Geometry of Biological Time* (Springer, New York 1980).
326. S.H. Strogatz, I. Stewart, *Sci. Am.* **269**, 102 (1993).
327. L. Glass, *Nature* **410**, 277 (2001).
328. G. Laurent, *Trends Neurosci.* **19**, 489 (1996); K. MacLeod, G. Laurent, *Science* **274**, 976 (1996).
329. R. Ritz, T.J. Sejnowski, *Curr. Opin. Neurobiol.* **7**, 536 (1997).
330. E. Niebur, S.S. Hsiao, K.O. Johnson, *Curr. Opin. Neurobiol.* **12**, 190 (2002).
331. I. Bar-Gad, H. Bergman, *Curr. Opin. Neurobiol.* **11**, 689 (2001).
332. J.M. Jørgensen, Å. Flock and J. Wersäll, *Z. Zellforsch* **130**, 362 (1972).
333. L.A. Wilkens, D.F. Russell, X. Pei, C. Gurgens, *Proc. Roy. Soc.* B **264**, 1723 (1997).
334. D.F. Russell, L.A. Wilkens, *F. Moss, Nature* **402**, 291-294 (1999).
335. A.B. Neiman, D.F. Russell, *J. Neurophysiol.* **92**, 492 (2004).
336. A. Neiman, X. Pei, D.F. Russell, W. Wojtenek, L.A. Wilkens, F. Moss, H.A. Braun, M.T. Huber, K. Voigt, *Phys. Rev. Lett.* **82** 660 (1999).
337. A. Neiman, D.F. Russell, *Phys. Rev. Lett.* **88**, 138103 (2002).
338. L. Glass, M.C. Mackey, *From Clocks to Chaos: The Rhythms of Life* (Princeton University Press, Princeton 1988).
339. A. Babloyantz, A. Deslexhe, *Biol. Cybernetics* **58**, 203 (1988).
340. P.I. Saparin et al., *Phys. Rev.* E **54**(1), 1 (1996).
341. A.N. Pavlov, N.B. Janson, *Izv. VUZ. Appl. Nonlinear Dyn.* **5**(1), 91 (1997).
342. V.S. Anishchenko et al., *Discrete Dyn. Nature Soc.* **4**, 201 (2000).
343. V.S. Anishchenko et al., *Int. J. Bifurcation Chaos* **10**(10), 2339 (2000).
344. C. Schäfer, M.G. Rosenblum, H. Abel, J. Kurths, *Phys. Rev.* E **60**, 857 (1998).
345. C. Schäfer, M.G. Rosenblum, J. Kurths, H.H. Abel, *Nature* **392**(6672), 239 (1998).
346. M. Rosenblum et al., *IEEE Eng. Med. Biol. Mag.* **17**(6), 46 (1998).
347. A.V. Gaponov-Grekhov, M.I. Rabinovich, I.M. Starobinetz, *Sov. Tech. Phys. Lett.* **39**, 561 (1984).
348. I.S. Aranson, A.V. Gaponov-Grekhov, M.I. Rabinovich, *J. Exp. Theor. Phys.* **89**, 92 (1985) (in Russian).
349. A.V. Gaponov-Grekhov, M.I. Rabinovich, Auto-structures. chaotic dynamics of ensembles. In: *Nonlinear Waves Structures and Bifurcations* (Nauka, Moscow 1987) p. 7 (in Russian).
350. L.A. Bunimovich, Ya.G. Sinai, *Nonlinearity* **1**, 581 (1988).
351. O. Rudzick, A. Pikovsky, *Phys. Rev.* E **54**, 5107 (1996).
352. M.I. Rabinovich, *Phys. Uspekhi* **125**, 10 (1987).
353. S.H. Strogatz, C.M. Marcus, R.E. Mirollo, R.M. Westervelt, *Physica* D **36**, 23 (1989).
354. K. Wiesenfeld, C. Bracikowski, G. James, R. Ray, *Phys. Rev. Lett.* **65**, 1749 (1990).
355. H.G. Winful, L. Rahman, *Phys. Rev. Lett.* **65**, 1575 (1990).
356. T. Kapitaniak, L.O. Chua, *Int. Bifurcation Chaos* **4**, 477 (1994).
357. C.W. Wu, L.O. Chua, *IEEE Trans. Circuit. Syst.* **43**, 161 (1996).
358. H. Sompolinsky, D. Golob, D. Kleinfeld, *Phys. Rev.* A **43**, 6990 (1991).
359. D. Hansel, H. Sompolinsky, *Phys. Rev. Lett.* **68**, 718 (1992).
360. W. Freeman, *Int. J. Bifurcation Chaos* **2**, 451 (1992).
361. S.K. Han, C. Kurrer, Y. Kuramoto, *Phys. Rev. Lett.* **75**, 3190 (1995).
362. P.C. Matthews, R.E. Mirollo, S.H. Strogatz, *Physica* D **51**, 293 (1991).
363. J.F. Heagy, L.M. Pecora, T.L. Carroll, *Phys. Rev. Lett.* **74**, 4185 (1994).
364. K. Kaneko, *Physica* D **23**, 436 (1986).

365. S. Watanabe, H.S.J. van der Zant, S.H. Strogatz, T.P. Orlando, *Physica* D **97**, 429 (1996).
366. S.P. Kuznetsov, *Sov. Tech. Phys. Lett.* **9**, 94 (1983) (in Russian).
367. A. Pikovsky, J. Kurths, *Phys. Rev.* E **49**, 898 (1994).
368. A. Morelli et al., *Phys. Rev.* E **58**, 128 (1998).
369. P. Grassberger, *Phys. Rev.* E **59**, R2520 (1999).
370. A.W. Hübler, E. Luscher, *Naturwissenschaft* **76**, 67 (1989).
371. E.A. Jackson, *Physica* D **50**, 341 (1991).
372. E.A. Jackson, *Phys. Lett.* A **151**, 478 (1990).
373. E. Ott, C. Grebogi, J.A. Yorke, *Phys. Rev. Lett.* **64**, 1196 (1990).
374. T. Shinbrot, C. Grebogi, E. Ott, J.A. Yorke, *Nature* **363**, 411 (1993).
375. K. Pyragas, *Phys. Lett.* A **170**, 421 (1992).
376. W.L. Ditto, S.N. Rauseo, M.L. Spano, *Phys. Rev. Lett.* **65**, 3211 (1990).
377. T. Shinbrot, E. Ott, C. Grebogi, J.A. Yorke, *Phys. Rev. Lett.* **65**, 3215 (1990).
378. J. Singer, Y. Wang, H. Bau, *Phys. Rev. Lett.* **66**, 1123 (1991).
379. F.J. Romeiras, C. Grebogi, E. Ott, W.P. Dayawasn, *Physica* D **58**, 165 (1992).
380. T. Shinbrot, W. Ditto, C. Grebogi, E. Ott, M. Spano, J.A. Yorke, *Phys. Rev. Lett.* **68**, 2868 (1992).
381. T. Shinbrot, E. Ott, C. Grebogi, J.A. Yorke, *Phys. Rev.* A **45**, 4165 (1992).
382. A. Garfinkel, M. Spano, W. Ditto, J. Weiss, *Science* **257**, 1230 (1992).
383. V. Petrov, V. Gaspar, J. Masere, K. Showalter, *Nature* **361**, 240 (1993).
384. Y.C. Lai, C. Grebogi, *Phys. Rev.* E **49**, 1094 (1994).
385. S.J. Schiff, K. Jerger, D.H. Duong, T. Chang, M.L. Spano, W.L. Ditto, *Nature* **360**, 615 (1994).
386. S. Hayes, C. Grebogi, E. Ott, A. Mark, *Phys. Rev. Lett.* **73**, 1781 (1994).
387. M. Spano, E. Ott, *Phys. Today* **48**, 34 (1995).
388. V. In, W.L. Ditto, *Phys. Rev.* E **51**, 2689 (1995).
389. E. Ott, M. Spano, Controlling chaos. In: *Chaotic, Fractal and Nonlinear Signal Processing, AIP Conf. Proc.*, vol. 375 R.A. Katz, ed. (AIP Press, New York 1995) p. 92.
390. E. Baretto, C. Grebogi, *Phys. Rev.* E **52**, 3553 (1995).
391. L. Poon, C. Grebogi, *Phys. Rev. Lett.* **75**, 4023 (1995).
392. G. Chen, X. Dong, *From Chaos to Order, Perspectives, Methodologies, and Applications* (World Scientific, Singapore 1998).
393. A.L. Fradkov, A.Yu. Pogromsky, *Introduction to Control of Oscillations and Chaos* (World Scientific, Singapore 1999).
394. K. Judd, A. Mees, K.L. Teo, T. Vincent (eds.), *Control and Chaos: Mathematical Modeling* (Birkhauser, Boston 1997).
395. T. Kapitaniak, *Chaos for Engineers: Theory, Applications, and Control* (Springer, New York 1998).
396. M. Lakshmanan, K. Murali, *Chaos in Nonlinear Oscillators: Controlling and Synchronization* (World Scientific, Singapore 1996).
397. M.J. Ogorzalek, *IEEE Trans. Circuit. Syst. I. Fundam. Theor. Appl.* **40**, 693 (1993).
398. G. Chen, X. Dong, *Int. J. Bifurcation Chaos* **3**, 1363 (1993).
399. Y.C. Lai, C. Grebogi: *Phys. Rev.* E **47**, 2357 (1993).
400. T.C. Newell, P.M. Alsing, A. Gavrielides, V. Kovanis, *Phys. Rev.* E **49**, 313 (1994).
401. M. Bernardo, *Int. J. Bifurcation Chaos* **6**, 557 (1995).
402. J.A.K. Suykens, P.F. Curran, L.O. Chua, *Int. J. Bifurcation Chaos* **7**, 671 (1996).
403. J.H. Peng, E.J. Ding, M. Ding, W. Yang, *Phys. Rev. Lett.* **76**, 904 (1996).
404. G. Malescio, *Phys. Rev.* E **53**, 2949 (1996).

405. C.K. Duan, S.S. Yang, *Phys. Lett.* A **229**, 151 (1997).
406. J. Yang, G. Hu, J. Xiao, *Phys. Rev. Lett.* **80**, 496 (1998).
407. H. Gang, Q. Zhilin, *Phys. Rev. Lett.* **72**, 68 (1994).
408. D. Auerbach, *Phys. Rev. Lett.* **72**, 1184 (1994).
409. I. Aranson, H. Levine, L. Tsimring, *Phys. Rev. Lett.* **72**, 2561 (1994).
410. R.V. Sole, L.M. Prida, *Phys. Lett.* A **199**, 65 (1995).
411. D. Battogtokh, A. Mikhailov, *Physica* D **90**, 84 (1996).
412. R.O. Grigoriev, M.C. Cross, H.G. Schuster, *Phys. Rev. Lett.* **79**, 2795 (1997).
413. M.K. Ali, J. Fang, *Discrete Dyn. Nat. Soc.* **1**, 179 (1997).
414. P. Parmananda, Yu. Jiang, *Phys. Lett.* A **231**, 159 (1997).
415. G. Hu, Z. Qu, K. He, *Int. J. Bifurcation Chaos* **5**, 901 (1995).
416. G.V. Osipov, V.D. Shalfeev, *IEEE Trans. Circuit. Syst.* **42**, 687 (1995).
417. G.V. Osipov, V.D. Shalfeev, J. Kurths, Asimmetry of coupling in a chain of discrete phase-locked loops as a reason of chaos suppression. In: *Proc. of Int. Conference on Control of Oscillations and Chaos*, F.L. Cernousko and A.L. Fradkov eds. (Inst. Problems of Mech. Engineering of RAS, St. Petersburg 1997).
418. V.V. Astakhov, A.V. Shabunin, V.S. Anishchenko, *Izv. Vuzov. Appl. Nonlinear Dyn.* **7**, 3 (1999) (in Russian).
419. V.V. Astakhov, A.V. Shabunin, P.A. Stalmakhov, A.V. Klimshin, *Izv. Vuzov. Appl. Nonlinear Dyn.* **8**, 91 (2000) (in Russian).
420. L.-Y. Cao, Y.-C. Lai, *Phys. Rev.* E **58**, 382 (1998).
421. N. Platt, E.A. Spiegel, C. Tresser, *Phys. Rev. Lett.* **70**, 279 (1993).
422. V.V. Astakhov, A.V. Shabunin, A.N. Silchenko, G.I. Strelkova, V.S. Anishchenko, *Radiotech. Electron.* **42**, 320 (1997) (in Russian).
423. V.V. Astakhov, A.N. Silchenko, G.I. Strelkova, A.V. Shabunin, V.S. Anishchenko, *Radiotech. Electron.* **41**, 1323 (1996) (in Russian).
424. P.L. Kapitsa, *J. Exp. Theor. Phys.* **21**, 588 (1951) (in Russian).
425. P.L. Kapitsa, *Phys. Uspekhi* **44**, 7 (1951).
426. V.V. Astakhov, V.S. Anishchenko, A.V. Shabunin, Stabilization of in-phase motions in symmetrically coupled oscillators by parametric perturbations. In: *Proc. of 1-st Int. Conf. on Control of Oscillations and Chaos*, vol. 2, F.L. Chernousko and A.L. Fradkov, eds. (Inst. Problems of Mech. Engineering of RAS, St. Petersburg 1997) p. 374.
427. V.V. Astakhov, A.V. Shabunin, *Izv. Vuzov. Appl. Nonlinear Dyn.* **5**, 15 (1997) (in Russian).
428. V.V. Astakhov, V.S. Anishchenko, A.V. Shabunin, *IEEE Trans. Circuit. Syst.* **42**, 352 (1995).
429. V.V. Astakhov, V.S. Anishchenko, G.I. Strelkova, A.V. Shabunin, Controlling spatiotemporal chaos in one- and two-dimensional coupled logistic map lattices. In: *Chaotic, Fractal and Nonlinear Signal Processing, AIP Conf. Proc.*, vol. 375 R.A. Katz, ed. (AIP Press, New York 1995) p. 104.
430. I. Waller, R. Kapral, *Phys. Rev.* A **30**, 2047 (1984).
431. K. Kaneko, *Physica* D **34**, 1 (1989).
432. J.P. Crutchfield, K. Kaneko, Phenomenology of spatiotemporal chaos. In: *Directions in Chaos*, H. Bai-lin, ed. (World Scientific, Singapore 1987) p. 272.
433. N.H. Packard, J.P. Crutchfield, J.D. Farmer, R.S. Shaw, *Phys. Rev. Lett.* **45**, 712 (1980).
434. F. Takens, Detecting strange attractors in turbulence. In: *Dynamical Systems and Turbulence*, eds. *Lecture Notes in Mathematics*, vol. 898 D. Rang and L.S. Young, (Springer, Berlin, Heidelberg 1980) p. 366.

435. R. Mane, On the dimension of the compact invariant sets of certain nonlinear maps. In: *Dynamical Systems and Turbulence*, Lecture Notes in Mathematics, vol. 898 D. Rang and L.S. Young, eds. (Springer, Berlin Heidelberg 1980) p. 230.

436. J. Stark, D.S. Broomhead, M.E. Davies, J. Huke, Takens embedding theorems for forced and stochastic systems. In: *Nonlinear Analysis, Theory, Methods and Applications, Proc. of 2nd Congress of Nonlinear Analysis*, vol. 30 (Elsevier, Amsterdam 1997) p. 5303.

437. J.D. Farmer, J.J. Sidorowich, *Phys. Rev. Lett.* **59**, 845 (1987).

438. M. Casdagli, *Physica* D **35**, 335 (1989).

439. H.D.I. Abarbanel, R. Brown, J.B. Kadtke, *Phys. Rev.* A **41**, 1782 (1990).

440. P. Grassberger, I. Procaccia, *Phys. Rev. Lett.* **50**, 346 (1983).

441. P. Grassberger, I. Procaccia, *Physica* D **9**, 189 (1983).

442. A. Wolf, J.B. Swift, H.L. Swinney, J.A. Vastano, *Physica* D **16**, 285 (1985).

443. J.P. Eckmann, S.O. Kamphorst, D. Ruelle, D. Gilberto, Phys. Rev. A **34**, 4971 (1986).

444. J. Cremers, A. Hübler, *Z. Naturforschung* A **42**, 797 (1987).

445. J.P. Crutchfield, B.S. McNamara, *Complex Syst.* **1**, 417 (1987).

446. T. Sauer, J.A. Yorke, M. Casdagli, *J. Stat. Phys.* **65**, 579 (1991).

447. D.S. Broomhead, G.P. King, *Physica* D **20**, 217 (1986).

448. M.T. Rosenstein, J.J. Collins, C.J. De Luca, *Physica* D **73**, 82 (1994).

449. T.S. Parker, L.O. Chua, *Practical Numerical Algorithms for Chaotic Systems* (Springer, New York 1989) p. 191.

450. A.M. Fraser, H.L. Swinney, *Phys. Rev.* A **33**, 1134 (1986).

451. W. Liebert, H.G. Schuster, *Phys. Lett.* A **142**, 107 (1989).

452. J.P. Eckmann, D. Ruelle, *Physica* D **56**, 185 (1992).

453. J.L. Breeden, N.H. Packard, *Int. J. Bifurcation Chaos* **4**, 311 (1994).

454. G.B. Mindlin, H.G. Solari, M.A. Natielo, R. Gilmore, X.-J. Hou, *J. Nonlinear Sci.* **1**, 147 (1991).

455. H.C. Tuckwell, *Introduction to Theoretical Neurobiology*, vols. 1 and 2 (Cambridge University Press, Cambridge 1988).

456. Task Force of the European Society of Cardiology and the North American Society of Racing and Electrophysiology, Circulation **93**, 1043 (1996).

457. A. Sherman, *Bull. Math. Biol.* **56**, 811 (1994).

458. T. Sauer, Reconstruction of integrate-and-fire dynamics. In: *Nonlinear Dynamics and Time Series*, C. Cutler and D. Kaplan eds. Fields Institute Communications, Vol. 11 (American Mathematical Society, Providence 1997) p. 63.

459. T. Sauer, *Phys. Rev. Lett.* **72**, 3911 (1994).

460. R. Castro, T. Sauer, *Phys. Rev.* E **55**, 287 (1997).

461. R. Castro, T. Sauer, *Phys. Rev. Lett.* **79**, 1030 (1997).

462. D.M. Racicot, A. Longtin, *Physica* D **104**, 184 (1997).

463. R. Hegger, H. Kantz, *Europhys. Lett.* **38**, 267 (1997).

464. N.B. Janson, A.N. Pavlov, A.B. Neiman, V.S. Anishchenko, *Phys. Rev.* E **58**, R4 (1998); A.N. Pavlov, O.V. Sosnovtseva, E. Mosekilde, V.S. Anishchenko, *Phys. Rev.* E **61**, 5033 (2000); A.N. Pavlov, O.V. Sosnovtseva, E. Modekilde, V.S. Anishchenko, *Phys. Rev.* E **63**, 036205 (2001).

465. D. Gabor, *J. IEEE Lond.* **93**, 429 (1946).

466. P. Panter, *Modulation, Noise, and Spectral Analysis* (McGraw-Hill, New York 1965).

467. J.L. Breeden, A. Hübler, *Phys. Rev.* A **42**, 5817 (1990).

468. R. Brown, N.F. Rulkov, E.R. Tracy, *Phys. Rev.* E **49**, 3784 (1994).

469. M.J. Bünner, M. Popp, Th. Meyer, A. Kittel, U. Rau, J. Parisi, *Phys. Lett.* A **211**, 345 (1996).

470. E.J. Kostelich, J.A. Yorke, *Physica* D **41**, 183 (1990).
471. G. Gouesbet, J. Maquet, *Physica* D **58**, 202 (1992).
472. G. Gouesbet, C. Letellier, *Phys. Rev.* E **49**, 4955 (1994).
473. K. Judd, A. Mees, *Physica* D **82**, 426 (1995).
474. E. Baake, M. Baake, H.G. Bock, K.M. Briggs, *Phys. Rev.* A **45**, 5524 (1992).
475. C.S.M. Lainscsek, F. Schürrer, J. Kadtke, *Int. J. Bifurcation Chaos* **8**, 905 (1998).
476. H. Haken, *Information and Self-Organization*, 2nd ed. (Springer, Berlin 1999).
477. O.L. Anosov, O.Ya. Butkovskii, Yu.A. Kravtsov, E.D. Surovyatkina, Predictable nonlinear dynamics: advances and limitations. In: *Chaotic, Fractal and Nonlinear Signal Processing*, AIP Conf. Proc., vol. 375 R.A. Katz. ed. (AIP, New York 1995) p. 71.
478. J. Kadtke, *Phys. Lett.* A **203**, 196 (1995).
479. M. Kremliovsky, J. Kadtke, M. Inchiosa, P. Moore, *Int. J. Bifurcation Chaos* **8**, 813 (1998).
480. J.S. Brush, Classifying transient signals with nonlinear dynamic filter banks. In: *Chaotic, Fractal and Nonlinear Signal Processing*, AIP Conf. Proc., vol. 375 R.A. Katz, ed. (AIP, New York 1995) p. 145.
481. N.B. Janson, A.N. Pavlov, V.S. Anishchenko, *Int. J. Bifurcation Chaos* **8**, 825 (1998).
482. A. Babloyantz, A. Deslexhe, *Biol. Cybern.* **58**, 203 (1988).
483. U. Parlitz, L.O. Chua, L. Kocarev, K.S. Halle, A. Shang, *Int. J. Bifurcation Chaos* **2**, 973 (1992).
484. U. Parlitz, *Phys. Rev. Lett.* **76**, 1232 (1996).
485. U. Parlitz, L. Kocarev, *Int. J. Bifurcation Chaos* **6**, 581 (1996).
486. H.D.I. Abarbanel, P.S. Linsay, *IEEE Trans. Circuit. Syst.* **40**, 643 (1993).
487. A.S. Dmitriev, Application maps with stored information in CDMA communication systems. In: *Proc. of the 1-st Int. Conf. on Control of Oscillations and Chaos*, vol. 2, F.L. Chernousko and A.L. Fradkov, eds. (Inst. Problems of Mech. Engineering of RAS, St. Petersburg 1997) p. 211.
488. S.O. Starkov, S.V. Yemetz, Digital communication systems, using chaos. In: *Proc. of the 1-st Int. Conf. on Control of Oscillations and Chaos*, vol. 2, F.L. Chernousko and A.L. Fradkov, ed. (Inst. Problems of Mech. Engineering of RAS, St. Petersburg 1997) p. 207.
489. D.A. Gribkov, V.V. Gribkova, Yu.I. Kuznetsov, A.G. Rzhanov, Gloval dynamical modeling of time series and application to restoration of broadband signal characteristics. In: *Chaotic, Fractal and Nonlinear Signal Processing*, AIP Conf. Proc., vol. 375, R.A. Katz, ed. (AIP, New York 1995) p. 181.
490. V.S. Anishchenko, A.N. Pavlov, *Phys. Rev.* E **57**, 2455 (1998).

# 3. Stochastic Dynamics

## 3.1 Stochastic Resonance

### 3.1.1 Introduction

The word "noise" is ordinarily associated with the term "hindrance". It was traditionally considered that the presence of noise can only make the operation of any system worse. There are well-known classical radio physical problems related to limitations of the sensitivity of amplifiers and a finiteness of the pulse bandwidth of oscillators due to the presence of natural and technical noise [1–3] (cf. Sect. 1.3).

In contrast, it has recently been established that noisy sources in nonlinear dynamical systems (DS) are able to induce completely new regimes that cannot be realized without noise. These effects were called noise-induced transitions [4], one beautiful example of which are noise-induced self-sustained oscillations [5]. The diversity and complexity of these transitions in nonlinear DS raised the following question: Does noise always bring disorder to a system's behavior, or are there cases when noise enhances the degree of order in a system or evokes improvement of its performance? The answer to this question is clear: Yes. Various studies have convincingly shown that in nonlinear systems increasing noise can induce new, more ordered behavior. Quite unexpected it can lead to the formation of more regular temporal and spatial structures, increase the degree of coherence, cause the amplification of weak signals accompanied by growth of their signal-to-noise ratio (SNR) or induce directed motions in systems with vanishing mean external forces. In other words, noise can play a constructive or beneficial role in nonlinear dynamics.

Stochastic resonance (SR) is one of the most shining and relatively simple examples of this nontrivial behavior in nonlinear systems under the influence of noise. The term was introduced in [6–8] in 1981–1982 in a study exploring a model proposed to explain the almost periodic recurrences of the Earth's ice ages. The temporal evolution of Earth's climate was modeled as the motion of an overdamped particle in a symmetric double-well potential driven by a periodic force. The two states of the system corresponded either to an ice period or to a warm climate on Earth, both being stable due to reflection or adsorption of energy, respectively, with respect to the present coverage of the

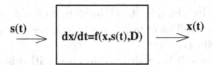

**Fig. 3.1.** General scheme of a stochastic resonator: A nonlinear noisy system with time scale depending on the noise intensity $D$ is forced by an ordered signal $s(t)$. The output becomes similar (ordered) to the input for optimally selected noise values

Earth by ice. The periodic force refers to tiny oscillations of the eccentricity of Earth's orbit changing the energy balance with a period of about $10^5$ years.

Estimations have shown that the actual amplitude of this periodic force is far too small to cause the drastic climate change. The possibility of switchings was achieved by introducing additional random forces into their model arising from fluctuations of the atmosphere. Not surprisingly, as Brownian particles hop in a double-well force field from one stable state to the other, atmospheric fluctuations induce transitions in the climate from stable cold to warm periods and vice versa. But remarkably, as the fundamental result of their model the authors found a sequence of temporally ordered transitions. The climate almost followed the vanishingly small external periodic perturbation for an assumed finite noise strength of the atmosphere.

The general scheme of SR is depicted in Fig. 3.1. It stands for phenomena for which the ordered response of a system with respect to weak input signals can be significantly increased by appropriately tuning the noise intensity to an optimal but nonvanishing value. Quantitatively, SR implies that some integral characteristics of the output, such as, e.g., the sharp peak in the output power spectrum located exactly at the frequency of the input (spectral power amplification, SPA), the SNR or input/output cross-correlation measures have well-marked maxima at a finite noise level. Also, the entropy-based measure of disorder attains a minimum, exemplifying the increase in noise-induced order [9].

Today SR is a well-known behavior of nonlinear stochastic dynamics [10–18]. It has been found and studied in a large variety of different physical systems, namely, in the electronic Schmitt trigger [19], in a ring laser [20], in magnetic systems [21], in passive optical bistable systems [22], in systems with electronic paramagnetic resonance [23], in experiments with Brownian particles [24], in experiments with magneto-elastic ribbons [25], in a tunnel diode [26], in superconducting quantum interference devices (SQUIDs) [27], and in ferromagnetics and ferroelectrics [28–30]. SR has been observed in chemical systems [31–33]. Finally, the most exciting applications of SR are in various biological systems. SR was observed on the level of single sensory neurons [34–36] and even on the level of ion channels [37]. Additionally, SR has been studied in human psychophysics [38] and experiments on animal behavior [39, 40].

SR was realized in bistable [6, 41] and monostable oscillators [42] and excitable dynamics as well as in nondynamical threshold systems [43–45] and in chaotic systems [46, 47].

In this chapter we will be mainly concerned with bistable models. In the following chapters we discuss various aspects of SR in noisy excitable systems and in noisy media or coupled networks of stochastic resonators.

### 3.1.2 Stochastic Resonance: Physical Background

The physical mechanism of SR is rather evident and simple. Contrary to a linear system, by definition a nonlinear system always possesses different frequencies or different time scales. Application of noise, even of an additive one, may populate different modes with different probability, which is different from traditionally studied linear systems. Hence, a change in the level of noise might change the temporal behavior of the nonlinear dynamics.

The secret of SR consist in the dependence of a single time scale on the intensity of the noise. Specifying to bistable overdamped dynamics, one controls the switching events between the two metastable states when tuning the noise level. We have studied this escape problem in detail in Sect. 1.2.6.

Indeed, the motion of a Brownian particle in a system with a symmetric double-well potential $U(x)$ is qualitatively characterized by two time scales. The first defines relaxations of fluctuations in the linear regime around the stable fixed points, called the intrawell or local dynamics. In the case of deep wells and noise that is not too strong, this time scale is independent of noise.

The second time scale characterizes the mean time of barrier crossings, also called the global dynamics. As outlined (see Sec. 1.2.6) it corresponds to the inverse of the rate of escape from a stable state. In the case of white noise a strong dependence on noise is expressed by the Arrhenius law [48, 49]:

$$r_{\mathrm{K}} = a_0 \exp(-\Delta U_0/D).$$

The prefactor $a_0$ is given by the curvature of the potential wells. At the same time it defines the relaxation rate in the linear regime near one of the fixed points. This relaxation time is always small compared with the global dynamics. Separation of the two time scales and hence a nonlinear regime dependent strongly on the noise level is achieved if the barrier height $\Delta U_0$ is large compared with the noise intensity $D$.

The amplitude of the periodic force $A$ is assumed to be sufficiently small. Transitions between the potential wells should still be excluded in the absence of noise. Nevertheless, at a first glance the presence of a small periodic force induces a small modulation of the barrier height, $\Delta U \simeq \Delta U_0 + A \sin \Omega t$ (Fig. 3.2), which results because of the exponential dependence in nonlinear modulation of the rates. Figure 3.3 presents the time series and the power spectrum of a reduced two-state-approximation. As a result, the power spectrum of the output signal shows a delta-peak at the modulation frequency

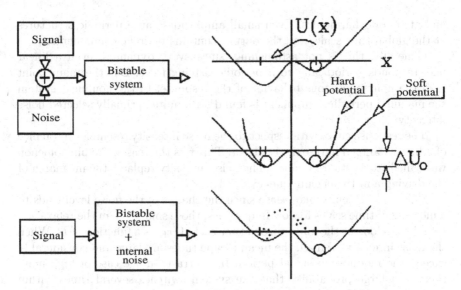

**Fig. 3.2.** A bistable potential with weak periodic modulation. The shape of the potential can vary, for example, as shown by the "hard" and "soft" curves. The particle symbolized by the *ball* can overcome the potential barrier $\Delta U_0$ only in the

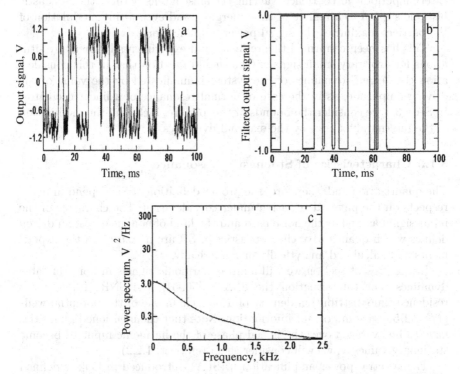

**Fig. 3.3.** (a) A time series at the output of a bistable system; (b) a similar time series which has been "two-state filtered"; (c) the power spectrum of the filtered signal

and at its odd harmonics. Even small amplitudes cause periodic structures at the global time scale, and the output contains periodic components.

Generally, this is not astonishing, since every dynamics, even a linear one, responds periodically to a periodic driving force. But the main point is in the nonmonotonous behavior of the response. In the nonlinear system the maximal periodic component is found for a finite optimally selected noise intensity.

The response curve with respect to the noise intensity resembles resonance of a linear damped harmonic oscillator. This was the reason the phenomenon was called stochastic resonance. The noise intensity replaces the mismatch of the driving and the eigenfrequency.

This replacement makes sense since the change of the noise level leads to a merging of time scales for low frequencies, those smaller than the relaxation rate. Let us look at the unperturbed time scale $1/r_K$ as a function of $D$. When the noise intensity is small, the mean escape times are large and considerably exceed the modulation signal period. In contrast, in the case of high noise there is a large probability that the system switches several times within one signal period. For moderate noise intensity, one can ensure a regime for which the mean barrier crossing time is close to the modulation signal period. Switching events will occur on average with the time scale of the external periodic force. Hence, varying the noise intensity tunes the stochastic bistable system to a regime of time merging, and even a tiny modulation of the barriers maximizes the signal power amplification.

This has been confirmed theoretically and experimentally for many situations for arbitrary small signals. One should mention that in this consideration the "eigen"-dynamics of the system is unaffected by the value of the driving amplitude, as in the case of normal resonance of a linear oscillator. In Sect. 3.3 we consider the bounds of this picture, including the modulation of the internal dynamics by the external dynamics.

### 3.1.3 Characteristics of Stochastic Resonance

The quantization and, therefore, the proper definition of SR depend in many respects on the physical situation under consideration. The character of the input signal as well as the noise used and the kind of nonlinear system driven defines which quantitative characteristics of SR are measured in the experiments and calculated analytically or numerically.

In the present section we will address harmonic signals in noisy bistable dynamics. For this situation the SPA $\eta$ [50, 51], the SNR [19] and the residence-time distribution density of a particle in one of the potential wells $p(\tau)$ [52, 53] give important information. Note that the mentioned quantities have to be averaged over the initial phase of the harmonic input to become stationary values as we will consider them (cf. Sect. 1.2.2).

The spectral power amplification $\eta$ (SPA), introduced in [50], is defined as the ratio between the signal powers at the output and at the input, i.e.

$$\eta = P_{\text{out}}/P_{\text{in}} \, . \tag{3.1}$$

In the case of weak harmonic input of frequency $\Omega$, the output power spectrum of the process $x(t)$ can be decomposed into the signal part and the noise background, $G_{x,x}^{(0)}(\omega)$. The signal part is represented by the delta-functions centered at the frequencies $\pm\Omega$: $\delta(\omega - \Omega) + \delta(\omega + \Omega)$. In this case $P_{\text{out}}$ is the integrated power stored in delta peaks at frequencies $\pm\Omega$ in the power spectrum of the output. The input signal carries the total power $P_{\text{in}} = A^2$.

In experiments the signal power is determined as an integral of the power spectral density over the range of measured frequencies. In experiments with harmonic signals, the frequency range corresponds to $\Omega \pm \Delta\omega$, where the frequency bin $\Delta\omega$ determines the frequency resolution of a measurement. The power spectral density at the signal frequency $S(\Omega)$ is represented in the form of regular and noisy components. Thus, the signal power can be calculated by the corresponding subtraction, i.e. $P_s = 2[G_{x,x}(\Omega) - \langle G_{x,x}^{(0)}\rangle]\Delta\omega$, where $\langle G_{x,x}^{(0)}\rangle$ is the noise spectral density averaged over the neighboring frequency bins around $\Omega$.

We use a definition of the signal-to-noise ratio (SNR) as the ratio of the total signal power to the noise power spectrum [12]. In the case of a harmonic signal it reads

$$SNR = \frac{1}{G_{x,x}^{(0)}(\Omega)} \, 2 \int_{\Omega-\Delta\omega}^{\Omega+\Delta\omega} G_{x,x}(\omega)\mathrm{d}\omega \, , \tag{3.2}$$

where the noise background $G_{x,x}^{(0)}$ is calculated at the frequency of the signal, $\omega = \Omega$. In analytical calculations we have to apply a limit of $\Delta\omega \to 0$ to the expression for the SNR. In practice, the value of $G_{x,x}^{(0)}(\Omega)$ is calculated by averaging $G_{x,x}^{(0)}(\omega)$ over the neighboring frequency bins around the signal frequency. Results can be presented using either linear (units) or logarithmic (dB) scales.

As a result of random switchings, the output of a stochastic bistable system without considering the intrawell dynamics can be represented by a stochastic telegraph signal (see Fig. 3.3b). The residence time for one of the potential wells is a random quantity whose probability density $p(\tau)$ in the absence of modulation shows evidence of an exponentially decreasing function [52]. When a modulation signal is added, the probability density becomes structured and contains a series of Gaussian-like peaks centered at $\tau = nT_s/2$, $n = 1, 3, 5, \ldots$, where $T_s$ is the modulation signal period. The maxima of $p(\tau)$ decay exponentially with $n$. In the regime of SR the peak of $p(\tau)$ at $\tau = T_s/2$ is the largest: the switchings between the potential wells are in phase with the external periodic signal, and the mean residence time $\langle \tau \rangle$ is closest to the half signal period. The description of SR based on the residence-time distributions, therefore, reflects a synchronization of system switchings by an external periodic force [53, 54]. Since the statistical properties of a telegraph signal depend on the noise intensity, the probability density structure can be controlled by noise variation. In this connection we have an alternative approach to analyzing the mechanisms of SR, based on the studies of residence-time statistics.

Characteristics and properties of SR must undoubtedly depend on the structure of signals applied to a nonlinear system. This concerns in equal degree both information and noisy signals. The modulation signal can be harmonic, can consist of several frequencies or represents a narrow-band stochastic process [55], or even be purely noisy. The stochastic force can be similar to white noise but can also have a finite correlation time and a bounded spectrum [56,57]. Depending on the signal properties, noise and the particular properties of nonlinear systems, SR is characterized by specific measures. At the same time without regard to the system characteristics and the structure of signals, SR is determined by its generic property of increasing the degree of order in the output signal for some optimal level of noise [9, 15, 58].

### 3.1.4 Response to a Weak Signal. Theoretical Approaches

Consider an overdamped bistable oscillator. It has become canonic for studying SR if additively driven by a periodic force. In this case the equation of motion in dimensionless variables reads

$$\dot{x} = x - x^3 + A \cos(\Omega t + \varphi_0) + \sqrt{2D}\, \xi(t). \tag{3.3}$$

It describes the overdamped motion of a Brownian particle in a double-well potential $U_0(x) = -x^2/2 + x^4/4$, driven by white Gaussian noise $\xi(t)$ with the intensity $D$ and periodic force $A \cos(\Omega t + \varphi_0)$. The $\varphi_0$ is a randomly distributed initial phase. The corresponding Fokker–Planck equation (FPE) for the probability density $p(x, t|x_0, t_0; \varphi_0)$ is

$$\frac{\partial p}{\partial t} = -\frac{\partial}{\partial x}\{[x - x^3 + A \cos(\Omega t + \varphi_0)]\, p\} + D\, \frac{\partial^2 p}{\partial x^2}. \tag{3.4}$$

This equation can be also written in the operator form

$$\frac{\partial p}{\partial t} = [\mathcal{L}_0 + \mathcal{L}_{\text{ext}}(t)]\, p, \tag{3.5}$$

where $\mathcal{L}_0 = -\frac{\partial}{\partial x}(x - x^3) + D\frac{\partial^2}{\partial x^2}$ is the unperturbed Fokker–Planck operator $(A = 0)$ and $\mathcal{L}_{\text{ext}}(t) = -A \cos(\Omega t + \varphi_0)\partial/\partial x$ refers to the periodic perturbation.

The rapid progress of SR studies has caused the development of a general theory of stochastic diffusion processes with periodically varying coefficients of drift and diffusion. Such a theory was proposed in [51, 59, 60] and is an extension of the Floquet theory to the case of FPE with periodic coefficients. One of the main conclusions of this theory is that the probability in the asymptotic limit $t_0 \to -\infty$ can be expanded into a Fourier series

$$p_{\text{asy}}(x, t; \varphi_0) = \sum_{n=-\infty}^{\infty} p_n(x) \exp\left[i\left(n\Omega t + \varphi_0\right)\right]. \tag{3.6}$$

Time inhomogeneity is removed by averaging over an equally distributed initial phase with the result $\langle p_{\text{asy}} \rangle_{\varphi_0} = p_0(x)$. The periodic nonstationarity can also be resolved by extending the system phase space using as an additional variable $\theta = \Omega t + \varphi_0$. Formally the process becomes stationary in $x, \theta$ whose stationary density agrees with previous results if integrated over $\theta$.

The asymptotic but time-dependent mean $\langle x(t) \rangle$ or the nonlinear response of the stochastic system to the harmonic force can be found from (3.6) to be

$$\langle x(t) \rangle_{\text{asy}} = \sum_{n=-\infty}^{\infty} M_n \exp\left[i\left(n\Omega t + \varphi_0\right)\right]. \tag{3.7}$$

Therein $M_n$ are complex-valued amplitudes connected with the multiple frequencies of the external force frequency. They depend on the noise intensity $D$, the signal frequency $\Omega$ and the amplitude $A$. The output power at the fundamental frequency is defined by the $|M_1|^2$ at positive and negative $\Omega$. Hence, the SPA $\eta$ according to (3.1) reads [50]

$$\eta = \left(\frac{2|M_1|}{A}\right)^2. \tag{3.8}$$

Analytical expressions for the amplification coefficient can be derived via some approximations. One of the main ones is a weak signal approximation when the response can be considered to be linear. Other approximations impose some restrictions on the signal frequency. We shall further study two approximate theories of SR, namely, the two-state theory proposed in [61] and the linear response theory [62–65].

**Two-state Theory of Stochastic Resonance.** The two-state theory, or the adiabatic theory [61], was the first theoretical description of SR. Since this theory is simple and elegant, it is used in the majority of studies on SR.

Let us approximate the global dynamics of a symmetric bistable system with attractor position at $\pm x_m$ by a two-state dynamics with discrete variables $\sigma(t) = \pm x_m$. Setting $\sigma = +x_m$ we define:

$$p(x, t | x_0, t_0) = p(\sigma | \sigma_0, t_0)\,\delta(x - x_m) + p(-\sigma, t | \sigma_0, t_0)\,\delta(x + x_m). \tag{3.9}$$

Herein $p(\sigma, t | \sigma_0, t_0)$ are the conditional probabilities of residing in one of the states $\sigma$ at time $t$ which satisfy normalization conditions.

Furthermore, by introducing the transition probabilities per unit time of switchings from one state to the other, $W_\sigma(t)$, we arrive at the simplest master equation [cf. (1.65)]:

$$\frac{dp(\sigma, t | \sigma_0, t)}{dt} = -W_\sigma(t)\,p(\sigma, t | \sigma_0, t) + W_{-\sigma}(t)\,p(-\sigma, t | \sigma_0, t). \tag{3.10}$$

Using the normalization this set of equations can be solved analytically for a given initial density.

The following transition probabilities make sense for a periodically driven bistable dynamics with assumed low frequencies compared to the local dynamics:

$$W_\sigma(t) = r_K \exp\left(-\frac{A\sigma}{D}\cos\Omega t\right).\tag{3.11}$$

Without external forcing ($A = 0$) the transition probabilities coincide with the Kramers rate $r_K$, which is in the case of (3.3)

$$r_K = \frac{1}{\sqrt{2}\,\pi}\exp\left(-\frac{1}{4D}\right).\tag{3.12}$$

The ansatz (3.11) is a Kramers rate with periodically modulated $\Delta U_{\text{eff}} = \Delta U_0 \pm A x_m \cos\Omega t$ and $\Delta U_0 = 1/4$. Sufficiently small amplitudes $\Delta U_0 - A \ll D$ still yield time scale separation between the local and global dynamics $r_K \ll 1/(\sqrt{2}\pi)$.

The given conditional probabilities allow calculation of the autocorrelation function. In order to simplify the explanation [61] we assume a weak signal, in detail $A x_m \ll D$, which introduces a lower limit on $D$. In this case one can expand the expressions (3.11) in a Taylor series of $A$. Retaining the linear term the time-dependent conditional probabilities are approximately

$$p(\sigma, t|\sigma_0, t_0) = \frac{1}{2}\left[\exp[-2r_K(t-t_0)]\left(2\delta_{\sigma_0,\sigma} - 1 - \frac{2r_K A x_m \cos(\Omega t_0 + \psi)}{D\sqrt{4r_K^2 + \Omega^2}}\right)\right.$$
$$\left. +1 + \frac{2r_K A x_m \cos(\Omega t + \psi)}{D\sqrt{4r_K^2 + \Omega^2}}\right],\tag{3.13}$$

where $\psi = -\arctan(\Omega/2r_K)$.

Of great importance is the mean value characterizing the system response. From (3.9) one obtains the conditioned first moment $\langle x(t)|x_0, t_0\rangle = \int xp(x, t|x_0, t_0)\,dx$. In the asymptotic limit $t_0 \to -\infty$, it follows the periodic but phase-shifted response,

$$\langle x(t)\rangle_{\text{asy}} = A_1(D)\cos[\Omega t + \psi(D)].\tag{3.14}$$

The amplitude $A_1(D)$ as well as the phase shift $\psi(D)$ depend on the intensity of the noise. Explicitly they read

$$A_1(D) = \frac{A x_m^2}{D}\frac{2\,r_K(D)}{\sqrt{4r_K^2(D) + \Omega^2}},\tag{3.15}$$

$$\psi(D) = -\arctan\frac{\Omega}{2\,r_K(D)},\tag{3.16}$$

where we have pointed out the noise dependence of the Kramers rates.

As in (3.8) the SPA in the case of a harmonic output is expressed in terms of the resulting amplitude (see 3.1):

$$\eta = \frac{4r_{\mathrm{K}}^2 x_m^4}{D^2(4r_{\mathrm{K}}^2 + \Omega^2)}. \tag{3.17}$$

With (3.12) and (3.17) it follows that the SPA attains a single maximum as a function of noise intensity $D$.

We proceed similarly to find the conditioned autocorrelation function $\langle x(t+\tau)x(t)|x_0, t_0\rangle$ and its asymptotic limit for $t_0 \to -\infty$. However, by virtue of periodic modulation of the transition probabilities, it depends not only on the time shift $\tau$ but also on the explicit time $t$. Thus, again one needs to perform an additional averaging over the period of the external force, a procedure which is equivalent to averaging over an equally distributed random initial phase. After Fourier transformation the expression for the spectral density $G_{\sigma,\sigma}(\omega)$ has the form

$$G_{\sigma,\sigma}(\omega) = G_{\sigma,\sigma}^{(0)}(\omega) + \frac{\pi}{2} A_1^2(D)\left[\delta(\omega - \Omega) + \delta(\omega + \Omega)\right]. \tag{3.18}$$

It contains two items, namely, a periodic one represented by the $\delta$-functions with weight proportional to the SPA, and a Lorentzian-like background $G_{\sigma,\sigma}^{(0)}(\omega)$:

$$G_{\sigma,\sigma}^{(0)}(\omega) = \frac{4r_{\mathrm{K}} x_m^2}{4r_{\mathrm{K}}^2 + \omega^2}\left(1 - \frac{A_1^2(D)}{2x_m^2}\right). \tag{3.19}$$

As seen from the last expression, the background is represented by the sum of the unperturbed spectrum and an additional term of order $A^2$. The appearance of the additional term is due to the signal which suppresses the noise background. It arises due to the application of Parseval's theorem in the two-state theory.

By computing the SNR according to (3.2) this term should be neglected in the linear approximation. Therefore, within this limit the SNR for the two-state model is

$$\mathrm{SNR} = \pi\left(\frac{Ax_m}{D}\right)^2 r_{\mathrm{K}}. \tag{3.20}$$

It exhibits a single maximum at noise value $D = D_{\max}^{\mathrm{SNR}} = \Delta U/2 = 1/4$. Surprisingly, this last expression does not reflect the dynamical explanation of the time-scale merging in the case of maximal amplification as explained above. This circumstance is included in the SPA only. Approximately we find in the low- and high-frequency limits for the SPA, respectively,

$$\eta_{\Omega \to 0} = \frac{4x_m^4}{D^2}, \qquad \eta_{\Omega \to \infty} = \frac{4r_{\mathrm{K}}^2 x_m^4}{D^2\,\Omega^2}. \tag{3.21}$$

The outgoing curve $\eta(D)$ due to (3.17) is always located below the two limits. The coincidence of both limits may serve to approximate the location of the maximal SPA yielding $2r_{\mathrm{K}}^2 = \Omega$, with the resulting $D_{\max}^{\mathrm{SPA}}$ explicitly expressing the dynamical origin [12].

**Linear Response Theory of Stochastic Resonance.** In case of a weak external force $f(t)$ the response of a nonlinear stochastic system $\langle x(t) \rangle$ in the asymptotic limit is determined by the linear integral relation [62, 63, 66]

$$\langle x(t) \rangle_{\mathrm{asy}} = \langle x \rangle_{\mathrm{st}} + \int_{-\infty}^{\infty} \kappa(t - \tau, D) \, f(\tau) \, \mathrm{d}\tau. \tag{3.22}$$

Here $\langle x \rangle_{\mathrm{st}}$ is the mean value of the unperturbed state variable $[f(t) = 0]$; later on we set $\langle x \rangle_{\mathrm{st}} = 0$. The function $\kappa(t)$ in (3.22) is called the response function and the assumption made is called the linear response theory (LRT).

The relation (3.22) accounts for arbitrary weak perturbations but one can consider a harmonic driving force without loss of generality. Since $\kappa(s) = 0$ for $s < 0$ the system response is expressed through the susceptibility $\chi(\omega)$, which is the Fourier transform of the response function. Formally, it reads

$$\langle x(t) \rangle = A \, |\chi(\Omega)| \, \cos\left(\Omega t + \psi\right), \tag{3.23}$$

with the phase shift $\psi$ found from

$$\psi = -\arctan \frac{\mathrm{Im} \, \chi(\Omega)}{\mathrm{Re} \, \chi(\Omega)}. \tag{3.24}$$

Consequently, the SPA is found as

$$\eta = |\chi(\Omega)|^2. \tag{3.25}$$

Likewise in the two-state approximation there is a noise background $G_{x,x}^{(0)}(\omega)$ with neglected signal suppression, and the periodic output contributes to the spectral density. The latter gives $\delta$-peaks with the SPA as weight, resulting in [cf. (3.18)]

$$G_{x,x}(\omega) = G_{x,x}^{(0)}(\omega) + \frac{\pi}{2} A^2 \, |\chi(\Omega)|^2 \, [\delta(\omega - \Omega) + \delta(\omega + \Omega)]. \tag{3.26}$$

The power at $\omega = \Omega > 0$ is the sum of the multipliers in front of the $\delta$-peaks at $\pm\Omega$. The SNR in LRT is in agreement with (3.2)

$$\mathrm{SNR} = \frac{\pi A^2 |\chi(\Omega)|^2}{G_{x,x}^{(0)}(\Omega)}, \tag{3.27}$$

and the susceptibility $\chi(\omega)$ occurs as the main quantity in all expressions.

LRT has delivered an elegant solution to determine $\chi(\omega)$. The solution can be realized via the fluctuation–dissipation theorem (FDT). As first shown by Kubo for systems in equilibrium [66], the response function in the case of weak perturbations is connected with the correlation function of the *unperturbed* system. Later on, this important connection was generalized by Hänggi and Thomas [63] to a wide class of stochastic processes where the perturbation operator in the corresponding master operator is of the gradient type as in

(3.4). Therefore, LRT should be applicable for SR, too, which indeed was proven by a lot of studies on different problems. But first we sketch the application to the overdamped bistable system (3.3).

The response function $\kappa(t)$ is connected with the autocorrelation function $c_{x,x}^{(0)}(t)$ of the unperturbed system as follows [50, 51]:

$$\kappa(t) = -\frac{\Theta(t)}{D} \frac{\mathrm{d}}{\mathrm{d}t} c_{x,x}^{(0)}(t), \tag{3.28}$$

where $\Theta(t)$ is the Heaviside function. Hence, the FDT relates two principally different processes: the statistical properties in the perturbed state with the linear response to an external driving force. The determination of the susceptibility requires an expression for the unperturbed autocorrelation function $c_{x,x}^{(0)}(\tau, D)$.

Unfortunately, so far there is no exact determination, even for the well-studied bistable systems. However, there exist several good approximate solutions. The most precise approach is based on the expansion of the Fokker–Planck operator in terms of eigenfunctions [41, 67]. The correlation function can be represented by a series $g_j \exp(-\lambda_j \tau)$, where $\lambda_j$ are the eigenvalues of the Fokker–Planck operator and $g_j$ are the coefficients which are computed by averaging the corresponding eigenfunctions over the unperturbed equilibrium distribution.

In the simplest case, when calculating the correlation function one only may take into account the smallest nonvanishing eigenvalue $\lambda_m$, which is related to the Kramers rate of escape from a potential well:

$$\lambda_m = 2\,r_\mathrm{K} = \frac{\sqrt{2}}{\pi} \exp\left(-\frac{1}{4D}\right). \tag{3.29}$$

Then the correlation function and the spectral density simplify to

$$c_{x,x}^{(0)}(\tau, D) \approx \langle x^2 \rangle_\mathrm{st} \exp(-\lambda_m \tau), \quad G_{x,x}^{(0)}(\omega) = \frac{2\lambda_m \langle x^2 \rangle_\mathrm{st}}{\lambda_m^2 + \omega^2}, \tag{3.30}$$

where $\langle x^2 \rangle_\mathrm{st}$ is the stationary second cumulant. Such an approximation corresponds to the two-state approach and only takes into account the global dynamics.

Improvements with the inclusion of the intrawell dynamics in the correlation function can be made by adding an additional exponential term in (3.30). It describes fast fluctuations within the potential wells. In this case the correlation function reflects both the global dynamics (factor $\lambda_m$) and the local intrawell dynamics [51]:

$$c_{x,x}^{(0)}(\tau, D) = g_1 \exp(-\lambda_m \tau) + g_2 \exp(-\alpha \tau), \tag{3.31}$$

where $\alpha$ stands for the relaxation rates in the potential minima. For the particular example (3.3), $\alpha = 2$. The coefficients $g_{1,2}$ in (3.31) are determined from the expression for the correlation function and its derivative for $\tau = 0$ and are equal to [51]

$$g_1 = \langle x^2 \rangle_{\text{st}} - g_2, \quad g_2 = \frac{\lambda_m \langle x^2 \rangle_{\text{st}}}{\lambda_m - \alpha} + \frac{\langle x^2 \rangle_{\text{st}} - \langle x^4 \rangle_{\text{st}}}{\lambda_m - \alpha}. \quad (3.32)$$

The susceptibility in the two-state or single-exponent approximation reads

$$\chi(\omega, D) = \frac{1}{D} \frac{\lambda_m \langle x^2 \rangle_{\text{st}}}{\lambda_m^2 + \omega^2} (\lambda_m - i\omega), \quad (3.33)$$

and with regard to the intrawell dynamics

$$\chi(\omega, D) = \frac{1}{D} \left( \frac{g_1 \lambda_m^2}{\lambda_m^2 + \omega^2} + \frac{g_2 \alpha^2}{\alpha^2 + \omega^2} \right) - i\omega \left( \frac{g_1 \lambda_m}{\lambda_m^2 + \omega^2} + \frac{g_2 \alpha}{\alpha^2 + \omega^2} \right). \quad (3.34)$$

With the derived susceptibilities one may readily find expressions for the SPA (3.25) and the SNR (3.27). For the single-exponent approximation, the SPA and the SNR coincide with corresponding expressions in the two-state theory ((3.17,3.20):

$$\eta(\Omega, D) = \frac{1}{D^2} \frac{(\langle x^2 \rangle_{\text{st}} \lambda_m)^2}{\lambda_m^2 + \Omega^2}, \quad \text{SNR} = \frac{\pi A^2}{2D^2} \langle x^2 \rangle_{\text{st}} \lambda_m. \quad (3.35)$$

Taking into account the intrawell dynamics yields the following expressions [51]:

$$\eta(\Omega, D) = \frac{(g_1 \lambda_m)^2 (\alpha^2 + \Omega^2) + (g_2 \alpha)^2 (\lambda_m^2 + \Omega^2) + 2 g_1 g_2 \alpha \lambda_m (\alpha \lambda_m + \Omega^2)}{D^2 (\lambda_m^2 + \Omega^2)(\alpha^2 + \Omega^2)}, \quad (3.36)$$

$$\text{SNR} = \frac{\pi A^2}{2D^2} \frac{(g_1 \lambda_m)^2 (\alpha^2 + \Omega^2) + (g_2 \alpha)^2 (\lambda_m^2 + \Omega^2) + 2 g_1 g_2 \alpha \lambda_m (\alpha \lambda_m + \Omega^2)}{g_2 \alpha (\lambda_m^2 + \Omega^2) + g_1 \lambda_m (\alpha^2 + \Omega^2)}. \quad (3.37)$$

As seen from Fig. 3.4a, the two approximations yield similar results in the range of maximal amplification at optimal noise. Differences arise for small noise intensities. In the single-exponent theory $\eta$ vanishes when $D \to 0$. The inclusion of the intrawell dynamics corrects this limit. The time scale of the global dynamics becomes exponentially large. During the duration of many external periods, the system is unable to respond on the global scale. Hence, the amplification behaves as $1/(\alpha^2 + \Omega^2)$, as in a linear system with the relaxation time $1/\alpha$.

When the driving frequency decreases, the maximum of the amplification shifts to the area of smaller noise intensities. But in case of smaller noise, the modulation of the rates (not of the barrier!) becomes effectively stronger and, therefore, the amplification factor itself increases. Figure 3.4b shows the

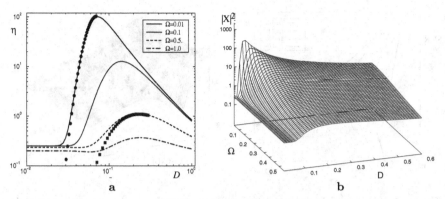

**Fig. 3.4.** (a) Spectral power amplification $\eta$ (3.36) as a function of the noise intensity $D$ for different values of the external periodic signal frequency; the dependences (3.35) obtained without regard to the intrawell dynamics are shown by *circles* and *squares*. (b) Susceptibility versus the noise intensity $D$ and signal frequency $\Omega$

amplification coefficient (3.36) as a function of the noise intensity and driving frequency. As follows from Fig. 3.4, the amplitude–frequency characteristic rises in the low-frequency range. This property follows from the nature of the fluctuations in the bistable systems. The bistable stochastic resonator is a low-frequency device. The noise enhances slow dynamics at the global scale, much smaller compared to the intrawell dynamics. The fluctuation spectrum is almost located in the low-frequency range with a width determined by the Kramers rate, which is limited from above by the relaxation time in the wells.

The SNR (3.37) as a function of noise intensity is displayed in Fig. 3.5. As seen, the SNR diverges for small values of $D$. This can be explained by the contribution of periodically modulated local dynamics inside the potential wells [61]. For sufficiently low driving frequencies the SNR achieves its maximum at $D \approx 1/8$ (note $\Delta U = 1/4$) and is, practically, independent on $\Omega$. However, with increasing $\Omega$, the SR effect disappears as the SNR becomes a monotonically decaying function of the noise intensity. Thus we see that the two SR measures, namely, the amplification coefficient and the SNR, again demonstrate different behavior. The SPA possesses a maximum even at high frequencies, while the SNR displays the SR effect only for frequencies low enough.

### 3.1.5 Array-Enhanced Stochastic Resonance

An often met physical situation is networks of coupled systems. Compared to the former case the coupling strength appears as a new control parameter. The question immediately following is how coupling affects SR.

Of course there will be different answers for the various systems. But in a series of papers it was shown that when stochastic bistable dynamics are coupled and driven in parallel the SPA and the SNR can be enhanced

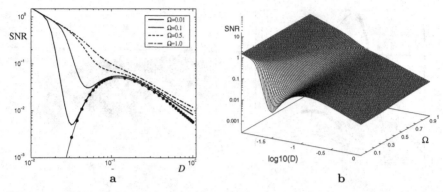

**Fig. 3.5.** (a) Signal-to-noise ratio (3.37) as a function of the noise intensity $D$ for different values of the external periodic signal frequency (the dependence (3.33) obtained with neglecting the intrawell dynamics is shown by *circles*. (b) SNR versus $D$ and $\Omega$)

[68–72, 75, 76]. Lindner et al. [77] introduced the notion of "array-enhanced stochastic resonance" for this coupling-induced increase in the SNR achieving a maximum at a finite coupling strength. Likewise for large noise a strong coupling lets the response fall off. But, apart from the noise strength in a single stochastic resonator, the coupling strength controls SR as well.

**Coupling Enhanced Response in Spin-chains.** As an easy way to illustrate this topic we present results from an analysis of a chain of ferromagnetically coupled two-state resonators [78,79]. Thereby Glauber's model of a stochastic Ising model [80] is widely employed.

In fact, the coupled magnetic spin system represents a good candidate for SR, as was shown in [81–83]. Coupling creates barriers for the spin flipping. Therefore, if a periodic force is applied for a given coupling strength, the temperature has to be chosen optimally to achieve the best periodic response of the system. But an increase of the coupling strength weakens monotonously the value of the peak in the spectrum; thus, weak coupling shows the best performance.

Alternatively, coupled bistable elements may be modeled as a connected chain of two-state resonators, where a barrier $\Delta U$ still exists for the single uncoupled element. In each element the periodically modulated but noise-dependent expression (3.11) has to be used. Coupling between these locally bistable elements can be introduced á la Glauber, which favors with $\gamma > 0$ a parallel ($\gamma < 0$: antiparallel) alignment of the states.

Then the rates for a transition $\sigma_i \to -\sigma_i$ of the $i$th spin in a chain are

$$W_i(\sigma_i) = r_{\rm K}(D)\left(1 - \sigma_i \frac{A}{D}\cos(\Omega t + \varphi_0)\right)\left(1 - \frac{\gamma}{2}(\sigma_{i-1} + \sigma_{i+1})\sigma_i\right). \quad (3.38)$$

These rates define the dynamics of the chain and should be inserted in the master equation

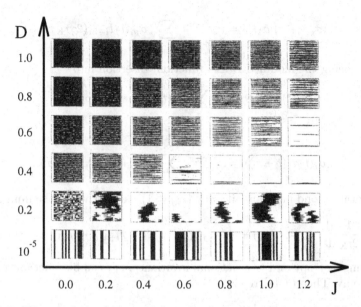

**Fig. 3.6.** Numerical simulations of the spin chain according to transition probabilities (3.38). In each panel a chain of one hundred elements with running time bottom-up is presented. Dark and white regions correspond to spin up and spin down, respectively [9, 84]

$$\dot{p}(\bar{\sigma}) = \sum_k W_k(-\sigma_k)\, p(\ldots, -\sigma_k, \ldots) - \sum_k W_k(\sigma_k)\, p(\ldots, \sigma_k, \ldots) \quad (3.39)$$

for the probability function $p(\bar{\sigma}, t)$ to find the chain in a particular configuration $\bar{\sigma} = (\ldots, \sigma_{k-1}, \sigma_k, \sigma_{k+1}, \ldots)$ at time $t$. We further introduce the spin coupling parameter $J$ by $\gamma = \tanh 2J/D$, giving without perturbation the stationary distribution of an Ising model.

The stochastic process can be easily simulated. Numerically generated realizations [84] (see also [11]) are presented in Fig. 3.6. A best periodic response of the chain is obtained if $\Delta U = 0.25$ near $D \propto 0.5$ and $J \propto 0.6$.

From the master equation the SPA can be determined, and the correlation function in the unperturbed case has been given by Glauber [80]. Both expressions in turn are obtained via the equation derived for the conditioned average from (3.39). One finds

$$\frac{1}{2r_K}\frac{\mathrm{d}}{\mathrm{d}t}\langle \sigma_i(t)|\bar{\sigma}\rangle = -\langle \sigma_i(t)|\bar{\sigma}\rangle + \frac{\gamma}{2}\left[\langle \sigma_{i+1}(t)|\bar{\sigma}\rangle + \langle \sigma_{i-1}(t)|\bar{\sigma}\rangle\right]$$

$$+ \frac{A}{D}\left(1 - \frac{\gamma}{2}(r_{i-1,i} + r_{i,i+1})\right)\cos(\Omega t + \varphi_0). \quad (3.40)$$

with the time-dependent $r_{i,j}(t) = \langle \sigma_i(t)\,\sigma_j(t)\rangle$.

Let us briefly report the unperturbed ($A = 0$) case. The solution of (3.40) is ($\tau \geq 0$)

$$\langle \sigma_i(t+\tau)|\bar{\sigma}(t)\rangle = e^{-2r_\mathrm{K}\tau} \sum_{m=-\infty}^{\infty} \sigma_m(t)I_{i-m}(2\gamma r_\mathrm{K}\tau), \tag{3.41}$$

with $I_n$ being the modified Bessel function. Insertion into

$$\langle \sigma_i(t)\sigma_j(t+\tau)\rangle = \sum_{\bar{\sigma}} \langle \sigma_j(t+\tau)|\bar{\sigma}(t)\rangle \sigma_i p(\bar{\sigma},t), \tag{3.42}$$

yields

$$\langle \sigma_i(t)\sigma_j(t+\tau)\rangle = e^{-2r_\mathrm{K}\tau} \sum_{m=-\infty}^{\infty} r_{i,m}(t)I_{i-m}(2\gamma r_\mathrm{K}\tau). \tag{3.43}$$

For the second moments, if $i \neq j$ ($r_{i,i} = 1$), one derives in the limit $A = 0$

$$\frac{1}{2r_\mathrm{K}}\frac{\mathrm{d}}{\mathrm{d}t}r_{i,j} = -2r_{i,j} + \frac{\gamma}{2}\big(r_{i,j-1} + r_{i,j+1} + r_{i-1,j} + r_{i+1,j}\big). \tag{3.44}$$

Assuming isotropy and translational invariance, $r_{i,j}$ can be functions of $\Delta = |i - j|$ only. Then (3.44) reads

$$\frac{1}{2r_\mathrm{K}}\frac{\mathrm{d}}{\mathrm{d}t}r_\Delta = -2r_\Delta + \gamma\big(r_{\Delta-1} + r_{\Delta+1}\big) \tag{3.45}$$

and $r_0 = 1$. In the asymptotic stationary limit it gives a recurrence relation, which is solved by $r_\Delta = \rho^\Delta$, where $\rho$ is given by

$$\rho^2 - 2\gamma^{-1}\rho + 1 = 0. \tag{3.46}$$

The physically relevant solution is [80]

$$\rho = \big(1 - \sqrt{1-\gamma^2}\big)\gamma^{-1} = \tanh(J/T). \tag{3.47}$$

Ending this brief passage, one is able to write down the stationary correlation function of Glauber's spin chain for arbitrary $\tau$:

$$\langle \sigma_i(t)\sigma_j(t+\tau)\rangle = e^{-2r_\mathrm{K}|\tau|} \sum_{m=-\infty}^{\infty} \eta^{|i-j+m|}I_m(2\gamma r_\mathrm{K}|\tau|). \tag{3.48}$$

The Fourier transform of this expression defines the noisy background spectrum.

The conditioned averages are calculated asymptotically with $t_0 \to -\infty$. Since $r_{i,j}$ in (3.40) is multiplied by $A/D$, it will be sufficient in LRT to take their asymptotic expressions from (3.47). Then in the asymptotic limit solutions $\langle \sigma_i(t)\rangle_\mathrm{asy}$ of the conditioned averages can be written as follows:

$$\langle \sigma_i(t)\rangle_\mathrm{asy} = A_1(D)\cos\big(\Omega t + \varphi_0 + \psi(D)\big). \tag{3.49}$$

The amplitude $A_1(D)$ of these oscillations reads

$$A_1(D) = \frac{A}{D}\frac{2r_\mathrm{K}\sqrt{1-\gamma^2}}{\sqrt{\Omega^2 + [4\,r_\mathrm{K}\,(1-\gamma)]^2}}, \tag{3.50}$$

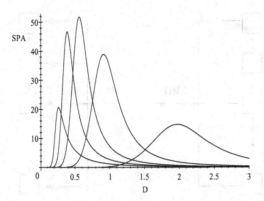

**Fig. 3.7.** SPA versus $D$ for several coupling parameters, $J$. Left to right: $0, 0.25, 0.5, 1.0, 2.5$ ($\Omega = 0.02$)

and the phase shift $\psi(D)$ is given by

$$\tan \psi(D) = \frac{\Omega}{2r_K(1-\gamma)}. \tag{3.51}$$

The response of the local spin embedded in the chain can be transferred to the two-state result (3.15) by replacing the rate as $r_K \to r_K(1-\gamma)$ and scaling the full expression by $\sqrt{(1+\gamma)/(1-\gamma)}$.

The SPA is presented in Fig. 3.7 and can be brought into the shape

$$\eta = \frac{\eta_s}{1 + \dfrac{\Omega^2}{4\,r_K^2\left[1-\tanh\left(\frac{2J}{D}\right)\right]^2}}, \tag{3.52}$$

with

$$\eta_s = \frac{1}{D^2}\exp\left(\frac{4J}{D}\right) \tag{3.53}$$

being the static response of the chain on a constant force. With increasing coupling $J$, this static response grows, whereas the second factor in (3.52) displays the dynamic inability to follow the signal. The larger the frequency the smaller the weight of the delta peak in the spectrum. Both dependences result in a bell-shaped curve for the SPA, since the static prefactor increases linearly in $\gamma$ whereas the dynamic response decreases with $\gamma^{-2}$.

A discussion of the SNR should distinguish between two principal arrangements of the output shown in Fig. 3.8. For the summed output

$$M(t) = \sum_i^N \sigma_i(t) \tag{3.54}$$

the noise part of the spectrum can be calculated explicitly [80] in the limit of an infinitely long chain, $N \to \infty$. The SNR per element monotonously decreases with increasing coupling

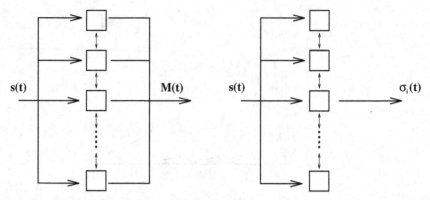

**Fig. 3.8.** Different schemes of coupled resonators. Whereas in the left scheme for the global output the SNR decreases monotonously, with increasing coupling the response of a single element in the coupled chain exhibits optimal coupling for the best SNR

$$\text{SNR}_M = \pi \frac{A^2}{D^2} r_K \sqrt{1 - \gamma^2}. \tag{3.55}$$

Alternatively, in Fig. 3.9 the SNR of a single element with the coupled chain is presented for different couplings. Array-enhanced response can be seen. With moderate coupling a single element inside the chain exhibits a larger SNR compared to the uncoupled resonator $J = 0$, which is the result of the McNamara–Wiesenfeld theory (3.33).

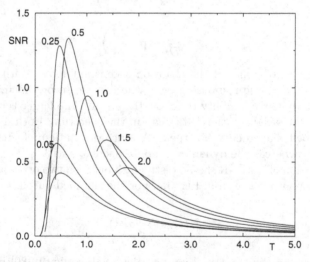

**Fig. 3.9.** SNR of a single element embedded in an infinitely long chain versus $D$ for different couplings $J$ ($\Omega = 0.02$)

Estimates of the SNR of a single element can be given for the low- and high-frequency limits:

$$\mathrm{SNR}_{\Omega \to 0} = \mathrm{SNR}_0(1 + \gamma)^2 = \mathrm{SNR}_0 \left( \tanh\left(\frac{2J}{D}\right) + 1 \right)^2, \qquad (3.56)$$

$$\mathrm{SNR}_{\Omega \to \infty} = \mathrm{SNR}_0 \sqrt{1 - \gamma^2} = \mathrm{SNR}_0 \cosh^{-1}\left(\frac{2J}{D}\right), \qquad (3.57)$$

where $\mathrm{SNR}_0$ is the SNR with vanishing coupling identical to the SNR of the two-state model. Since the SNR for finite frequencies can never exceed the low-frequency expression, we are able to find the upper limit of array-induced SR. The coupling-induced improvement of the SNR is bounded by the factor 4 compared to the uncoupled element.

**Stochastic Resonance of Front Motion.** A typical situation in systems with two stable phases or in bistable diffusively coupled dynamics is front propagation. The front represents a wave of transition between the two states dividing the two phases by a narrow moving interface. A lot of problems in high-energy physics, structural phase transitions, chemical dynamics and magnetization dynamics can be analyzed in the framework of this model [85–91].

Let a bistable reaction–diffusion system identical with a Landau–Ginzburg equation,

$$\tau\frac{\partial u(r,t)}{\partial t} = f(u) + r_0^2\frac{\partial^2 u(r,t)}{\partial r^2} = 2u(r,t)(1 - u(r,t)^2) + r_0^2\frac{\partial^2 u(r,t)}{\partial r^2}, \quad (3.58)$$

be the starting point of the analysis. The density of the bistable medium is labelled $u(r,t)$. It possesses two homogeneous stable fixed points $u_{1,3} = \pm 1$ and an unstable one $u_2 = 0$. In (3.58) $\tau$ defines the characteristic time scale of evolution, $r_0$ is the diffusion length.

If boundary conditions $u(r \to \pm\infty) = u_{1,3}$ are applied a solitary front of transition from $u_1$ to $u_3$ will be established [86, 90, 91]. A narrow interface of thickness $l$ of order $r_0$ divides two regions with homogeneous $u_1$ and $u_3$ and is moving with velocity $c$. A general expression for the velocity has been derived [92]:

$$c = \frac{\int_{u_1}^{u_3} f(u)\mathrm{d}u}{\tau \int_{-\infty}^{+\infty} \left(\frac{\partial u^0}{\partial r}\right)^2 \mathrm{d}r}. \qquad (3.59)$$

where $u^0$ is the solution of (3.58) with the applied boundary conditions. With $f(u)$ being a polynomial of third degree, this solution is well known [93]:

$$u^0(r,t) = u_1 + (u_3 - u_1)\frac{1}{1 + \exp\left[(r - ct)/l\right]}. \qquad (3.60)$$

The velocity can be calculated, $c = \frac{r_0}{\tau}(u_3 + u_1 - 2u_2)$, and the interface thickness is $l = r_0/(u_1 - u_3)$. For the special choice in (3.58) the front will be at rest, $c = 0$, $u^0 = u^0(r) = \tanh(r/r_0)$, and the width of the interface $l$ is given by diffusion length via $l = r_0/2$.

The solution (3.60) is stable with respect to fluctuations of the stationary front profile. It is marginally stable to shifts of the front position. Both circumstances are expressed by the eigenvalue spectrum when considering the linear stability of the front solution. As a result one obtains one vanishing eigenvalue, belonging to the Goldstone mode $u_r^0 = \partial u^0/\partial r$. The other discrete eigenvalue and the continuous spectrum describing front deviations are well separated from zero [86, 94].

Therefore, not surprisingly an additionally applied noise first affects the position of the front, which is without a restoring force. The velocity (3.59) becomes a stochastic time-dependent process $c_\eta = c(\eta(t))$ with given properties dependent on the applied noise. Assuming weak noise there might be a noise-induced shift and nonvanishing second moments, the latter yielding diffusion of the front position [95–98]. Therefore, in a first approximation the front position behaves like a Brownian particle.

The second ingredient to control the position of the front is stationary inhomogeneities, for example, those of rate coefficients entering into (3.58). They can pin the front on centers or amplify the motion out of repelling regions. Reducing again the description to that of the front position, one obtains a picture of a particle moving in an external force field.

As a "canonical" example, we introduce a double-well-shaped inhomogeneity, which should model two adjacent attracting inhomogeneities. This kind of inhomogeneity is typical for a variety of physical situations [99–102]. Taking additional noise, we therefore obtain a Kramers problem for the front position. Noise generates transitions of the interface between the two inhomogeneities. The rate of those will be of the Arrhenius type for sufficiently strong inhomogeneities and if the distance between the two centers is large compared to the front thickness.

The situation displayed above represents a good prototype for SR in distributed systems if subject to temporal periodic forces. It was studied in [103, 104] and also called *solitonic stochastic resonance* [105]. Since a motion of fronts is often accompanied with radiation it should be experimentally accessible. One can imagine that the ordered hopping of fronts between pinning centers can be employed as a technical device for transmitting periodic signals and could deliver experimental data about the existing defects in the present physical system.

The two adjacent regions attracting the front are introduced by an inhomogeneous term in (3.58). We assume a localized double well

$$h(r) = \epsilon \left( -\delta^2 r^2/2 + r^4/4 - 2\delta^4 \right), \quad |r| < 2\delta \tag{3.61}$$

and 0 otherwise. The parameter $\delta$ stands for the distance between the two wells and $\epsilon$ determines the depth. To achieve stability of the front profile, the strength of the inhomogeneities $\varepsilon\delta^4$ should be small, i.e., $(2\varepsilon\delta^4 \ll 1)$.

To have stochastic resonance of the front motion the model is completed by adding Gaussian white noise and periodic forces. In summary we obtain

$$\tau\frac{\partial u}{\partial t} = 2u(1-u^2) + r_0^2\frac{\partial^2 u}{\partial r^2} + 2h(r)u + A\cos\left(\Omega t + \varphi_0\right) + \sqrt{2D}\xi(t). \quad (3.62)$$

**Fronts as Brownian Particles.** The motion of fronts in dissipative bistable media has many similarities to the motion of solitons [86,106–109]. As well as solitons the solitary behavior of fronts has proven to be robust with respect to slowly varying external fields and spatial perturbations having characteristic length scales larger than the interface thickness of the front solution [110]. For relevant experiments, see, e.g., [111].

The front position $R(t)$ may be defined by the location of the unstable value, i.e., $u\big(R(t),t\big) = u_2$. Derivation with respect to time yields

$$\frac{\partial}{\partial t}u(R,t) + \frac{\partial}{\partial R}u(R,t)\,\dot{R} = 0. \quad (3.63)$$

Hence for the front velocity $\dot{R}$ it is sufficient to know the derivatives for $r = R(t)$.

Let $\phi = r - R(t)$ be the relative coordinate in the system of a resting front. Multiplying (3.62) by the Goldstone mode $u_r^0(\phi)$ with assumed stationarity of the front $u(r,t) \to u^0(\phi)$ is one of several techniques used to reduce the dynamics to $R(t)$. Then in a first approximation all functions depend on time via $R(t)$. For sharp interfaces $(u_r(\phi))^2$ is nonzero at the position of the front $r = R$ or $\phi = 0$ only, and one can approximately write

$$\int_{-\infty}^{\infty} dr\,\frac{\partial u^0}{\partial t}\frac{\partial u^0}{\partial r} = \int_{-\infty}^{\infty} dr\,\frac{\frac{\partial u^0(r,t)}{\partial t}}{\frac{\partial u^0(r,t)}{\partial r}}\left(\frac{\partial u^0}{\partial r}\right)^2 \approx -\dot{R}\int_{-\infty}^{\infty} dr\,(u_r^0)^2.$$
$$(3.64)$$

As a result (3.62) and (3.64) yield the following first approximations:

$$\tau\dot{R} = \frac{2\int_{-\infty}^{+\infty} dr\, u_r^0(\phi)u^0 h(r)}{\int_{-\infty}^{+\infty} dr\,(u_r^0)^2} + \tilde{A}\cos(\Omega t + \varphi_0) + \sqrt{2\tilde{D}}\eta(t), \quad (3.65)$$

with rescaled noise intensity $\tilde{D} = 9r_0^2 D/4$ and amplitude $\tilde{A} = 3r_0 A/2$; we can recall that the deterministic velocity vanishes in the considered case, $c = 0$. The first item may be formulated as an external force,

$$\tau\dot{R} = -\frac{dU_{r_0}(R)}{dR} + \tilde{A}\cos(\Omega t + \varphi_0) + \sqrt{2\tilde{D}}\eta(t). \quad (3.66)$$

with potential still dependent on $R(t)$ by $\phi(t)$ and defined by

$$-\frac{dU_{r_0}(R)}{dR} = \frac{3}{2}\int_{-\infty}^{+\infty} h(r)\,\mathrm{sech}^3(\phi/r_0)\sinh(\xi/r_0)dr. \quad (3.67)$$

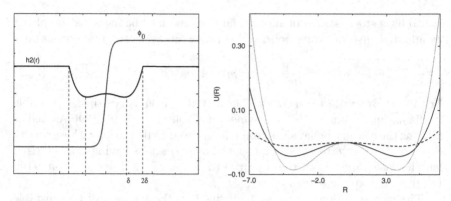

**Fig. 3.10.** (a) Sketch of the front in the inhomogeneous potential $h(r)$, see (3.61);
(b) The potential $U(R)$ for $\delta = 4$: $r_0 = 0.5$ (*dashed line*), $r_0 = 1.0$ (*solid line*),
$r_0 = 1.5$ (*dotted line*)

This dependence can be calculated explicitly. Replacing in (3.65) the derivative $u_r^0(r - R)$ by $-u_R^0(r - R)$, this derivation can be brought out of the
integral. As a result the potential reads

$$U_{r_0}(R) = \frac{3r_0^2}{2} \int_{-\infty}^{+\infty} h(r) \frac{1}{2r_0} \operatorname{sech}^2[(r - R)/r_0] dr, \tag{3.68}$$

with constant contributions omitted.

For $r_0 \to 0$ the second expression under the integral is known to be a
representation of Dirac's delta function, $\operatorname{sech}^2[(r - R)/r_0]/2r_0 \to \delta(r - R)$.
In the limit of vanishing $r_0$, it yields $U_0(R) = 3r_0^2 h(R)/2$. The potential can
also be defined for nonvanishing $r_0$. Substituting in (3.68) $x = (r - R)/r_0$,
one is able to obtain ($R < 2\delta$)

$$U_{r_0}(R) = \frac{3}{2} \epsilon r_0^2 \left[ -\frac{1}{2} \left( \delta^2 - \frac{3r_0^2 \pi}{12} \right) R^2 + \frac{R^4}{4} \right]. \tag{3.69}$$

Since the inhomogeneous perturbation term $h(r)$ has a double-well shape,
the potential $U_{r_0}(R)$ achieves a double-well structure as long as $r_0$ is smaller
than $2\delta/\sqrt{\pi}$. If $R > 2\delta$ the potential is constant. The potential is shown for
different $r_0$ in Fig. 3.10.

**The Signal-to-Noise Ratio.** The front behaves similarly to an overdamped
Brownian particle moving in a periodically modulated double-well potential.
We first estimate the mean escape rate from a metastable state of the potential (3.69) in the absence of the periodic field. For weak noise we use Kramers
formula (1.214), yielding

$$r_K = \frac{r_0^2}{2\sqrt{2}\tau\pi} \varepsilon \left( 4\delta^2 - \pi r_0^2 \right) \exp \left[ -\frac{1}{6D} \tau \epsilon \left( \delta^2 - \frac{\pi}{4} r_0^2 \right)^2 \right]. \tag{3.70}$$

Insertion into the SNR expression from the two-state theory (3.20) leads to

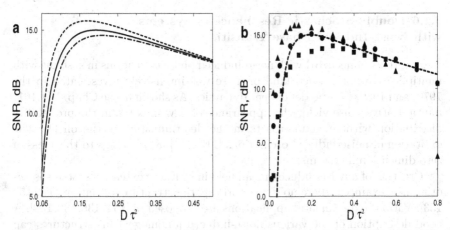

**Fig. 3.11.** (a) The SNR for three different values of $r_0$: 1.5 (dashed); 1.0 (solid); 0.5 (dash-dotted); (b) Numerically determined SNR versus $D$ for different values of coupling strength: $r_0^2 = 1.5$ (*triangles*), $r_0^2 = 1$ (*circles*), $r_0^2 = 0.5$ (*squares*). The *dashed line* represents the theoretical estimation from adiabatic theory for $r_0 = 1$. Other parameters value are $r_0 = 1.0$, $D = 0.2$, $\delta = 4$, $\varepsilon = 0.0005$, $A = 0.01$, $\Omega = 0.1$ and $\tau = 0.05$

$$\text{SNR} = \frac{A^2\tau^2}{6D^2\sqrt{2}}\varepsilon\left(4\delta^2 - \pi r_0^2\right)\exp\left(-\frac{\tau\varepsilon}{6D}\left[\delta^2 - \frac{\pi}{4}r_0^2\right]^2\right). \qquad (3.71)$$

In Fig. 3.11a we present the theoretical curve. The maximum of the SNR is located at

$$D_{\max} = \frac{1}{12}\tau\varepsilon\left(\delta^2 - \frac{\pi}{4}r_0^2\right)^2, \qquad (3.72)$$

which shifts with increasing coupling $r_0$ to smaller values of $D$. Similar to the case of array-enhanced SR the maximum increases with stronger coupling.

We note a good qualitative agreement with the results of the numerical simulations (Fig. 3.11b). The weak dependence of the SNR on the coupling strength $r_0$ results from the increase of the interface of the front (see (3.60)] if $r_0$ is becoming larger. Fronts with finite interfaces are more sensitive to fluctuations. The main result is a shift of the minimum of the effective potential $U_{r_0}(R)$ if the interface thickness comes closer to the distance of the minimum of the external inhomogeneity $\delta$. Hence, for the front it becomes easier to overcome the barrier between the two inhomogeneities. Therefore, the mean escape rate increases with the increase in the coupling strength [see e.g., (3.70)]. As a consequence the matching condition for SR (matching of the mean escape time and a half-period of the periodic force) can be achieved for a smaller noise intensity where the SNR peaks at higher values. Therefore, the output SNR should increase with an increase in the coupling strength.

## 3.1.6 Doubly Stochastic Resonance in Systems with Noise-Induced Phase Transition

This section connects SR with noise-induced phase transitions in systems with multiplicative noise. Such kind of noise was intensively investigated in the 1970s and 1980s in time-dependent dynamics. As shown in the Chap. 1 (1.161) multiplicative noise yields the appearance of new maxima in the probability distribution without counterparts in the deterministic description [4]. The excitation of noise-induced oscillations [112–114] also belongs to this class of zero-dimensional systems.

Creation of a noise-induced mean field in spatially extended systems is another new example where noise supports spatial structure formation [98,115–133]. Various experimental applications are proposed in [134–136,138,139]. A good description of the various noise-induced inhomogeneous structures can be found in [98] and references therein.

It is well known that a noise-induced bistability in zero-dimensional systems is unable to serve as a stochastic resonator. The effective barriers defining the escape rates between the two states are too small to respond significantly to the variation of a signal and an external noise. However, as was reported in [141], bistable noise-induced states in extended systems possess time scales sufficiently separated from relaxation times. This is due to an existing strong interaction between the coupled or neighboring cells in the medium, which also serves as one of the conditions for a noise-induced phase transition.

We briefly sketch a periodically driven system with two noise-induced phases. An additive noise controls the escape time between the two states. The SR occurring is called doubly stochastic resonance since the phases are due to the multiplicative noise and the additive noise exhibits the SR.

**Noise-Induced Phase Transitions.** First we address the occurrence of noise-induced phase transitions. For this purpose a nonlinear lattice of coupled stochastic overdamped oscillators is studied [120,126,131]:

$$\dot{x}_i = f(x_i) + \sqrt{2D_m}g(x_i)\xi_i(t) + \frac{\gamma}{2d}\sum_j(x_j - x_i) + \sqrt{2D_a}\zeta_i(t). \quad (3.73)$$

In this coupled set of Langevin equations $x_i(t)$ represents the state of the $i$th oscillator. $i = 1, ..., L^d$ labels $N = L^d$ cells of a cubic lattice with size $L$ in $d$ dimensions. The sum in (3.73) runs over nearest neighbors, with $\gamma$ being the strength of the coupling. $\xi_i(t)$ and $\zeta_i(t)$ are mutually uncorrelated Gaussian noise sources, with zero mean and uncorrelated both in space and time,

$$\langle\xi_i(t)\xi_j(t')\rangle = \delta_{i,j}\delta(t - t'), \quad \langle\zeta_i(t)\zeta_j(t')\rangle = \delta_{i,j}\delta(t - t'), \quad (3.74)$$

The first is multiplicative noise responsible for the phase transition; the second is the additive noise which controls the escape times between the two attracting states induced by the multiplicative noise.

The functions $f(x)$ and $g(x)$ are specified to be of the form [120]

$$f(x) = -x(1+x^2)^2, \qquad g(x) = 1+x^2. \tag{3.75}$$

We point out that this kind of dynamics does not create new maxima in a zero-dimensional system.

The model (3.73) can be handled analytically within mean-field theory [90, 98]. This mean-field approximation consists in the replacement of the neighboring states $x_j$ by the global average of the system $m$. In this way, one obtains a one-dimensional FPE parametrically dependent on $m$ and, hence, including a symmetry break. Assuming a quick relaxation to a value $m = \text{const.}$ the following steady-state probability distribution $P_{\text{st}}$ is obtained:

$$P_{\text{st}}(x, m) = \frac{C(m)}{\sqrt{D_m g^2(x) + D_a}} \exp\left(\int_0^x \frac{f(y) - \gamma(y - m)}{D_m g^2(y) + D_a} \, dy\right). \tag{3.76}$$

$C(m)$ is a normalization constant. Afterwards, the mean field $m$ is determined self-consistently by

$$m = \int_{-\infty}^{\infty} x P_{\text{st}}(x, m) \, dx, \tag{3.77}$$

which is the equation of definition of $m$ with dependences on the noise and coupling intensities.

Solving (3.77) numerically with respect to $m$, one determines transitions between ordered ($m \neq 0$) and disordered ($m = 0$) phases. Transition boundaries between different phases are shown in Fig. 3.12 and the corresponding dependence of the order parameter on $D_m$ is presented in Fig. 3.13. The influence of additive noise results in a shift of this line as shown. For $D_a = 0$ an increase in the multiplicative noise causes a disorder–order phase transition, which is followed by the reentrant transition to disorder [120]. In the ordered phase the system occupies one of two symmetric possible states with

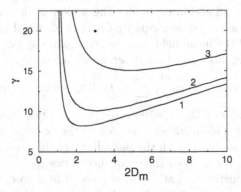

**Fig. 3.12.** Transition lines between ordered and disordered phases on the plane $(D_m; \gamma)$ for different intensities of additive noise: $D_a = 0$ (*1*), 0.5 (*2*), and 2.5 (*3*). A mean field different from zero exists above the curves shown. The *black dot* corresponds to $\gamma = 20$, $D_m = 1.5$

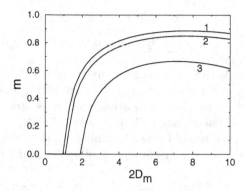

**Fig. 3.13.** The order parameter $|m|$ versus the intensity of multiplicative noise for $\gamma = 20$ and $D_a = 0$ (*1*), 0.5 (*2*), and 2.5 (*3*)

the mean fields $m_1 = -m_2 \neq 0$, depending on initial conditions. We underline that although the phase transition is created by multiplicative noise this phenomenon is quite different from the case of zero-dimensional systems and disappears with vanishing coupling.

**Doubly Stochastic Resonance.** Now let us turn to the problem of how the system (3.73) responds to periodic forcing:

$$\dot{x}_i = f(x_i) + \sqrt{2D_m}g(x_i)\xi_i(t) + \frac{\gamma}{2d}\sum_j(x_j - x_i) + \sqrt{2D_a}\zeta_i(t) + A\cos(\Omega t + \varphi_0),$$
(3.78)

with amplitude $A$, frequency $\Omega$ and random initial phase $\varphi$. In [141] a set of parameters $(D_m; \gamma)$ was taken within the region of two coexisting ordered states with nonzero mean field, in particular, values given by the dot in Fig. 3.12 were chosen. Simulations were performed on a two-dimensional lattice of $L^2 = 18 \times 18$ oscillators under the action of the harmonic external force. Runs were averaged over equally distributed initial phases.

Time series of the mean field and the corresponding periodic input signal are plotted in Fig. 3.14 for three different values of $D_a$. The current mean field is calculated as $m(t) = \frac{1}{L^2}\sum_{i=1}^{N}x_i(t)$. The series resembles a picture of SR. For small values of $D_a$ transitions between the two symmetric states $m_1$ and $m_2$ are rather seldom without relation to the external force. For medium noise the transitions occur with the periodicity of the external force. If $D_a$ is increased further, the ordered time sequence is destroyed.

Figure 3.14 illustrates that additive noise is able to optimize the signal processing in the system (3.78). In order to characterize this SR effect the SNR from the power spectral density $S(\omega)$ was computed. Its dependence on the intensity of the additive noise is shown in the Fig. 3.15; its dependence is also shown in a two-state approximation where $m(t)$ was approximately

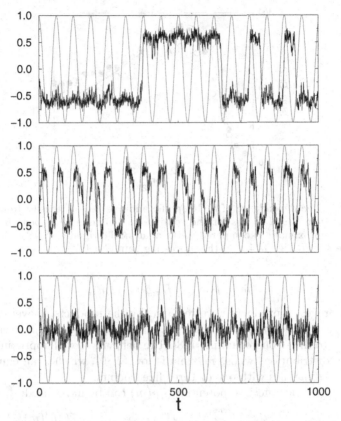

**Fig. 3.14.** Example of input/output synchronization. The time evolution of the current mean field (output) and the periodic external force $F(t)$ (input) for different intensities of additive noise (from top to bottom): $D_a = 0.005$, $0.5$ and $2.5$. If the intensity of the additive noise is close to their optimal value (middle row), hops occur with the period of the external force. The remaining parameters are $A = 0.1$, $\Omega = 0.1$, $\gamma = 20$ and $D_m = 1.5$

replaced by $m(t) = +1$ or $m(t) = -1$, respectively. Both curves exhibit the well-known bell-shaped dependence on $D_a$ typical for SR.

**The Effective Potential and the Signal-to-Noise Ratio.** Let us first address the situation without signal and additive noise when $A$ and $D_a$ vanish. In the case of strong coupling the first moment can be regarded as homogeneous and its time evolution is given by the drift part in the corresponding FPE:

$$\langle \dot{x} \rangle = \langle f(x) \rangle + D_m \langle g(x)g'(x) \rangle. \tag{3.79}$$

The mechanism of the noise-induced transition in coupled systems can be explained by means of a short-term evolution approximation [126]. Starting with an initial Dirac $\delta$-function the dynamics are as follows:

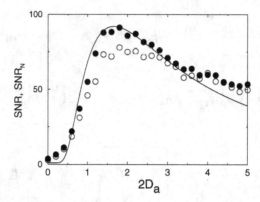

**Fig. 3.15.** The dependence of SNR on the additive noise intensity from numerical simulations (*solid circles* for the continuous dynamics and *open circles* for a two-state approximation) compared to the analytical estimation $SNR_N$ (*solid line*), parameters as in Fig. 3.14, and the processing gain $G = 0.7$

$$\dot{m} = f(m) + D_m g(m) g'(m), \qquad (3.80)$$

where averages of products were replaced by the products of averages. It is valid if $f(\langle x \rangle) \gg \langle \delta x^2 \rangle f''(\langle x \rangle)$. This required suppression of fluctuations is again due to the strong coupling [128]. The behavior is well approximated by Gaussian distributions whose maxima follow (3.80), which is an equation of dissipative dynamics with one or three fixed points.

In addition, an effective potential $U_{\text{eff}}(m)$ can be introduced:

$$U_{\text{eff}}(m) = U_0(m) + U_{\text{noise}} = -\int^m dm' \, f(m') - \frac{D_m g^2(m)}{2}, \qquad (3.81)$$

where $U_0(m)$ is a monostable potential. $U_{\text{noise}}$ represents the influence of the multiplicative noise and exhibits bimodality of the potential above a critical value of $D_m$ in this approximative picture. We can see the strong similarity of the potential (3.81) with the effective potential of a Mathieu–Kapiza pendulum [142].

Returning to the additively driven situation with small amplitudes $A$ and varying noise $D_a$ a typical SR situation arises. Let us look at a sufficiently large multiplicative noise. It implies that transition times between the two noise-induced states without additive noise are much larger than the period of the driving force $T_{\text{per}} = 2\pi/\Omega$. Of course, an increase in additive noise will reduce the escape times, and if they are of the order of the period SR occurs.

Application of the adiabatic two-state theory [61] neglecting again intrawell dynamics yields in LRT the SNR as

$$SNR_1 = \frac{\pi A^2}{D_a^2} r_K, \qquad (3.82)$$

where $r_K$ is the Kramers rate,

$$r_{\mathrm{K}} = \frac{\sqrt{U_{\mathrm{eff}}''(m_{\mathrm{min}})|U_{\mathrm{eff}}''(m_{\mathrm{max}})|}}{2\pi} \exp\left(-\frac{\Delta U_{\mathrm{eff}}}{D_a}\right), \qquad (3.83)$$

for surmounting the potential barrier $\Delta U_{\mathrm{eff}}$ by additive noise. Further, rescaling this value by the number of oscillators $N$ in the lattice [78] and taking into account the processing gain $G$ and the bandwidth $\Delta$ in the power spectrum [61], the $\mathrm{SNR_N}$ of the $N$ elements reads

$$\mathrm{SNR_N} = \mathrm{SNR_1}\frac{NG}{\Delta} + 1. \qquad (3.84)$$

This dependence is shown in the Fig. 3.15 and demonstrates good agreement with the results of the numerical simulations despite the rough approximation. Nearly exact agreement is found at the location of the maximum as well as for the quantitative values of the SNR.

It should be remarked that additive noise also changes the properties of the effectively bistable system (see Figs. 3.12 and 3.13) which leads to limitations. As a consequence, the amplitude of the output decreases, and even bistability might disappear for large noise intensities $D_a$. Experimental realizations of this interesting topic were proposed in [143].

### 3.1.7 Stochastic Resonance for Signals with a Complex Spectrum

In the majority of studies on SR the external force is a harmonic signal of small amplitude. Hence, a natural problem arises in studying the system response to multifrequency and noisy signals. This is especially important for biological and engineering applications. Signals recognized by living organisms are often noisy and may not contain strongly periodic components. An investigation of the system response to a quasiharmonic signal with a finite spectral linewidth, resulting from the fluctuation contribution, seems to be more realistic.

An advantage of LRT is that it can be naturally extended to the case of signals with a complex spectral composition [144]. The spectral density at the output takes the form

$$G_{x,x}(\omega) = G_{x,x}^{(0)}(\omega) + |\chi(\omega)|^2 G_{s,s}(\omega), \qquad (3.85)$$

where $G_{s,s}(\omega)$ is the spectrum of the signal. Below we shall discuss a series of examples which are of practical importance.

**Response of a Stochastic Bistable System to Multifrequency Signals.** Let us consider a weak signal possessing a discrete spectrum. The external force $s(t)$ can be represented in the form of a Fourier series

$$s(t) = A\sum_{k=1}^{M} b_k \cos \Omega_k t, \qquad (3.86)$$

where $Ab_k \ll 1$ are the small amplitudes of the harmonics, and $\Omega_k$ are their frequencies. According to LRT the system response is

$$\langle x(t) \rangle = A \sum_{k=1}^{M} b_k |\chi(\Omega_k, D)| \, \cos \left[ (\Omega_k + \psi_k) \, t \right]. \tag{3.87}$$

The response $\langle x(t) \rangle$ (3.87) contains the same spectral components (3.86) but with different amplitudes and phases. In (3.87) phase shifts $\psi_k$ for each harmonic are given via the susceptibility as

$$\psi_k(\Omega_k, D) = -\arctan \frac{\operatorname{Im} \chi(\Omega_k, D)}{\operatorname{Re} \chi(\Omega_k, D)}. \tag{3.88}$$

In accordance with LRT, SR measures, such as the SPA and the SNR, are determined as follows [51, 64]:

$$\eta(\Omega_k, D) = |\chi(\Omega_k, D)|^2, \tag{3.89}$$

$$\mathrm{SNR}(\Omega_k, D) = \frac{\pi (A \, b_k)^2 |\chi(\Omega_k, D)|^2}{G_{x,x}^{(0)}(\Omega_k, D)}. \tag{3.90}$$

The frequency dependence of the susceptibility provides frequency distortions of the output signal. The magnitudes of those distortions can be estimated by using the ratio of the amplitudes of different harmonics at the output and at the input:

$$E(\Omega_k, \Omega_j, D) = \frac{|\chi(\Omega_k, D)|}{|\chi(\Omega_j, D)|}. \tag{3.91}$$

As follows from (3.33-35), the overdamped bistable oscillator (3.3) represents an amplifier with low-frequency filtering of a signal at the output. The parameters of such an amplifier (the SPA and the SNR) are controlled by the intensity of an external noise. A question arises: Is it possible, using such a device, to provide an amplification of information- carrying signals (for instance, amplitude- and frequency-modulated signals) without significant distortions? This problem has been discussed in [145, 146], where an affirmative answer has been given to this question. If the signal is weak and if its effective frequency range does not exceed 25% of the carrier frequency, then all the frequency components of the signal will be amplified almost similarly and the output signal will contain practically no linear distortions [147]. Evidently, the requirements for undistorted amplification impose special restrictions on the frequency range of amplitude modulation (or on the index of frequency modulation), which determine an effective width of the frequency range of the information signal.

**Stochastic Resonance for Signals with a Finite Spectral Linewidth.** Actual periodic signals always possess a finite spectral linewidth due to the presence of amplitude and phase fluctuations of the oscillator. Will SR be observed for such signals and which features may it lead to if one takes into account the finite width of the spectrum? The answer to these questions is of great importance in practical applications.

As a model of a signal with a finite spectral line width we took so-called "harmonic noise" [148–154]. Harmonic noise is represented by a two-dimensional Ornstein–Uhlenbeck process and is governed by the system of two stochastic differential equations (SDE):

$$\dot{s} = y, \quad \dot{y} = -\Gamma y - \Omega^2 s + \sqrt{2\varepsilon\Gamma}\,\xi_2(t), \tag{3.92}$$

where $\xi(t)$ is a Gaussian white noise, $\langle\xi_2(t)\xi_2(t')\rangle = \delta(t - t')$, $\Gamma$ is the parameter of dissipation, and $\varepsilon$ is the intensity of harmonic noise. The spectral density $G_{s,s}(\omega)$ is known and has a Lorenzian shape:

$$G_{s,s}(\omega) = \frac{2\varepsilon\,\Gamma}{\omega^2\Gamma^2 + (\omega^2 - \Omega^2)^2}. \tag{3.93}$$

For $\Omega > \Gamma/2$, the spectral density (3.93) possesses a peak at the frequency

$$\omega_{\mathrm{p}} = \sqrt{\Omega^2 - \Gamma^2/2} \tag{3.94}$$

and is characterized by the width $\Delta\omega_{\mathrm{in}}$ determined at the half height of the peak maximum:

$$\Delta\omega_{\mathrm{in}} = \sqrt{\omega_{\mathrm{p}}^2 + \Gamma\omega_1} - \sqrt{\omega_{\mathrm{p}}^2 - \Gamma\omega_1}, \quad \omega_1 = \sqrt{\Omega^2 - \Gamma^2/4}. \tag{3.95}$$

Another quantity characterizing the spectrum (3.93) is the quality factor of the spectrum. It is defined as the ratio of the peak frequency to the spectral width $\Delta\omega$:

$$Q_{\mathrm{in}} = \frac{\omega_{\mathrm{p}}}{\sqrt{\omega_{\mathrm{p}}^2 + \Gamma\omega_1} - \sqrt{\omega_{\mathrm{p}}^2 - \Gamma\omega_1}}. \tag{3.96}$$

Let us turn again to the bistable overdamped oscillator that is governed by the SDE [55]:

$$\dot{x} = x - x^3 + \sqrt{2D}\,\xi_1(t) + s(t) \tag{3.97}$$

with $\xi_1(t)$ being independent of $\xi_2(t)$. SDE (3.92) and (3.97) represent a three-dimensional Markovian process, for which the following FPE for the probability density $p(x, s, y, t)$ is valid:

$$\frac{\partial p}{\partial t} = -\frac{\partial}{\partial x}[(x - x^3 + s)\,p] - \frac{\partial}{\partial s}(y\,p) + \frac{\partial}{\partial y}[(\Gamma s + \Omega^2 y)\,p]$$

$$+ D\frac{\partial^2 p}{\partial x^2} + \varepsilon\Gamma\frac{\partial^2 p}{\partial y^2}. \tag{3.98}$$

The previous case (3.4) corresponds to a random process inhomogeneous in time, while the FPE (3.98) describes a homogeneous random process, which has the stationary statistical characteristics.

In a weak signal approximation $\varepsilon \ll 1$, and according to LRT the expression for the spectral density at the output is

$$G_{x,x}(\omega) = G_{x,x}^{(0)}(\omega, D) + |\chi(\omega, D)|^2\,G_{s,s}(\omega). \tag{3.99}$$

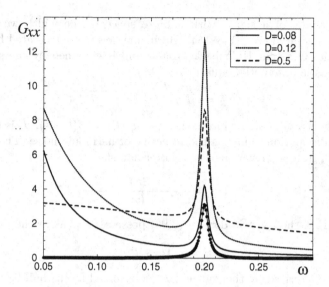

**Fig. 3.16.** The spectral power density at the output and at the input (*circles*) for different values of the noise intensity. The parameters of the harmonic noise are $\varepsilon = 10^{-3}$, $\Omega = 0.2$ and $\Gamma = 0.008$

We take into consideration only the global dynamics of switchings between the potential wells in the SR regime. With this, the susceptibility and the correlation function of the unperturbed system are given by (3.33) and (3.30), respectively. Hence, the spectral density at the output of the bistable system reads

$$G_{x,x}(\omega) = \frac{2\langle x^2 \rangle_{\mathrm{st}} \lambda_{\mathrm{m}}}{\omega^2 + \lambda^2} \left( 1 + \frac{\langle x^2 \rangle_{\mathrm{st}} \lambda_{\mathrm{m}}}{D^2} \frac{\varepsilon \Gamma}{\omega^2 \Gamma^2 + (\omega^2 - \Omega^2)^2} \right). \qquad (3.100)$$

As seen from Fig. 3.16, the spectrum of the output signal also attains a maximum at a certain frequency $\omega_{\mathrm{m}}$. It is clear that SR is realized for harmonic noise, i.e., the peak height is maximal at a certain optimal noise intensity, $D$.

Let us explore in more detail the properties of the output spectrum. Since the amplitude–frequency properties of the bistable oscillator significantly depend on the noise intensity, the frequency $\omega_{\mathrm{m}}$ at which a peak appears in the output spectrum and the spectral width $\Delta\omega_{\mathrm{out}}$ also become functions of $D$. To quantify the parameters of the output spectrum we introduce the quantity $R$ as the ratio between the spectral widths at the output $\Delta\omega_{\mathrm{out}}$ and at the input $\Delta\omega_{\mathrm{in}}$:

$$R(D) = \frac{\Delta\omega_{\mathrm{out}}(D)}{\Delta\omega_{\mathrm{in}}}. \qquad (3.101)$$

This quantity is determined from the expression for the spectral density (3.100).

The relative spectral linewidth $R$ possesses its minimum at a certain noise intensity, as seen from Fig. 3.17a. The magnitude of this intensity depends on

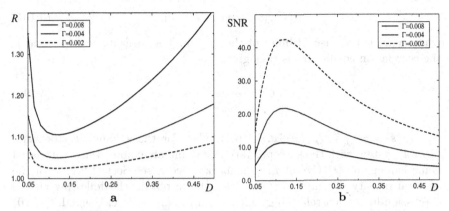

**Fig. 3.17.** (a) Relative spectral linewidth $R$ and (b) the SNR (3.103) as functions of the noise intensity for different values of $\Gamma$. The parameters of the harmonic noise are $\varepsilon = 10^{-3}$ and $\Omega = 0.2$

the damping coefficient $\Gamma$. When $\Gamma$ is decreased these dependences disappear. For a signal in the form of harmonic noise, a certain ambiguity appears in the SNR definition, because the signal itself is a narrow-band random process. Let us introduce another characteristic, namely, the product of the quality factor $Q_\text{out}$ and the peak height $h$:

$$C(D) = Q_\text{out}(D) \cdot h(D), \qquad (3.102)$$

which is called the measure of coherence. The dependence of the coherence measure $C(D)$ on the noise intensity has a characteristic maximum typical for the SR phenomenon. The amplification of SR is achieved by increasing the quality factor of the input signal.

In this case the SNR can be defined, for instance, as the ratio of the peak height to the value of the unperturbed spectral density at the peak frequency:

$$\text{SNR} = 1 + \frac{\langle x^2 \rangle_\text{st} \lambda_\text{m}}{D^2} \frac{\varepsilon \Gamma}{\omega_\text{m}^2 \Gamma^2 + (\omega_\text{m}^2 - \Omega^2)^2}. \qquad (3.103)$$

Its dependence on noise intensity $D$ shown in Fig. 3.17b demonstrates a maximum typical for SR.

**Aperiodic Stochastic Resonance.** The notion of stochastic resonance can be extended to purely stochastic signals. This has been done in [155], and the effect was termed *aperiodic* stochastic resonance(ASR). In this case, when the spectral density of the signal does not possess any peaks, the input–output cross-correlation functions must be used rather than conventional measures of SR, e.g., the SPA and the SNR.

For a weak aperiodic signal $s(t)$, LRT can be used for ASR [156]. In the limit of LRT the spectral density at the output $G_{x,x}(\omega)$ is given by (3.85), while the cross-spectrum $G_{s,x}(\omega)$ between input and output is defined as

$$G_{s,x}(\omega) = \chi(\omega)G_{s,s}(\omega), \qquad (3.104)$$

where $\chi(\omega)$ is the susceptibility of the system. The measure of ASR, that is the correlation coefficient, is thus given by

$$C = \frac{\langle s\,x \rangle}{\sqrt{\langle x^2 \rangle \langle s^2 \rangle}}, \qquad (3.105)$$

where $\langle s\,x \rangle = \int_0^\infty G_{s,x}(\omega)d\omega$. In particular, for the symmetric stochastic bistable system with the susceptibility (3.33) and an exponentially correlated Gaussian signal $s(t)$ (Ornstein-Uhlenbeck-process, see Sect. 1.2.3) with the spectral density $G_{s,s}(\omega) = \gamma Q/(\gamma^2 + \omega^2)$, we obtain the following simple expression for the correlation coefficient (in the limit of weak signal, $Q \to 0$) [156]:

$$C = \frac{\lambda_{\mathrm{m}} Q}{D(\gamma + \lambda_{\mathrm{m}})}, \qquad (3.106)$$

where $\lambda_m$ is due to (3.29). The coefficient of correlation (3.106) possesses a single maximum as a function of noise intensity $D$ and therefore reflects ASR.

**Enhancement of Stochastic Resonance by Additional Noise.** In [157, 158] the application of two different signals onto the noisy dynamics has aimed at controlling SR. As it was shown two harmonic inputs with the same frequency and both modulating the system's threshold are able to enhance a variety of different behaviors. A change of the phase shift between the two signals evokes either enhancement or suppression of the response.

The known ability to manipulate the rates of a stochastic threshold system by dichotomic noise [159, 160] opens another possibility of controlling SR. In this case two uncorrelated inputs are used [161]. A first quick telegraph signal affects the rates of the noisy nonlinear system, causing a better or worse amplification of the second harmonic signal at lower frequencies.

In fact, the model possesses two statistically independent noise sources: (i) "thermal" white noise, which is responsible for stochastic switching between metastable states, and (ii) dichotomic noise, influencing the switching events between states of the system. The magnitude of dichotomic noise is always small, so that it cannot induce transitions by itself, and the presence of thermal noise is always necessary.

Analysis in a bistable system can be performed by reducing it to again two symmetric states $\sigma(t) = \pm 1$, which represent position of a particle in the right or left well of a bistable potential with barrier $\Delta U$. Dichotomic noise $\lambda(t) = \pm 1$ with magnitude $B$ and flipping rate $0 < \gamma \ll a_0$ modifies the transition rate [162, 163]:

$$W_0(\sigma, \lambda) = a_0 \exp\left(-\frac{\Delta U + \sigma\lambda B}{D}\right). \qquad (3.107)$$

We obtained a 4-state system with two states each for the output and input. It is described by the master equation

$$\frac{d}{dt} p(\sigma\,\lambda) = -W_0(\sigma, \lambda)\,p(\sigma, \lambda) + W_0(-\sigma, \lambda)\,p(-\sigma, \lambda) + \gamma\,[p(\sigma, -\lambda) - p(\sigma, \lambda)].$$

(3.108)

The next step is to add a weak periodic signal $s(t) = A\cos(\Omega t + \varphi_0)$. For sufficiently slow ($\Omega < \gamma \ll a_0$) and weak $A \ll \Delta U - B$ the harmonic force leads to the following time-dependent rates in LRT [161]:

$$W(\sigma, \lambda) = W_0 \exp\left(-\frac{A\sigma}{D}\cos(\Omega t + \varphi_0)\right) \approx W_0\left(1 - \frac{A\sigma}{D}\cos(\Omega t + \varphi_0)\right).$$

(3.109)

The autocorrelation function with an amplitude-independent noise part and signal part of order $\propto (A/D)^2$ can be obtained from the equations for the conditioned mean and for the cross-correlation function $\langle\sigma(t)\lambda(t')\rangle$. Both have to be solved with the initial conditions $\langle\sigma(t)^2\rangle = 1$ and the stationary correlator [162]

$$\langle\sigma\lambda\rangle_{\text{stat}} = \frac{a_2 - a_1}{a_1 + a_2 + 2\gamma}.$$

(3.110)

In the latter expression we have introduced the abbreviations ($a_0 = 1$)

$$a_{1,2} = \exp\left[-(\Delta U \pm B)/D\right].$$

(3.111)

This autocorrelation function yields after averaging over the initial phase $\varphi_0$ and after a Fourier transform the power spectrum $G_{\sigma,\sigma}(\omega) = G_{\sigma,\sigma}^{(0)}(\omega) + A^2\pi\eta\delta(\omega - \Omega)$ [161] with the background

$$G_{\sigma,\sigma}^{(0)}(\omega) = 4\frac{a_1 + a_2}{(a_1 + a_2)^2 + \omega^2}\left(1 + \frac{(a_2 - a_1)^2}{4\gamma^2 + \omega^2}\right) - 4\frac{(a_2 - a_1)^2}{(a_1 + a_2 + 2\gamma)(4\gamma^2 + \omega^2)}$$

(3.112)

and the SPA

$$\eta = \frac{1}{D^2}\left(a_1 + a_2 - \frac{(a_2 - a_1)^2}{(a_1 + a_2 + 2\gamma)}\right)^2 \frac{1}{(a_1 + a_2)^2 + \Omega^2}.$$

(3.113)

The SPA and the SNR are presented in Fig. 3.18 versus thermal noise intensity for fixed values of the flipping rate $\gamma$ and the signal frequency $\Omega$ but for different values of the magnitude of dichotomic noise. We immediately conclude that both the SPA and the SNR are enhanced for a large enough $B$ in comparison with the conventional case, when dichotomic noise is absent ($B = 0$). The optimal value of the noise intensity which maximizes SPA and SNR shifts towards smaller values with increasing magnitude of dichotomic noise. Moreover, the behavior of the SPA and the SNR versus thermal noise intensity is qualitatively different from the conventional case, as both measures possess two maxima.

An interpretation of the enhancement as well as of the two maxima can be given by looking at the mean escape rate between the two states when perturbed by a dichotomic process [159, 160]. For this purpose it is useful to consider two limiting cases. For a fast dichotomic noise the escape is mainly

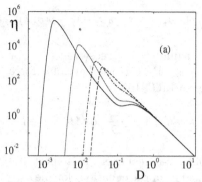 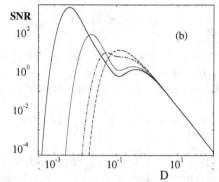

**Fig. 3.18.** (a) SPA and (b) SNR versus noise intensity $D$ for different values of the magnitude of dichotomic noise $B$: $B = 0$ (*dashed line*), $B = 0.1$ (*dash-dotted line*), $B = 0.2$ (*dotted line*) and $B = 0.24$ (*solid line*). Other parameters are $\gamma = 0.1$ and $\Omega = 0.001$

governed by the larger rate $a_2$ with the lowered barrier $\Delta U - B$. At lower barriers the merging condition between the input frequency shifts to a smaller noise level, with a better performance, as shown previously for the two-state model. That is why SR is greatly enhanced. This explains the first enhanced maximum in the SPA and SNR. In contrast, for vanishingly slow dichotomic noise and large noise intensity with $\gamma \ll a_1$, $a_2$ the system performs many transitions in a static asymmetric potential during one round-trip of the dichotomic noise. Forward and backward transitions occur together with a common rate $\propto \exp(-\Delta U/D)/\cosh(B/D) \approx a_1$. This explains the second maximum. But it is known that SR is gradually suppressed with the increase in asymmetry of a bistable system and that the position of the maximum does not depend on signal frequency [164].

### 3.1.8 Stochastic Resonance in Chaotic Systems with Coexisting Attractors

One of the typical properties of systems demonstrating dynamical chaos is the coexistence of different attractors in phase space [165]. Basins of attraction of coexisting attractors are separated by separatrix hypersurfaces in the phase space. Without external noise a phase trajectory belongs to either one or the other attractor depending on the initial conditions. The influence of external noise leads to the appearance of random switchings between coexisting attractors of the system. The statistics of these switchings is defined by the properties of the noise and the DS.

A theoretical consideration of the influence of external noise on the regimes of dynamical chaos is available in the limits of small [166, 167] and large [168] Gaussian noise. The theory of random perturbations of DS [169] is based on the notion of a quasipotential and has been extended to systems with complex dynamics [170–172]. Suppose that a DS has an attractor and for the latter there exists an invariant probability measure. Let the system be subjected to a weak Gaussian noise with intensity $D$ and be described by the set of SDE

$$\dot{x}_i = f_i(x) + \xi_i(t), \qquad i = 1, \ldots, N, \tag{3.114}$$

where $\langle \xi_i(t)\xi_j(0) \rangle = 2D\,\delta_{i,j}\delta(t)$. Then the stationary probability density $p(x)$ can be written via the quasipotential (or nonequilibrium potential) $\Phi(x)$ as follows:

$$p(x) \propto \exp\left(-\frac{\Phi(x)}{D}\right). \tag{3.115}$$

The quasipotential, being an analog of the free energy for the nonequilibrium stationary state [173], depends only on the state variables and the parameters of the system and not on the noise intensity $D$. The quasipotential takes its minimal values on the attractor. If a few attractors coexist in the phase space of the system $\Phi(x)$ possesses local minima corresponding to these attractors. In this case for a weak noise we may formulate the Kramers problem of escape from the basin of attraction of an attractor. For $D \ll 1$, the motion of a system involves a slow time scale related to the mean escape time from the basin of attraction of the attractor. The dependence of the mean escape time upon the noise intensity is characterized by an exponential law $\tau \propto \exp(\Delta\Phi/D)$. If additionally to external noise we apply a weak periodic signal to the system, which cannot evoke transitions between attractors in the absence of noise, the phenomenon of SR should be observed [46, 47].

A principally different effect called *deterministic stochastic resonance* has been recently revealed for systems with chaotic dynamics [46, 47]. It is known that with the variation of control parameters of a nonhyperbolic system a crisis of its attractors may occur. As an example we can mention here the phenomenon of two attractors merging, leading to a "chaos–chaos" dynamical intermittency [174]. In this case a phase trajectory spends a long time on each of the merged attractors and rarely makes irregular transitions between them. We note that such random-like switchings occur in the absence of external noise and are controlled via the deterministic law [175, 176]. For the systems with "chaos–chaos" intermittency, the mean residence time $T_i$ of a phase trajectory to be on an attractor obeys a universal scaling law [175, 177]:

$$T_i \propto (a - a_{\mathrm{cr}})^\gamma, \tag{3.116}$$

where $a$ is a control parameter, $a_{\mathrm{cr}}$ is its bifurcation value at which a crisis occurs and corresponds to the onset of intermittency and $\gamma$ is the scaling exponent. Hence, in this case the control parameter plays the role of noise intensity. It controls a slow time scale and, consequently, the spectral properties of the system [177]. If the system is driven by a slow periodic signal, then via variation of its control parameter, one can obtain a situation in which the driving period and the mean time of switching from one attractor to the other one coincide, i.e. the conditions of SR are realized. Note that

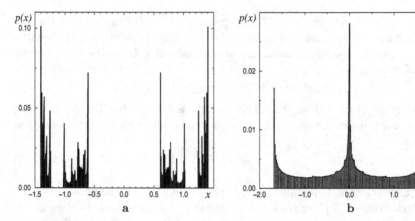

**Fig. 3.19.** Stationary probability distribution density of trajectories on the attractors of map (3.117) with $A = D = 0$, $b = 10$ for the parameter $a$ values: **(a)** 2.5 and **(b)** 2.84

the regimes of dynamical intermittency exhibit an exponential sensitivity to external noise [176, 177], which makes a realization of the conventional SR possible.

We shall illustrate deterministic SR using a simple example of a discrete system demonstrating a crisis of chaotic attractors:

$$x_{n+1} = (ax_n - x_n^3)\exp(-x_n^2/b) + A\sin\Omega n + \sqrt{D}\xi(n). \qquad (3.117)$$

The system (3.117) is a one-dimensional cubic map driven by a weak periodic signal ($A \ll 1$) and a delta-correlated noise of intensity $D$. The exponential term is introduced to prevent escape of phase trajectories to infinity.

Let us describe the behavior of the unperturbed map ($A = D = 0$). At $a < a_{\mathrm{cr}} = 2.839\ldots$, two chaotic attractors are separated by the saddle point $x_n = 0$ and coexist in the phase space of the system. At $a = a_{\mathrm{cr}}$, these attractors merge with the onset of "chaos–chaos" dynamical intermittency, as seen in Fig. 3.19.

First we study the conventional SR before the crisis $a < a_{\mathrm{cr}}$, when the switching events occur due to external noise. As seen from the numerical results in Fig. 3.20a, both the SPA and the SNR pass through a maximum at optimal noise levels. The simulations have confirmed that the Kramers rate is close to the external signal frequency at the optimal noise intensity $D = D_{\mathrm{opt}}$.

Now we exclude noise ($D = 0$) and consider the system reaction to periodic disturbance in the regime of "chaos–chaos" intermittency, when the parameter $a$ is slightly larger than its critical value, $a > a_{\mathrm{cr}}$. According to simulations, the mean switching frequency monotonically increases with increasing parameter $a$. Figure 3.20b shows the dependences of the SPA and the SNR on $a$. The graphs illustrate the effect of deterministic SR, i.e., maxima

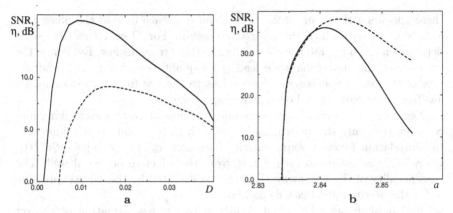

**Fig. 3.20.** The SNR (*solid line*) and the SPA (*dashed line*) as functions (**a**) of the noise intensity $D$ for $a = 2.5$, $A = 0.05$, $\Omega = 0.1$ and (**b**) of the control parameter $a$ for $D = 0$, $A = 0.005$, $\Omega = 0.1$

of the SPA and the SNR can be obtained by tuning the control parameter in the range $2.85 < a < 2.88$, when the Kramers rate matches the external signal frequency.

The effects described above have a generic character, which was verified by numerical simulations of various discrete and flow systems demonstrating both noise-induced intermittency and a crisis [46]. As an example we discuss SR in the Lorenz model [178] in the regime of noise-induced intermittency. The Lorenz system is a suitable model for studying the influence of noise, since noise sources can be included into the equations of motion using fluctuation–dissipation relations [179]. The ergodicity of a stochastic Lorenz model has been proved in [180]. Furthermore, in the system parameter space there are regions of the Lorenz attractor and a nonhyperbolic attractor [181]. The stochastic Lorenz model is described by the following system of SDE:

$$
\begin{aligned}
\dot{x} &= -\sigma x + \sigma y + \xi_1(t), \\
\dot{y} &= -y + r x - x z + \xi_2(t), \\
\dot{z} &= -b z + x y + \xi_3(t), \\
\langle \xi_i(t)\xi_j(0) \rangle &= D\,\delta_{i,j}\,\delta(t).
\end{aligned}
\tag{3.118}
$$

With standard parameter values $\sigma = 10$, $r = 210$, $b = 8/3$, corresponding to the existence of a nonhyperbolic attractor, noise induces "chaos–chaos" intermittent behavior of switchings between two symmetric attractors [182].

In the absence of noise, two symmetric attractors are realized in the system for different initial conditions. When $D \neq 0$, these attractors merge. For small noise intensities the phase trajectory is retained on each of the attractors for a long time and rarely makes transitions between them. As a result, there appears a slow time scale – the mean residence time of the phase trajectory to be on each of the merged attractors. For $D \ll 1$ this time scale depends on the noise intensity according to the Arrhenius law. Evidently, the presence of the slow time scale leads to a qualitative change in the structure of the power spectrum which evolves to the low-frequency domain. Its low-frequency part has a Lorentzian shape [182].

Neglecting the local chaotic dynamics on these attractors and taking into consideration only the switchings between them (two-state approximation), the correlation function exponentially decreases: $c_{x,x}(\tau) \propto \exp\left(-2r(D)\right)$, where $r(D)$ is the mean escape rate from the effective potential well. The susceptibility of the Lorenz system can be estimated by the same expression as for the overdamped bistable oscillator.

Add a weak periodic signal $A \sin(\Omega t)$ to the first equation of the set (3.118). In this case the spectral density will contain a delta-peak at the signal frequency.

The simplest estimate for the SNR at the system output is

$$\text{SNR} \propto \frac{r(D)}{D^2} = \frac{1}{D^2} \exp\left(-\frac{\Delta\Phi}{D}\right), \tag{3.119}$$

where $\Delta\Phi$ is the barrier height of the corresponding effective potential.

Numerical calculations of the SNR based on (3.118) and (3.119) are presented in Fig. 3.21. As seen from the figure, the expression (3.119) is rather

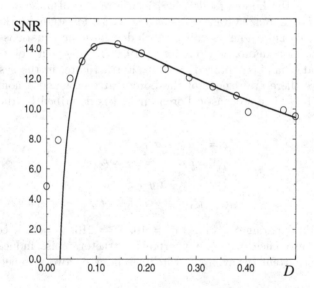

**Fig. 3.21.** SNR as a function of the noise intensity for the Lorenz system at $A = 1.0, \Omega = 0.1$. The numerical results are shown by *circles*; the *solid line* corresponds to the results of an approximation using (3.119) with $\Delta\Phi = 0.24$

universal and describes the SNR for simple bistable systems as well as for systems with complex dynamics. Note that the Lorenz model can be transformed into the form of a bistable oscillator with inertial nonlinearity [165, 183]:

$$\ddot{u} + \gamma \dot{u} + u^3 + (v-1)u = 0, \quad \dot{v} = h(\beta u^2 - \alpha v), \tag{3.120}$$

where $\gamma = (1+\sigma)/\sqrt{\sigma(r-1)}$, $h = (r-1)^{-1/2}$, $\beta = (2\sigma - b)/\sqrt{\sigma}$ and $\alpha = b/\sqrt{\sigma}$. For large values $r \gg 1$, variable $v(t)$ is slow and after excluding it from (3.120) we arrive at the equation for a bistable inertial oscillator no longer containing the nonlinear inertial term:

$$\ddot{u} + \gamma \dot{u} + \frac{\alpha + \beta}{\alpha} u^3 - u = 0. \tag{3.121}$$

Thus, in the limit of large $r$ the Lorenz system can be described in terms of a bistable oscillator. This fact guarantees the existence of SR.

### 3.1.9 Analog Simulation

SR has now been repeatedly observed in a large variety of experiments. The very first experimental verification of the SR phenomenon was realized in an electronic circuit, in a simple Schmitt trigger [19], and in a bistable ring laser [20]. Since then, SR has been verified experimentally in different bistable systems [58]. In the majority of the experimental works the dependences of the SNR on the noise intensity at the system output were measured and then compared with theoretical conclusions. This is obvious since the definition of SR is based on the shape of this dependence.

As has been shown above, the SR effect is also characterized by the resonant dependence of the amplification factor of the input signal on the noise intensity [see (3.17) and (3.35)]. In connection with this we consider the experimental results for the amplification coefficient of a harmonic signal of small amplitude with the variation of noise intensity in an electronic circuit modeling the classical bistable oscillator (3.3) [184–186]. A schematic of the electronic circuit corresponding to (3.3) is given in Fig. 3.22. The scheme consists of operational amplifiers $D1 - D4$, resistors $R_0 - R_8$, capacitor $C$ and two anti-phase connected diodes. The input signal in the form of a sum of harmonic $A\sin(2\pi ft)$ and noisy $\xi(t)$ signals is applied to resistor $R_1$. Both signals are fed into the simulation circuit by means of suitable voltage generators. The system response $x(t)$ is measured at the output of the scheme. The main point of this experiment is to approximate the nonlinear function which describes the shape of a bistable potential $U(x)$. The anti-phase connection of two identical diodes and a suitable selection of resistors in the scheme provided the possibility to model the characteristic which approximates with a high level of accuracy the classical bistable potential in the interval of state variable variation $-1.5 < x < 1.5$ (variable $x$ is normalized):

$$-\frac{dU(x)}{dx} = ax - bx^3. \tag{3.122}$$

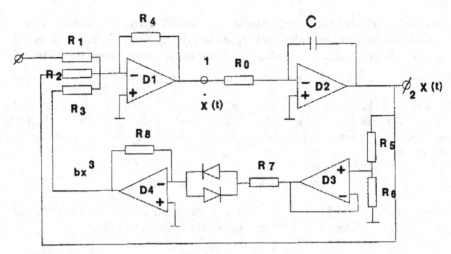

**Fig. 3.22.** Electronic analog for the overdamped bistable oscillator. Values of scheme elements are $R_0 - 5.11\,\text{k}\Omega$, $R_{1,4,6} - 10\,\text{k}\Omega$, $R_2 - 51.2\,\text{k}\Omega$, $R_3 - 2.2\,\text{k}\Omega$, $R_5 - 0.5\,\text{k}\Omega$, $R_7 - 1.1\,\text{k}\Omega$, $R_8 - 57\,\text{k}\Omega$, $C = 150$

In this equation the coefficient $a$ depends on the resistor $R_2$ values and the coefficient $b$ is determined by the values of resistors $R_3$, $R_5$ and $R_8$ [186].

To compare correctly the experimental and theoretical (3.17) results, one needs to ensure that the condition of the adiabatic approximation is fulfilled, i.e.,

$$2\pi f \ll \tau^{-1} = U''(x_{\min}), \tag{3.123}$$

where $\tau$ is the relaxation time of the oscillator (3.3). The external force period must be significantly larger than the relaxation time. In this case we can apply the two-state approach. The condition (3.123) imposes certain restrictions on the coefficients $a$ and $b$ in (3.122) and, consequently, on the resistors values. For the values indicated in Fig. 3.22, from (3.123) it follows that the external frequency must be bounded, e.g., $f \ll 12.5\,\text{kHz}$. This restriction was also found in experiments.

Theoretical and experimental dependencies of the amplification coefficient versus noise intensity and external frequency are in good qualitative agreement with each other but differ quantitatively due to the error of measurement of about $\pm 5\%$ (see Fig. 3.23a). It should be noted, however, that these differences exceed on average the indicated experimental error for at least two reasons: (i) the nonlinear characteristics of the scheme are different from the cubic polynomial (3.122), although special measurements have shown that this difference is minor in the operating range $|x| \leq 2$ (it does not exceed 1%); (ii) the influence of internal noise of the scheme has not been taken into account in experiments. The direct numerical simulation of the system (3.3) dynamics allows one to exclude the reasons mentioned above. Numerical results are shown in Fig. 3.23b in comparison with theoretical

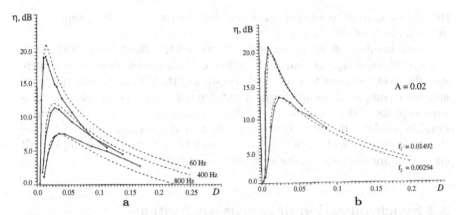

**Fig. 3.23.** The amplification coefficient as a function of the noise intensity, obtained experimentally (*solid line*) and theoretically (*dashed line*) (**a**) for different values of the input signal frequency and (**b**) from numerical simulation of the system (3.3) dynamics (*solid line*) and theoretically (*dashed line*). The amplitude of the input signal is $A = 20\,\text{mV}$; the normalized frequencies $f_1$ and $f_2$ correspond to 60 Hz and 400 Hz in the full-scale experiment

ones. As seen from the graphs, the differences between theoretical and calculated data are significantly less and are entirely within the error of numerical calculations, which is due to the finiteness of the time series used for their statistical processing.

### 3.1.10 Summary

The results presented in this section readily illustrate one of the interesting effects in nonlinear dynamics of stochastic systems – SR as a noise-induced transition in bistable systems. In the regime of SR certain characteristics of a system can be optimized by varying the noise intensity. Undoubtedly, the quantitative SR measures depend on the structure of the external signal, noise statistics and dynamical properties of the bistable system. But the physical essence of the phenomenon principally does not change. One of the characteristic times of the system has to be a nonlinear function of the noise intensity. In this case it is evident that by varying the noise intensity one can control the system response to an external force and obtain an optimal result.

An understanding of the physical mechanism of SR allows one to first assume and then experimentally demonstrate the possibility of realizing a similar effect in deterministic chaotic bistable systems. The mean switching time in the regime of "chaos–chaos" intermittency in such systems is a nonlinear function of the system control parameter which plays the role of noise intensity. By varying the control parameter, one can control the response of the chaotic system to an external force and observe an SR-like phenomenon in the absence of external noise.

The presented results are generally illustrated by using the simplest model of a one-dimensional overdamped oscillator. This model allows one to provide the most complete theoretical description of the SR effect, perform direct numerical computations and set up a suitable full-scale experiment. The comparison of theoretical, numerical and experimental results for the overdamped bistable oscillator demonstrates surprisingly good agreement. One may hope that these results will serve as a great stimulus for further studying the SR effect in more complex nonlinear stochastic systems.

## 3.2 Synchronization of Stochastic Systems

### 3.2.1 Introduction

One of the main aims in nonlinear dynamics is a generalization of the notion of *synchronization* to driven or coupled chaotic and stochastic systems. Intuitively we understand synchronization as an adjustment of rhythms of coupled systems due to their interaction. This adjustment is indicated by various measures. It is accompanied by a suppression of some oscillation modes, by a shrinkage of the possible excursions of phase trajectories in the phase space and results in the reduction of a number of degrees of freedom which are necessary to describe the system dynamics.

In general, there are two different origins for oscillating modes in simple noisy dissipative dynamics. First, oscillations exist due to the internal deterministic dynamics of the system or due to external periodic forcing. The effect of noise in this case is just broadening of the spectral peaks of damped or self-sustained deterministic periodic motion so that the coherence of oscillations is usually degraded by noise. Noise also destroys synchronization for this type of system. A representative of this type of DS is the van der Pol oscillator.

Second, oscillations are evoked by noise. For example, one can consider an overdamped system with one or a few attractors at finite distance. These systems does not possess deterministic natural frequencies but solely relaxation times. As in the previous case the noisy forces enhance a continuous spectrum corresponding to a variety of stochastic oscillatory motions. The essential dynamics of these type of systems is the returns over energetic barriers. The return time depends on the level of noise. When a weak periodic signal is applied, SR predicts an amplification of the weak periodic signal for an optimal noise intensity (see Sect. 3.1). Intuitively the system at the optimal noise should exhibit stochastic returns which are in phase with the signal.

How does that picture merge with the notion of *synchronization*? Linear response theory (LRT) has not brought sufficient insight into the discussion of this topic. There are, however, a few other approaches giving an indication that during the amplification of the switchings other motions are suppressed. One of them is the nonlinear response theory where the signal suppresses the noise background [187]. Another hint for the occurrence of synchronization is based on the waiting-time densities [188] and is called *bona fide resonance* [54]. The peak over the first harmonics undergoes a maximum, decreasing the probability for faster and slower processes.

A treatment closely following the theory of nonlinear oscillations has given evidence that synchronization accompanies SR [162, 163, 189–191]. It is based on the introduction of an instantaneous phase for the stochastic trajectories as also proposed for chaos synchronization. Such a description opens up the possibility of returning to synchronization as a phenomenon of phase relation between interacting systems.

In this section it will be shown that the stochastic switching processes *lock* to periodic forces. We discuss the criteria, mechanisms, and regularities of this effect. Since the dynamics is fully induced by the acting noise and is not destroyed over a large range of intensities, we call it *stochastic synchronization*.

### 3.2.2 Synchronization and Stochastic Resonance

Let us consider again the overdamped bistable system prioritized in studies of SR. Adding a Gaussian white noise changes the situation qualitatively compared to the noiseless case. The stochastic system has a new physical feature: Under the influence of noise the particle begins to perform jumps from one deterministically attracting states to the other one and back. Despite the fact that the jumps occur according to a Poissonian waiting-time distribution one can call these returns stochastic self-sustained oscillations. The random oscillations are permanently generated, and their statistical properties in the asymptotic limit do not depend on its initial conditions, i.e. they perform a stationary process being invariant to a time shift.

Then it is natural to pose the question of synchronization. Is it possible to synchronize the random oscillations in the bistable dynamics by an additional harmonic force? If yes, what are the features of this effect?

**Analytical Signal Approach to the Overdamped Kramers Oscillator.** To answer the questions posed above and to prove the existence of synchronization in the periodically driven random oscillators, we consider again the overdamped Kramers oscillator and apply the analytic signal approach described in Chap. 1.

The stochastic differential equation (SDE) of periodically driven overdamped Kramers oscillator is given by

$$\dot{x} = \alpha x - \beta x^3 + \sqrt{2D}\,\xi(t) + A\,\cos(\Omega t + \varphi_0). \qquad (3.124)$$

Without periodic driving this system has no natural deterministic frequency. At the same time, the stochastic dynamics of this bistable system is characterized by the noise-controlled mean return time (see Sect. 1.2.6). In the frequency domain this time scale defines the mean switching frequency of the system. The added periodic signal with amplitude $A$ represents an external "clock" with frequency $\Omega$ and initial phase $\varphi_0$. (Later on, we shall perform averaging over an equally distributed phase $\varphi_0$.)

Suppose that $\alpha$, $\beta > 0$ and for the moment $\varphi_0 = 0$ in (3.124) and the periodic modulation amplitude $A$ is always sufficiently small, so that transitions do not occur without noise. Furthermore, we will suppose that the modulation frequency is low as compared to the intrawell relaxation rates.

We introduce the analytic signal $w(t) = x(t) + iy(t)$, where $y(t)$ is the Hilbert transform of the original process $x(t)$:

$$y(t) = H[x] = \frac{1}{\pi} \int_{-\infty}^{\infty} \frac{x(\tau)}{t - \tau} d\tau = \frac{1}{\pi} \int_{0}^{\infty} \frac{x(t - \tau) - x(t + \tau)}{\tau} d\tau. \quad (3.125)$$

The instantaneous amplitude and phase are defined as the absolute value and the argument of the complex function $w(t)$, respectively:

$$w(t) = R(t) \exp[i\Phi(t)]. \quad (3.126)$$

The direct application of the analytical signal concept to (3.124) gives the following SDE for the analytical signal $w(t)$:

$$\dot{w} = \alpha w - \frac{\beta}{4}(3 R^2 w + w^3) + \psi(t) + A e^{i\Omega t}, \quad (3.127)$$

where $\psi(t) = \xi(t) + i\eta(t)$ is the analytical noise with $\eta(t)$ being the Hilbert transform of $\xi(t)$. From (3.127) it is easy to derive the SDE for the instantaneous amplitude and phase:

$$\dot{R} = \alpha R - \frac{\beta}{2} R^3 [1 + \cos^2(\phi + \Omega t)] + A \cos\phi + \xi_1(t),$$

$$\dot{\phi} = -\Omega - \frac{A}{R} \sin\phi - \frac{\beta}{4} R^2 \sin[2(\phi + \Omega t)] + \frac{1}{R} \xi_2(t), \quad (3.128)$$

where $\phi(t) = \Phi(t) - \Omega t$ is the instantaneous phase difference. The noise sources $\xi_{1,2}(t)$ are defined by the following expressions:

$$\xi_1(t) = \xi(t) \cos\Phi + \eta(t) \sin\Phi,$$
$$\xi_2(t) = \eta(t) \cos\Phi - \xi(t) \sin\Phi. \quad (3.129)$$

As seen, the second SDE in (3.128), describing the evolution of the phase difference, is similar to the equation for the phase of a synchronized van der Pol oscillator with noise (1.268). However, in contrast with the van der Pol case, the term corresponding to the natural frequency in the frequency mismatch is absent. Instead, the item $\Omega$ occurs singularly in (3.128). This is another indication that the system has no deterministic time scales. The corresponding rotation term is hidden in (3.128) and will occur after averaging.

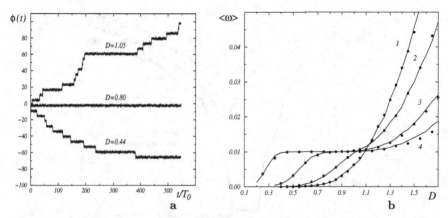

**Fig. 3.24.** (a) Instantaneous phase difference $\phi$ versus time for the indicated values of the noise intensity for $A = 3$; (b) the mean frequency (*solid line*) and the mean switching frequency (*circles*) versus noise intensity for different values of the driving amplitude: $1 - A = 0$, $2 - A = 1$, $3 - A = 2$, and $4 - A = 3$. Other parameters are $\alpha = 5$, $\beta = 1$ and $\Omega = 0.01$ [191]

**Phase Synchronization of Switchings by the Periodic Signal.** Numerically computed phase differences versus time defined by means of the analytic signal concept are shown in Fig. 3.24a for different values of the noise intensity. The slope of the curves gives the difference between the instantaneous frequency of $x$ and $\Omega$. As seen there exists an optimal noise level for which the frequencies converge, i.e., the slope vanishes. Additionally, for this noise and the selected amplitude of the driving the phase is locked during the observation time.

Deviations from this optimal noise intensity lead to phase differences, and the appearance of the phase slips, which leads to a systematic nonvanishing slope of the curves. Resulting mean frequencies defined via $\langle \omega \rangle = \lim_{T \to \infty} \frac{1}{T} \int_0^T \frac{d\Phi(t)}{dt} \, dt$ are presented in Fig. 3.24b for different values of the amplitude of the signal $A$ and versus noise intensity $D$. These curves demonstrate locking of the mean frequency of the output $\langle \omega \rangle$ for optimal noise. It was firstly reported in [190] for the case of the stochastic Schmitt trigger (see Sect. 3.2.3).

Fig. 3.24b illustrates both the presence of a threshold for synchronization and the broadening of the synchronization region with growing amplitudes of the modulating force. In the absence of the signal the mean frequency grows monotonically in accordance with Kramers rate for the overdamped case multiplied by $\pi$. For larger amplitudes starting about $A \propto 1$ the dependence $\langle \omega \rangle$ versus $D$ has a twist where $\langle \omega \rangle$ does not depend on $D$. A further increase in the amplitude leads to the broadening of the synchronization region [190, 191].

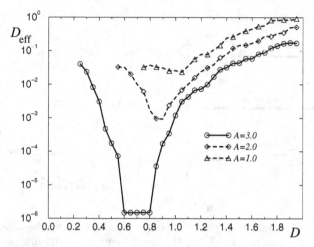

**Fig. 3.25.** The effective diffusion coefficient as a function of the noise intensity for the indicated values of the driving amplitude; other parameters are $\alpha = 5$, $\beta = 1$ and $\Omega = 0.01$ [191]

As outlined above the occurrence of the frequency entrainment is accompanied by at least a small number of phase slips. Therefore, not only the mean frequency is locked but the mean phase difference is minimized, too. In accordance with the introduced definition (see Sect. 1.3.5) one might call $x$ effectively synchronized with the input external periodic force.

Next, we prove that one has to calculate the effective diffusion constant

$$D_{\text{eff}} = \frac{1}{2} \frac{d}{dt} \left[ \langle \phi^2(t) \rangle - \langle \phi(t) \rangle^2 \right] \tag{3.130}$$

using phases defined via the analytical signal. As we already discussed in Chap. 1, the effective diffusion measures the quality of synchronization: the smaller the effective diffusion constant, the longer the durations of phase-locking epochs. Results from computer simulations are presented in Fig. 3.25. With the growth of the amplitude $A$ the presence of the minimum in the dependence of $D_{\text{eff}}$ versus noise intensity becomes strongly pronounced. The quantitative values of $D_{\text{eff}}$ testify that phase and frequency locking takes place over about $10^3$ periods $T_1 = 2\pi/\Omega$ of the driving signal.

Therefore, the presence of the effective synchronization at the basic tone was verified for the standard example of SR. It manifests itself in the mean frequency locking by the external signal and minimal diffusion coefficients. It is an important fact that the introduction of noise into the system leads to the ordering of its phase dynamics: the increase in the noise intensity causes the deceleration of the phase difference diffusion.

**Two-State Description of Synchronization.** Interestingly enough, the time-dependent rates (3.11) of the two-state model treated by McNamara and Wiesenfeld [61] can be recast to show an explicit dependence on the phase difference $\phi(t)$ between the dichotomic output phase $\Phi_{\text{out}}(t) = N(t)\pi$ and the input phase $\Phi_{\text{in}}(t) = \Omega t$

$$\phi(t) = \Phi_{\text{out}}(t) - \Phi_{\text{in}}(t) \tag{3.131}$$

Applying a trigonometric theorem it yields

$$\cos(\Phi_{\text{out}})\cos(\Phi_{\text{in}}) = \frac{1}{2}\left[\cos(\phi) + \cos(2\Phi_{\text{out}} - \phi)\right] = \cos(\phi) \tag{3.132}$$

since the phase of the output $\Phi_{\text{out}}$ takes over multiple values of $\pi$, only. Thus we find that the rates of output switches of the two-state dynamics 3.11 reads in dependence on the unwrapped phase difference $\phi$ [163]

$$W_\phi^{\text{out}}(t) = r_{\text{K}}(D)\exp\left(-\frac{A}{D}\cos[\phi(t)]\right), \tag{3.133}$$

with $A$ being the amplitude of the input.

From this point on we will consider dichotomic signals for the input as well as already done for the output. This restriction defines the $2 \times 2$-state system [see (3.107) and later] if the phase difference is taken to be $\text{mod}(2\pi)$. Here the unwrapped phase is a sequence

$$\phi(t) = k(t)\pi \tag{3.134}$$

and $k(t)$ is defined by a simple Markovian birth and death process, $k \to k+1$ being output switches and $k \to k-1$ changes of the input, respectively. We briefly denote $p_k(t)$ as the probability of observing a phase difference $k\pi$ at time $t$ (conditioned by $\phi_0 = 0$ at time $t_0 = 0$).

Hence, the stochastic dynamics is governed by the following master equation:

$$\frac{\partial p_k(t)}{\partial t} = W_{k+1}^{\text{in}}p_{k+1} - W_k^{\text{in}}p_k + W_{k-1}^{\text{out}}p_{k-1} - W_k^{\text{out}}p_k, \tag{3.135}$$

There are two states for the output and, respectively, for the input. We are left with the two noise-dependent rates for changes of the output

$$W_{k_{\text{even}}}^{\text{out}} = a_1 = r_{\text{K}}(D)\exp\left(-\frac{A}{D}\right), \quad W_{k_{\text{odd}}}^{\text{out}} = a_2 = r_{\text{K}}(D)\exp\left(\frac{A}{D}\right) \tag{3.136}$$

corresponding to the in-synchrony and out-of-synchrony transitions due to output switchings $k \to k+1$, respectively.

The first two items on the right-hand side standing for backwardly running $k$ if the input changes. They are independent on $k$. In the case of dichotomic periodic driving it reads

$$W_k^{\text{DPP}}(t, \varphi_0) = \pi \sum_{n=-\infty}^{\infty} \delta\left(t - \frac{n\pi + \varphi_0}{\Omega}\right). \tag{3.137}$$

Here, the nonstationary rate depends explicitly on the initial phase $\varphi_0$ of the signal. Averaging over this initial phase eventually yields a stationary process with

$$\left\langle W_k^{\text{DPP}} \right\rangle_{\varphi_0} = \frac{1}{2\pi} \int\limits_0^{2\pi} W_{\pm}^{\text{DPP}}(t, \varphi_0) \, \mathrm{d}\varphi_0 = \Omega. \tag{3.138}$$

A similar rate is used in the case of aperiodic driving by a stationary dichotomic Markovian process. With $\gamma$ being the inverse mean time $\tau$ of a single stochastic input change, the mean angular velocity reads

$$W_k^{\text{DMP}} = \frac{2\pi}{2\tau} = \pi\gamma. \tag{3.139}$$

Both stationary rates are constants which allows a common analysis of their mean frequency. A common treatment is impossible when calculating the effective phase diffusion coefficient since second moments of the DPP and DMP differ even in the stationary limit [163].

**Noise-Induced Frequency Locking.** From the master equation (3.135) we can compute the mean frequency $\langle \dot{\phi} \rangle$ and find

$$\left\langle \frac{\mathrm{d}}{\mathrm{d}t}\phi \right\rangle = -\langle \omega_{\text{in}} \rangle + \langle \omega_{\text{out}} \rangle = -\langle \omega_{\text{in}} \rangle + \frac{\pi}{2}(a_1 + a_2) - \frac{\pi}{2}(a_2 - a_1)\langle \cos(\phi) \rangle, \tag{3.140}$$

with $\langle \omega_{\text{in}} \rangle = \gamma\pi$ for the DMP and $\langle\langle \omega_{\text{in}} \rangle\rangle_{\varphi_0} = \Omega$ for the DPP.

Equation (3.140) is in close analogy to the averaged Adler equation (1.268). By comparison we see that the frequency mismatch corresponds to

$$\Delta = \pi \frac{a_1 + a_2}{2} - \langle \omega_{\text{in}} \rangle, \tag{3.141}$$

whereas the difference $\pi(a_2 - a_1)/2$ is the analogue of the nonlinearity parameter $\mu$. Note that now both parameters are noise dependent, which explains the fact that frequency locking is noise induced. Moreover, we stress the fact that the dependence of $\mu$ on the ratio $A/D$ is nonlinear.

The first item in (3.141) defines the required frequency of the driven stochastic bistable dynamics. It increases monotonously with growing noise, starting from zero in the case of absent noise. With vanishing amplitudes $a_1 = a_2 = r_K$. Therefore, in analogy to oscillators the Kramers rate multiplied by $\pi$ can be called the *eigenfrequency* of the stochastic oscillations in the two-state picture.

In general, the term $\langle \cos\phi \rangle$ is time dependent and its evolution is determined by the initial preparation. However, the asymptotic value is unique and can be derived as the stationary value of the related master equation

$$\lim_{t \to \infty} \langle \cos(\phi) \rangle = \left\langle \sigma^{\text{stat}} \right\rangle = \frac{a_2 - a_1}{2\dfrac{\langle \omega_{\text{in}} \rangle}{\pi} + a_1 + a_2}. \tag{3.142}$$

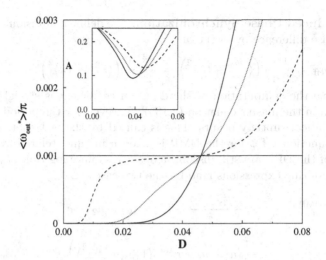

**Fig. 3.26.** The mean frequency of the output [cf. (3.140)] in units of $\pi$ for the DMP with $\gamma = 0.001$ and for the DPP with $\Omega/\pi = 0.001$ for three signal amplitudes: $A_1 = 0$, i.e., vanishingly small, (*solid line*), $A_2 = 0.1$ (*dotted line*) and $A_3 = 0.2$ (*dashed line*). For sufficiently large amplitudes (with $\alpha_0 = 1$ and $\Delta U = 0.25$), a plateau around $\gamma = \Omega/\pi$ is formed which – in connection with a suitable definition – gives rise to the tongue like structures shown in the inset

As was shown in [162] this value equals the stationary correlation coefficient of the input and output of the $2 \times 2$-state model.

In Fig. 3.26 we show the asymptotic mean output frequency with (3.142) inserted,

$$\left\langle \omega_{\text{out}}^{\text{stat}} \right\rangle = \frac{\pi}{2} \left( a_1 + a_2 - \frac{(a_2 - a_1)^2}{2\dfrac{\langle \omega_{\text{in}} \rangle}{\pi} + a_1 + a_2} \right), \qquad (3.143)$$

in units of $\pi$ as a function of noise intensity $D$ for the DMP with $\gamma = 0.001$ and for three amplitudes $A_{1,2,3} < \Delta U = 0.25$. For sufficiently large amplitudes we see the formation of a plateau around the frequency of the input signal. This picture resembles the frequency-locking graph in noisy self-sustained oscillators. The range of noise intensities yielding values belonging to this plateau – a criterion which we pinpoint by the demand of a small slope – in dependence on the signal amplitude $A$ shapes the tongue-like structures shown in the inset of Fig. 3.26. These results identically apply to the DPP when setting $\Omega/\pi = \gamma$. We note that similarities to nonlinear oscillators would become complete if plotting with respect to the noise-dependent eigenfrequency $\pi r_{\text{K}}$ instead of noise intensity.

**Noise-Induced Phase Synchronization.** We define a diffusion coefficient of the phase difference by virtue of

$$D_{\text{eff}} = \frac{1}{2}\frac{\mathrm{d}}{\mathrm{d}t}\left(\langle\phi^2\rangle - \langle\phi\rangle^2\right) = \frac{\pi^2}{2}\frac{\mathrm{d}}{\mathrm{d}t}\left(\langle k^2\rangle - \langle k\rangle^2\right). \tag{3.144}$$

In principle the computation of the diffusion coefficient is straightforward, starting from the master equation (3.135). However, for the periodic driving a mathematical subtlety occurs. This is caused by the fact that, while the mean of functions of $\phi$ for the DMP is smooth in time, related averages in the case of the DPP are still discontinuous across jumping times $t_n$ [163].

The resulting expressions read for the two cases

$$D_{\text{eff}}^{\text{DMP}} = \frac{\pi^2}{2}\left(\gamma + \frac{\langle\omega_{\text{out}}^{\text{stat}}\rangle}{\pi} - (2\gamma - (a_1 + a_2))\,\langle\sigma^{\text{stat}}\rangle^2\right.$$
$$\left. -\frac{1}{2}(a_2 - a_1)\langle\sigma^{\text{stat}}\rangle(1 + \langle\sigma^{\text{stat}}\rangle^2)\right),$$

$$D_{\text{eff}}^{\text{DPP}} = \frac{\pi^2}{2}\left(\frac{\langle\omega_{\text{out}}^{\text{stat}}\rangle}{\pi} - \left(2\frac{\Omega}{\pi} - (a_1 + a_2)\right)\langle\sigma^{\text{stat}}\rangle^2\right.$$
$$\left. -\frac{1}{2}(a_2 - a_1)\langle\sigma^{\text{stat}}\rangle(1 + \langle\sigma^{\text{stat}}\rangle^2) + \frac{\Omega}{\pi}\langle\sigma^{\text{stat}}\rangle\right).$$

Its general structure is $D_{\text{eff}} = D_{\text{in}} + D_{\text{out}} - D_{\text{coh}}$. Therein the last coherence term scales with $\langle\sigma^{\text{stat}}\rangle^n$ ($n = 1, 2, 3$). Only this term can reduce the effective phase difference diffusion for the optimal noise intensity.

Effective phase synchronization reveals itself by the extended duration of locking episodes. We can determine the average duration of a locking episode $\langle T_{\text{lock}}\rangle$ from the ansatz

$$\langle\phi^2\rangle = \left\langle\frac{\mathrm{d}\phi}{\mathrm{d}t}\right\rangle^2\langle T_{\text{lock}}\rangle^2 + 2D_{\text{eff}}\langle T_{\text{lock}}\rangle = \pi^2. \tag{3.145}$$

From this we find

$$\langle T_{\text{lock}}\rangle = D_{\text{eff}}\left\langle\frac{\mathrm{d}\phi}{\mathrm{d}t}\right\rangle^{-2}\left[\sqrt{1 + \left(\frac{\pi}{D_{\text{eff}}}\left\langle\frac{\mathrm{d}\phi}{\mathrm{d}t}\right\rangle^2\right)} - 1\right]. \tag{3.146}$$

In Fig. 3.27 we plot the dependence of $\langle T_{\text{lock}}\rangle$ on the noise intensity $D$ for the DMP and DPP. In both plots we find huge maxima around noise intensities where frequency locking also occurs, indicating the region of effective phase synchronization. Let us note that the onset of phase synchronization is triggered by a sharp increase in $\langle\sigma^{\text{stat}}\rangle$. However, even beyond the region of effective phase synchronization the correlator maintains a large value; thus, it is not an equivalent measure.

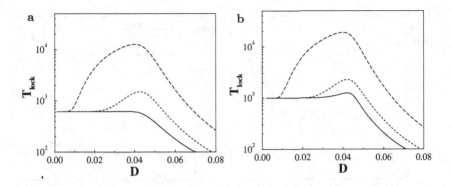

**Fig. 3.27.** The average duration of locking episodes $\left\langle T_{\text{lock}} \right\rangle$ [cf. (3.146)] for (a) the DMP and (b) the DPP both for signal amplitudes: $A_1 = 0$ (*solid line*), $A_2 = 0.1$ (*dotted line*) and $A_3 = 0.2$ (*dashed line*). Pronounced maxima with logarithmic scale occur around noise intensities which coincide with the frequency-locking condition (cf. Fig. 3.26). For the DPP we find a maximum even for $A \to 0+$ around the noise intensity at which the Kramers rate coincides with the signal frequency because the phase diffusion coefficient increases rather slowly

### 3.2.3 Forced Stochastic Synchronization of the Schmitt Trigger

In this section we describe an experimental study of noise-induced synchronization in a stochastic Schmitt trigger driven by a weak periodic signal. The Schmitt trigger is one of the experimentally and theoretically best investigated examples of nonlinear stochastic dynamics exhibiting SR [19, 192].

Results of experimental investigations and numerical simulations of the mean switching frequency for the Schmitt trigger are presented and discussed in [190]. Noise with a cutoff frequency $f_c = 100\,\text{kHz}$ and a periodical signal with frequency $f_0 = 100\,\text{Hz}$ were applied to the Schmitt trigger with a threshold voltage $\Delta U = 150\,\text{mV}$.

The mean switching frequency, $\left\langle f \right\rangle = \lim_{M \to \infty} \frac{1}{M} \sum_{k=1}^{M} \frac{\pi}{t_{k+1} - t_k}$, is calculated from the random telegraph signal of the output, which is recorded by a computer. The results of measurements are shown in Fig. 3.28a. For a weak signal, the dependence of the mean frequency on the noise intensity obeys Kramers law, i.e., it exponentially grows when the noise intensity increases. With increasing signal amplitude, the dependence is qualitatively different. There appears a range of noise intensities in which the mean frequency practically does not change with increasing noise and, within the limits of experimental accuracy, coincides with the signal frequency. Thus, the effect of mean switching frequency locking is observed [190].

Repeating the measurements of the mean frequency for different values of the amplitude and phase of the signal one may obtain synchronization regions in the parameter plane "noise intensity – amplitude of periodic force". Inside these regions the mean frequency matches the signal frequency. Synchroniza-

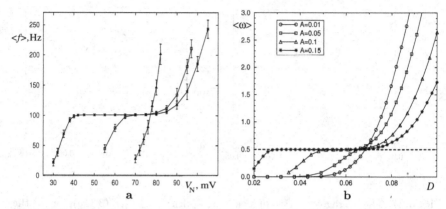

**Fig. 3.28.** Mean frequency of the output signal of the Schmitt trigger obtained (a) experimentally and (b) from numerical simulation of the system (3.147) as a function of the noise intensity for different signal amplitudes: $A = 0\,\text{mV}$ ($\triangle$), $A = 60\,\text{mV}$ ($\square$), $A = 100\,\text{mV}$ ($\star$) with $K = 0.2$ and $\Omega = 0.5$

tion regions resembling Arnold's tongues are shown in Fig. 3.29. As seen from this figure, there is a threshold amplitude, $A_{\text{th}}$, of the external signal beyond which the synchronization of switchings is observed. When the threshold is achieved, the periodic signal starts to effectively control the stochastic dynamics of switchings. With increasing signal frequency, the synchronization is adversely affected: the synchronization regions shrink, and the threshold values of the signal amplitude increase (see Fig. 3.29).

Figure 3.30 illustrates the residence-time probability density $p(t)$ of the trigger in one of the metastable states for different noise intensities. For a weak noise ($V_{\text{N}} = 35\,\text{mV}$), $p(t)$ exhibits peaks centered at odd multiples of the half signal period. Inside the synchronization region ($V_{\text{N}} = 70\,\text{mV}$) the mean residence time in one of the states coincides with the half period and the residence-time distribution density possesses a single well-marked peak at $t = T_0/2$. For strong noise ($V_{\text{N}} = 115\,\text{mV}$), beyond the synchronization region, the mean time for switchings is much less than the half period. In one period the system switches repeatedly from one state to another and a peak corresponding to short switching times appears. The peak centered at the half signal period is smeared, and coherence of the output signal is destroyed.

The Schmitt trigger is modeled by the equation

$$x(t + \Delta t) = \text{sign}\left[K\,x(t) - D\,y(t) - A\sin(\Omega t)\right], \tag{3.147}$$

where $K = 0.2$ characterizes the operating threshold of the trigger, $y(t)$ is an Ornstein–Uhlenbeck process correlation time, $\tau_{\text{c}} = 10^{-2}$ and unit intensity.

The dependence of the mean switching frequency of the trigger versus noise intensity obtained by numerical simulation of (3.147) are shown in Fig. 3.28b and completely verify the experimental results presented in

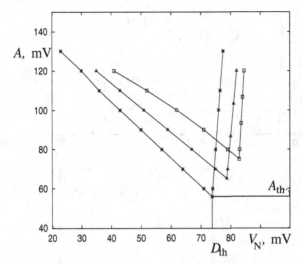

**Fig. 3.29.** Synchronization regions for the Schmitt trigger (full-scale experiment) for different signal frequencies: $f_0 = 100\,\mathrm{Hz}\,(\star)$, $f_0 = 250\,\mathrm{Hz}\,(\triangle)$, and $f_0 = 500\,\mathrm{Hz}\,(\square)$

Fig. 3.28a. The dynamics of the phase difference resulting from definition (1.286) for $A = 0.1$ is pictured in Fig. 3.31a. At an optimal noise level ($D = 0.06$), the phase difference remains constant over the observation time. Beyond the synchronization region ($D = 0.04$, $D = 0.08$), the phase jumps are much more frequent. This leads to fast phase diffusion and, consequently, to the lack of synchronization.

If the signal amplitude is small ($A \leq A_{\mathrm{th}}$), the effect of mean switching frequency locking is not observed. Calculation results of the instantaneous phase difference also indicate the lack of phase synchronization, i.e., there are relatively short locking segments, and time series of the instantaneous phase difference display random-walk-like behavior [191].

The numeric results of the diffusion coefficient are shown in Fig. 3.31b. The dependence $D_{\mathrm{eff}}(D)$ is characterized by a minimum that results from the effect of stochastic synchronization.

From the findings presented above it follows that an external periodic signal of sufficient amplitude synchronizes the stochastic dynamics of switching events. This phenomenon is accompanied by instantaneous phase and mean frequency locking.

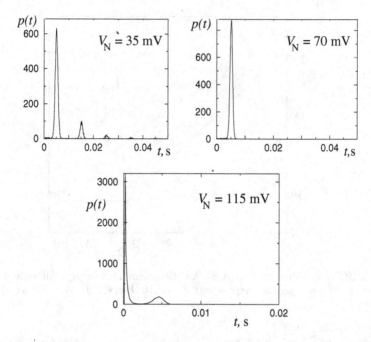

**Fig. 3.30.** Residence-time probability density for the Schmitt trigger in one of the states for different noise intensities. Signal amplitude $A = 100\,\mathrm{mV}$, and signal frequency $f_0 = 100\,\mathrm{Hz}$

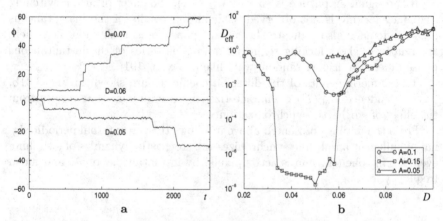

**Fig. 3.31.** (a) Time dependencies of the instantaneous phase difference $\phi$ with $A = 0.1$ and $\Omega = 0.5$ for the indicated values of the noise intensity; (b) diffusion coefficient of the phase difference $D_{\mathrm{eff}}$ of the Schmitt trigger versus noise intensity for different values of the signal amplitude for $K = 0.2$ and $\Omega = 0.5$

## 3.2.4 Mutual Stochastic Synchronization
## of Coupled Bistable Systems

Now consider the simplest case of two symmetrically coupled overdamped bistable oscillators which are governed by the following SDE [70, 189]:

$$\dot{x} = \alpha\, x - x^3 + \gamma\,(y - x) + \sqrt{2D}\,\xi_1(t),$$
$$\dot{y} = (\alpha + \Delta)\, y - y^3 + \gamma\,(x - y) + \sqrt{2D}\,\xi_2(t). \qquad (3.148)$$

In (3.148) the parameter $\alpha$ determines the Kramers frequency of the first subsystem without coupling, $\Delta$ is the parameter of detuning of the second system with respect to the first, and $\gamma$ is the coupling coefficient. The white noise sources $\xi_1(t)$ and $\xi_2(t)$ are assumed to be statistically independent: $\langle \xi_i(t)\xi_j(t+\tau) \rangle = \delta_{i,j}\,\delta(\tau)$. The latter means that for $\gamma = 0$ the stochastic processes $x(t)$ and $y(t)$ in the subsystems will also be statistically independent. We also assume the intensities of uncorrelated noises to be identical, e.g., $D_1 = D_2 = D$.

The Fokker–Planck equation (FPE) for the probability density $p(x, y, t)$, corresponding to the SDE (3.148), has the form

$$\frac{\partial p}{\partial t} = -\frac{\partial}{\partial x}\left[\alpha x - x^3 + \gamma(y - x)\right]p \qquad (3.149)$$
$$- \frac{\partial}{\partial y}\left[(\alpha + \Delta)y - y^3 + \gamma(x - y)\right]p + D\left(\frac{\partial^2 p}{\partial x^2} + \frac{\partial^2 p}{\partial y^2}\right).$$

In this case the stationary solution $p(x, y)$ of the FPE can be found analytically as the coefficients of drift and diffusion of FPE (3.149) obey the potential conditions [67, 193]. The stationary probability density is

$$p(x, y) = C \exp\left(-\frac{U(x, y)}{D}\right), \qquad (3.150)$$

$$U(x, y) = -\alpha\frac{x^2}{2} + \frac{x^4}{4} - (\alpha + \Delta)\frac{y^2}{2} + \frac{y^4}{4} + \frac{\gamma}{2}(x - y)^2,$$

where $C$ is the normalization constant. Hence, the SDE (3.148) and the FPE (3.149) describe overdamped Brownian motion in a two-dimensional potential $U(x, y)$. The fact that the system is potential one (comp (1.163)) facilitates its bifurcation analysis. In this case qualitative changes in the probability density structure (the change in number of extrema) correspond to the bifurcations of the deterministic system (see Sect. 1.2.5).

Since individual subsystems are bistable, we assume the processes $x(t)$ and $y(t)$ to be synchronized if the transitions between metastable states in the subsystems occur at the same times. Figure 3.32 displays time series of the processes in the subsystems for two values of the coupling coefficient. The value of detuning parameter $\Delta = -0.5$ represents faster motion in the second subsystem $y(t)$. For weak coupling the processes in the subsystems are nonsynchronized. By increasing the coupling coefficient the process in

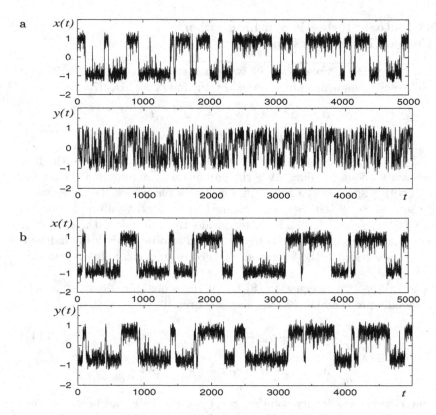

**Fig. 3.32.** Time series $x(t)$ and $y(t)$ of the system (3.148) for (**a**) $\gamma = 0.05$ and (**b**) $\gamma = 0.5$. Other parameters are $\alpha = = 1.0$, $\Delta = -0.5$ and $D = 0.07$

the second subsystem is being slowed down, and for a sufficient value of $\gamma$ the processes in the subsystems become coherent.

To quantify the synchronization process we use the coherence function defined by the expression

$$\Gamma^2(\omega) = \frac{|G_{x,y}(\omega)|^2}{G_{x,x}(\omega)\,G_{y,y}(\omega)}, \tag{3.151}$$

where $G_{x,y}(\omega)$ is the mutual spectral density of the processes $x(t)$, $y(t)$, while $G_{x,x}(\omega)$ and $G_{y,y}(\omega)$ are the power spectra of $x(t)$ and $y(t)$, respectively.

The coherence function obtained from numeric simulation of the SDE (3.148) is shown in Fig. 3.33a for different values of the coupling coefficient. With increasing coupling coefficient $\gamma$ the stochastic processes $x(t)$, $y(t)$ become coherent in the low-frequency range, which corresponds to the Kramers frequencies of the subsystems.

We now explore the evolution of the Kramers frequencies of the subsystems as one varies the coupling coefficient. For $\gamma = 0$, the natural time scales

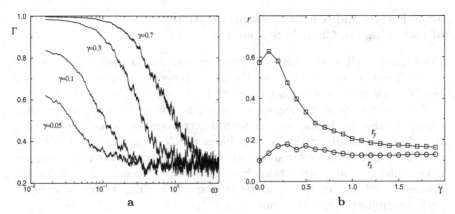

**Fig. 3.33.** (a) Coherence function $\Gamma(\omega)$ for the system (3.148) for different values of the coupling coefficient, and (b) the mean switching frequencies $r_{x,y}$ as functions of the coupling coefficient for parameters $\alpha = 1.0$, $\Delta = -0.5$ and $D = 0.1$

of the subsystems are represented by the Kramers rates (frequencies) $r_x$, $r_y$ of escape from a metastable state:

$$ r_x = \frac{\sqrt{2}\alpha}{\pi} \exp\left(-\frac{\alpha^2}{4D}\right), \quad r_y = \frac{\sqrt{2}(\alpha + \Delta)}{\pi} \exp\left(-\frac{(\alpha + \Delta)^2}{4D}\right). \quad (3.152) $$

Consider the mean switching frequencies in each subsystem, taking into account the coupling between them. These quantities can be obtained numerically by fixing the time moments $t_i^x$, $t_j^y$ when the processes $x(t)$ and $y(t)$ intersect the zero axis in a positive direction, respectively:

$$ r_{x,y} = \lim_{N \to \infty} \frac{1}{N} \sum_{i=1}^{N} \frac{2\pi}{t_{i+1}^{x,y} - t_i^{x,y}}. \quad (3.153) $$

The mean switching frequencies versus $\gamma$ are shown in Fig. 3.33b. As can be seen, with increasing coupling strength the mean frequencies of the partial subsystems come closer together.

The coherence function characterizes only a linear dependence and contains no information about the phases of the processes. The instantaneous phase of the processes in the subsystems can be introduced on the basis of switching times $t_i^x$, $t_j^y$ as introduced in section 1.3.5. Thus we define

$$ \Phi_x(t) = 2\pi \frac{t - t_i^x}{t_{i+1}^x - t_i^x} + 2\pi i, \quad t_i^x < t < t_{i+1}^x, $$

$$ \Phi_y(t) = 2\pi \frac{t - t_j^y}{t_{j+1}^y - t_j^y} + 2\pi j, \quad t_j^y < t < t_{j+1}^y. \quad (3.154) $$

Numerical simulation of the dynamics of instantaneous phase difference $\phi(t) = \Phi_x(t) - \Phi_y(t)$ shows the effect of phase locking for $\gamma > 1$: the phase difference remains constant over long time intervals.

### 3.2.5 Forced and Mutual Synchronization of Switchings in Chaotic Systems

Deterministic SR is realized in chaotic systems in the regimes of "chaos–chaos" intermittency. In this case the mean frequency of irregular switching events is governed by a system's control parameter (see Sect. 3.1).

If, from the physical viewpoint, these phenomena have a deep physical generality, then in chaotic systems with strong interactions one must observe the effect of synchronization of switchings which is qualitatively equivalent to that described above for stochastic bistable systems. Our recent studies have confirmed this assumption. For illustrative purposes we shall further discuss forced and mutual synchronization of switching events in systems with deterministic chaotic dynamics [9, 194].

**Forced Synchronization.** Consider the one-dimensional cubic map

$$x_{n+1} = (ax_n - x_n^3)\exp(-x_n^2/b) + A\sin(\Omega n) \qquad (3.155)$$

in the regime of "chaos–chaos" intermittency when $a > a_{cr} = 2.839\ldots$ Suppose that the map (3.155) is driven by a periodic force with a large enough amplitude $A$, and explore the system dynamics in the absence of external noise. In this case the switching process is principally nonlinear, and its statistics essentially depends on the parameter $a$. Applying the two-state method we evaluate the evolution of the residence-time probability density $p(\tau)$ in a single potential well as the parameter $a$ is varied. $\tau$ is $n$ given in units of the period of the signal $2\pi/\Omega$. The results are shown in Fig. 3.34 and agree qualitatively with the data obtained for the Schmitt trigger (Fig. 3.30). At an optimal value $a = 8.34$ the distribution possesses a single Gaussian-like peak with a maximum at $\tau = 0.5$. This means that the mean switching frequency between chaotic attractors coincides with the external force frequency.

The effect of mean switching frequency locking by the external signal is illustrated in Fig. 3.35a for different values of parameter $a$. With increasing signal amplitude, the synchronization region expands as expected. As seen from Fig. 3.35b, this region represents a typical synchronization zone, as for the Schmitt trigger (Fig. 3.29), and also demonstrates the presence of a threshold for the synchronization. Hence, the effect of forced synchronization of the switching frequency in a deterministic chaotic system is confidently observed and proves to be equivalent to the effect of noise-induced stochastic synchronization.

The effect of mean switching frequency locking is universal and manifests itself in a wide class of DS with intermittency. As an example we study the nonlinear Chua's system realizing the regime of dynamical intermittency [146]. Chua's circuit is described by the following set of equations:

$$\dot{x} = \alpha[y - h(x)],$$
$$\dot{y} = x - y + z, \qquad (3.156)$$
$$\dot{z} = -\beta y + F(t),$$

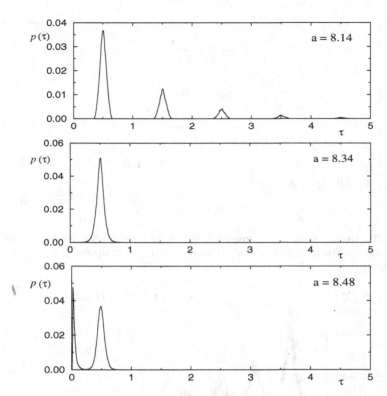

**Fig. 3.34.** The residence-time probability density to be in one of the potential wells of the map (3.117) for three different values of the control parameter $a$ and given frequencies of the external driving. The behaviour remembers on the dynamics of a stochastic Schmitt trigger (see Fig. 3.30). $\tau$ is given in units of the period of the external force

where $h(x) = m_1 x + 0.5(m_0 - m_1)(|x + 1| - |x - 1|)$ is the piecewise-linear characteristic of the system with fixed parameters $m_0 = -1/7$, $m_1 = 2/7$, and $F(t) = A \cos(\Omega t)$ is the external periodic force.

As was shown in [46], the switching process in the chaotic bistable system (3.156) can be either noise-induced or of a purely dynamical nature. We again apply the two-state approach. Let parameter $\beta$ be fixed, namely, $\beta = 14.286$. The dynamics of the system depends on $\alpha$ as well as on the amplitude $A$ and the frequency $\Omega$ of the external signal. Without the driving signal ($A = 0$) an intermittency is realized for $\alpha \approx 8.8$. With increasing $\alpha$, when $\alpha > 8.8$, the mean switching frequency monotonically increases. We choose the amplitude value $A = 0.1$ when the system response to the external force is, in principle, nonlinear. The results of numerical calculations of the return-time probability density $p(\tau)$ are shown in Fig. 3.36 for different values of $\alpha$ [195]. One can see a surprising similarity to the data obtained for the Schmitt trigger (Fig. 3.30) an to the data displayed in Fig. 3.34. For a certain value, $\alpha = 8.8325$, the

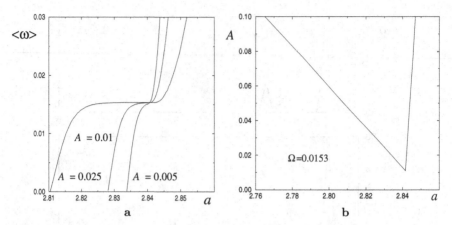

**Fig. 3.35.** (a) Mean switching frequency versus parameter $a$ for different values of the signal amplitude, and (b) the synchronization region of the switching frequency for map (3.155)

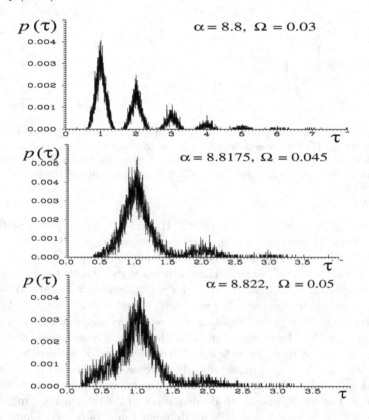

**Fig. 3.36.** Return-time probability density for Chua's circuit for different values of the parameter $\alpha$ in (3.156) and the external signal frequency $\Omega$. The time $\tau$ is scaled in units of the period of the external driving $T = 2\pi/\Omega$

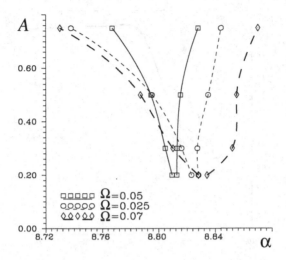

**Fig. 3.37.** Synchronization regions for Chua's system in dependence on the amplitude of the external signal $A$ and on $\alpha$ from (3.156) for the three frequencies $\Omega$ of the signal

probability density $p(\tau)$ has a single Gaussian-like peak near $\tau = 1$. This means that the mean return frequency coincides with the frequency of the external force. The effect of forced synchronization which manifests itself in the mean switching frequency locking by the periodic signal is registered.

Figure 3.37 illustrates synchronization regions obtained in full-scale experiments with Chua's circuit. These regions are qualitatively similar to Arnold's tongues, as in the case of the Schmitt trigger (Fig. 3.29). The only difference is that, as the signal frequency increases, the threshold of the synchronization practically does not change and the width of the synchronization regions is increased. These differences are caused by the nonlinear properties of the system (3.156) and do not relate to the nature of the observed phenomenon. More detailed calculations testify that the effect of mean switching frequency synchronization in systems (3.155) and (3.156) corresponds to the effect of phase synchronization, the latter being completely equivalent to the cases of the Schmitt trigger and the overdamped oscillator considered above.

**Mutual Synchronization.** To illustrate mutual synchronization of switching events we examine the dynamics of two coupled Lorenz models [196]:

$$\dot{x}_1 = \sigma(y_1 - x - 1) + \gamma(x_2 - x_1),$$
$$\dot{y}_1 = r_1 x_1 - x_1 z_1 - y_1,$$
$$\dot{z}_1 = x_1 y_1 - z_1 b,$$
$$\dot{x}_2 = \sigma(y_2 - x_2) + \gamma(x_1 - x_2),$$  (3.157)
$$\dot{y}_2 = r_2 x - 2 - x_2 z_2 - y_2,$$
$$\dot{z}_2 = x_2 y_2 - z_2 b.$$

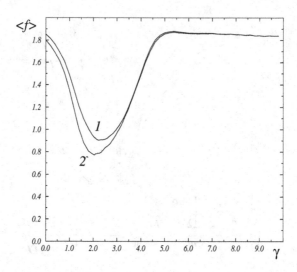

**Fig. 3.38.** Mean switching frequencies $\langle f_1 \rangle$ (*1*) and $\langle f_2 \rangle$ (*2*) for two coupled non-indentical Lorenz models (3.157) versus the coupling strength. Other parameters: $\sigma = 10$, $r_1 = 28.8$, $r_2 = 28$ and $b = 8/3$

We choose the parameters to be: $\sigma = 10$, $r_1 = 28.8$, $r_2 = 28$ and $b = 8/3$, when the Lorenz attractor is realized in each subsystem [181]. The Lorenz attractor in an individual system may be treated as a generalized bistable oscillator where irregular switching events occur with a mean frequency controlled by the parameter $\gamma$ [182]. The introduction of coupling ($\gamma > 0$) must cause the changes of the mean switching frequencies in each of the subsystems and lead to the effect of mutual synchronization of switchings.

We perform numerical simulation using the two-state approximation. The calculation results show that the mean switching frequencies $\langle f_1 \rangle$ and $\langle f_2 \rangle$ practically coincide for a coupling coefficient $\gamma > 5$ (Fig. 3.38). Moreover, numerical simulation of the dynamics of the instantaneous phase difference between processes $x_1(t)$ and $x_2(t)$ has confirmed the effect of mutual synchronization, e.g., for $\gamma > 5$, the phase difference approaches zero at times significantly exceeding the mean switching time [196]. Therefore, the effect of mutual synchronization takes place for strong coupling between two chaotic bistable oscillators.

### 3.2.6 Stochastic Synchronization of Ensembles of Stochastic Resonators

The effect of SR can be significantly enhanced if an array of coupled bistable systems is taken instead of a single one (see Sect. 3.1.5). It has been found out that at optimal values of the noise intensity and the coupling coefficient, the SNR in an array attains its maximal level, demonstrating an array-enhanced SR effect. First we consider the case of extremely weak coupling when the interaction between individual stochastic resonators can be neglected.

The model under consideration is schematically shown in Fig. 3.39 and contains $N$ subsystems, each demonstrating SR. Each resonator $SR_k$ is subjected to the same input signal $s(t)$, which may be aperiodic, and includes an internal Gaussian noise source $\xi_k(t)$. In addition, the internal noises in the subsystems are statistically independent. The outputs of the elements converge onto a summing center, giving a collective response $x_M(t)$:

$$x_M(t) = \frac{1}{N} \sum_{k=1}^{N} x_k(t). \tag{3.158}$$

This model is widely used in practice as the simplest method for increasing the signal-to-noise ratio (SNR). When the number of elements in the array is sufficiently large ($N \gg 1$), the internal noises at the collective output disappear due to averaging, and the SNR increases proportionally to the number of elements. This model is truly nontrivial because each element is treated as a stochastic resonator. Hence, besides enhancement of the SNR [197], it is also possible to significantly amplify the signal. Synchronization of an ensemble by an external signal is also of great interest. This model seems to be generic for a number of applications and, in particular, for a simple network of sensory neurons [198] and a model of ion channels [199]. A similar model, where self-sustained systems were used as individual elements, has been studied in [200, 201].

**Linear Response Theory for Arrays of Stochastic Resonators.** In the case of a weak signal $s(t)$ the statistical properties of the response of a single stochastic system can be calculated via LRT. The problem is to compute the spectral characteristics of the collective response $x_M(t)$ [156].

Denote the spectral density at the output of the $k$th element by $G_{k,k}(\omega)$, the cross-spectral density of the $k$th and $m$th elements by $G_{k,m}(\omega)$, and the spectral density at the summing output by $G_{M,M}(\omega)$. The spectral density $G_{M,M}(\omega)$ is derived immediately from (3.158) as follows:

$$G_{M,M}(\omega) = \frac{1}{N^2} \left( \sum_{k=1}^{N} G_{k,k}(\omega) + \sum_{k=1}^{N} \sum_{\substack{m=1 \\ k \neq m}}^{N} G_{k,m}(\omega) \right). \tag{3.159}$$

In absence of the signal one has $G_{k,m}(\omega) = 0$ by virtue of the statistical independence of the internal noises in the elements. Each element of the

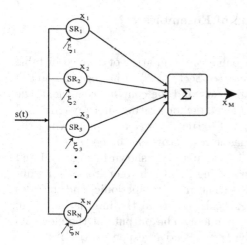

**Fig. 3.39.** Block scheme of the studied model. $SR_k$ are the stochastic resonators, $s(t)$ is a weak external signal, $\xi_k(t)$ are the internal noises in array elements, and $x_M$ is the collective output of the ensemble

array has a known susceptibility $\chi_k(\omega, D)$, where $D$ is the intensity of the internal noise. For the spectral density at the output of the $k$th element we have

$$G_{k,k}(\omega) = G_{k,k}^{(0)}(\omega, D) + |\chi_k(\omega, D)|^2\, G_{s,s}(\omega), \qquad (3.160)$$

where $G_{k,k}^{(0)}(\omega, D)$ is the spectral density of the $k$th element in absence of a signal, and $G_{s,s}(\omega)$ is the spectral density of the input signal. The cross spectral density $G_{k,m}(\omega)$ is determined as [156]

$$G_{k,m}(\omega) = \chi_k^*(\omega, D)\,\chi_m(\omega, D)\,G_{s,s}(\omega), \qquad (3.161)$$

where the symbol * denotes complex conjugation. In the absence of the signal $G_{k,m}(\omega) = 0$. Substituting (3.153-154), into (3.159) we obtain the spectral density of the collective output:

$$G_{M,M}(\omega) = \frac{1}{N^2}\sum_{k=1}^{N} G_{k,k}^{(0)}(\omega, D) + \frac{G_{s,s}(\omega)}{N^2}\sum_{k=1}^{N}\sum_{m=1}^{N}\chi_k^*(\omega, D)\chi_m(\omega, D).$$

$$(3.162)$$

The cross-spectral density of the collective output and the input signal $G_{s,M}(\omega)$ is

$$G_{s,M}(\omega) = \frac{G_{s,s}(\omega)}{N}\sum_{k=1}^{N}\chi_k(\omega, D). \qquad (3.163)$$

The relations obtained allow us to determine all the necessary SR measures, in particular, the coherence function and the SNR. To simplify our analysis we consider an array of identical elements with susceptibility $\chi(\omega, D) \equiv \chi_k(\omega, D)$ and unperturbed spectral density $G_{x,x}^{(0)}(\omega, D) \equiv G_{k,k}^{(0)}(\omega, D)$. For this case we have the following expressions for the spectral densities

$$G_{M,M}(\omega) = \frac{1}{N} G_{x,x}^{(0)}(\omega, D) + |\chi(\omega, D)|^2\, G_{s,s}(\omega), \qquad (3.164)$$

$$G_{s,M}(\omega, D) = \chi(\omega, D)\, G_{s,s}(\omega). \tag{3.165}$$

In the limit of large $N$, the first item in (3.164), being responsible for the internal fluctuations in the elements of the array, becomes vanishingly small and the whole ensemble behaves as an equivalent *linear* system with the transfer function $\chi(\omega, D)$.

Let a weak input signal $s(t)$ be the sum of the periodic and noise components, i.e., $s(t) = A \sin \Omega t + n(t)$. The SNR at the input is fixed:

$$\mathrm{SNR_{in}} = \frac{\pi A^2}{G_{n,n}(\Omega)}, \tag{3.166}$$

where $G_{n,n}(\omega)$ is the spectral density of the noisy component. The spectral density of the collective output will also consist of a noisy background and a delta-peak corresponding to the periodic part of the signal:

$$G_{M,M}(\omega) = \frac{1}{N} G_{x,x}^{(0)}(\omega, D) + |\chi(\omega, D)|^2 \left[ G_{n,n}(\omega) + \pi A^2\, \delta(\omega - \Omega) \right]. \tag{3.167}$$

The SNR of the collective output, $\mathrm{SNR_{out}}$, can be easily derived from the last expression. However, of much interest is the ratio of the SNR at the output to the SNR at the input:

$$\eta = \frac{SNR_{\mathrm{out}}}{SNR_{\mathrm{in}}} = 1 - \frac{G_{x,x}^{(0)}(\Omega, D)}{G_{x,x}^{(0)}(\Omega, D) + N\, |\chi(\Omega, D)|^2\, G_{n,n}(\Omega)}. \tag{3.168}$$

This ratio is always less than 1, unless $N$ tends to infinity. In the latter case input and output SNRs coincide. Although, as follows from the analysis above, *the SNR at the output of the ensemble of stochastic resonators cannot be improved compared to the SNR at the input*, the periodic component of the signal can be significantly amplified by $|\chi(\Omega, D)|$ times.

Estimation of the number of elements necessary to achieve a given ratio $\eta$ is of practical importance. This number can easily be found from (3.168):

$$N = \frac{G_{x,x}^{(0)}(\Omega, D)}{|\chi(\Omega, D)|^2\, G_{n,n}(\Omega)} \frac{\eta}{1 - \eta}. \tag{3.169}$$

Consider again the array of bistable stochastic oscillators subjected to the input signal $s(t) = n(t) + A \sin \Omega t$, where $n(t)$ is the Gaussian colored noise with spectral density $G_{n,n}(\omega) = \gamma Q/(\omega^2 + \gamma^2)$. We obtain

$$\eta = 1 - \frac{\Omega^2 + \gamma^2}{D^2\,(\Omega^2 + \gamma^2) + N\, \lambda_m\, \langle x^2 \rangle_{\mathrm{st}}\, Q\, \gamma} \tag{3.170}$$

and

$$N = \frac{D^2\,(\Omega^2 + \gamma^2)}{\lambda_m\, \langle x^2 \rangle_{\mathrm{st}}\, Q\, \gamma} \frac{\eta}{1 - \eta}. \tag{3.171}$$

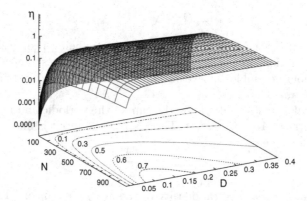

**Fig. 3.40.** The ratio $\eta$ (3.170) for the ensemble of stochastic bistable resonators; $Q = 10^{-3}$, $\gamma = 1.0$, $A = 0.1$ and $\Omega = 0.1$. The lines of constant levels are defined by (3.171)

where $N$ is the number of elements in the array necessary for calculating the ratio $\eta$.

The results are summarized in Fig. 3.40, where the ratio $\eta$ is shown as a function of the internal noise intensity $D$ and the number of elements $N$. In this figure one can also see the lines of constant levels, which can be defined by the expression (3.171). It is clear that the dependence $N(D)$ possesses a minimum at a certain optimal noise intensity. Hence, at an optimal noise level the required value of the ratio $\eta$ can be reached for a minimal number of elements in the array. Besides, Fig. 3.40 reflects the SR effect without tuning, i.e., for large $N$ the ratio $\eta$ practically does not depend on $D$.

**Synchronization of an Ensemble of Stochastic Resonators by a Weak Periodic Signal. Numerical Simulation.** In the limit of an infinite number of elements in the array and in the absence of external noise, the collective response of the array to a weak periodic signal is a periodic function of time:

$$x_M^\infty(t) = A\,|\chi(\Omega, D)|\,\cos{(\Omega t + \psi)}, \qquad (3.172)$$

where $\psi$ is the phase shift defined by

$$\psi = -\arctan\frac{\mathrm{Im}\,\chi(\Omega, \mathrm{D})}{\mathrm{Re}\,\chi(\Omega, \mathrm{D})}. \qquad (3.173)$$

Of great interest are the situations when the number of elements is large but finite. Such cases can be often encountered in biology, for example, populations of neuroreceptors [198] and arrays of ion channels in cell membranes [199]. The following question arises: How much does the process at the collective output of the ensemble reflect the input signal? The previous sections have answered this question only in terms of the averaged characteristics, such as the SNR and the coherence function. As has been shown above, for a single stochastic resonator, synchronization of the output and input is possible for a large enough amplitude of periodic signal. The collective output of an array of stochastic resonators can be synchronized by an arbitrary weak periodic signal for optimally tuned internal noise [202].

We start by discussing the results of numerical simulation. As an example we refer to an ensemble of Schmitt triggers:

$$x_k(t + \Delta t) = \text{sgn} \left[ K\, x_k(t) - \xi_k(t) - \eta(t) - A \sin \Omega t \right], \qquad (3.174)$$

where $K$ is the operating threshold of the triggers, $\xi_k(t)$ is the internal noise in the $k$th element, and $\eta(t)$ is a weak external noise. In the numerical simulations with $K = 0.2$, the internal noise was assumed to be Gaussian exponentially correlated noise with correlation time $\tau_c = 0.01$ and intensity $D$. The external noise was also Gaussian colored noise with the same correlation time and intensity $Q = 0.03$ (OUP). We chose the amplitude of the periodic signal $A = 0.03$ and the frequency $\Omega = 0.5$, at which the phenomenon of stochastic synchronization did not occur in a single element. The numerical simulation was carried out using a CRAY super-computer and a SUN-Ultra/4 workstation.

To quantify the synchronization of the collective output we use the notions of the mean frequency and the instantaneous phase. The mean frequency $\langle \omega \rangle$ is shown in Fig. 3.41a as a function of the internal noise intensity for different numbers of elements $N$ in the array. For a single element $N = 1$, the dependence $\langle \omega \rangle(D)$ is exponential. However, with an increasing number of elements the exponential law is violated, and for a large enough $N$ ($N \geq 100$) mean frequency locking takes place. This effect is similar to the synchronization of a single bistable resonator as the signal amplitude increases (see Figs. 3.24a and

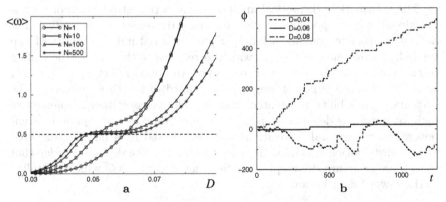

**Fig. 3.41.** (a) Mean frequency $\langle \omega \rangle$ of the collective output for the ensemble of Schmitt triggers as a function of the internal noise intensity $D$ for different numbers of elements $N$ in the ensemble. The frequency of the external signal is $\Omega = 0.5$; (b) time series of the phase difference between the collective output and the periodic input for an ensemble of 500 Schmitt triggers for different values of the internal noise intensity

3.29). But in our case the signal amplitude is small and the synchronization can be achieved by increasing the number of elements in the array.

The synchronization effect is also verified by calculations of the instantaneous phase difference of the collective output and the periodic signal. The results are shown in Fig. 3.41b. At an optimal internal noise intensity ($D \approx 0.06$), the phase difference remains constant over long time intervals. We already noted earlier that the mean value of these time intervals can be estimated using the effective diffusion coefficient $D_{\text{eff}}$ of the phase difference. As the calculations have shown, the effective diffusion coefficient is minimal for an optimal noise intensity. When the number of elements $N$ in the array increases, the absolute value of the diffusion coefficient decreases, and the optimal intensity $D$ of internal noise shifts to the range of smaller values.

The synchronization phenomenon described above is observed for any elements demonstrating SR with variation of the internal noise, including the elements represented by neuron models.

### 3.2.7 Stochastic Synchronization as Noise-Enhanced Order

One of the major motivations for SR research is the idea of gaining information through an optimally tuned stochastic bistable filter. The most appropriate measure describing the information transmission through a bistable system is the spectrum of Shannon conditional entropies [203,204]. In contrast to other measures (a linear version of transinformation), used in [35,205] and quantifying the degree of linear dependence between the input and the output of the system, the hierarchy of Shannon conditional entropies [206,207] characterizes correlations of all higher orders and in the limit is considered to be a measure of the order (disorder) in the system.

The information–theoretical analysis requires the introduction of a symbol alphabet corresponding to the stochastic dynamics of the system. For bistable stochastic systems, a binary alphabet is natural. It consists of two symbols, for instance, '0' and '1', which correspond to the state of the system to the left and to the right with respect to the barrier. Let $\mathbf{i}_n = i_1, \ldots, i_n$ be a binary subsequence of length $n$ or a word of length $n$ ($n$-word). The stationary probability (estimated from its relative repetition frequency) of such an $n$-word is denoted by $p(\mathbf{i}_n)$. If a sequence contains a periodic component, then temporal correlations will be reflected by a highly structured $n$-word distribution function. In order to quantify the degree of order that rules these structures we employ the Shannon entropy [206] which is applied to the $n$-word distribution

$$H_n = - \sum_{(\mathbf{i}_n) \in \{0,1\}^n} p(\mathbf{i}_n) \log_2 p(\mathbf{i}_n). \tag{3.175}$$

The $n$-block entropy $H_n$ is interpreted as the average information necessary to predict the appearance of the $n$-word $(i_1, \ldots, i_n)$.

The conditional or dynamical [207, 208] entropies are introduced for $n = 1, 2, \ldots$ in the following way:

$$h_n = H_{n+1} - H_n = \langle - \sum_{i_{n+1}} p(i_{n+1}|\mathbf{i}_n) \log_2 p(i_{n+1}|\mathbf{i}_n) \rangle_{(\mathbf{i}_n)}, \qquad (3.176)$$

where $\langle \rangle$ indicate averaging over the prehistory $\mathbf{i}_n$. This definition is supplemented by "the initial condition" $h_0 = H_1$. In (3.176), $p(i_{n+1}|\mathbf{i}_n)$ denotes the appearance probability for the symbol $i_{n+1}$ conditioned by the $n$ preceding symbols $\mathbf{i}_n$. The dynamical entropies $h_n$ are interpreted as the average information necessary to predict the symbol $i_{n+1}$ (or gained after its observation) with given prior knowledge of $\mathbf{i}_n$. In other words, $h_n$ characterizes the uncertainty in prediction of the next symbol in a sequence $\mathbf{i}_n$. This amount of information is usually decreased by correlations between symbols in a sequence. The limit $h_n$ when $n \to \infty$, i.e.,

$$h = \lim_{n \to \infty} h_n, \qquad (3.177)$$

is named the source entropy [209]. The source entropy determines the minimal amount of information necessary to predict the next symbol in a sequence with given knowledge of the whole prehistory of the process.

We apply the information–theoretical approach to the experimental data obtained for the Schmitt trigger (see Sect. 3.2.3). Binary random sequences generated by a Schmitt trigger were stored in a computer via an analog-to-digital converter. Simultaneously we recorded the input sequences (the signal and the signal plus noise) which were represented by 0's and 1's depending on their sign. In all experiments the length of sequences was $15\,000 \Delta t$, where $\Delta t$ is the sampling step. The optimal sampling step was chosen to be approximately a twelfth of the signal period: $\Delta t = T_0/12 \approx 8.33 \times 10^{-4}$ s. We chose *the regime of synchronization* of stochastic switchings of the trigger when the mean switching frequency is locked. This regime occurs for a signal amplitude $A = 100$ mV.

It is reasonable to suggest that a symbolic sequence generated by the Schmitt trigger will be maximally ordered in the regime of stochastic synchronization or SR. Hence, one may expect the following scenario for the source entropy behavior: For very weak noise, when the trigger switching events are very rare, the sequence is characterized by a large redundancy, and the entropy is small. With increasing noise, the entropy should increase and then decrease, attaining a minimum due to SR, and rise again when the dynamics of the system is fully controlled by noise.

The picture described above was completely verified by the calculations performed from experimental data [203, 204]. All entropy measures were computed by averaging over 20 time series of length $1500 \Delta t$. The results are shown in Fig. 3.42a. The curves in this figure display a well-marked minimum around the expected noise intensity. Thus, the predictability of the output sequences can be maximized by tuning the noise intensity. This im-

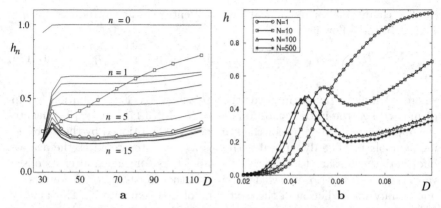

**Fig. 3.42.** (a) Dynamical entropies $h_n$ ($n = 0, 1, \ldots, 15$) versus noise intensity. $h_6(D)$ at the trigger input is shown by *squares*; $h_6(D)$ at the output is shown by *circles*. (b) Source entropy of a binary sequence generated by an ensemble of Schmitt triggers versus internal noise intensity for different numbers of elements in the ensemble

portant effect cannot be principally observed at the output of conventional linear filters.

The increase of predictability implies an enhancement of ordering in the output sequence. With application to SR, entropies reflect an amplification of a periodic component of the output signal, and for a certain optimal noise level we may speak of *noise-enhanced order* in time. The most ordered state means that a maximal amount of switching events takes place during a time equal to half the signal period, and the output is characterized by the longest correlations.

Note that the minimum of the dependence of the source entropy on the noise intensity is observed only for large-enough amplitudes of the periodic signal, when the synchronization phenomenon of trigger stochastic switchings occurs. In the case of a weak signal, when the response of a stochastic system to the signal is basically linear, the entropy monotonically grows with increasing noise and tends to 1 in the limit of high noise level [203, 204].

The fact that the degree of order is enhanced in the regime of SR is also verified when analyzing the collective output of an ensemble of stochastic resonators driven by a weak periodic signal. In order to calculate the source entropy of the collective output of an array, we introduce the symbolic description

$$u(t) = \begin{cases} 0, & \text{if } x_M(t) < 0, \\ 1, & \text{if } x_M(t) \geq 0. \end{cases} \tag{3.178}$$

The calculated results for the source entropy $h$ are shown in Fig. 3.42b. For a single element $N = 1$, the entropy monotonically increases with increasing internal noise intensity $D$ until it saturates. When the number of elements goes up, the behavior of the entropy qualitatively changes. For weak internal noise, the residence times are exponentially large and the symbolic sequence generated by the array is characterized by a high redundancy. As a result, the entropy is close to zero. With increasing $D$, the source entropy rises and reaches a maximum at the internal noise intensity corresponding to the boundary of the synchronization region, when the mean frequency is locked. Starting from this value the source entropy falls, approaches a minimum at a certain optimal noise intensity, and finally rises again. Hence, with increasing internal noise intensity the collective output of the array of stochastic resonators becomes more ordered. We emphasize that in contrast to the synchronization of a single element by a periodic signal with large amplitude, in the case of an ensemble of stochastic resonators the entropy decreases for weak signals and, hence, single elements remain nonsynchronized. Extensive numerical investigations have shown that the values of the noise intensity which minimize the source entropy are the same as the optimal noise level maximizing the output spectral power amplification of a single element.

### 3.2.8 Summary

The results presented in this section give a positive answer to the main question formulated in Sect. 3.2.1. The synchronization effect of switching events takes place in noisy bistable systems. It is principally nonlinear (it can be observed only for finite force amplitudes and for sufficiently strong coupling) and can be described on the basis of a generalized knowledge of the phase synchronization theory. The influence of noise which causes random switchings requires the introduction of the notion of effective synchronization when the time over which the phase difference is locked is finite and depends on the noise intensity.

Knowledge of stochastic synchronization can be extended with reference to ensembles of interacting bistable oscillators. Moreover, by analogy with the SR phenomenon, the problem of switching synchronization can be stated and solved as applied to deterministic chaotic systems with a generalized type of bistable behavior.

As in the classical oscillation theory, the effects of both external (forced) synchronization and mutual synchronization of switching events are realized in bistable stochastic and chaotic systems. The important result is that in the SR regime it is the synchronization effect that leads to an enhancement of the degree of order (self-organization) when the entropy of the output signal has a well-marked minimum. This means that both synchronization and self-organization in bistable systems in SR regimes are realized for sufficiently strong interaction only. These phenomena are principally nonlinear and cannot by studied in the frame of linear approximation, in particular, on the basis of LRT.

## 3.3 The Beneficial Role of Noise in Excitable Systems

So far our attention has been paid to the properties of noisy nonlinear dynamics with a relatively unstructured behavior. In systems with SR the main origin of structure is located outside the system. Ordered behavior is observed as a response to periodic or random correlated inputs. In this section we consider another situation where noise enhances coherence of inherent oscillatory modes of the system.

### 3.3.1 Coherence Resonance Near Bifurcations of Periodic Solutions of a Dynamical System

The SR regime allows one to optimize the degree of coherence of the input and output signals by choosing an optimal noise intensity to control the switching events of a bistable system. Similar phenomena can be observed in DS with noise but without bistability. In this case the system must be oscillatory and the corresponding Fokker–Planck operator must have complex eigenvalues. Moreover, noise can be used to enhance coherence of oscillations even without external periodic signals. An example of such a phenomenon has been described in [210], where a limit cycle was enhanced near a Hopf bifurcation but before the deterministic dynamics exhibits periodic cycles. This effect became more pronounced when colored noise was applied [211, 212].

In [213] a noise-induced fluctuation peak in the spectrum of a nonlinear oscillator with small friction was investigated. Noise-induced coherent motion has also been observed in a self-sustained oscillator near the saddle-node bifurcation [214–216] and in neuron models [217, 218]. In these studies the fluctuation spectrum peak possesses optimal characteristics, that is the peak has maximal height and minimal width, at a certain optimal noise level. This phenomenon is very similar to SR and may be called *coherence resonance* [219]. In the present section, we study coherence resonance near the local bifurcations of periodic solutions, and we substantiate the fact that a simple mechanism of this effect is universal for a wide class of DS [220].

Consider a situation in which noise perturbes a nonlinear DS near a bifurcation of its periodic solution. Assume that a noise-induced peak of height $h$ appears in the spectral density at the frequency $\omega_p$ so that a noisy precursor of the bifurcation can be observed [221]. The following question arises: What will happen with the spectrum shape when the noise intensity is varied? To answer this one should take into account two competitive effects: (i) when the noise level is increased, the system phase trajectories go far away from the stable periodic trajectory (this is the reason for induced oscillations at the frequency $\omega_p$) and the peak height $h$ grows (the growth of $h$ must be bounded due to nonlinearity); (ii) since amplitude and phase fluctuations increase, the noise growth leads to a spreading of the spectrum, e.g., the relative spectral width $W(D) = 1/Q = \Delta\omega/\omega_p$ grows or its quality factor $Q$ decreases. The spreading of the spectral line makes difficult detection of the noise-induced peak and destroys the coherent motion at the frequency $\omega_\mathrm{p}$.

It is reasonable to suppose that there exists an optimal noise level at which the peak in the spectrum has a sufficient height and is narrow enough so that it can be clearly observed on the noise background.

To quantify the degree of coherence of a system, introduce into consideration the quantity $C(D)$ [214]:

$$C(D) = h(D)/W(D) = Q(D) \cdot h(D), \tag{3.179}$$

where $Q(D)$ is the quality factor of the peak. We show below that the coherence measure $C(D)$ attains its maximum at a certain noise level.

Consider the logistic map perturbed by noise:

$$x_{n+1} = 1 - ax_n^2 + \sqrt{D}\xi_n, \tag{3.180}$$

where $a$ is the control parameter and $D$ is the intensity of white noise $\xi(n)$. Without noise the cascade of period-doubling bifurcations $2^k$ takes place for $a = a_k$: $a_1 = 0.75$, $a_2 = 1.25$, $a_3 = 1.368099$, ... For $a = 1.24$ and in the absence of noise ($D = 0$) the map (3.180) has a stable period-2 fixed point, and a delta-peak is observed in the spectrum at the frequency $\omega_0 = \pi$. In the presence of noise there appears a noisy precursor of the bifurcation, a period-4 cycle. In the spectrum one can distinguish an additional characteristic peak at the subharmonic frequency $\omega_\mathrm{p} = \pi/2$ [222]. With increasing $D$ this peak becomes more and more pronounced over the noise background. However, a further increase in the noise intensity broadens the peak so that it cannot be practically observed over the noise background.

The dependence of the coherence degree $C(D)$ is shown in Fig. 3.43a and confirms the presence of coherence resonance in the system. As seen from Fig. 3.43b, its relative width $W$ increases linearly with $D$ [1]. At the same time, the peak height grows linearly for small noise intensity, then its growth is slowed down and, finally, is saturated. Therefore, the dependences $h(D)$ and $W(D)$ can be approximated by the following expressions:

$$W(D) \propto W_0 + D, \quad h(D) \propto 1 - \exp(-\alpha D), \tag{3.181}$$

where $W_0$ and $\alpha$ are certain constants. The competition between an increase in the peak height and the relative spectral width yields the maximal coherence:

$$C(D) \propto \frac{1 - \exp(-\alpha D)}{W_0 + D}. \tag{3.182}$$

We are also interested in the behavior of noisy precursors of the bifurcation when varying the control parameter. Introduce the parameter of supercriticality $\varepsilon = a_k - a$, where $a_k$ is the bifurcation point. For any noise intensity $D$ the dependence of the coherence measure on supercriticality $\varepsilon$ can be described by the law $C(\varepsilon) \propto \varepsilon^{-3}$. This description shows agreement with theoretical predictions in [221]. Hence, the effect must be enhanced

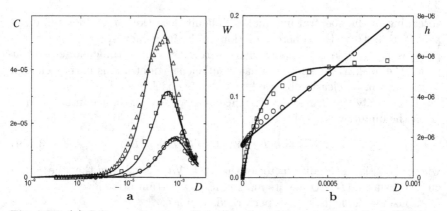

**Fig. 3.43.** (a) Coherence measure $C$ as a function of the noise intensity for different values of the control parameter $a$ of the logistic map: 1.2 (○); 1.22 (□); 1.23 (△). The *solid lines* shows the results of approximation by formula (3.182). (b) The dependences $h(D)$ (right scale, □) and $W(D)$ (left scale, ○) for $a = 1.23$

when one approaches the bifurcation point. Note that the optimal noise intensity $D_{opt}$ is shifted to large values when parameter $a$ moves away from the bifurcation point (see Fig. 3.43a). Numerical simulations have shown that the optimal noise intensity and the parameter of supercriticality are related linearly, i.e., $D_{opt} \propto \varepsilon$. A qualitatively similar behavior has also been observed for subsequent period-doubling bifurcations. Indeed, the same effect can be observed for other precursors, for example, near the Hopf bifurcation [220].

### 3.3.2 Coherence Resonance in Excitable Dynamics

**FitzHugh–Nagumo Dynamics.** An interesting system which exhibits coherence resonance is the FitzHugh–Nagumo (FHN) model. It is a simplified version of the well-known Hodgkin–Huxley neuron model, which accounts for the main essentials of the regenerative firing mechanisms in a nerve cell [223, 224]. On the other hand, this model is a representative of the so-called activator–inhibitor system. Its dynamics is given by

$$\varepsilon \, \dot{x} = x - x^3 - y, \qquad \dot{y} = \gamma x - y + b, \qquad (3.183)$$

where $x$ is the fast variable (activator or voltage variable), $y$ is the slow variable (inhibitor or recovery variable) and $\varepsilon$ is the ratio of the activator to inhibitor time scales. These equations can be re-written in an oscillatory form as

$$\varepsilon \, \ddot{x} = (1 - \varepsilon - 3\,x^2) \, \dot{x} + (1 - \gamma) \, x - x^3 - b. \qquad (3.184)$$

The FHN system possesses a rather complex bifurcation behavior with multiple fixed points and limit cycles. For our purposes we will focus on the simplest bifurcation of limit cycle birth. The parameters $b$ and $\gamma$ determine the position

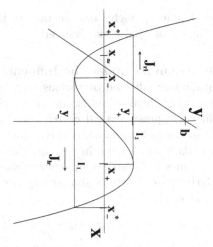

**Fig. 3.44.** Null clines of the system and particular points used in the calculations. The system has one stable fixed point on the left branch of the cubic null cline. The currents $J_l, J_r, J_{rl}$ and $J_{lr}$, used in the approximation, flow on the left and right side of the cubic function and on the straight lines $l_1$ and $l_2$, respectively

of the fixed points and a single fixed point is realized if $(\gamma-1)^3/27+b^2/4 > 0$. If $4\varepsilon\gamma > (\varepsilon + 1 - 3x_0^2)^2$ this fixed point at $x_0$ possesses complex eigenvalues and undergoes a Andronov–Hopf bifurcation if $3x_0^2 + \varepsilon - 1$ changes its sign.

The behavior of the FHN model can be understood in terms of the null clines of $x$ and $y$ (cf. Fig. 3.44), i.e., the cubic function in $x$ with maximum $P_{\max} = (x_+, y_+)$ and minimum $P_{\min} = (x_-, y_-)$ and the straight line in $y$. The former consists of three distinct regions, separated by $P_{\max}$ and $P_{\min}$: the stable left and right branches and the unstable middle branch. By choosing proper values of $\gamma$ and $b$, both null clines intersect only once on the left stable branch (excitable regime), providing a single fixed point in $(x_{\text{fix}}, y_{\text{fix}})$. In the original work by FitzHugh [223] this point corresponds to the resting state of the nerve cell, while points on the right branch belong to the excited state and the points largely above the fixed point on the left-hand side to a refractory state.

We note that the voltage variable possesses a much shorter time scale than the recovery variable ($\varepsilon \ll 1/\gamma$), i.e., the system is forced to relax quickly to the $x$-null cline. Since the middle branch is unstable, the motion of the system is restricted to a narrow region around the left and right branches and the two connecting lines between them.

The deterministic system started at an appropriate initial state (for instance with $y < y_-$) will make one long excursion in the phase space. First the trajectory quickly reaches the right branch, moving along this branch upwards until it reaches its top; afterward it switches to the left branch, where it relaxes into the fixed point (which takes an infinite time). Then in the voltage variable one "spike" or "pulse" is observed.

**FitzHugh–Nagumo Dynamics under the Influence of Noise.** In reality, neurons are permanently affected by various sources of different kinds of noise, e.g., the fluctuating opening and closing of ion channels within the membrane of the cell, noisy presynaptical currents and fluctuations of the distinct conductivities in the system, to name only a few. As a source of fluctuations we include additive white noise in the equation of $y$. The location of the noise in the dynamics is rather unimportant because the noise shifts the null clines in a relatively qualitatively similar way. We also introduce the adiabatically slow signal $s$.

$$\epsilon \dot{x} = x - x^3 - y + s, \qquad \dot{y} = \gamma x - y + b + \sqrt{2D}\,\xi(t). \qquad (3.185)$$

Later on we set, without loss of generality, $s = 0$. The sensitivity of the FHN model with respect to adiabatically slow signals in the current can be expressed through the dependence on $b$. Transformation of the recovery variable $y \rightarrow y - s$ (neglecting time derivatives of $s$) leads to a modified $b \rightarrow b - s$.

The fluctuations cause a sequence of stochastic excitations. The spike train and its properties have attracted much interest in recent studies on the constructive role of noise in nonlinear systems. For instance, if signals are input into the dynamics, SR [225–229] as well as synchronization with the input signals [162,231] can be observed. Bursting behavior, as in real neurons, takes place if additionally harmonic noise is implied in the system [232].

Similar to bistable dynamics the application of stationary noise in excitable systems evokes a new time scale, the mean excitation time of the occurrence of new spikes. Its value and properties depend significantly on the characteristics of the applied noise, and hence noise controls that time scale. On the other hand, the inherent recovery time of relaxation into the fixed point after firing does not vary much with the action of the noise.

Consequently, excitable systems driven by white noise exhibit the phenomenon of *coherence resonance* [219]. Intuitively, the neuron is unable to fire during the recovery state, and if it becomes excited at the same time scale nearly oscillatory behavior is observed. Since the excitation time depends strongly on the noise, an optimal noise level with respect to the regularity of the spike train exists.

**Quantification of Coherence Resonance.** With $D \neq 0$ and starting in the fixed point, fluctuations allow the system after a typical activation (excitation) time to overcome the inherent threshold at $(x_-, y_-)$ and – as in the deterministic case – to perform the excursion in the phase space, returning back to the vicinity of the fixed point. In the course of time a stochastic spike train of the voltage variable is generated (cf. Fig. 3.45), which on one hand can be characterized by the pulse rate measured by time averaging:

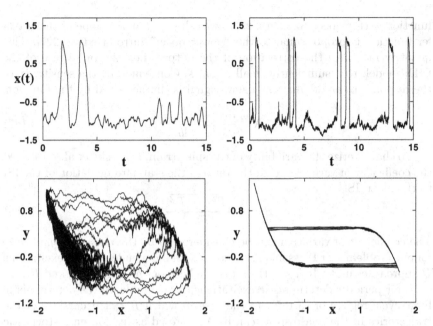

**Fig. 3.45.** Simulations of the stochastic differential equations. Voltage variable versus time and the corresponding trajectory in phase space for $\gamma = b = 1.5, D = 0.3$. Left: $\varepsilon = 10^{-1}$; right: $\varepsilon = 10^{-4}$

$$r = \lim_{T \to \infty} \frac{N}{T}. \tag{3.186}$$

On the other hand, it can be characterized by the mean time between two pulses (mean interspike interval (ISI)),

$$\langle T \rangle = \lim_{N \to \infty} \frac{1}{N} \sum_{i=1}^{N} T_i,$$

where $T_i$ is the time between the $i$th and $(i+1)$th spike.

Both measures give the same information, since in the long-term limit

$$r = \lim_{N \to \infty} \left( \frac{T_0}{N} + \frac{1}{N} \sum_{i=1}^{N} T_i + \frac{T_{N+1}}{N} \right)^{-1} = \frac{1}{\langle T \rangle}$$

with $T_0$ and $T_{N+1}$ being the time intervals until the first and after the last spike, respectively.

As shown above a slow signal $s$ decreases the value of the parameter $b$. Therefore, the dependence of the rates on $b$ can be thought of as a transfer function with respect to $s$. In [229] it was shown that the slope of the transfer function is proportional to the "power norm" introduced in [227]. This quantity measures the correlation of the output, i.e., the pulse rate of the FHN model, to a sufficiently small signal $s$. Consequently, the sensitivity of the response to an adiabatically slow signal is characterized by the function

$$\lambda = \frac{dr}{ds} = -\frac{dr}{db}. \tag{3.187}$$

To characterize the variability of the spike train, Pikovsky et al. [219] used the coefficient of variation, e.g., the ratio of the standard deviation of the ISI to the mean ISI:

$$R = \frac{\sqrt{\langle T^2 \rangle - \langle T \rangle^2}}{\langle T \rangle}. \tag{3.188}$$

The coefficient of variation can be considered to be the *noise-to-signal ratio* for our problem. For Poissonian sequences with independent single excitations $R$ approaches unity. If $R < 1$ the sequence becomes more regular, and $R$ vanishes for periodic deterministic excitations, for example, in the deterministic limit cycle regime of (3.185). In the case in which $R < 1$ excursions of the trajectories in the phase space can be interpreted as motion on a stochastic limit cycle [123, 210].

This effect has been studied numerically by different authors [219,233,234] and experimentally in an excitable electronic circuit [235]. In addition, numerical simulations with coupled neurons were performed, and the existence of optimally selected noise intensities and coupling coefficients were proven for a synchronously oscillating (ordered) response to the coupled network [123, 133, 236].

We would like to mention that the occurrence of ordered sequences of excitations could also be discussed with respect to spectral measures or correlation functions [210]. For our purposes the quantity (3.188) is advantageous since it requires the first two moments of the ISI distribution. Both moments will be available using the approach discussed below for arbitrary noise intensity.

**Fokker–Planck–Equation Analysis.** Since the FHN model is a nonpotential system [237, 238], even the analytical solution of the corresponding stationary FPE is still a nontrivial problem. A particular analytical solution was reported in [239], assuming a special relationship between the parameters of the applied noise and those of the dynamics. The authors of [240] proposed a perturbative approach to nonpotential systems in the bistable regime of an FHN model.

We consider the FHN model in the limit of a fast voltage variable (activator) and with a noise source in the recovery variable (inhibitor) [241]. In this limit, following a method by Melnikov [242], we find analytically the stationary marginal probability density of the nonpotential system and the rate of spikes in the voltage variable (pulse rate) [241].

The corresponding FPE for the probability density $P(x, y)$,

$$\partial_t P = -\frac{1}{\varepsilon}\partial_x(x - x^3 - y)P + \partial_y(y - \gamma x - b + D\partial_y)P, \qquad (3.189)$$

cannot be solved analytically, even in the stationary case. Also not available are the marginal densities

$$\rho(x) = \int\limits_{-\infty}^{+\infty} dy P(x, y)\,, \ p(y) = \int\limits_{-\infty}^{+\infty} dx P(x, y),$$

which reveal how an ensemble of independent neurons is distributed over the excited and resting states or the recovery variable, respectively.

However, all these functions may be achieved by time averaging of a simulation of (3.185). We have used a simple Euler procedure with time step $\Delta t$, which is two orders of magnitude smaller then $\varepsilon$, and shall compare the analytical findings with numerical results.

With a look at the simulations at $\varepsilon = 10^{-4}$ in Fig. 3.45 one notes that the dynamics becomes effectively one dimensional in the limit $\varepsilon \to 0$. In that case at least $p(y), \rho(x), r, dr/db, T$ and $R$ can be calculated analytically by solving two one-dimensional FPE connected by appropriate boundary conditions and showing a constant flux through the system.

In this limit the FHN system closely resembles a Schmitt trigger driven by an Ornstein–Uhlenbeck process (OUP) [19, 242]. The recovery variable $y$ replaces the input variable of a Schmitt trigger centered around $\gamma x(t) - b$ as can be seen in Fig. 3.46.

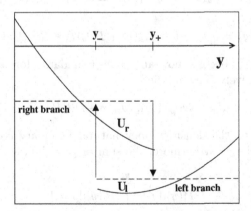

**Fig. 3.46.** Schematic of the system for $\varepsilon \to 0$. Depicted are the effective potentials $U_l(y), U_r(y)$ (*solid lines*) as functions of the slow inhibitor $y$ for $\gamma = b = 1.5$ and the absorption points $y_+, y_-$; the *dashed lines* indicate the two branches. Although the scheme resembles a Schmitt trigger, note that in this case the output depends on the slow variable $y$ and is not constant

On the other hand, there are fundamental differences between the excitable system and a Schmitt trigger. First and most importantly, the excited state is a dynamics without a fixed point. Its effective potential does not possess a minimum. Hence, the excited state is left even without noise. Second, the output is not a binary value. $x(y)$ depends strictly on the value of the slow variable. In addition, the dynamics at both branches are highly nonlinear.

**Probability Density.** For small $\varepsilon$ the voltage variable $x$ relaxes quickly toward one of the stable branches of the null cline $y = x - x^3$, where $x$ obeys the inverse of the cubic function on the left- or right-hand side, respectively,

$$x_{\mathrm{l}}(y) = 3y_- \cos\left(\frac{1}{3}\arccos(y/y_+)\right),$$

$$x_{\mathrm{r}}(y) = 3y_+ \cos\left(\frac{1}{3}\arccos(y/y_-)\right). \tag{3.190}$$

In the limit $\varepsilon \to 0$ the motion is restricted to these two lines. The two-dimensional Markovian system *separates into two one-dimensional subsystems* exchanging probability by currents $J_{\mathrm{rl}}, J_{\mathrm{lr}}$ infinitely quickly via the straight lines $l_1, l_2$ (see Fig. 3.44). In this limit there is no finite probability on these lines. Thus we obtain two coupled FPE; this was introduced for the first time by Melnikov [242], who considered the stochastic Schmitt trigger.

The FPE of the two systems do not only contain the usual drift and diffusion terms, but additional sources and sinks of probability, changing the probabilities at $y_+$ or $y_-$, respectively. They read

$$\partial_t P_{\mathrm{l}}(y) = \partial_y(y - b - \gamma x_{\mathrm{l}}(y) + D\partial_y)P_{\mathrm{l}} + J_{\mathrm{rl}}\delta(y - y_+)$$

$$\tag{3.191}$$

$$\partial_t P_{\mathrm{r}}(y) = \partial_y(y - b - \gamma x_{\mathrm{r}}(y) + D\partial_y)P_{\mathrm{r}} + J_{\mathrm{lr}}\delta(y - y_-),$$

and $P_{\min}$ and $P_{\max}$ become now absorbing boundaries for the left and right branches, respectively. This implies

$$J_{\mathrm{lr}} = D\partial_y P_{\mathrm{l}}(y), \quad J_{\mathrm{rl}} = -D\partial_y P_{\mathrm{r}}(y), \tag{3.192}$$

while in $y \to \pm\infty$ the densities obey natural boundary conditions on the respective branches. Furthermore, the sum of probabilities on both sides is conserved:

$$\int_{y_-}^{\infty} P_{\mathrm{l}}(y)\mathrm{d}y + \int_{-\infty}^{y_+} P_{\mathrm{r}}(y)\mathrm{d}y = 1. \tag{3.193}$$

In steady state the currents have to be constant and coincide with each other and with the pulse rate introduced in (3.186):

$$J_{\mathrm{lr}} = J_{\mathrm{rl}} = r. \tag{3.194}$$

With (3.186-187) one finds the solutions of the coupled FPE (3.191) [241]:

$$P_l(y) = \frac{r}{D} \exp\left(-U_l(y)/D\right) \int_{y_-}^{y} dz \exp\left(U_l(z)/D\right) \cdot \Theta(y_+ - z), \quad (3.195)$$

$$P_r(y) = \frac{r}{D} \exp\left(-U_r(y)/D\right) \int_{y}^{y_+} dz \exp\left(U_r(z)/D\right) \cdot \Theta(z - y_-), \quad (3.196)$$

with the effective potentials $U_l(y), U_r(y)$ explicitly given by

$$U_l(y) = \frac{(y-b)^2}{2} - \gamma\frac{x_l(y)}{4}[3y - x_l(y)],$$

$$U_r(y) = \frac{(y-b)^2}{2} - \gamma\frac{x_{rmr}(y)}{4}[3y - x_r(y)]. \quad (3.197)$$

Taking into account the change of the volume element, the marginal density $\rho(x)$ is given by $\rho(x) = p(y)\left|\frac{dy}{dx}\right|$. It does not exhibit any contribution between $x_{min}$ and $x_{max}$, since, as assumed in our approach, there is no probability on the straight lines between the branches. In contrast, simulations at finite $\varepsilon$ provide a small amount of probability within that range, which becomes comparably small by a decrease in $\varepsilon$ in the *logarithmic* plot (Fig. 3.47). The density around the maxima agrees fairly well with the numerical data.

**The Pulse Rate.** The pulse rate is obtained by (3.193) as follows [241]:

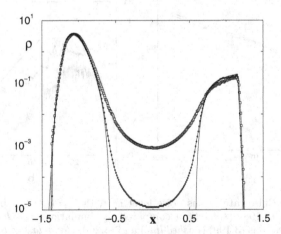

**Fig. 3.47.** Density $p(y)$, simulations at different values of $\varepsilon$ compared to the approximation. Log plot of $\rho(x)$, simulations at different values of $\varepsilon$ compared to the approximation. Parameters are $D = 0.1, \gamma = 1.5, b = 1.5$ and $\varepsilon = 10^{-3}$ (*squares*) and $\varepsilon = 10^{-5}$ (*circles*). Note that the right branch is due to noise-induced excursions

$$r = D \left[ \int\limits_{y_-}^{y_+} du \int\limits_{u}^{\infty} dv \ \exp \left( \frac{U_l(u) - U_l(v)}{D} \right) \right.$$

$$\left. + \int\limits_{y_-}^{y_+} du \int\limits_{-\infty}^{u} dv \ \exp \left( \frac{U_r(u) - U_r(v)}{D} \right) \right]^{-1} . \tag{3.198}$$

For extremely small $D$ this may be simplified to an Arrhenius-like shape:

$$r \approx U_l''(y_{\text{fix}}) \sqrt{\frac{\Delta U_l}{\pi D}} \exp \left( -\frac{\Delta U_l}{D} \right), \quad D \ll \Delta U_l. \tag{3.199}$$

For large $D$ one finds the rate obeying

$$r \approx \frac{\sqrt{2D}}{4y_+ \sqrt{\pi}}, \quad D \gg \Delta U_l. \tag{3.200}$$

The pulse rate and its dependencies on the noise intensity and on the slope of the second null cline $\gamma$ are compared to numeric simulations in Fig. 3.48. For sufficiently small $\varepsilon$ the approximation gives satisfactory agreement. At finite $\varepsilon$ one notices that (3.198) overestimates the rate. Two reasons are seen for these deviations: obviously the transition between the two branches takes a finite time; additionally, a finite $\varepsilon$ originates subthreshhold-oscillations, diminishing the rate as well.

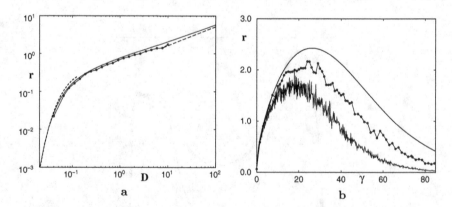

**Fig. 3.48.** (a) Pulse rate versus noise level $D$, with $\gamma = 0.8$, $b = 0.9$, $\varepsilon = 10^{-4}$. Approximation (3.198) (*solid line*) compared to simulations (*circles*) and to the simplified expressions (3.199) (*dashed line*) and (3.200) (*long-dashed line*). (b) Pulse rate versus the slope $\gamma$, while $D = 0.1$ and the fixed point is fixed to $x_{\text{fix}} = -0.8$, therefore $b = b(\gamma)$. Approximation (3.198) (*solid line*) compared to simulations with $\varepsilon = 10^{-4}$ (*thin solid line*) and $\varepsilon = 10^{-5}$ (*circles*)

**Aperiodic Stochastic Resonance in the FitzHugh–Nagumo Model.**
An increase in the parameter $b$ changes the position of the fixed point, inducing an enlargement of the distance to the threshold. Therefore, growing $b$ diminishes monotonously the pulse rate. This rate acts as a transfer function for adiabatically slow signals.

The slope of the transfer function $r$ at a finite $b$ is found by

$$\frac{dr}{db} =$$

$$-\frac{\int\limits_{y_-}^{y_+} du \left[\int\limits_u^\infty dv\,(v-u)\exp\left(\frac{U_l(u)-U_l(v)}{D}\right) + \int\limits_{-\infty}^u dv\,(v-u)\exp\left(\frac{U_r(u)-U_r(v)}{D}\right)\right]}{\left(\int\limits_{y_-}^{y_+} du \left[\int\limits_u^\infty dv\,\exp\left(\frac{U_l(u)-U_l(v)}{D}\right) + \int\limits_{-\infty}^u dv\,\exp\left(\frac{U_r(u)-U_r(v)}{D}\right)\right]\right)^2}.$$

and exhibits a nonmonotonous behavior with respect to $D$ as shown in Fig. 3.49. The latter is a fingerprint of *aperiodic stochastic resonance*. The slope is in a linear approximation proportional to the "power norm", taking account of the cross-correlation between an adiabatically slow signal $b = b_0 - b_1(t)$ and the output firing rate [229]. The value of the noise $D_{max}$ where the maximum occurs can be estimated as

$$D_{max} \approx 2(2 - \sqrt{3}) \cdot (\Delta U_l)^2 \qquad (3.201)$$

and is indicated in Fig. 3.49 by an arrow.

**Noise-to-Signal Ratio.** The mean time between two spikes, i.e., the mean ISI is given by the sum of the passage times from injection to absorption point on each branch. Therefore in estimating the time sequence of the spikes one deals with the classical mean first passage time (MFPT) problem and can use standard formulas [193].

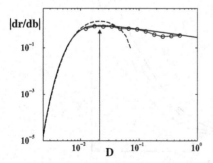 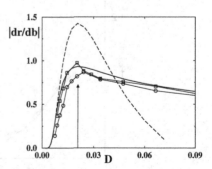

**Fig. 3.49.** Absolute value of the derivative of the pulse rate with respect to $b$ versus noise strength $D$ for $b = 0.7$ and $\gamma = 1.0$. Expression (3.201) (*solid line*) and and small noise approximation (*dashed line*) compared to simulations with $\varepsilon = 10^{-3}$ (*circles*). The *arrow* indicates the maximum calculated by (3.201)

The first and second moments on both branches are statistically independent and obey

$$\langle T_{\mathrm{l}}(y_+)\rangle = \frac{1}{D} \int\limits_{y_-}^{y_+} du \, \exp\left(\frac{U_{\mathrm{l}}(u)}{D}\right) \int\limits_{u}^{\infty} dv \, \exp\left(-\frac{U_{\mathrm{l}}(v)}{D}\right), \qquad (3.202)$$

$$\langle T_{\mathrm{l}}^2 \rangle = \frac{2}{D} \int\limits_{y_-}^{y_+} du \, \exp\left(\frac{U_{\mathrm{l}}(u)}{D}\right) \int\limits_{u}^{\infty} dv \, \exp\left(\frac{-U_{\mathrm{l}}(v)}{D}\right) \langle T_{\mathrm{l}}(v)\rangle, \quad (3.203)$$

and analogously for $\langle T_{\mathrm{l}}\rangle, \langle T_{\mathrm{l}}^2\rangle$ with appropriate integration boundaries for
the right branch and potentials according to (3.197). Equation (3.202) proves
again the equality of current (pulse rate) and the inverse of the sum of the
passage times, i.e., the mean ISI.

With (3.195-196) the noise-to-signal ratio (3.188) can be calculated as
follows:

$$R(D) = \frac{\sqrt{\langle T_{\mathrm{l}}^2 \rangle + \langle T_{\mathrm{r}}^2 \rangle - \langle T_{\mathrm{l}}\rangle^2 - \langle T_{\mathrm{r}}\rangle^2}}{\langle T_{\mathrm{l}} + T_{\mathrm{r}}\rangle}. \qquad (3.204)$$

This quantity exhibits a minimum with respect to the noise strength
(Fig. 3.50), indicating a coherent (i.e., most regular) spike train for a particular noise level. Results of the simulations, depicted in the same figure, confirm
the analytical findings, particularly the rough position of the maximum as
well as its depth. It is remarkable that a finite $\varepsilon$ deepens the minimum and
shifts it toward larger values of $D$. Because there is no doubt that the coherence is destroyed for larger values of $\varepsilon$, there should be a critical value of $\varepsilon$
for which $R$ becomes minimal.

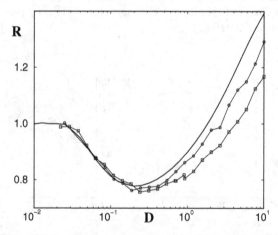

**Fig. 3.50.** Relative fluctuations versus noise strength $D$, with $\gamma = 0.8$ and $b = 0.9$.
Approximation (*solid line*) compared to simulations with $\varepsilon = 10^{-3}$ (*squares*) and
$\varepsilon = 10^{-4}$ (*circles*)

### 3.3.3 Noise-Enhanced Synchronization
### of Coupled Excitable Systems

In the previous section we described how the inherent oscillatory regimes of an excitable system can be controlled by externally applied noise: the coherence of inherent oscillations can be maximized at a nonzero noise level, while the mean frequency of oscillation (e.g., the mean firing rate) is a function of noise intensity. What would be the synchronization effects in a system of coupled excitable elements, each of which demonstrating the phenomenon of coherence resonance? First results have shown that coupled identical FHN elements exhibit collective oscillations, indicated by a well-developed peak in the spectrum [133]. In this section we describe mutual synchronization of locally coupled nonidentical FHN oscillators [236]. This discrete network of diffusely coupled oscillator mimics a noisy excitable media which is of high interest in biology, chemistry and physics. It is described by the following set of stochastic differential equations:

$$\varepsilon \dot{x}(t, n) = x - \frac{x^3}{3} - y + \gamma \sum_{n'} [x(t, n') - x(t, n)],$$

$$\dot{y}(t, n) \;\; = x + a(n) + \sqrt{2D}\, \xi(t, n), \qquad\qquad (3.205)$$

where $x(t, n)$ and $y(t, n)$ are again the fast and slow variables, respectively. For the one-dimensional case these variables are defined on a chain $n = 1, \ldots, N$, while in the two-dimensional case $x$ and $y$ are defined on a square lattice. The sum over the neighbors stands for the discrete Laplace operator in one and two dimensions, modeling the local interactions with coupling strength $\gamma$. The parameter $a(n)$ depends on the spatial variable $n$ and is assumed to be a uniformly distributed random variable. In this way we simulate a network of nonidentical FHN elements. Further, we assume stochastic forcing by Gaussian white noise $\xi$, statistically independent in space and with zero mean $\langle \xi(t, n)\, \xi(t + \tau, m) \rangle = \delta_{m,n}\, \delta(\tau)$.

The number of parameters in the model can be reduced by introducing the spacing of the lattice, $l$, and then scaling it as $l = \sqrt{\gamma}\, l_0$. Then the coupling factor in front of the Laplacian becomes one, but the noise intensity changes. As a result the effect of the noise and the dependence on the coupling strength can be discussed with regard to a common parameter $Q = D/\sqrt{\gamma^d}$, where $d$ equals 1 or 2 for the one- or two-dimensional case, respectively. For example, strong coupling decreases the action of the noise and the large noise case corresponds to the weak coupling limit. That is why in the following we fix $\gamma$ and use the noise intensity as a control parameter.

We expect that for coupling that is strong enough the firing events of particular elements will be synchronized. In our numerical simulations we fixed $\varepsilon = 0.01$ and $\gamma = 0.05$, while the activation parameters $a(n)$ are random numbers distributed uniformly on $[1.03, 1.1]$. This leads to a distribution of spiking times if noise is applied. We also use a free boundary and random initial conditions. In the absence of noise any initial state of the system evolves to an equilibrium state. Depending on noise intensity $D$, for a sufficiently large value of the coupling strength three basic types of space time behavior can be observed. For a small noise centers of excitation are nucleated very seldom in random positions in the medium, giving rise to propagating target waves. Collapse of such waves cannot exhibit stable spiral waves since the velocity of the waves at the intersection is always directed outside of the intersecting region. Therefore no new open spirals may occur. However, in the case of parametric noise the propagating fronts may locally backfire small directed spots which break propagating excitations and make spirals possible [243, 244]. In this case different cells in the medium are correlated only on the short time scale of the mean time of wave propagation, and there is no synchronization between distant cells.

For a large noise strength, the nucleation rate is very high and the medium is represented by stochastic firing cells. However, for an optimal noise intensity the medium becomes phase coherent: firings of different and distant cells occur almost in phase. Those three cases are shown in Fig. 3.51.

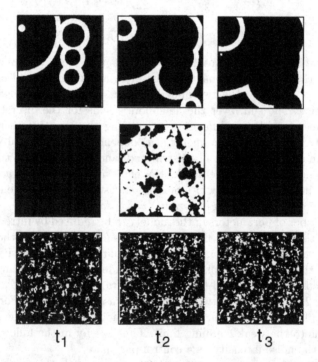

**Fig. 3.51.** Snapshots of a two-dimensional $200 \times 200$ system (3.205) for three moments in time $t_1 < t_2 < t_3$. The white color corresponds to the excited states. For the optimal noise intensity (second raw) the medium exhibits collective noise induced oscillations. First raw: $D = 1.1\,10^{-4}$, second raw: $D = 3.12\,10^{-4}$ and third raw: $D = 5\,10^{-3}$

At the optimal noise level the whole medium oscillates nearly periodically (see the middle row in Fig. 3.51). Finally, the case with large noise is represented by randomly flushing clusters. The same behavior has been observed in a model of the visual cortex [245].

To study this phenomenon of noise-induced global oscillations we describe this effect for the one-dimensional case in terms of phase synchronization. We introduce the instantaneous phase $\Phi(t, n)$ of the $n$th element using the analytic signal representation. We choose the central cell in the media ($n = N/2$) as a reference element and then calculate the phase differences $\phi(t, k) = \Phi(t, N/2) - \Phi(t, N/2 + k)$, $k = -N/2, \ldots, N/2$. Our numerical simulations have shown that for the optimal noise level the phases of different oscillators are locked during the time of computations. In the case of large distances between oscillators the phase fluctuations do indeed grow. Nevertheless, the phase difference is still bounded during long periods of time in a certain range. For non-optimal noise intensities a partial phase synchronization with randomly occurring phase slips can be observed only between neighboring elements. For larger distances the diffusion of the phase differences becomes very strong and synchronization breaks down.

In our case an appropriate measure of stochastic synchronization is the cross-diffusion coefficient defined as

$$D_{\text{eff}}(k) = \frac{1}{2} \frac{d}{dt} \left[ \langle \phi^2(t, k) \rangle - \langle \phi(t, k) \rangle^2 \right].$$    (3.206)

This quantity describes the spreading in time of an initial distribution of the phase difference between the $(N/2)$th element and all other elements. If this diffusion constant decreases, a longer phase-locking epoch appears and, therefore, phase synchronization becomes stronger. A single measure is obtained by averaging $D_{\text{eff}}(k)$ over the spatial distance:

$$D_{\text{eff}} = \frac{1}{N} \sum_{k=-N/2}^{N/2} D_{\text{eff}}(k).$$    (3.207)

The dependence of this averaged effective cross-diffusion constant versus noise intensity is shown in Fig. 3.52; it demonstrates a global minimum at a nonzero noise level. Thus, phase synchronization can be enhanced by tuning the noise intensity.

Synchronization is also defined as a frequency-locking effect. In the case of a stochastic excitable system one must use the mean frequencies $\langle \omega(n) \rangle = \langle \dot{\Phi}(t, n) \rangle$ of the oscillators [1]. Due to the given distribution of $a(n)$, the elements in the network have different randomly scattered frequencies for vanishing coupling. We have numerically built the distribution of the mean frequencies, calculated for every element across the network, $P(\langle \omega \rangle)$, for different noise intensities. The results are shown in Fig. 3.53. A remarkable effect of noise-enhanced space–time synchronization can be seen from this figure. For the optimal noise intensity, when the phases of different oscillators are locked for long periods of time, the mean frequencies are entrained

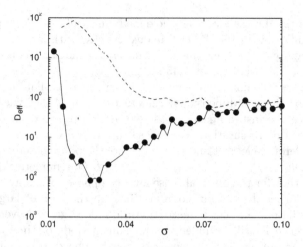

**Fig. 3.52.** The averaged effective cross-diffusion constant versus noise intensity. The *dashed line* corresponds to the uncoupled array ($\gamma = 0$)

and the distribution of the mean frequencies becomes extremely narrow. For nonoptimal noises the mean frequencies show rather wide distributions, indicating the lack of synchronization. Figure 3.53 clearly indicates noise-induced space–time ordering in the system based on synchronization mechanism. This behavior can be quantified further by calculating the mean square deviation, shown in Fig. 3.54, of the mean frequencies averaged over the network, which shows a deep minimum at the same optimal noise intensity as the effective diffusion constant [236].

The mechanism of noise-induced synchronization is rooted in the behavior of a single uncoupled element. The noise-induced oscillations are most coherent at a nonzero noise intensity, and the quality factor of the noise-induced peak in the power spectrum is maximal. In this regime the mean firing rate (or the mean frequency) of the system approaches the peak frequency in the power spectrum. In the case of weak noise the mean firing rate depends exponentially on the control parameter $a$ ($a > 1$). However, with the increase in noise the dependence of the mean frequency on $a$ becomes very weak. That is why with the increase in noise from a very low level the mismatch between characteristic frequencies of elements in the coupled array decreases, providing better conditions for mutual synchronization. On the other hand, the noise-induced oscillations becomes more coherent. These effects will tend to facilitate synchronization among elements in the network. Large noise, however, destroys again the coherence of local stochastic oscillations (the frequency and phase fluctuations grow rapidly) and also leads to the destruction of spatially coherent structures. The optimal noise intensity at which synchronization is most pronounced depends on the range of the

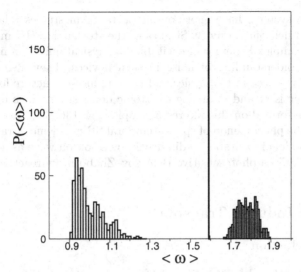

**Fig. 3.53.** Distribution of the mean frequencies of oscillators for the three values of noise. From left to right: $D = 1.1\,10^{-4}$, $D = 3.12\,10^{-4}$ and $D = 5\,10^{-3}$

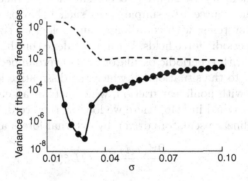

**Fig. 3.54.** Variance of the mean frequencies of oscillators for the indicated values of noise variance

distribution of activation parameters $a(n)$: with the increase in the range of disorder, the optimal noise intensity shifts towards smaller values.

### 3.3.4 Summary

We have presented analytical and numerical results for the FHN model driven by white noise [241]. In the limit of a fast voltage variable, we could calculate the stationary probability densities as well as the mean ISI and the variance of the ISI [219]. Analytical study has revealed that the pulse rate exhibits a maximum versus the slope of the recovery null cline and has proven the existence of a minimum of the coefficient of variation (noise-to-signal ratio).

Excitable systems have enormous importance in studies related to ion channels and neuronal activity. Studies of the stochastic FHN model which mimics the complex behavior of excitable biophysical units can nevertheless help in the understanding of noise in such devices. They also bridge the gap between research in biophysics and noise-induced structure formation in nonlinear chemistry and in analog electronic devices, for which experiments have been performed on this interesting topic [134–136, 138, 139].

In [122] the phenomenon of spatio-temporal SR was demonstrated numerically in a model of excitable media driven by a soliton wave as a signal and experimentally in a photosensitive Belousov–Zhabotinsky reaction.

## 3.4 Noise-Induced Transport

### 3.4.1 Introduction

Most mechanical and heat engines perform a directed motion as a periodic repetition of a temporal sequence in response to an external force or energy supply. This periodicity is required to permanently extract work from the applied forces and sources. In simple cases such behavior can be mapped onto mechanical systems, with coordinates and velocities characterizing the motion in given periodic force fields. With dependence on the value of friction one is left either with a periodic oscillator or its overdamped pendant.

Let us specify to the situations in which the dissipative force is linear. A counter example with nonlinear friction, which acts as an energy pump into the system, was studied in [246]. Here we look at the overdamped motion of conservative nonlinear oscillators driven by external forces and given by

$$\dot{x} = -\frac{\partial V(x,t)}{\partial x} + \sqrt{2D}\xi(t)\,. \tag{3.208}$$

The potential consists of two parts $V(x,y) = U(x) + W(x,t)$. The first is stationary and periodic, $U(x + L) = U(x)$, with one or more extrema in the period. The second describes the influence of the external driving scaling with $F_0$. In the case of an additive force $W = -F_0\, x\, y(t)$ with a given temporal variation $y(t)$ different from and generally uncorrelated with the Gaussian white noise $\xi(t)$.

The existence of directed motion in equilibrium is forbidden by the second law of thermodynamics if no bias is applied to the system, i.e., $F_0 = 0$. First Smoluchowski [247] and later on Feynman [248] pointed out that even rectifying configurations are unable to extract work from equilibrium fluctuations.

The situation qualitatively changes if $F_0$ is set different to zero in (3.208) and, therefore, perturbed externally and brought out of equilibrium. Then the existence of directed fluxes is general if translational invariance in space or in time is broken and as a result detailed balance is violated. Which symmetry

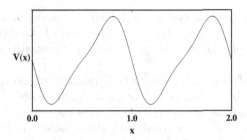

**Fig. 3.55.** Typical sawtooth-like potential $U(x)$ with period $L = 1$ and broken reflection symmetry.

classes of forces excite directed motion in dependence on the properties of the potential $U(x)$ is discussed intensely in [249].

We will consider stochastic problems with translational invariance in time. Then, noise and fluctuations achieve a fundamental role in two cases. First, the external driving is random in time, for example, an Ornstein–Uhlenbeck process (OUP) (see Sect. 1.2.3) with correlation time $\tau$ or a dichotomic process. Second, the driving force is too small in their finite-valued elongation, and the presence of white noise $\xi(t)$ makes it possible to escape over maxima of $U(x)$ only.

With the assumed stationary and symmetric noise [all odd moments of $y(t)$ vanish], a symmetry-breaking ingredient will be required to select a directed motion of Brownian particles. It will be the broken reflection symmetry of the periodic potential, i.e., $U(x) \neq U(-x)$ for arbitrary values $x$ over one period. This broken symmetry is usually introduced by the choice of a periodic but asymmetric potential, a so-called ratchet, washboard or sawtooth potential (see Fig. 3.55).

As will be shown in this section unbiased but nonthermal fluctuations and a broken reflection symmetry are sufficient prerequisites for the molecular transport mechanism. Such devices have recently attracted interest [250–281] in view of novel mass separation techniques [282–286] and of potential biological transport applications, such as muscle contraction [267,277] or the motion of actin on myosin [287], kinesin molecules on microtubuli [288], biotransport [287–296] and transmembrane transport [297–301]. In addition, there are reports on physical experiments proving that nonequilibrium fluctuations are able to induce directed motion [302–307]. A generalization of the considered effects to fluxes of magnetic phases in SQUIDs [308] or of chemical species in configurational spaces is possible, as was shown in [256]. It is worth mentioning that noise-induced directed current has implicitly been observed experimentally in one-dimensional organic conductors as early as in the late 1970s [309] via the occurrence of a finite stop-voltage.

Depending on the system under consideration the coordinate may be a continuous spatial axis or a discrete reaction coordinate. Transport in spatially continuous systems can be described by means of a Fokker–Planck equation (FPE) [67]. Alternatively, for the case of discrete events master equations are employed, for example, for birth-and-death processes defined by rates [310].

Within both frameworks the central quantity of interest, the stationary current, can be formulated rigorously. Additionally, the consideration of diffusion around the averaged motion gives more precise information. A lot of investigations are concerned with the possible efficiency of such devices [248, 311] which it will not be discussed here. We propose looking at the quality of the directed transport by the relation between conductive and convective transport expressed by the mean flux over diffusive motion over one period [312].

### 3.4.2 Flashing and Rocking Ratchets

One has to mention that already ancient water pumps have made use of situations with broken reflection symmetry. Nowadays ratchet-like devices with deterministic dynamics are widely used as rectifiers [313]. Here the new point under consideration are objects in ratchets if noise or fluctuations are applied. They are to be used to induce diffusional motion or to escape over existing barriers and model thermal noise, stochastic external forces or in general, stochastic changes of the reaction of the object in the ratchet.

At present two different stochastic prototypes of ratchets are subject to investigation [259, 261]. In the first prototype external temporal force fields are applied to mechanical oscillators with thermal noise. The fields vanish on average. Investigations evolve along the theory of stochastic mechanical oscillators including rate theory. This class of systems is the so-called rocking ratchets [255]. Additive forces are periodic or random with nonwhite or non-Gaussian correlations.

The second class is characterized by a switching between different profiles of the potential energy. They are called flashing ratchets and Brownian motion in this temporally changing potentials is considered. Such fluctuating or periodically changing potentials, for example, may be caused from externally driven chemical reactions which result in configurational changes of the considered molecules [254, 256]. In theoretical minimal models the temporal variation does not affect the periodicity of the potential. It is motivated because chemical reactions can modify the potential landscapes but as scalar processes they do not induce directed forces. In most biophysical applications flashing ratchets are considered as birth-and-death processes with temporally varying rates.

Not every ratchet can be integrated in that system. An early investigation by Büttiker [314] proved the existence of a particle flux if the phase of multiplicative noise in periodic systems is shifted relative to the periodic potential. In addition, instead of varying potentials, temperature or noise intensities might be varied from outside [315].

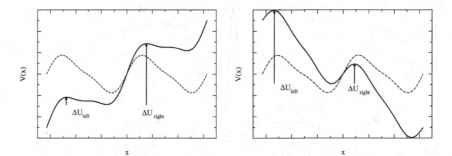

**Fig. 3.56.** Brownian motion in rocked ratchet. *Solid lines* show the potentials $V(x, t) = U(x) - F_0 y(t)$ for $y$-values with different sign. The *dashed lines* indicates the unperturbed $U(x)$ for comparison. *Arrows* indicate the different energetic barriers for a stochastic escape to the left or right, respectively. In the case of adiabatically slow driving $y(t)$ the mean transport is directed to the left

**Brownian Motion in Rocking Ratchets.** The main working principle of a rocking ratchet is easily understood and is illustrated in Fig. 3.56. The potential $U(x)$ for different $y(t)$ is turned by the part $W(x, t) = -F_0 y(t)$ around the origin of coordinates. If the maximal elongation is sufficiently small, the energetic barriers to the left and right do not disappear and without application of noise the particles are unable to move to the next minimum. With noise stochastic escapes over the remaining barriers are possible. Escape rates define the time scale of these stochastic motions to the neighboring minima of the potential.

As seen in Fig. 3.56 due to the broken reflection symmetry, the barriers have different magnitude despite a symmetric driving $y(t)$. The left barrier in the left plot is remarkably smaller than the right barrier in the right plot. Therefore, particles have a preferred direction and move with higher probability to the left.

The process is strictly noise induced in the case of dichotomic or periodic driving if the figures present maximal elongations of $F_0 y$. For vanishingly small additive noise the motion effectively disappears. For large intensities the difference between the barriers becomes insignificant. There exist medium noise values for which the differences amplified in the corresponding Arrhenius factors come into operation and an averaged motion to the left takes place.

Described here is the picture of overdamped motions when the driving changes at scales slowly compared to all other times. It should be mentioned that faster variations from outside or underdamped particles can exhibit a more complex behavior with one or more current reversals when noise is varied.

**Brownian Motion in Flashing Ratchets.** In the most simple case the temporal variation of the periodic potential consists in a switching on–off process of potential. As a continuous process this can be described by the Langevin equation

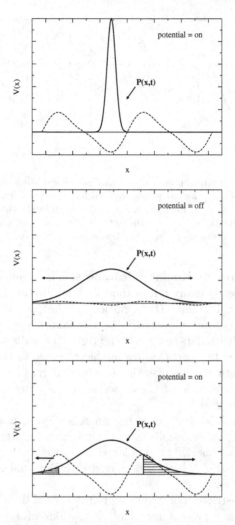

**Fig. 3.57.** Three subsequent steps of the evolution of probability density for the position in a flashing ratchet. Initially the density is concentrated near the minimum. If by external perturbation the potential is switched off the particle can freely diffuse. After switching on again differently sized portions (*hatched*) of the tails of the density go to the left and right neighbors, respectively. In the case of adiabatically slow switchings the mean transport is directed to the right. *Dashed lines* show the flashing potential

$$\dot{x} = -\frac{\partial}{\partial x} U(x)\,y(t) + \sqrt{2D}\,\xi(t), \tag{3.209}$$

and $y(t) = 1, 0$ is a dichotomic periodic or Markovian process. This situation is depicted in Fig. 3.57, and we draw a small amplitude of the "off" state for ease of comprehension.

The transport in the case of slow variation of $y(t)$ can be understood quickly as in the previous case. Being localized in the "on" state of the potential near a minimum of the potential, particles start to diffuse in the "off" state. After switching the potential due to the asymmetry "on" again a larger part of the probability is cut at the right tail compared to that at the left one. These hatched portions move to neighboring minima of the potential, inducing an effective flux to the right.

Again in the case of vanishing $D$ the particles will not diffuse and no net transport occurs. Also for a large noise the difference between the hatched regions is negligible. Hence, an optimal noise takes care of the asymmetry and realizes the effective flux.

On the other hand, for finite noise $D \neq 0$, slow switchings let the particles diffuse over long regions and the asymmetry does not count. Similarly, no transport is realized if the switchings are very fast. The particles simply have no time to freely diffuse. Therefore there exists a moderate switching time which exhibits a maximal current.

### 3.4.3 The Adiabatic Approach

Analytical results for mean values of fluxes can be given in discrete approximations, as will be shown below. In the case of a continuous coordinate, solutions are available if the driving $y(t)$ is either slowly or quickly changing [255, 259], for piece-wise linear dynamics [256, 260] and if the ratchet is forced by white shot noise [316, 317]. Here the first situation will be presented for rocked ratchets with an additive force in concrete considerations. Slowly varying forces imply that the system is able to adiabatically follow the variations of the external perturbation. Then in the asymptotic limit $t \to \infty$ analytic expressions can be given. The analysis follows modified approaches as given in Sects. 1.2.6 and 1.3.4.

The FPE corresponding to (3.208) reads

$$\frac{\partial}{\partial t} P(x,t) = -\frac{\partial}{\partial x}\left(-\frac{\partial V(x,t)}{\partial x} P(x,t)\right) + D \frac{\partial^2}{\partial x^2} P(x,t). \qquad (3.210)$$

From the periodicity of $U(x) = U(x + L)$ it follows that the asymptotic density $P_{\mathrm{asy}}(x,t)$ will be periodic as well. Then (3.210) with $\partial P_{\mathrm{asy}}/\partial t = 0$ has to be solved only in the interval $[0, L]$ with periodic boundary conditions. Its solution implicitly depends on time via $y(t)$.

The expression under the derivative of the right-hand side of (3.210),

$$J(x,t) = -\frac{\partial V(x,t)}{\partial x} P(x,t) - D \frac{\partial}{\partial x} P(x,t), \qquad (3.211)$$

can be interpreted as the instantaneous probability flux, which must be independent of $x$ in the asymptotic limit $J(x,t) \to J_{\mathrm{asy}}(t)$. Multiplication of (3.210) by $x$ and integration over the support gives the mean velocity

$$\langle \dot{x}(t) \rangle_L = \int\limits_0^L J(x,t)\,\mathrm{d}x \tag{3.212}$$

giving asymptotically $\langle \dot{x} \rangle_L = L J_{\mathrm{asy}}(t)$. As in problems of stochastic resonance for periodic driving the mean velocity has to be averaged over one period. In contrast, for stochastic stationary $y(t)$ the asymptotic values are time independent but should be averaged over the stationary density of the driving.

The asymptotic solution of the FPE (3.210) valid for slow variation of $y(t)$ reads [1]

$$P_{\mathrm{asy}}(x,y) = P^0(x,y)\left( C(\dot{y}) - \frac{J_{\mathrm{asy}}(y)}{D} \int\limits_0^x \mathrm{d}x' \,\frac{1}{P^0(x',y)} \right) \tag{3.213}$$

with $P^0(x,y) = \exp[-\Phi(x,y)]$ where $\Phi(x,t) = V(x,y)/D$. The flux $J_{\mathrm{asy}}(y)$ and $C(y)$ are adiabatically taken integration constants dependent on time via the temporal driving $y$. One relation between them follows from the periodic boundary condition:

$$J_{\mathrm{asy}}(y) = C(y)\,D\left( 1 - \frac{P^0(0,y)}{P^0(L,y)} \right)\left( \int_0^L \mathrm{d}x\,\frac{1}{P^0(x,y)} \right)^{-1}. \tag{3.214}$$

Taking into account normalization $C(y)$ is

$$C(y) = \left( \int_0^L \mathrm{d}x\,P^0(x,y) \int_x^{x+L} \mathrm{d}x'\,\frac{1}{P^0(x',y)} \right)^{-1} \int_0^L \mathrm{d}x\,\frac{1}{P^0(x,y)}. \tag{3.215}$$

Insertion gives after some straightforward manipulations the mean asymptotic velocity [255]:

$$\frac{\langle \dot{x}(y) \rangle_L}{L\,D} = \left( 1 - \frac{P(0,y)}{P(L,y)} \right)\left( \int_0^L \mathrm{d}x\,P^0(x,y) \int_x^{x+L} \mathrm{d}x'\,\frac{1}{P^0(x',y)} \right)^{-1}. \tag{3.216}$$

This velocity still has to be averaged over the external force:

$$\langle \langle \dot{x} \rangle_L \rangle_y = \int \mathrm{d}y\,\langle \dot{x}(y) \rangle_L\,\rho(y), \tag{3.217}$$

with the given distribution $\rho(y)$. Explicitly in the case of periodic driving with $y(t) = A\cos(\omega t + \varphi_0)$, it yields

$$\langle \langle \dot{x} \rangle_L \rangle_y = \frac{1}{2\pi} \int_0^{2\pi} \mathrm{d}\varphi_0\,\langle \dot{x}(A\cos(\omega t + \varphi_0)) \rangle_L, \tag{3.218}$$

which is not dependent on $\omega$ as a result of the adiabatic approach.

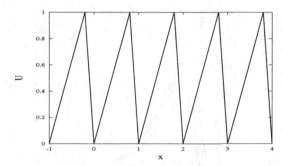

**Fig. 3.58.** Typical sketch of the piece-wise linear potential

In the application of this approach we study a piece-wise linear dynamics [130]:

$$f(x) = -\frac{\partial}{\partial x} U(x) = \begin{cases} a_1 & \text{for } nL < x < nL + x_1, \\ -a_2 & \text{for } nL - x_2 < x < nL, \end{cases} \tag{3.219}$$

driven by a periodic additive force with amplitude $A$ and small frequency. The potential is shown in Fig. 3.58. $U_0 = a_1 x_1 = a_2 x_2$ is the height of the potential barrier, $x_1 + x_2 = L$. In the small amplitude case

$$\max y(t) \ll \frac{a_1 a_2}{a_2 - a_1} \min\left(1, \frac{D}{U_0}\right), \tag{3.220}$$

one finds approximately the averaged velocity [318]

$$\overline{\langle \dot{x} \rangle} \approx \frac{U_0^2(a_2 - a_1)A^2}{8D^2 a_1 a_2 \sinh^2(U_0/2D)} \left(\frac{U_0^2}{4D^2 \sinh^2(U_0/2D)} + \frac{U_0}{2D \tanh(U_0/2D)} - 2\right) \tag{3.221}$$

which vanishes for large noise and in the absence of noise.

Examples of $\overline{\langle \dot{x} \rangle}/A^2$ on $D/U_0$ approximated by evaluations of the corresponding integrals [318] are shown in Fig. 3.59 for different $A$. For $A < \min(a_1, a_2)$ and $A > \min(a_1, a_2)$ the physical situation changes. In the first case a maximum is observed with respect to $D/U_0$ which becomes smaller for increasing amplitudes $A$. In the second case the normalized mean velocity tends to a certain finite value as $D/U_0 \to 0$, which was calculated within the theory of noise-free vibration transport [281, 313]. For $A < 0.5$ the dependences found are almost coincident with small amplitude approximation (3.221).

One has to remark that periodically rocked ratchets were solved numerically in [255] by using the matrix continued fraction (MCF) technique. For low frequencies the numerical results coincide with the adiabatic approximation. Otherwise, in the case with high frequencies current reversals arise which cannot be obtained in the adiabatic theory.

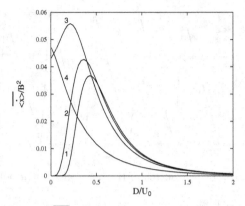

**Fig. 3.59.** Mean velocity $\overline{\langle \dot{x} \rangle}/A^2$ on $D/U_0$ in the adiabatic theory. $a_1 = 1.25$, $a_2 = 5$, $x_1 = 0.8$, $x_2 = 0.2$, and $A = 0.1$ (*curve 1*), $A = 1$ (*curve 2*), $A = 2$ (*curve 3*) and $A = 5$ (*curve 4*) [130]

### 3.4.4 The Overdamped Correlation Ratchet

In most applications in biophysics the particle moves in a highly viscous medium. Therefore, the overdamped situation has general importance. In [250] it was proposed that an OUP with finite correlation time serving as additive force can induce directed motion of Brownian particles in a ratchet potential. An analysis of this noise-induced motion was given in [266]. Here some results should be summarized.

Let us look at

$$\dot{x} = \frac{1}{\gamma}\left(-U'(x) + F + y(t) + \sqrt{2D}\xi(t)\right), \tag{3.222}$$

giving the overdamped dynamics for a Brownian particle subject to high friction $\gamma$.

Therein $y(t)$ stands for the exponentially correlated random force

$$\langle y(t)y(t')\rangle = \frac{Q}{\tau}\exp\left(-\frac{|t-t'|}{\tau}\right). \tag{3.223}$$

$\tau$ is the correlation time and $Q$ its intensity (see Sect. 1.2.3. We are interested in the mean velocity $\langle \dot{x} \rangle$ or the steady-state probability current $J_0 = \langle \dot{x} \rangle/L$, and its dependence on the noise parameters, the loading force and the friction. Unfortunately, the corresponding FPE for the Markovian dynamics in the two-dimensional phase space $x, y$ cannot be solved explicitly.

Equations (3.215-216) can be treated by way of computer simulation using a fast algorithm developed by Fox et al. [149,319,320]. A more advanced technique in determining the flux by solving high-dimensional FPE was developed by Risken, Jung and Bartussek using MCF [67,321]. In that semianalytical approach the stationary nonequilibrium density and the flux is expanded in a set of appropriate eigenfunctions. It yields an infinitely large algebraic system

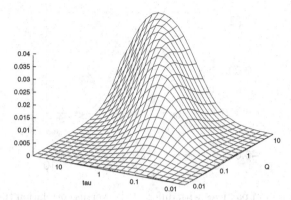

**Fig. 3.60.** Particle current versus correlation time and noise strength $Q$. Parameters: $\gamma = 1.0$ and $D = 0.1$ [320]

of equations for the amplitudes of the eigenfunctions which converges best if represented as MCF.

Results [320] of this approach using about 50 eigenfunctions show the influence of the noise parameters $Q$ and $\tau$. Let first $F$ be zero and $\gamma = 1$ and we choose the ratchet potential as in Fig. 3.55:

$$U(x) = \frac{1}{(2\pi)} \left[ \sin(2\pi x) + 0.25 \sin(4\pi x) \right]. \qquad (3.224)$$

The dependence of the mean velocity on the noise strength and the correlation time is given in Fig. 3.60. It possesses maxima with respect to $Q$ and $\tau$. The current is zero for vanishing (white noise) and infinite correlation time if $\langle y^2 \rangle = 0$. Taking $\tau$ to be fixed yields a maximum in $Q$ since vanishing as well as infinitely large amplitudes destroy the weak ratchet effect. Current reversal does not appear for this simple potential but was shown for more complex potentials with three Fourier modes [266] or in the case of more complicated driving [154, 265, 269].

A particularly appealing feature of Brownian motors is their ability to separate particles of differing friction strength or mass [261]. The overdamped model allows a discussion of the influence on friction. Variation of friction models different sizes of particles by Stokes law. With increasing $\gamma$ the mobility decreases as well as the impact of the white noise. The first reduces the mean velocity; the second lessens the velocity in most parts of the parameter regime.

A possible separation of particles with different size [322] includes an additional constant force $F \neq 0$ in (3.222). This force should be directed against the preferred direction of the ratchet. For finite correlation times the ratchet effect overwhelms the effect of the force and the particles move "uphill". The correlation times will be different for different friction, which means that large particles still follow the bias, whereas the smaller ones move uphill. MCF results confirm these expectations. In a region around $\tau = 0.1$,

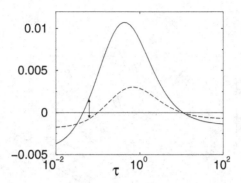

**Fig. 3.61.** Current of particles with different size versus correlation time for a finite bias. Parameters: $D = 0.1$, $Q = 0.2$, $F = -0.005$, $\gamma_1 = 1.0$ and $\gamma_2 = 2.0$

indicated by the arrow in Fig. 3.61, the weak currents of two particles have different size.

### 3.4.5 Particle Separation by Ratchets Driven by Colored Noise

A more detailed discussion about properties of particles can be gained from studies taking care of mass dependence of the flux. Investigations on such devices including dynamical inertial effects can be found in a few studies, only, considered in [322–327].

As will be seen, inertial ratchets possess current reversals as a function of the mass of the particle. That is why these ratchets are ideally suited to separate particles of differing masses, and thus allow for the conceptual operation of molecular shuttles [325]. Therein an inertial Brownian carrier is able to move back and forth massive cargo along pre-assigned routes.

To start, the model for the inertia ratchet which was studied in [327] is presented. Let $x$ and $v$ denote the space coordinate and velocity of the particle of mass $m$, with $\gamma$ being the viscous friction strength. The stochastic dynamics of the ratchet then reads

$$\dot{x} = v, \quad m\dot{v} = -\gamma v - U'(x) + F + y(t) + \sqrt{2D}\xi_1(t). \qquad (3.225)$$

Here, $y(t)$ is the colored, unbiased noise (OUP) and models stochastic non-equilibrium forces and $\xi_1(t)$ accounts for thermal fluctuations of strength $D$. The scaling

$$\tilde{t} = \frac{t}{t_0} \; , \quad \tilde{x} = \frac{x}{x_0} \; , \quad \tilde{v} = \frac{v t_0}{x_0} \; , \quad \tilde{y} = \frac{y V_0}{x_0} \; , \quad V(\tilde{x}) = \frac{U(x)}{V_0} \; , \quad \tilde{F} = \frac{x_0 F}{V_0}$$

leads to a dimensionless formulation of the dynamics in a potential $V$ with $V(\tilde{x}) = V(\tilde{x} + 1)$. We choose $t_0 = \gamma x_0^2 / V_0$ to obtain a dimensionless friction coefficient equal to one. Rescaled mass and noise parameters are given as

$$\mu = \frac{mV_0}{x_0^2\gamma^2} \ , \ \ \tilde{D} = \frac{D}{V_0\gamma} \ , \ \ \tilde{Q} = \frac{Q}{V_0\gamma} \ , \ \ \tilde{\tau} = \frac{\tau V_0}{\gamma x_0^2}.$$

The new dimensionless dynamics reads

$$\dot{x} = v, \ \ \mu\dot{v} = -v - V'(x) + F + y(t) + \sqrt{2D}\xi_1(t), \ \ \dot{y} = -\frac{y}{\tau} + \frac{\sqrt{2Q}}{\tau}\xi_2(t); \quad (3.226)$$

tildes are omitted here and hereafter.

The quantity of foremost interest is the mean velocity $\langle v \rangle$ or the steady-state probability current of immersed Brownian particles. We are interested in its dependencies on the noise parameters, but particularly on the particle mass, and present results for the potential (3.224).

The three-dimensional Markovian dynamics can be treated by several methods. An adiabatic approach was performed in [326] and first contributions to an expansion $\propto 1/\gamma$ with respect to high friction were given in [325]. It can be studied by means of analytical approximation schemes such as a unified colored noise approximation (UCNA) or a path integral approach [327]. Numerically, it can be investigated either by direct computer simulations of (3.226) or by applying again the MCF method to the corresponding FPE.

**Unified Colored Noise Approximation.** The objective in the UCNA is to formulate an approximate Markovian description of a generally intractable non-Markovian dynamics [328]. This approximation has been developed for overdamped stochastic dynamics driven by OUP [329] in a common way valid for small and large correlation times $\tau$. Refined and generalized approaches put forward in [149, 330] were applied in [154, 266] to colored-noise-driven directed transport.

First, a nonlinear coordinate transformation of $x, v, y$ to linearly uncouple stochastic variables is performed. In a second step, a Markovian description for the coordinate $x$ only is achieved via separation of time scales for these new variables, admitting the adiabatic elimination of the "fast" ones.

Adapting this general line to (3.226) yields a single Langevin equation with white noise sources. One finds expressions for small correlation times $\tau$ and, simultaneously, for a strongly overdamped dynamics $\mu \to \infty$. In the Stratonovich interpretation the following was found [320]:

$$\dot{x} = \frac{1}{g(x)}\left(-V'(x) + F + \sqrt{2(Q + D)}\,\xi(t)\right), \quad (3.227)$$

where the state- and mass-dependent function $g(x)$ reads

$$g(x) = 1 + \frac{d}{dx}\frac{\tau Q\left[V'(x) - F\right]}{(D + Q)(1 + \mu/\tau) + \tau V''(x)}. \quad (3.228)$$

Afterward, the steady-state probability current $J$ can be calculated following standard approaches [1]. With $\langle v \rangle = \langle \dot{x} \rangle = J$ we find

$$\langle \dot{x} \rangle = \frac{L\,(Q+D)\,\left[1 - \exp\left(\Phi(1)/(Q+D)\right)\right]}{\int\limits_0^1 \mathrm{d}x\, g(x)\, \exp\left(-\Phi(x)/(Q+D)\right) \int\limits_x^{x+1} \mathrm{d}y\, g(y)\, \exp\left(\Phi(y)/(Q+D)\right)}.$$

(3.229)

In this expression the effective potential

$$\Phi(x) = \int_0^x g(y)\,[V'(y) - F]\,\mathrm{d}y \qquad (3.230)$$

occurs. In the white noise limit $\tau \to 0$ it follows from (3.228) that $g(x) \to 1$. The current $\langle \dot{x} \rangle_{\tau=0}$ thus vanishes accordingly to [309, 331] if $F \neq 0$, independent of the mass $\mu$. Otherwise, that is for $0 < \tau < \infty$, the current is generically nonzero for nonsymmetric potentials $V(x)$, even for $F = 0$. The asymptotic behavior of (3.229) for small $\tau$ and zero load $F = 0$ is obtained as

$$\langle \dot{x} \rangle = -\frac{\hat{\tau}^2\,Q}{A(0)(D+Q)^2} \int_0^1 V'(y)V''(y)^2\,\mathrm{d}y, \qquad (3.231)$$

$$A(F) = \int\limits_0^1 \mathrm{d}x \int\limits_x^{x+1} \mathrm{d}y\, \exp\left([V(y) - V(x) + (x-y)F]/(D+Q)\right). \quad (3.232)$$

The $\mu$ dependence of the UCNA result (3.229) can be completely absorbed into the renormalized correlation time $\hat{\tau} = \tau/(1 + \mu/\tau)$.

A comparison of the UCNA-results with results of numeric simulations of the corresponding Langevin equations 3.225 is made in Fig. 3.62 for different $\mu$ values. The maximum of the approximation shifts in agreement with the simulations towards larger $\tau$ with increasing mass. The maximal value of the current in the UCNA is unaffected by mass due to the mentioned dependence through $\hat{\tau}$.

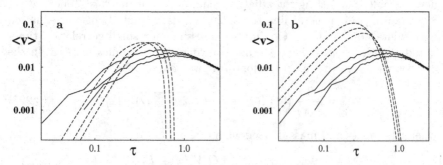

**Fig. 3.62.** Comparison of (**a**) the UCNA and (**b**) the path-integral method with numeric simulations. *Solid lines* are from simulations of the Langevin-equations (3.225). *Dashed lines* correspond to the analytic approximations **a** (3.232) and **b** (3.235)

**Path Integral Approach for Inertia Ratchets.** As in quantum mechanics a reformulation of stochastic dynamics yields a compact representation by path integrals [332]. Within the restriction of weak thermal noise the current of the Brownian particles can be approximated by a rate description:

$$\langle \dot{x} \rangle = k_+ - k_-, \tag{3.233}$$

where $k_+$ ($k_-$) is the noise-activated hopping rate to the next right (left) neighboring well with an Arrhenius-like dependence $k_\pm = \zeta_\pm \exp - \Delta\Phi_\pm/D$. The $\Delta\Phi_\pm$ stands for effective barriers independent of $D$, with $x^{\#}$ being the position of the local maximum of $V(x) - xF$, and $x_+$ and $x_- = x_+ - 1$ the locations of the corresponding neighboring minima to the right and left, respectively. In the case of small $\tau$ [333] the explicit results read [327]

$$\Delta\Phi_\pm(\tau) = \Delta\Phi_\pm^{(0)} + \Delta\Phi_\pm^{(1)}(\tau) \tag{3.234}$$

$$= \frac{V(x^{\#}) - V(x_\pm) + (x_\pm - x^{\#})F}{1 + R} + \tau^2 \frac{R}{(1 + R)^2} \int_{-\infty}^{\infty} \ddot{q}_\pm^2(t)\, dt$$

and $R = Q/D$. The functions $q_\pm(t)$ are saddle-point trajectories found from the noise-free dynamics $\mu\ddot{q}_\pm(t) = -\dot{q}_\pm(t) - V'(q_\pm(t)) + F$ with initial conditions at the maximum with vanishing velocity and ending in one of the possible minima as $t \to \infty$.

The prefactors $\zeta_\pm$ can be approximated in the zeroth order $\zeta(\tau) \simeq \zeta(\tau = 0)$ with the effective noise $(D + Q)$ [334]. We thus infer that

$$\langle \dot{x} \rangle = k_+^0 \left[ \exp\left( - \Delta\Phi_+^{(1)}/D \right) - \exp\left( - [\Delta\Phi_-^{(1)} + LF/(1 + R)]/D \right) \right], \tag{3.235}$$

where $k_+^0 = k_+(\tau = 0)$, which corresponds to escape over the barrier $\Delta\Phi_+^{(0)}$ with two white noise sources [cf. (1.213)].

It is interesting to note that $\Delta\Phi_+^{(1)} - \Delta\Phi_-^{(1)}$ can change its sign depending on $\mu$ in the low-damping regime. Therefore, (3.235) predicts a reversal of the current direction in the underdamped case ($\mu \gg 1$), as will be confirmed by the MCF analysis. However, the reversal occurs in the energy-diffusion-limited regime, where (3.235) fails qualitatively. In this case long excursions over multiple barriers appear which are not taken into account by (3.233). Their occurrence is given in good approximation by the rates $k_\pm$ and depends on the asymmetry of the ratchet potential; it might serve as an indicator for a reversal of the current.

The path integral prediction (3.235) was compared with numerical simulations (see Fig. 3.62). Apart from its absolute value, which is better estimated by the UCNA, the mean flux converges satisfactorily up to about $\tau = 0.5$. The shift with respect to $\tau$ and the change of the maximum with increasing inertia $\mu$ are qualitatively described.

**Matrix Continued Fraction Analysis.** The MCF analysis starts by expanding the steady-state density $P(x, v, y)$ in Hermité-functions along $v$ and $y$, and Fourier modes in the coordinate $x$, and solves algebraic equations for

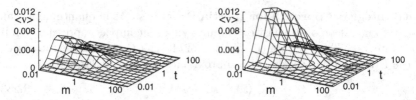

**Fig. 3.63.** Currents from the MCF analysis with respect to mass and correlation time: (a) $Q/D < 1$ and (b) $Q/D > 1$ where $Q$ and $D$ are the intensities of the stochastic external force (OUP) or of the white noise in 3.225

the amplitudes. Precise values for currents calculated for two different values of $Q$ are depicted in Fig. 3.63 with the intensity $Q$ chosen smaller and larger than $D$.

Finite inertia causes a complex behavior of the current as can be deduced from all following figures. The current vanishes for $\mu \to \infty$ as well as for small and large $\tau$. Starting with small values of $\mu$ one observes in a certain range of $\tau$ a novel unexpected increase in the current with respect to $\mu$. A global maximum with respect to both $\mu$ and $\tau$ appears for finite mass not predicted by the analytical approximations.

A further increase ($\mu \approx 0.1$) results in a rapid decrease in the current. The maximum with respect to $\tau$ for a given $\mu$ shifts toward larger values, in agreement with the UCNA and the path integral approximation. This drop of the current is understandable. Fluctuations and the forces have a weaker effect on a larger mass, increasing the mass the particle motion is slowing down; this can be compensated by a larger correlation time $\tau$, which explains the shift of the maximum.

In the strongly underdamped case at arbitrary $R$, negative velocities appear for moderate $\tau$. For moderate values of $\mu$ a double reversal of the velocity is obtained with respect to the correlation time Fig. (3.64a). Negative currents are observed for a finite interval of $\tau$. For larger $\mu$ the current exhibits

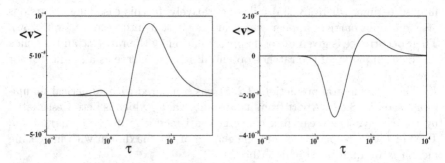

**Fig. 3.64.** Current reversals with respect to correlation time $\tau$. (a) Moderate values of $\mu$ with multiple current reversals, (b): heavy particles with a single reversal

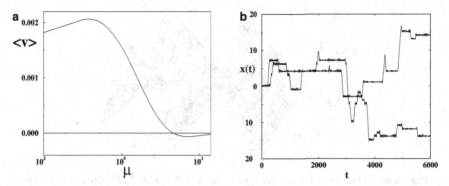

**Fig. 3.65.** Currents reversal for underdamped particles from (a) MCF analysis and (b) simulations in the underdamped regime (see text)

a single reversal only (Fig. 3.64b). But one sees in Fig. 3.65 immediately that the absolute values of the reversed currents are rather small compared to the maximal current for small masses.

The current reversal in the underdamped case was verified by simulations [327]. A time span for two individual trajectories demonstrates the effect of strong inertia (Fig. 3.65b). One can generally distinguish between "running states" and "locking states" in the underdamped case. But, as was checked numerically and supported by the path integrals, the reason for the reversal arises from the asymmetry of the potential. It is expressed by the inverted difference of rates $k_+$ and $k_-$ for transitions from the locked state into the left- or right-running state.

### 3.4.6 Two-Dimensional Ratchets

The majority of theoretical ratchets models address one-dimensional periodic structures. In these devices the time-dependent forces act in parallel to the motion of the induced flows. On the other hand, most of experiments on transport in sawtooth-like potentials address two-dimensional devices. For example, a directed motion of particles was observed in the 2D "Christmas tree"-like structure of obstacles in [302, 303, 307].

The motivation for constructing such systems comes from the need for an effective technique to separate macromolecules [283–285, 335, 336]. It has been found that 2D systems with constant forces in the presence of obstacles with a ratchet-like geometry can excite a transport orthogonally to the applied force. Moreover, if the shape of the obstacles is symmetric in the direction of the applied force the induced flux is an even function of the driving force [336]. This feature makes the phenomenon even more interesting for practical applications since the flux does not depend from the direction of the force and hence also periodic forcing can be applied [335].

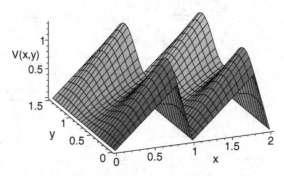

**Fig. 3.66.** The 2D potential $V(x, y)$ with $A = 0.3$, $L = 1$, $F = 0$ and $k = 0.8$

**Noise-Induced Motion Transverse to the Applied Force.** We look at the following physical situation [337]: The symmetry of the 2D sawtooth-like potential is broken in the $x$-direction. A constant bias acts in the $y$-direction and drives the system into nonequilibrium. Periodic boundary conditions are applied, in both $x$ and $y$.

A potential $V(x, y)$ which fulfills our requirements in $x \in [0, 1]$, $y \in [0, L]$ is shown in Fig. 3.66. Explicitly we will use in calculations

$$V(x, y) = V_0(x, y) - Fy = \{1 - A\sin(2\pi y/L)\}U(x) - Fy. \qquad (3.236)$$

$U(x)$ is a 1D piecewise linear ratchet potential of unit height with the parameter of asymmetry $k \in (0, 1)$ if $k = 1/2$ then $U(x)$ is symmetric. $U(x)$ is periodically modulated in the $y$-direction with an amplitude $A$. The linear term stands for the constant bias which drives the system out of equilibrium.

The dynamics is given by the dimensionless overdamped Langevin equations,

$$\dot{x} = f_x(x, y) + \sqrt{2D}\xi_x(t), \qquad \dot{y} = f_y(x, y) + \sqrt{2D}\xi_y(t), \qquad (3.237)$$

and the Gaussian white noise $\xi_i$ is independent in both components. $f_i(x, y)$ are the force fields generated by the potential (3.236). From these equations the 2D Smoluchowski equation for the evolution of the probability density $p(x, y, t)$ can be immediately formulated.

Using the finite element method (FEM) [337,338] the Smoluchowski equation was numerically solved in the asymptotic case with periodic boundary conditions $p^s(x, y) = p^0(x + 1, y)$, $p^s(x, y) = p^s(x, y + L)$. The stationary density $p^S(x, y)$ contains all of the relevant information for determining the stationary flows in the system, namely

$$J_x(x, y) = f_x(x, y)p^s(x, y) - D_x\frac{\partial}{\partial x}p^s(x, y),$$

$$J_y(x, y) = f_y(x, y)P^0(x, y) - D_y\frac{\partial}{\partial y}P^0(x, y). \qquad (3.238)$$

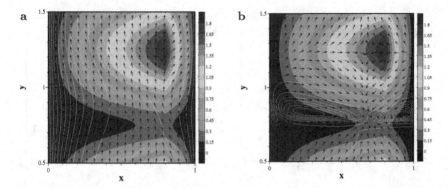

**Fig. 3.67.** Vector field of the probability current $[J_x(x, y), J_y(x, y)]$. *Arrows* and *lines* show the direction of the flow, which at the background is a potential without tilt. **(a)** Isotropic diffusion with $D = 0.2$, $A = 0.8$, $L = 1$, $F = 5$ and $k = 0.8$. **(b)** Complex local fluxes in a system with $D_x = 0.3$ and $D_y = 0.01$. $F = 1$, $L = 1$, $k = 0.8$ and $A = 0.8$

The total transport over one period in the $x$-direction can be quantified by determining the average velocity $\langle v_x \rangle$. Taking into account the normalization condition, the average velocity is expressed by the total current via

$$\langle v_x \rangle = J_x^{\text{tot}} = \int_0^1 \int_0^L J_x(x, y) \mathrm{d}x \, \mathrm{d}y. \tag{3.239}$$

Nonzero tilt $F \neq 0$ induces mean flows in the system. As well as the expected flow in the $y$-direction, a current appears in the $x$-direction. The vector field of the probability flux density is presented in Fig. 3.67. Directions of the flow on the entire plane $[0, 1] \times [0, L]$ are drawn by arrows of unit length. One sees that the $x$-component of the local current adopts positive and negative values.

The main contribution to the total current comes from the flow through the saddle point of the potential ($x = 0.8$ and $y = 0.75$). Globally a flux to the left side dominates. Fig. 3.68a presents the averaged velocity. Due to the symmetry of $V_0(x, y)$ in the $y$-direction the flow does not change replacing $F \to -F$. The current $J_x^{\text{tot}}(F)$ is $O(\propto F^2)$ for small $F$ [336]. Increasing $F$ induces a negative flow, and $J_x^{\text{tot}}$ reaches its maximal absolute value at $F \simeq 5.5$. It is near the critical value of the force, above which the potential monotonously decays in the $y$-direction. Larger values of $F$ suppress the effect.

Monte-Carlo simulations of the Langevin equations (3.237) have been included in Fig. 3.68a and confirm the findings. Good agreement with independent numerical sources ensures the validity of the algorithms used. However,

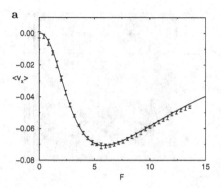
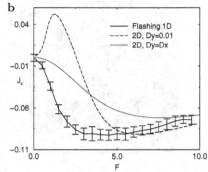

**Fig. 3.68.** (a) Total flux in the $x$-direction versus driving force $F$. Error bars depict the results of Monte-Carlo simulations. (b) Comparison of the 1D flashing ratchet with the 2D ratchet with symmetric and nonsymmetric diffusion constant. The remaining parameters are $D_x = 0.3$, $k = 0.8$ and $A = 0.8$

it was underlined in [337] that the Monte-Carlo simulations are in general less precise and require much more computing power compared to the FEM solution of the Smoluchowski equation.

**Comparison with One-Dimensional Flashing Ratchets.** The considered 2D ratchet system can be interpreted as an extension of the concept of the 1D flashing ratchet. The particles moving in the $y$-direction underly a periodic modulation in the same way as a particle in a periodically flashing 1D potential. This correspondence becomes exact if the dynamics of the second Langevin equation (3.237) can be replaced by $\dot{y} = F$.

The situation can be modeled by introducing anisotropic noise in the $x$- and $y$-directions, $D_x$ and $D_y$, respectively. In Fig. 3.68b three situations with such kind of noise are compared: the 1D flashing ratchet $D_y = 0$ and the 2D ratchet with isotropic diffusion ($D_x = D_y$) and with anisotropic diffusion ($D_x \gg D_y = 0.01$).

First, one sees for forces larger than a threshold ($F \simeq 5$) the values of the currents become similar. If $F$ is of the order of 10 then both $D_y = 0.3$ and $D_y = 0.01$ are negligible and a situation similar to that for a periodically flashing potential is realized.

However, for smaller $F$ the behavior of the 2D systems changes and differs from the simple model. The 2D ratchet with isotropic diffusion gives a smaller current than the flashing one. It is due to the fact that the motion of the particle in the $y$-direction is significantly disturbed by forces coming from the potential $V_0$, and a regime with a constant mean velocity along $y$ is left. This motion is hindered by high barriers, and particles can move forwardly in $y$ only, escaping barriers by thermal activation.

In the anisotropic case, $D_x \gg D_y$, the behavior is even more complex. An unexpected phenomenon takes place: for small values of $F$ the total current becomes positive, which is qualitatively different from the behavior of the

1D flashing ratchet. In this situation the local probability current indicates a complicated dynamics of the Brownian particles (see Fig. 3.68b). The current reversal is accompanied with the occurrence of circulating flows. The vector field now has multiple singularities.

We underline the fact that such circulations appear only in the case of anisotropic diffusion. They persist for vanishing forces $F$ but without total current. Oppositely, higher values of $F$ remove this complex pattern of the local current and the flow becomes laminar again.

### 3.4.7 Discrete Ratchets

This section presents results concerning directed transport with discrete jumps. Especially biophysical situations with particle numbers and the reduction of the dynamics of macromolecules or ion channels to a small set of discrete configurations require such consideration. We have included in this section basic models which can be generalized to applied situations.

The goal is to study a simple ratchet given by specific rates of transition $W_\sigma(i \to i \pm 1)$ for the discrete events. The suffix $\sigma = \pm 1$ stands for external temporal driving. As will be seen, even basic systems allow the discussion of interesting phenomena such as the relation between directed transport and diffusion, disordered ratchets and coupled ratchets with the ability to structure formation.

**The Minimal Discrete Flashing Ratchet.** A broken reflection symmetry with zero flux in equilibrium requires at least three states over one period [339]. Subject to periodic boundary conditions the mass action law is obeyed if multiplication over forward rates equals the product of backward rates. Then, in the case of two states obviously symmetric rates obey the mass action law and periodicity only. Hence, ratchet models with two states include external forces and motion even in equilibrium [268].

Let us assume that the transition rates between the different states $i = 1, 2, 3$ can be represented as exponentials and a common prefactor scales time. One obtains periodicity and broken reflection symmetry if two rates of forward jumps $i \to i + 1$ are smaller than 1; in contrast the remaining is larger than 1. It will be sufficient to take

$$W_\sigma(1 \to 2) = W_\sigma(2 \to 3) = k_\sigma \tag{3.240}$$

and

$$W_\sigma(3 \to 4) = k_\sigma^{-2} = 1/[W_\sigma(1 \to 2) \times W_\sigma(2 \to 3)], \tag{3.241}$$

and by periodicity state "4" equals state "1". The backward jumps will be defined as the inverse

$$W_\sigma(i \to i - 1) = \frac{1}{W_\sigma(i - 1 \to i)}. \tag{3.242}$$

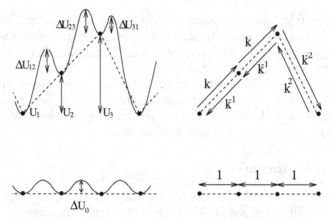

**Fig. 3.69.** The two alternating potentials $\sigma = +$ (upper) and $\sigma = -$ (lower) together with periodic boundary conditions $(1 \leftrightarrow 3)$ motivate the selected choice of transition rates

Flashing or rocking of the ratchet implies a temporal driving expressed by temporal variations of the rates. Here, specifying to the simplest flashing mode, $\sigma = 1$ corresponds to potential "on" and $\sigma = -1$ to potential "off". One may assume two alternating set of rates (cf. Fig. 3.69) between the system switches with rate $\gamma$.

With the notation used, $\sigma = 1$ stands for rates $k < 1$, and for $\sigma = -1$ we set simply $k = 1$. In this situation simple diffusion remains. Hence, we are left with two basic constants – $k$ and the switching rate $\gamma$ – setting in the following the period $L = 1$.

With periodic boundary conditions a six-state model, as introduced in [312, 339], is a minimal discrete flashing ratchet. Stochastic switchings between the two potentials are modeled by assuming $\sigma(t)$ is a dichotomic process specified by the switching rate $\gamma$. The evolution of the time-dependent probability $P_\sigma(j, t)$ to meet a single particle at position $j$ experiencing the potential $\sigma$ is defined by the following master equation:

$$\frac{\partial P_\sigma(j, t)}{\partial t} = W_\sigma(j+1 \to j) P_\sigma(j+1, t) + W_\sigma(j-1 \to j) P_\sigma(j-1, t)$$
$$- \left[ W_\sigma(j \to j+1) + W_\sigma(j \to j-1) \right] P_\sigma(j, t) - \gamma P_\sigma(j, t) + \gamma P_{-\sigma}(j, t).$$
$$(3.243)$$

The transition rates $W_\sigma(i \to j)$ can be represented by considering two alternating potentials, as shown in Fig. 3.69, and defining

$$k_\sigma = \exp\left( -\frac{\sigma(t)+1}{2} \, \Delta U/D \right). \tag{3.244}$$

The product of rates thus transforms to the sum over exponents of the piece-wise linear segments, which should be zero. Setting the potential in the switched-off case to zero is justified by the scaling of time.

One may even discuss the origin of nonequilibrium with the aid of this figure [263]. Assumption of constant rates $\gamma$ between different potential landscapes independently of the value of the potential at different $i$ serves as a source of nonequilibrium. Equilibrium would require specific rates $\gamma_i$ with Arrhenius-like dependence.

The master equation (3.243) for the six states can be cast in form of a matrix equation:

$$\dot{P} = WP, \tag{3.245}$$

with the vector $P$ comprising the six states and $W$ being the transition matrix. As can be easily seen by summing all the lines of the matrix, the rank of $W$ is less than six. It expresses conservation of probability. However, the normalization condition provides the independent sixth equation.

The stationary solution of (3.245) with $\dot{P} = 0$ can thus be achieved analytically by solving linear algebraic equations. The current $J(i,t)$ between states $i$ and $i+1$ reads

$$
\begin{aligned}
J(i,t) = \; & [\, W_{+1}(i \to i+1)\, P_{+1}(i,t) - W_{+1}(i+1 \to i)\, P_{+1}(i+1,t)\,] \\
& + [\, W_{-1}(i \to i+1)\, P_{-1}(i,t) - W_{-1}(i+1 \to i)\, P_{-1}(i+1,t)\,] \,.
\end{aligned}
\tag{3.246}
$$

Of course, the stationary current $J_0$ if $t \to \infty$ does not depend on the location, i.e., $J_0 = J(1) = J(2) = J(3)$. It is depicted in Fig. 3.70 versus the switching

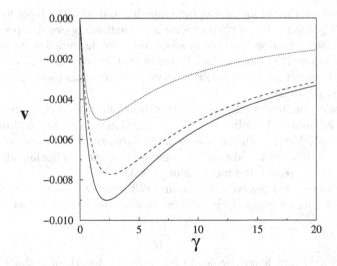

**Fig. 3.70.** The stationary current for varying switching rate $\gamma$ and for three different transition rates: $k = 0.1$ (*dashed line*), $k = 0.2$ (*solid line*) and $k = 0.4$ (*dotted line*); note that the maximum means vanishing current, whereas the minimum around $\omega \approx 2$ and $k \approx 0.2$ corresponds to a maximum current directed to the left

rate $\gamma$ and on the transition rate $k$. The nonmonotonic structure of the curves becomes transparent when considering the limits. For a vanishing switching rate the statistical problem corresponds to two uncoupled ensembles with the two different potentials, respectively. Since in every ensemble no net transport occurs due to detailed balance there is, evidently, no flux taken the both ensembles together. Alternatively, for very high switching rates the current tends towards zero because then the time spent in the switched-off state $\sigma = -1$ is much to small to reach a neighboring barrier by diffusion. In between there exists an optimal switching rate, with which diffusion to the right minimum but not to the left one is most effective.

Setting the rate to $k = 1$ means choosing a flat potential in the state $\sigma = +1$ as well. However, then the spatial asymmetry and, as a consequence, also the current vanish. The case for which $k \to 0$ corresponds to raising the barriers in the switched-on state beyond all limits. The concentration of the distribution in the minimum cannot be overcome by diffusion; hence the current vanishes. Again, in between there exists an optimal transition rate.

**Envelope Description of Discrete Ratchets.** When considering a chain of identical ratchets one may ask for the evolution of some initially sharp distribution. On a coarse-grained scale one would expect the distribution envelope to drift with the mean velocity in the direction determined by the asymmetry of the potential and at the same time to broaden due to diffusion [312, 323, 340].

The observation of an enveloping Gaussian was also the basis for an approximating description applied to a dichotomic flashing overdamped ratchet [272]. The same approach has been adapted to the discrete 3-state model introduced above [312], having the advantage that velocity and diffusion coefficient can be analytically related to system parameters as given in a different context in [341, 342].

Diffusion counteracts the desired transport of a molecular system, which requires the cargo to be delivered at its destination reliably, i.e., within a small time interval. Large diffusion means large variance of the times of arrival, i.e., low reliability. In addition, in separation devices diffusion affects the efficiency of the separation mechanism [284, 335].

The competition between drift $v$ and diffusion $D$ in advection–diffusion problems is often expressed by a dimensionless number, the Peclet number:

$$Pe = \frac{|v|L}{D} . \tag{3.247}$$

Here $L$ is a typical length scale, in our case the length of a single ratchet element. The larger the Peclet number the more net drift predominates over diffusion. The break-even point, beyond which drift wins over diffusion, defines a critical length $L_c = 2D/v$. Demanding this length to be not larger than the length of a ratchet unit $a$ requires Peclet numbers not smaller than 2.

A systematic way to account for this envelope evolution starts from a gradient expansion ansatz:

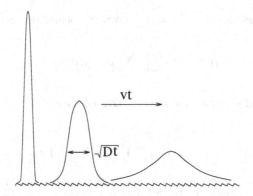

**Fig. 3.71.** Envelope approximation of Brownian motion in a ratchet

$$P_\sigma(x,t) = \sum_{n=0}^{\infty} p_\sigma^n(x)\partial_x^n \mathcal{P}(x/\lambda,t) . \tag{3.248}$$

Here, the time-independent functions $p_\sigma^n(x)$ reflect the periodicity of the potentials, whereas the function $\mathcal{P}(x/\lambda,t)$ describes the envelope which evolves in time on some coarse-grained scale $\lambda \gg L$ (see Fig. 3.71). The refined structure of the exact solution which possesses sharp bends is accounted for by the higher derivatives of the envelope $\partial_x^n \mathcal{P}(x/\lambda,t)$.

The envelope evolution on the coarse-grained scale $\lambda$ is reflected by an effective FPE:

$$\partial_t \mathcal{P}(x/\lambda,t) = \partial_x(v - D\partial_x)\mathcal{P}(x/\lambda,t) , \tag{3.249}$$

which is valid only in an approximated sense in powers of $1/\lambda$. The parameters $v$ and $D$ are the drift velocity and diffusion coefficient of the spreading Gaussian. Their relation to the outgoing dynamics is made by inserting the ansatz (3.248) into the basic dynamical equation, the master equation (3.243).

This connection can be obtained by the following expansion:

$$P(x,\sigma,t) = \sum_{n=0}^{\infty} p^{(n)}(x,\sigma)\partial_x^n \mathcal{P}(x/\lambda,t) . \tag{3.250}$$

It consists of an infinite set of periodic functions $p^{(n)}$ of period $L$. Small $n$ terms describe the smooth components, and, hence, one obtains only the first functions $p^{(0)}$, $p^{(1)}$ and $p^{(2)}$ to be involved in the envelope description. Indeed, equating terms of order $1/\lambda^n$ yields

$$[W + \Gamma]p^{(0)} = 0, \tag{3.251}$$

$$[W + \Gamma]p^{(1)} = -(v - \hat{V})p^{(0)}, \tag{3.252}$$

$$[W + \Gamma]p^{(2)} = (D - \hat{T})p^{(0)} - (v - \hat{V})p^{(1)}, \tag{3.253}$$

with $\hat{V}$ and $\hat{T}$ being two operators (matrices) involving rates $w(n \to m,\sigma)$ and $\gamma$ (see [312]). Equation (3.251) reveals that $p^{(0)}$ is the stationary solution of (3.243). Taking traces and obeying correct normalization,

$$\text{Tr}\{p^{(n)}\} = \sum_{\sigma=-1}^{+1} \sum_{i=1}^{3} p^{(n)}(i,\sigma) = \delta_{n,0}, \tag{3.254}$$

yields the desired relations for the effective coefficients:

$$v = \text{Tr}\{\hat{V}\, p^{(0)}\} \tag{3.255}$$
$$D = \text{Tr}\{\hat{T}\, p^{(0)}\} - \text{Tr}\{\hat{V}\, p^{(1)}\}\,. \tag{3.256}$$

The fact that the function $p^{(2)}$ is not involved becomes transparent when taking the traces of (3.244-246). Due to the norm-conserving property of the operator $[W + \Gamma]$ traces of the left sides vanish identically. The functions $p^{(0)}$ and $p^{(1)}$ are obtained by solving (3.251) and (3.252), respectively. All manipulations necessary to yield (3.255) and (3.256) can be done in the framework of linear algebra. They can be performed by an algebraic computer program. The resulting expressions are rather long and cannot be simplified. However, they still grant the benefit of being analytically exact and explicit. In particular, exact limits can be evaluated.

**Diffusion versus Drift.** Without going into the details of the analogous calculations we cite the final expressions for the velocity $v$ and the diffusion coefficient $D$ [312]:

$$v = \frac{1}{3} \sum_{\sigma=\pm} \sum_{i=1,2,3} [W_\sigma(i \to i+1) - W_\sigma(i \to i-1)] p_\sigma^0(i)\,, \tag{3.257}$$

$$D = \frac{1}{6} \sum_{\sigma=\pm} \sum_{i=1,2,3} [W_\sigma(i \to i+1) + W_\sigma(i \to i-1)] p_\sigma^0(i)$$
$$- \frac{1}{3} \sum_{\sigma=\pm} \sum_{i=1,2,3} [W_\sigma(i \to i+1) - W_\sigma(i \to i-1)] p_\sigma^1(i)\,. \tag{3.258}$$

We note that the velocity $v$ is identical to the stationary current $J_0$ as could be expected. The stationary solution $p_\sigma^0(i)$ as well as $p_\sigma^1(i)$ can be achieved by solving two linear equation systems. The necessary algebra was done employing a computer algebra program (maple). The explicit formula for $v(k,\gamma)$ and $D(k,\gamma)$ comprise an abundance of terms. Nevertheless, the value of this model system lies in its analytical accessibility.

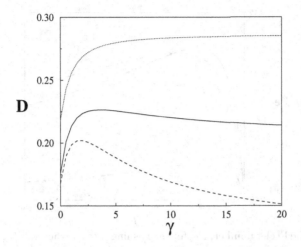

**Fig. 3.72.** The diffusion coefficient of the envelope as a function of the switching rate $\gamma$ for three different transition rates: $k = 0.1$ (*dashed line*), $k = 0.2$ (*solid line*), $k = 0.4$ (*dotted line*); note that the stationary current and the diffusion coefficient simultaneously peak for $\gamma \approx 2$

In addition to the dependence of the velocity on the switching rate $\gamma$ depicted in Fig. 3.70 we plot the diffusion constant $D(k, \gamma)$ for the same range of $\gamma$ and parameters $k$ in Fig. 3.72. It can be seen that the diffusion coefficient becomes independent of $\gamma$ for $k = 1$, which is obvious since this corresponds to the case of flat potentials in each of the two alternating cases. There its value is $D(k = 1, \gamma) = 1/3$. With increasing $\gamma$ the diffusion becomes more and more restricted because the frequently switched-on potential always contracts the distribution in the minimum. For very small $k$, i.e., rather steep slopes, a local maximum for nonvanishing $\gamma$ occurs. When viewing this result in the context of an effective separation device the maximum current condition for $k \approx 0.2$ and $\gamma \approx 3$ unfortunately is connected with rather high diffusion; in fact, the maximum absolute value of $v$ and the relative maximum of $D$ nearly coincide.

In Fig. 3.73 we depict the related Peclet numbers, $Pe$. We see that Peclet numbers never reach values of the order of 1. This basic example clearly demonstrates that diffusion effects are far from negligible. We mention that the same qualitative result was found in the analysis of a discrete rocking ratchet by formulating appropriate rates in the master equation.

Therefore as a general feature it is found that maximal drift is linked with rather high diffusion [312]. The limitation by diffusion imposes on the transport efficiency can be considered quantitatively in terms of the dimensionless Peclet number.

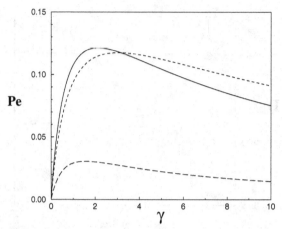

**Fig. 3.73.** The Peclet number, $Pe$, for the flashing ratchet when varying the flashing rate $\gamma$ for three different asymmetry parameters: $k = 0.1$ (*dashed line*), $k = 0.2$ (*solid line*) and $k = 0.5$ (*long-dashed line*)

### 3.4.8 Sawtooth-like Media

The previous section was devoted to a single sawtooth with a periodic boundary condition. Obviously, such kinds of ratchet models describe also stationary fluxes in potential landscapes consisting of many sawteeth which flip coherently between the two states. At any place on the landscape the potential is either switched "on" or it is switched "off". Hence, giving a single sawtooth an individual number of its switching state $\sigma_i$, where $i$ is the position of a single sawtooth; the previous model assumes $\sigma_i = \sigma_j$ for arbitrary $i, j$. We call this situation a correlated ratchet.

Crossing now from a single ratchet to a chain consisting of many sawteeth, it is obviously not the exclusive situation for the switching possibilities. One possible way to generalize the problem is the inclusion of failures in the periodic arrangement of the chain of sawteeth [272]. Another approach supposes different spatial correlations between the states of the neighboring sawteeth [339]. Possible limiting cases are

(i)   correlated: $\sigma_{i+1} = \sigma_i$ $(\langle \sigma_{i+1} \sigma_i \rangle = 1)$;
(ii)  anticorrelated: $\sigma_{i+1} = -\sigma_i$ $(\langle \sigma_{i+1} \sigma_i \rangle = -1)$;
(iii) uncorrelated: $\langle \sigma_{i+1} \sigma_i \rangle = 0$.

The angular brackets denote averaging over the dichotomic noise. Case (i) corresponds to the coherent switching of all segments as formerly discussed. In the anticorrelated case (ii) again the chain adopts two states, only. Indeed, the potential has a period of two sawteeth. In the first state, for example, the left sawtooth is switched "on" and the right one is "off". Respectively, in the second state the right one is "on" and the left one is "off".

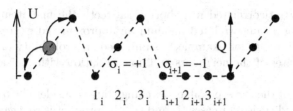

**Fig. 3.74.** In the uncorrelated ratchet every sawtooth $i$ switches "on" ($\sigma_i = +1$) and "off" ($\sigma_i = -1$) independently of its neighbors. The points should hint schematically at the three energy levels of the discrete model

Compared with former investigations a qualitatively new situation is the uncorrelated case (iii). The switching of each sawtooth is independent from its adjacent sawteeth, as illustrated in Fig. 3.74.

The analysis of the new situations is quite similar, as described below [339]. It can be reduced to the treatment of a single sawtooth. It is exact for the anticorrelated case. For the uncorrelated case we make a mean field assumption. The new point of the analysis consists in a proper formulation of the boundary conditions which connect the left-hand side of the sawtooth with the right-hand side and an adopted solution of the stationary master equation. That solution now depends on the selected switching rules by the corresponding states $\sigma_{i\pm1}$ of the neighboring sawteeth.

For the anticorrelated choice we have to apply antisymmetric boundary conditions. If the $i$th sawtooth is "on", the neighbor is "off" and vice versa. It means that the probability fluxes at $\sigma_i = 1$ have to be linked with those of $\sigma_{i\pm1} = -1$. It gives a simple picture for one sawtooth, the outwards pointing fluxes are the inputting fluxes at the other end of the sawtooth but into the state with different $\sigma_i$.

The balance of probability fluxes can be formulated in the same way for the uncorrelated case. But now these probability fluxes are random. They depend on the present state of the adjacent sawteeth. Let $J(\sigma_i - 1, \sigma_i)$ label the four possible fluxes between states $n = 3$ for the $(i - 1)$th sawtooth and $n = 1$ of the $i$th one and pointing from $i - 1 \to i$. These $J(\sigma_i - 1, \sigma_i)$ are given by

$$J(\sigma_{i-1}, \sigma_i) = W(3 \to 1, \sigma_{i-1})P^{\bar{\sigma}_{i-1}}(3, \sigma_{i-1}) - W(1 \to 3, \sigma_{i-1})P^{\bar{\sigma}_i}(1, \sigma_i), \tag{3.259}$$

with $P^{\bar{\sigma}_i}(n, \sigma_i)$ being the probability function for the state $n, \sigma_i$ of the $i$th element. It depends on the given environment $\bar{\sigma}_i = \{\sigma_{i-1}, \sigma_{i+1}\}$.

Therewith a general formulation of the boundary condition can be given as

$$J(\sigma_{i-1}, \sigma_i) = \frac{1}{2}(1 + \sigma_{i-1}\sigma_i)J(\sigma_i, \sigma_i) + \frac{1}{2}(1 - \sigma_{i-1}\sigma_i)J(-\sigma_i, \sigma_i); \tag{3.260}$$

for the general formulation at the right side between sawteeth with numbers $i$ and $i + 1$. In the case of correlated elements the second item vanishes; in contrast, for anticorrelated neighboring sawteeth the first item disappears. In the case of an uncorrelated medium an approximated expression can be found by averaging, for example, assuming equal probability of both "on"– and "off"–states of the neighbors and no correlation with the sawtooth considered.

Insertion of the three possible switching states considered closes the set of boundary conditions. For cases (i) and (ii) the product is replaced by $+1$ and $-1$, respectively. In both cases the dependence on $\bar{\sigma}_i$ is formal and defined by the present state $\sigma_i$. In the uncorrelated state we average over the possible states of the adjacent sawteeth. The product $\sigma_{i-1}\sigma_i$ vanishes in the average, and the dependence on $\bar{\sigma}_i$ is lost.

Figure 3.75 shows the stationary fluxes of the three cases for a given value of $k = 0.2$ [339]. For all switching rates $\gamma$ the absolute value of the flux for uncorrelated and anticorrelated situations increases compared to the usually considered correlated case. This remains valid qualitatively for all relevant values $0 < k < 1$. The larger flux results from the following: In the correlated case a diffusive transport of a particle between different segments takes place in the "off" case only. The particle has to stay at least the time $\tau = \gamma^{-1}$ of the "on" state within one segment. Hence, the transport is interrupted. On the contrary, in the other cases a transport might take place at any time. In the limiting anticorrelated case the particle is able to reach the next but one segment during two switching times $\tau$.

A stronger enhancement of the particle flux is observed if the sign of the potential is changed $[U(x) \rightarrow -U(x)]$ or, equivalently, by the replacement $k \rightarrow k^{-1}$. The fluxes are shown in Fig. 3.76. While in the correlated situa-

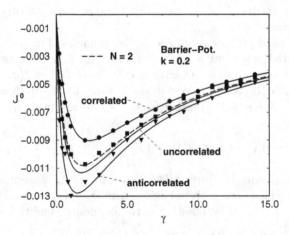

**Fig. 3.75.** Current $J$ versus switching rate $\gamma$ ($k = 0.2$) for correlated, uncorrelated and anticorrelated ratchets (from top to bottom). The *lines* are the analytical calculations described in the text, while the *points* are obtained by computer simulations ($N = 10$)

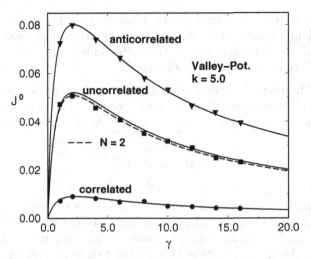

**Fig. 3.76.** The fluxes of particles if the potential is reversed

tion only the direction of the current changes, leaving the value of the flux unchanged; the uncorrelated and anticorrelated chains strongly enhance the absolute value of the flux. In the latter case the current increases nearly by one order of magnitude. This is shown in Fig. 3.76 for $k = 5.0$. This behavior is caused by the shape of the potential. While in the original $(U > 0)$, areas of constant plateaus (no force) are embedded by repelling segments, in the image $(U < 0)$, they are surrounded by attracting forces.

### 3.4.9 Making Spatial Structures Using Ratchets

The three states of the minimal ratchet can be interpreted as configurational states of a molecule undergoing a circular chemical reaction:

$$D^a \leftrightarrow D^s \leftrightarrow D^r \leftrightarrow D^a. \tag{3.261}$$

To realize flashing one additionally may imagine the action of external temporal perturbations switching between two states $\sigma(t) = \pm 1$ with a mean time $\tau$.

In particular, we will consider that the effect of the perturbation results in a temporal change of the rates of the transition between different configurations. As previously, let us suppose that for both states $\sigma$ the set of rates obey the mass action law. Hence, adiabatically slowly driven $(\tau \to \infty)$, the reaction approaches in both states an equilibrium given by detailed balance. In equilibrium the circular reaction has no net flux.

But with asymmetrically chosen rates the circular reaction represents a prototype of a ratchet. For medium switching times $\tau$ a clock-wisely directed flow of the reaction is induced. As will be seen this circular flow causes a nonequilibrium pattern formation [343]. Discarding the perturbation, $\tau \to \infty$, or if switching infinitely fast, $\tau \to 0$ the structure disappears. Hence, we presume and intend to show that the ratchets can be considered as the origin of nonequilibrium structure formation [344].

**Circular Chemical Reactions and Ratchets.** Let a second passive chemical obstacle $S$ be connected to the circular reaction. $D^a$ is able to react with the obstacles $S$, forming a complex $D^s$ when the obstacle is removed from the surface.

$$D^a + S \overset{\alpha^+}{\underset{\alpha^-}{\longleftrightarrow}} D^s. \tag{3.262}$$

$\alpha^+$ is the rate of the forward reaction, $\alpha^-$ the rate of the backward reaction.

The complex $D^s$ will be assumed to be the second state of the particles $D$. If this complex collides during its random walk with another obstacle $S$, the complex decays to a particle $D$ in its third state $D^r$ and an obstacle $S$, which is placed in the neighborhood of the colliding obstacle, i.e.,

$$D^s + S \overset{\beta^+}{\underset{\beta^-}{\longleftrightarrow}} D^r + 2S. \tag{3.263}$$

As in the first case (3.262) the reaction can be performed in the opposite direction. $\beta^+$ and $\beta^-$ denote the rates in the forward and backward directions, respectively.

In a last step the particles in state $D^r$ convert back to state $D^a$ with a certain rate $\gamma^+$ and become active again. The rate of the corresponding backward reaction is $\gamma^-$.

$$D^r \overset{\gamma^+}{\underset{\gamma^-}{\longleftrightarrow}} D^a. \tag{3.264}$$

The overall numbers $N$ of agents $D$ and $M$ of obstacles $S$ is conserved. Denoting by small letters the densities of $D$ and of $S$, respectively, the following equations are obtained:

$$\frac{1}{A} \int_A dA \ (d^a + d^s + d^r) = 1; \qquad \frac{1}{A} \int_A dA \ (s + d^s) = 1, \tag{3.265}$$

where $\int_A dA$ denotes the integral over the surface $A$ and summed densities are normalized to 1.

**External Driving: The Ratchet.** For the moment we simplify matters by setting $s = \text{const.}$ in this subsection. Let $\sigma(t)$ be the dichotomic Markov process. We suppose that all forward and backward rates depend on the actual realization of the two possible values of $\sigma$. For every $\sigma$, three kinetic equations for the reaction system can be formulated:

$$\dot{d}_\sigma^a = -(\alpha_\sigma^+ s + \gamma_\sigma^- + \tfrac{1}{\tau}) d_\sigma^a + \alpha_\sigma^- d_\sigma^s + \gamma_\sigma^+ d_\sigma^r + \tfrac{1}{\tau} d_{-\sigma}^a,$$

$$\dot{d}_\sigma^s = \alpha_\sigma^+ s\, d_\sigma^a - (\alpha_\sigma^- + \beta_\sigma^+ s - \tfrac{1}{\tau}) d_\sigma^s + \beta_\sigma^- s^2 d_\sigma^r + \tfrac{1}{\tau} d_{-\sigma}^s, \qquad (3.266)$$

$$\dot{d}_\sigma^r = \gamma_\sigma^- s\, d_\sigma^a + \beta_\sigma^+ s\, d_\sigma^s - (\beta_\sigma^- s^2 + \gamma_\sigma^+ + \tfrac{1}{\tau}) d_\sigma^r + \tfrac{1}{\tau} d_{-\sigma}^r.$$

Here $d_\sigma^{a,s,r}$ are the densities for the three states of $D$, and $\alpha_\sigma, \beta_\sigma, \gamma_\sigma$ the rates for a given $\sigma$. For constant $s$ the system (3.266) can be solved analytically in the stationary limit following (3.243) and (3.246). In detail we obtain the current given by

$$J = \langle \alpha_\sigma^+ s\, d_\sigma^a - \alpha_\sigma^- d_\sigma^s \rangle = \langle \beta_\sigma^+ s\, d_\sigma^s - \beta_\sigma^- s^2 d_\sigma^r \rangle = \langle \gamma_\sigma^+ d_\sigma^r - \gamma_\sigma^- d_\sigma^a \rangle,$$
$$(3.267)$$

where stationary densities have to be inserted and $\langle \ldots \rangle$ means summation over $\sigma$.

The current $J$ is shown in Fig. 3.77 as a function of the switching time $\tau$ for three densities of the obstacles $s$. With respect to $\tau$ a bell-shaped dependence is obtained. Fluxes vanish for slow and fast driving. Therefore, we obtain that for an optimal $\tau$ the chemical reaction is maximally forced by the external perturbation to approach a nonequilibrium state. A stationary, irreversible exchange in the forward direction exchange,

$$D^a \;\rightarrow\; D^s \;\rightarrow\; D^r \;\rightarrow\; D^a, \qquad (3.268)$$

takes place. The effective velocity of the reactions scales with the current $J$.

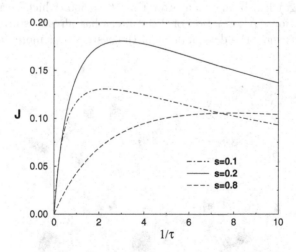

**Fig. 3.77.** Stationary current in the configuration space versus the mean switching time $\tau$ for three values of obstacle density $s$, $\alpha_{+1}^+ = \beta_{+1}^+ = 10$ and $\alpha_{-1}^+ = \gamma_{-1}^+ = 10$, other parameters according to $\alpha_\sigma^+ = 1/\alpha_\sigma^-$; $\beta_\sigma^+ = 1/\beta_\sigma^-$; $\gamma_\sigma^+ = 1/\gamma_\sigma^- = 1/(\alpha_\sigma^+ \beta_\sigma^+)$

**The Distributed System: Nonequilibrium Pattern Formation.** We call the $D$ molecules agents which may move diffusively on a two-dimensional plane. They are simulated as Brownian particles by solving numerically the Langevin equation [345, 346]:

$$\dot{x}_i = \sqrt{2\varepsilon}\,\xi_i^x, \qquad \dot{y}_i = \sqrt{2\varepsilon}\,\xi_i^y. \tag{3.269}$$

Here $\xi_i^{x,y}$ are Gaussian white noise sources and $i = 1, \ldots, N$ labels the number of the agents. Diffusive motion is modeled, with $\varepsilon$ being the diffusion coefficient of the agents $D$. In simulations we took $\varepsilon = 10^4$.

While the motion of the agents $D$ is continuous in space, the positions of the obstacles $S$ are bound to an underlying grid with box length $\Delta L = 1$, size $200 \times 200$ and periodic boundary conditions. Each cell of the grid can contain one or no obstacle, meaning that the obstacles have a volume in the size of the cell. Each time step we check whether a particle resides in a cell containing an obstacle or in a cell which does not contain an obstacle. If the particle is situated in a cell with an obstacle, the reactions which need an obstacle will be performed according to the given rates. Otherwise only reactions which do not need an obstacle can take place.

Initially the obstacles as well as the agents are distributed randomly on the surface with uniform probability. Application of temporal switchings of the dichotomic signal $\sigma$ results in an effective forward flux of the reaction coordinate. The system now shows cluster formation of obstacles $S$ in a special range of the switching time $\tau$. The sequence of snapshots in Fig. 3.78 demonstrates this for a switching time $\tau = 1.25$. The first snapshot (upper left image) has been taken after time $t = 500$, which corresponds to $100\,000$ iterations. A formation of dense areas has already started. The following images show the development of the system over more dense regions

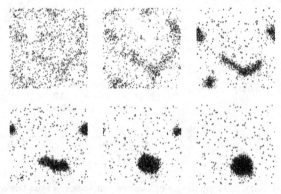

**Fig. 3.78.** Temporal sequence of forthcoming snapshots of a simulation with $\tau = 1.25$. Black dots indicate positions of obstacles $S$. Other parameters: $\varepsilon = 10^4$ and rates as in Fig. 3.77. From the upper left to the lower right: $t = 500, 1000, 2000, 4000, 8000, 9000$. At about $t = 9000$ a single stable cluster arose

up to several clusters, which end up in one single spherical cluster after $t = 9000$.

Simulations with different switching times $\tau$ demonstrate that this cluster formation of obstacles is most pronounced for maximal current $J$ at an optimal switching time. No clusters are formed in the case of vanishingly small currents for very high and very small switching times $\tau$, respectively.

**Ratchets Making Structures.** The origin of the structure formation is given by the nonequilibrium flux between different states of the agents $D$. To prove this fact let us look at the simplified model (3.268) including diffusion. For this system a directed flow with forward rates $\alpha^+, \beta^+, \gamma^+ \propto J(\tau)$ and vanishing backward rates $\alpha^- = \beta^- = \gamma^- = 0$ is considered. Addition of diffusion of the $D$ agents and of the obstacles $S$ yields

$$\dot{d}^a = \gamma^+ d^r - \alpha^+ s\, d^a + \varepsilon\Delta d^a \ , \ \dot{d}^s = \ \alpha^+ s\, d^a - \beta^+ s\, d^s + \varepsilon\Delta d^s,$$
$$\dot{d}^r = \beta^+ s\, d^s - \gamma^+ d^r + \varepsilon\Delta d^r \ , \ \dot{s} = \beta^+ s\, d^s - \alpha^+ s\, d^a + \varepsilon_0\varepsilon\Delta s, \tag{3.270}$$

where $\varepsilon_0 \ll 1$ is the ratio between the mobilities of the agents and the obstacles (a small mobility of $S$ should mimic the finite volume of the obstacles).

Equations (3.270) possess maximally three homogeneous solutions. Usual stability analysis proves either monostability for a positive stationary solutions $s_{\mathrm{stat}} > 0$ and $d_{\mathrm{stat}}^{a,s,r} > 0$ or bistability with an additional stable state located at $s_{\mathrm{stat}} = 0$.

The stable state of free obstacles with nonvanishing density $s_{\mathrm{stat}} \neq 0$ can be destabilized by diffusing agents $D$ in the case of a nonvanishing flux $J$. Indeed, allowing nonhomogeneous perturbations, one can prove the instability of this state by varying $J/(\varepsilon * k^2)$, with $k$ being the wave number of inhomogeneous perturbations. Interestingly, the region of instability is increased if the flux $J$ (3.267) of the three-state reaction is increased.

Analytic results of the stability analysis are shown in Fig. 3.79. The presented curve indicates critical values of $\varepsilon_0^{\mathrm{crit}}$ (the ratio of the mobility of agents and obstacles) plotted versus $J/(\varepsilon * k^2)$. Below the curve drawn the system amplifies inhomogeneous perturbations. Thus for a finite given $\epsilon_0$ an overcritical flux $J$ of the reaction is necessary for $\varepsilon * k^2 > 0$.

Increasing $\tau$ starting from a first critical value induces a required current $J$. Above a second critical $\tau$ the current again becomes small and no structure will be excited. For optimal overcritical $J$ slightly inhomogeneous distributions with finite wave numbers occur very soon (see Fig. 3.78). Later on, beyond our linear stability analysis the competition of those arising clusters leads to a small number of larger clusters and the system ends up with one large cluster consisting of all obstacles $S$. In this region the $s$-dependence of the rates cannot be neglected as we have done in the given approximation.

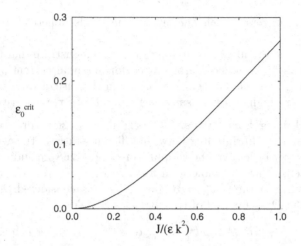

**Fig. 3.79.** Results of the stability analysis of the nonvanishing stable homogeneous state of (3.270). Below the plotted critical values of $\epsilon_0$ inhomogeneous perturbations are amplified. Rates $\alpha^+ = J\alpha^0$, $\beta^+ = J\beta^0$ and $\gamma^+ = J\gamma^0$, with $\alpha^0 = \beta^0 = \gamma^0 = 1$

### 3.4.10 Summary

In addition to the many possible biophysical applications these studies on noise-induced motion address fundamental problems in statistical physics. Violation of detailed balance in the kinetic equations is most clearly expressed within the frame of these investigations. Models as presented above allow the detailed inspection of the transition from equilibrium to nonequilibrium and the occurrence of a minimal order.

The most advanced studies about ratchets address a connection and combination with the other nonlinear stochastic phenomena studied in this chapter. Of course, synchronization as rotation of phases can be interpreted as an ordered flux in periodic landscapes and might be characterized by Peclet numbers [347]. Another direction was started was in [349–352]. Therein, noise-induced phase transitions and ratchets were interconnected as was done also with stochastic resonance (see Sect. 3.1.6) [141]. Such coupling of different noise-induced phenomena will deliver a lot of interesting and surprising results in the future and shed new light on the fascinating role of noise in nonlinear dynamics and dynamics far from equilibrium.

# References

1. R. L. Stratonovich, *Selected Problems of Fluctuation Theory in Radiotechnics* (Sov. Radio, Moscow 1961) (in Russian); *Topics in the Theory of Random Noise* (Gordon and Breach, New York, Vol. 1 1963, Vol. 2 1967).
2. S.M. Rytov, Ya.A. Kravtsov, V.I. Tatarskii, *Principles of Statistical Radiophysics* (Springer, Berlin, Heidelberg 1987).
3. A.N. Malakhov, *Fluctuations in Autooscillating Systems* (Nauka, Moscow, 1968) (in Russian).
4. W. Horsthemke, R. Lefever, *Noise-induced Transitions. Theory and Applications in Physics, Chemistry and Biology* (Springer, Berlin, Heidelberg, 1984).
5. P.S. Landa, A.A. Zaikin, JETP **84**, 197 (1997).
6. R. Benzi, A. Sutera, A. Vulpiani, J Phys. A: Math. Gen. **14**, L453 (1981).
7. R. Benzi, G. Parisi, A. Sutera, A. Vulpiani, Tellus **34**, 10 (1982).
8. C. Nicolis, Tellus **34**, 1 (1982).
9. V.S. Anishchenko, A.B. Neiman, F. Moss, L. Schimansky-Geier, Phys. Uspekhi **42**, 7 (1999).
10. F. Moss, D. Pierson, D. O'Gorman, Int. J. Bifurc. Chaos. **4**, 1383 (1994).
11. A. Bulsara A., L. Gammaitoni, Phys. Today **49**, 36 (1996).
12. L. Gammaitoni, P. Hänggi, P. Jung, F. Marchesoni, Rev. Mod. Phys. **70**, 223 (1998).
13. F. Moss, P.V.E. McClintock (eds.), *Noise in Nonlinear Dynamical Systems* Vols. 1–3 (Cambridge University Press, Cambridge 1990).
14. Proc. NATO Adv. Res. Workshop on Stochastic Resonance in Physics and Biology, J. Stat. Phys. **70** (1993).
15. Proc. Int. Workshop on Fluctuations in Physics and Biology: Stochastic Resonance, Signal Processing and Related Phenomena (1995). Nuovo Cimento D **17**.
16. L. Schimansky-Geier, T. Pöschel (eds.), *Stochastic Dynamics*, Lecture Notes on Physics, Issue 484 (Springer, Berlin, Heidelberg 1997).
17. J.B. Kadtke, A.R. Bulsara (eds.), *Applied Nonlinear Dynamics and Stochastic Systems Near the Millennium* (AIP Press, New York 1997).
18. J. A. Freund, T. Pöschel (eds.), *Stochastic Processes in Physics, Chemistry and Biology*, Lecture Notes on Physics, Issue 557 (Springer, Berlin, Heidelberg 1997).
19. S. Fauve, F. Heslot, Phys. Lett. A **97**, 5 (1983).
20. B. McNamara, K. Wiesenfeld, R. Roy, Phys. Rev. Lett. **60**, 2626 (1988).
21. A.N. Grigorenko, P.I. Nikitin , A.N. Slavin, P.Y. Zhou, J. Appl. Phys. **76**, 6335 (1994).
22. M.I. Dykman, A.L. Velikovich, G.P. Golubev, D.G. Luchinsky, S.V. Tsupikov, JETP Lett. **53**, 193 (1991).
23. L. Gammaitoni, M. Martinelli, L. Pardi, S. Santucci, Phys. Rev. Lett. **67**, 1799 (1991).
24. A. Simon, A. Libchaber, Phys. Rev. Lett. **68**, 3375 (1992).
25. M.L. Spano, M. Wun-Fogle, W.L. Ditto, Phys. Rev. A **46**, R5253 (1992).
26. R.N. Mantegna, B. Spagnolo, Phys. Rev. E **49**, R1792 (1994).
27. A.D. Hibbs, E.W. Jacobs, A.R. Bulsara, J.J. Bekkedahl, F. Moss, IL Nuovo Cimento **17**, 811 (1995).
28. A. Pérez-Madrid, J.M. Rubí, Phys. Rev. E **51**, 4159 (1995).
29. Z. Néda, Phys. Lett. A **210**, 125 (1996).
30. A.E. Dubinov, K.E. Mikheev, I.B. Nizhegorodtsev, Izv. Ross. Acad. Nauk. Ser. Fiz. **60**, 76 (1996).
31. D.E. Leonard, L.E. Reichl Phys. Rev. E **49**, 1734 (1994).

32. M.I. Dykman, T. Horita, J. Ross, J. Chem. Phys. **103**, 966 (1995).
33. W. Hohmann, J. Müller, F.W. Schneider, J. Phys. Chem. **100**, 5388 (1996).
34. J. Douglass, L. Wilkens, E. Pantazelou, F. Moss, Nature **365**, 337 (1993).
35. J.E. Levin, J.P. Miller, Nature **380**, 165 (1996).
36. J.J. Collins, T.T. Imhoff, P. Grigg, J. Neurophysiol. **76**, 642 (1996).
37. S. Bezrukov, I. Voydanoy, Nature **378**, 362 (1995).
38. J.J. Collins, T.T. Imhoff, P. Grigg, Nature **383**, 770 (1996).
39. D.F. Russell, L.A. Wilkens, F. Moss, Nature **402**, 291 (1999).
40. J. A. Freund, J. Kienert, L. Schimansky-Geier, B. Beisner, A. Neiman, D. F. Russell, T. Yakusheva, F. Moss, Phys. Rev. E **63**, 031910 (2001).
41. R.F. Fox, Phys. Rev. A **39**, 4148 (1989).
42. N.G. Stocks, N.D. Stein, P.V.E. McClintock, J. Phys. A **26**, L385 (1993).
43. Z. Gingl, L.B. Kiss, F. Moss, Europhys. Lett. **29**, 191 (1995).
44. P. Jung, Phys. Lett. A. **207**, 93 (1995).
45. S.M. Bezrukov, I. Vodyanoy, Nature **385**, 319 (1997).
46. V.S. Anishchenko, A.B. Neiman, M.A. Safonova, J. Stat. Phys. **70**, 183 (1993).
47. V.S. Anishchenko, M.A. Safonova, L.O. Chua, Int. J. Bifurc. Chaos **2**, 397 (1992).
48. H.A. Kramers, Physica **7**, 284 (1940).
49. P. Hänggi, P. Talkner, M. Borkovec, Rev. Mod. Phys. **62** 251, (1990).
50. P. Jung, P. Hänggi, Phys. Rev. A **144**, 8032 (1991).
51. P. Jung, Phys. Rep. **234** (1993).
52. T. Zhou, F. Moss, P. Jung, Phys. Rev. A **42**, 3161 (1990).
53. L. Gammaitoni, F. Marchesoni, E. Menichella-Saetta, S. Santucci, Phys. Rev. Lett. **62**, 349 (1989).
54. L. Gammaitoni, F. Marchesoni, S. Santucci, Phys. Rev. Lett. **74**, 1052 (1995).
55. A. Neiman, L. Schimansky-Geier, Phys. Rev. Lett. **72**, 2988 (1994).
56. P. Hänggi, P. Jung, C. Zerbe, F. Moss, J. Stat. Phys. **70**, 25 (1993).
57. A. Neiman, W. Sung, Phys. Lett. A **224**, 341 (1996).
58. F. Moss, Stochastic Resonance: From the Ice Ages to the Monkey Ear, In: Contemporary Problems in Statistical Physics, ed. by G.H. Weiss (SIAM, Philadelphia 1994), pp. 205.
59. P. Jung, P. Hänggi, Phys. Rev. A **41**, 2977 (1990).
60. P. Talkner, New J. Phys. **1**, 4 (1999).
61. B. McNamara, K. Wiesefeld, Phys. Rev. A **39**, 4854 (1989).
62. L.D. Landau, E.M. Lifshitz, *Statistical Physics* Part 1 (Pergamon Press, Oxford 1980).
63. P. Hänggi, H. Thomas, Phys. Rep. **88**, 207 (1982).
64. M.I. Dykman, R. Mannella, P.V.E. McClintock, and N.G. Stocks, JETP **52**, 141 (1990).
65. M.I. Dykman, H. Haken, Hu Gang, D.G. Luchinsky, P.V.E. McClintock, C.Z. Ning, N.D. Stein, and N.G. Stocks, Phys. Lett. A **180**, 332 (1993).
66. R. Kubo, M. Toda, N. Hashitsume, *Statistical Physics II*, Springer Series in Solid State Sciences, Vol. 31 (Springer, Berlin, New York 1985).
67. H. Risken, *The Fokker–Planck Equation* (Springer, Berlin, Heidelberg 1988).
68. P. Jung, U. Behn, E. Pantazelou, F. Moss, Phys. Rev. A **46**, R1709 (1992).
69. A. Bulsara, G. Schmera, Phys. Rev. E **47**, 3734 (1993).
70. A. Neiman, L. Schimansky-Geier, Phys. Lett. A **197**, 379 (1995).
71. M. Morillo, J. Gómez-Ordoñez, J. M. Casado, Phys. Rev. E **52**, 316 (1995).
72. F. Marchesoni, L. Gammaitoni, A. Bulsara, Phys. Rev. Lett. **76**, 2609 (1996).
73. F. Castelpoggi, H.S. Wio, Phys. Rev. E **57**, 5112 (1998).
74. N. Kuperman, H.S. Wio, G. Izus, Phys. Rev. E **57**, 5122 (1998).

75. I. Dikshtein, A. Neiman, L. Schimansky-Geier, J. Magn. Magn. Mater. **188**, 301 (1998).

76. R. Rozenfeld, L. Schimansky-Geier, Chaos, Solitons, Fractals **11**, 1937 (2000).

77. J. Lindner, B. Meadows, W. Ditto, M. Inchiosa, A. Bulsara, Phys. Rev. Lett. **75**, 3 (1995); ibid, Phys. Rev. E **53**, 2081 (1996).

78. L. Schimansky-Geier, U. Siewert, A. Glauber, "Dynamics Approach to Coupled Stochastic Resonators", In: *Stochastic Dynamics*, ed. by L. Schimansky-Geier, T. Pöschel, Lecture Notes on Physics, Issue 484 (Springer, Berlin, Heidelberg 1997), p. 245.

79. U. Siewert, L. Schimansky-Geier, Phys. Rev. E **58**, 2843–2852 (1998).

80. R.J. Glauber, J. Math. Phys. **4**, 294 (1963).

81. J.J. Brey, A. Prados, Phys. Lett. **A216**, 240 (1996).

82. Z. Néda, Phys. Rev. E **51**, 5315 (1995).

83. K.-T. Leung, Z. Néda, Phys. Rev. E **59**, 2730 (1999).

84. P. Ruszczynski, Diploma thesis (Humboldt-University, Berlin 1999).

85. V.G. Makhankov, Phys. Rep. **35**, 1 (1978).

86. J.F. Currie, J.A. Krumhansl, A.R. Bishop, S.F. Trullinger, Phys. Rev. B **22**, 477 (1980).

87. W. Ebeling, R. Feistel, *Physik der Selbstorganisation und Evolution*, (Akademie-Verlag, Berlin 1982).

88. R. Rajaraman, *Solitons and Instantons. An Introduction to Solitons and Instantons in Quantum Field Theory*, (North Holland, Amsterdam 1982).

89. R.K. Dold, J.C. Ellbeck, J.D. Gibson, H.C. Morris, *Solitons and Nonlinear Wave Equations* (Academic, London 1989).

90. H. Malchow, L. Schimansky-Geier, *Noise and Diffusion in Bistable Nonequilibrium Systems*, Teubner Texte zur Physik 5 (Teubner, Leipzig 1986).

91. A.S. Mikhailov, *Foundations of Synergetics I* (Springer, Berlin, Heidelberg 1989).

92. F. Schlögl, Z. Phys. **253**, 147 (1972).

93. E.W. Montroll, In: *Statistical Mechanics*, ed. by S.A. Rice, K.F. Freed, and J. C. Light (Chicago University Press, Chicago 1972).

94. L. Schimansky-Geier, H. Hempel, R. Bartussek, C. Zülicke Z. Phys. B Cond. Matter **96**, 417 (1995).

95. A.S. Mikhailov, L. Schimansky-Geier, W. Ebeling, Phys. Lett. A **96**, 453 (1983).

96. L. Schimansky-Geier, A.S. Mikhailov, W. Ebeling Ann. Phys. **40**, 277 (1983).

97. L. Schimansky-Geier, Ch. Zülicke, Z. Phys. B **82**, 157 (1991).

98. J. García Ojalvo, J.M. Sancho, *Noise in Spatially Extended Systems* (Springer, New York 1999).

99. I. Dikshtein, A. Neiman, Bull. Am. Phys. Soc. II **42**, 338 (1997).

100. I. Dikshtein, A. Neiman, L. Schimansky-Geier, "Spatio-Temporal Stochastic Resonance of Domain Wall Motion in Inhomogeneous Magnets", In: *Studies in Applied Electromagnetics and Mechanics*, ed. by: V. Kose and J. Sievert (IOS Press, Amsterdam 1998), p. 119.

101. A.N. Grigorenko, P.I. Nikitin, Appl. Surf. Sci. **92**, 466 (1996).

102. A.N. Grigorenko, G.V. Roshchepkin, Zh. Exp. Teor. Fiz. **112**, 628 (1997).

103. I. Dikshtein, A. Neiman, L. Schimansky-Geier Phys. Lett. A **246**, 259 (1998).

104. P. Ruszczynski, L. Schimansky-Geier, I. Dikshtein, Europhys. J. B **14**, 569–577 (2000).

105. J.A. Gonzáles, B.A. Mello, L.I. Reyes, L.E. Guerrero, Phys. Rev. Lett. **80**, 1361 (1998).

106. K.A. Gorshkov, L.A. Ostrovsky, Physica D **3**, 428 (1981).

107. D.J. Kaup, A.C. Newell, Proc. R. Soc. Lond. **A361**, 413 (1978).

108. M. Büttiker, R. Landauer, Phys. Rev. A **23**, 1397 (1994).
109. F. Marchesoni, Phys. Rev. Lett. **73**, 2394 (1994).
110. A counter example in a conservative dynamics was reported by F. Marchesoni and L. Vazquez, Physica D **14**, 273 (1985).
111. A. Kulka, M. Bode, H.-G. Purwins, Phys. Lett. A **203**, 33 (1995).
112. P. Landa, A. Zaikin, Phys. Rev. E **54**, 3535 (1996).
113. V. Berdichevsky, M. Gitterman, Phys. Rev. E **60**, 1494 (1999).
114. P.S. Landa, A.A. Zaikin, V.G. Ushakov, J. Kurths, Phys. Rev. E **61**, 4809 (2000).
115. A.S. Mikhailov, Phys. Lett. A **73**, 143 (1979).
116. A.S. Mikhailov, Z. Phys. B **41**, 277 (1981).
117. J. Garcíca-Ojalvo, J.M. Sancho, L. Ramirez-Piscina, Phys. Lett. A **168**, 35 (1992).
118. J. García-Ojalvo, A. Hernández-Machado, J.M. Sancho, Phys. Rev. Lett. **71**, 1542 (1993).
119. C. Van den Broeck, J.M.R. Parrondo, J. Amero, A. Hernández-Machado, Phys. Rev. E **49**, 2639 (1994).
120. C. Van den Broeck, J.M.R. Parrondo, R. Toral, Phys. Rev. Lett. **73**, 3395 (1994).
121. C. Van den Broeck, J.M.R. Parrondo, J. Armero, A. Hernández-Machado, Phys. Rev. E **49**, 2639 (1994).
122. P. Jung, G. Mayer-Kress, Phys. Rev. Lett. **74**, 2130 (1995).
123. Ch. Kurrer, K. Schulten, Phys. Rev. E **51**, 6213 (1995).
124. J. García-Ojalvo, J.M.R. Parrondo, J. M. Sancho, C. Van den Broeck, Phys. Rev. E **54**, 6918 (1996).
125. J.M.R. Parrondo, C. Van den Broeck, J. Buceta, E. Javier de la Rubia, Physica A **224**, 153 (1996).
126. C. Van den Broeck, J.M.R. Parrondo, R. Toral, R. Kawai, Phys. Rev. E **55**, 4084 (1997).
127. R. Müller, K. Lippert, A. Kühnel, U. Behn, Phys. Rev. E **56**, 2658 (1997).
128. C. Van den Broeck, "From Stratonovich Calculus to Noise-Induced Phase Transitions", In: *Stochastic Dynamics*, ed. by L. Schimansky-Geier, T. Pöschel, Lecture Notes on Physics, Issue 484 (Springer, Berlin, Heidelberg 1997), p. 7..
129. S. Mangioni, R. Deza, H.S. Wio, R. Toral, Phys. Rev. Lett. **79**, 2389 (1997).
130. P.S. Landa, A.A. Zaikin, L. Schimansky-Geier, Chaos, Solitons, Fractals **9**, 1367 (1998).
131. A. Zaikin, L. Schimansky-Geier, Phys. Rev. E **58**, 4355 (1998).
132. A.A. Zaikin, J. García-Ojalvo, L. Schimansky-Geier, Phys. Rev. E **60**, R6275 (1999).
133. H. Hempel, L. Schimansky-Geier, J. García-Ojalvo, Phys. Rev. Lett. **82**, 3713–3716 (1999).
134. S. Kadar, J. Wang, K. Showalter, Nature **391**, 770 (1998); F. Moss, Nature **391**, 743 (1998).
135. P. Jung, A. Cornell-Bell, F. Moss, S. Kadar, J. Wang, K. Showalter, haos **8**, 567 (1998).
136. J. Wang, S. Kadar, P. Jung, K. Showalter, Phys. Rev. Lett. **82**, 855 (1999).
137. I. Sendiña-Nadal, A. P. Muñuzuri, D. Vives, V. Pérez-Muñuzuri, J. Casademunt, L. Ramírez-Piscina, J. M. Sancho, and F. Sagués, Phys. Rev. Lett. **80**, 5437 (1998).
138. M. Löcher, D. Cigna, E.R. Hunt, Phys. Rev. Lett. **80**, 5212 (1998).

139. I. Sendiña-Nadal, S. Alonso, V. Pérez-Muñuzuri, M. Gómez-Gesteira, V. Pérez-Villar, L. Ramírez-Piscina, J. Casademunt, J.M. Sancho, F. Sagués, Phys. Rev. Lett. **84**, 2734 (2000).
140. S. Alonso, I. Sendiña-Nadal, V. Pérez-Muñuzuri, J. M. Sancho, and F. Sagués, Phys. Rev. Lett. **87**, 078302 (2001).
141. A. Zaikin, J. Kurths, L. Schimansky-Geier, Phys. Rev. Lett. **85**, 227 (2000).
142. L.D. Landau, E.M. Lisfschitz, *Theoretische Mechanik*, Lehrbuch zur Theoretischen Physik, Bd I (Akademie-Verlag, Berlin 1984).
143. A.A. Zaikin, K. Murali, J. Kurths, Phys. Rev. E **63**, 020103R (2001).
144. M. Dykman, D.G. Luchinsky, R. Mannella, P.V.E. McClintock, N.D. Stein, and N.G. Stocks, Nuovo Cimento D **17**, 661 (1995).
145. V.S. Anishchenko, A.B. Neiman, M.A. Safonova, I.A. Khovanov, "Multifrequency Stochastic Resonance", In: *Proc. of Euromech Colloquium on Chaos and Nonlinear Mechanics,* ed. by T. Kapitaniak and J. Brindley (World Scientific, Singapore 1995), p. 41.
146. V.S. Anishchenko, *Dynamical Chaos – Models and Experiments* (World Scientific, Singapore 1995).
147. V.S. Anishchenko, M.A. Safonova, L.O. Chua, Int. J. Bifurc. Chaos **4**, 441 (1994).
148. W. Ebeling, L. Schimansky-Geier, "Transition Phenemena in High-Dimensional Systems and Models of Evolution", In: *Noise in Nonlinear Dynamics Systems,* ed. by F. Moss, P.V.E. McClintock (Cambridge University Press, Cambridge 1990), p. 279.
149. L. Schimansky-Geier, C. Zülicke, Z. Phys. B **79**, 451 (1990).
150. J.J. Hesse, L. Schimansky-Geier, Z. Phys. B **84**, 467 (1991).
151. M.I. Dykman, P.V.E. McClintock, N.D. Stein, N.G. Stocks, Phys. Rev. Lett. **67**, 933 (1991).
152. M.I. Dykman, R. Mannella, P.V.E. McClintock, N.D. Stein, N.G. Stocks, Phys. Rev. E **47**, 3996 (1993).
153. S.J.B. Einchcomb, A.J. McKane, Phys. Rev. E **49**, 259–266 (1994).
154. R. Bartussek, P. Hänggi, B. Lindner, L. Schimansky-Geier Physica D **109**, 17 (1997).
155. J.J. Collins, C.C. Chow, T.T. Imhoff, Phys. Rev. E **52**, R3321 (1995).
156. A. Neiman, L. Schimansky-Geier, F. Moss, Phys. Rev. E **56**, R9 (1997).
157. L. Gammaitoni, M. Löcher, A. Bulsara, P. Hänggi, J. Neff, Phys. Rev. Lett. **82**, 4574 (1999).
158. M. Löcher, M.E. Inchiosa, J. Neff, A. Bulsara, K. Wiesenfeld, L. Gammaitoni, P. Hänggi, W. Ditto, Phys. Rev. E **62**, 317 (2000).
159. U. Zürcher, Ch. Doering, Phys. Rev. E **47**, 3862 (1993).
160. C. Van den Broeck, Phys. Rev. E **47** , 3862 (1993).
161. R. Rozenfeld, A. Neiman, L. Schimansky-Geier, Phys. Rev. E **62**, R3031 (2000).
162. A. Neiman, L. Schimansky-Geier, F. Moss, B. Shulgin, J. Collins, Phys. Rev. E **60** , 284 (1999).
163. J. Freund, A. Neiman, L. Schimansky-Geier, Europhys. Lett. **50**, 8 (2000).
164. M.I. Dykman, D.G. Luchinsky, P.V.E. McClintock, N.D. Stein, N.G. Stocks, Phys. Rev. A **46** , 1713 (1992).
165. V.S. Anishchenko, *Complex Oscillations in Simple Systems* (Nauka, Moscow, 1990) (in Russian).
166. Yu.I. Kifer, Izv. Acad. Nauk SSSR. Ser. Math. **38**, 1091 (1974).
167. E.B. Vul, Ya.G. Sinai, K.M. Khanin, Usp. Math. Nauk **39**, 3 (1984) (in Russian).
168. V.S. Anishchenko, W. Ebeling, Z. Phys. B **81**, 445 (1990).

169. M.I. Freidlin, A.D. Wentzell, *Random Perturbations of Dynamical Systems* (Springer, New York 1984).
170. R. Graham, T. Tel, Phys. Rev. A **31**, 1109 (1985).
171. R. Graham, "Macroscopic Potentials, Bifurcations and Noise in Dissipative Systems", In: *Noise in Nonlinear Dynamical Systems*, Vol. 1, ed. by F. Moss, P.V.E. McClintock (Cambridge University Press, Cambridge 1990), p. 225.
172. R. Graham, A. Hamm, T. Tel, Phys. Rev. Lett. **66**, 3089 (1991).
173. Yu.L. Klimontovich, *Statistical Physics* (Harwood Academic Publisher, New York 1986).
174. V.S. Anishchenko, Sov. Tech. Phys. Lett. **10**, 266 (1984).
175. V.S. Anishchenko, A. Neiman, Sov. Tech. Phys. Lett. **13**, 444 (1987).
176. V.S. Anishchenko, A. Neiman, Sov. Tech. Phys. **35**, 1 (1990).
177. V.S. Anishchenko, A. Neiman, L.O. Chua, Int. J. Bifurc. Chaos **4**, 99 (1994).
178. E.N. Lorenz, J. Atmos. Sci. **20**, 130 (1963).
179. Yu.L. Klimontovich (ed.), *Wave and Fluctuations Processes in Lasers* (Nauka, Moscow, 1974) (in Russian).
180. H.M. Ito, J. Stat. Phys. **35**, 151 (1984).
181. V.V. Bykov, L.P. Shilnikov, "About Boundary of Existence of a Lorenz-Attractor", In: (*Methods of the Qualitative Theory and the Theory of Bifurcations*, ed. by L.P. Shilnikov (GGU Publisher, Gorky 1989), pp. 151 (in Russian).
182. V.S. Anishchenko, A. Neiman, Sov. Tech. Phys. Lett. **17**, 510 (1991).
183. M.I. Rabinovich, Sov. Phys. Usp. **21**, 443 (1978).
184. T. Zhou, F. Moss, Phys. Rev. A. **41**, 4255 (1990).
185. V.S. Anishchenko, D.E. Postnov, I.A. Khovanov, B.V. Shulgin, Izv. Vuzov – Appl. Nonlinear Dyn. **3**, 16 (1995) (in Russian).
186. B.V. Shulgin, *Stochastic Resonance in Bistable Radio-physical Systems*, PhD thesis (Saratov State University, Saratov 1995) (in Russian).
187. V. Shneidman, P. Jung, P. Hänggi, Phys. Rev. Lett. **72**, 2682 (1994).
188. G.N. Milstein, M.V. Tretyakov, Physica D **140**, 244 (2000).
189. A. Neiman, Phys. Rev. E **49**, 3484 (1994).
190. B. Shulgin, A. Neiman, V.S. Anishchenko, Phys. Rev. Lett. **75**, 4157 (1995).
191. A. Neiman, A. Silchenko, V.S. Anishchenko, L. Schimansky-Geier, Phys. Rev. E **58**, 7118 (1998).
192. V.I. Melnikov, Phys. Rev. E **48**, 2481 (1993).
193. C.V. Gardiner, *Handbook of Stochastic Methods for Physics, Chemistry, and the Natural Sciences* (Springer, Berlin 1983).
194. V.S. Anishchenko, A. Neiman, A. Silchenko, I.A. Khovanov, Dyn. Stab. Syst. **14**, 211 (1999).
195. I.A. Khovanov, V.S. Anishchenko, Tech. Phys. Lett. **22**, 854 (1996).
196. V.S. Anishchenko, A.N. Silchenko, I.A. Khovanov, Phys. Rev. E **57**, 316 (1997).
197. J.J. Collins, C.C. Chow, T.T. Imhoff, Nature **376**, 236–238 (1995).
198. X. Pei, L. Wilkens, F. Moss, Phys. Rev. Lett. **77**, 4679 (1996).
199. P.C. Gailey, A. Neiman, J.J. Collins, F. Moss, Phys. Rev. Lett. **79**, 4701 (1997).
200. A.S. Pikovsky, Radiophys. Quantum Electron. **27**, 576 (1984) (in Russian).
201. A.S. Pikovsky, "Synchronization and Stochastization of Nonlinear Oscillations by External Noise", In: *Nonlinear and Turbulent Processes in Physics*, Vol. 3, ed. by R.Z. Sagdeev (Harwood Academic Publisher, New York 1984), p. 1601.

202. A. Neiman, F. Moss, L. Schimansky-Geier, W. Ebeling, "Synchronization in Ensembles of Stochastic Resonators", In: *Applied Nonlinear Dynamics and Stochastic Systems Near the Millenium,* ed. by J.B. Kadtke, A.R. Bulsara (AIP Press, Woodbury 1997), p. 151.
203. A. Neiman, B. Shulgin, V.S. Anishchenko, W. Ebeling, L. Schimansky-Geier, J. Freund, Phys. Rev. Lett. **76**, 4299 (1996).
204. L. Schimansky-Geier, J. Freund, A. Neiman, B. Shulgin, Int. J. Bifurc. Chaos **8**, 869 (1998).
205. A. Bulsara, A. Zador, Phys. Rev. E **54**, R2185 (1996).
206. W.J.A. Sloane, A.D. Wyner (eds.) *Claude Elwood Shannon: Collected Papers* (IEEE Press, New York 1993).
207. W. Ebeling, G. Nicolis, Chaos, Solitons, Fractals **2**, 635 (1995).
208. W. Ebeling, J. Freund, K. Rateitschak, Int. J. Bifurc. Chaos **6**, 611 (1996).
209. A.Ya. Khinchin, Usp. Math. Nauk **11**, 17 (1956) (in Russian).
210. W. Ebeling, H. Herzel, W. Richert, L. Schimansky-Geier, Zeitschrift für angewandte Mathematik und Mechanik **66**, 141 (1986).
211. K. Lekkas, L. Schimansky-Geier, H. Engel-Herbert, Z. Phys. B **70**, 517 (1988).
212. L. Schimansky-Geier, F. Moss, G. Schmera, "Influence of Colored Noise on Hopf-Bifurcating Systems", In: *Irreversible Processes and Selforganization*, ed. by W. Ebeling and H. Ulbricht, Teubner Series on Physics (Teubner, Leipzig 1989),p. 231.
213. S.M. Soskin, Physica A **155**, 401 (1989).
214. Hu Gang, T. Ditzinger, C.Z. Ning, H. Haken, Phys. Rev. Lett. **71**, 807 (1993).
215. T. Ditzinger, C.Z. Ning, Hu Gang, Phys. Rev. E **50**, 3508–3516 (1994)
216. B. Lindner, L. Schimansky-Geier, Phys. Rev. E **61**, 6103 (2000).
217. A. Longtin, A. Bulsara, F. Moss Phys. Rev. Lett. **67**, 656 (1991).
218. A. Longtin, Phys. Rev. E **55**, 868 (1997).
219. A. Pikovsky, J. Kurths, Phys. Rev. Lett. **78**, 775 (1997)
220. A. Neiman, P. Saparin, L. Stone, Phys. Rev. E **56**, 270 (1997).
221. K. Wiesenfeld, J. Stat. Phys. **38**, 1071 (1985).
222. A. Neiman, V.S. Anishchenko, J. Kurths, Phys. Rev. E **49**, 3801 (1994).
223. R.A. FitzHugh, Biophys. J. **1**, 445 (1961).
224. A.C. Scott, Rev. Mod. Phys. **47**, 487 (1975).
225. A. Longtin ,J. Stat. Phys. **70**, 309 (1993).
226. A. Longtin, Nuovo Cimento D **17**, 835 (1995).
227. J.J. Collins, C.C. Chow, T.T. Imhoff, Phys. Rev. E **52**, 3321 (1995).
228. X. Pei, K. Bachmann, F. Moss, Phys. Lett. A **206**, 61–65 (1995).
229. D.R. Chialvo, A. Longtin, J. Müller-Gerking, Phys. Rev. E **55**, 1798 (1997).
230. M.G. Vilar, R.V. Solé, J.M. Rubí, Phys. Rev. E **59**, 5920 (1999).
231. A. Longtin, D.R. Chialvo, Phys. Rev. Lett. **81**, 4012 (1998).
232. J.B. Baltanas, J.M. Casado, Physica D **122**, 231–240 (1998).
233. J.M. Casado, Phys. Lett. A **235**, 489-492 (1997).
234. S.R. Massanés, C.J.P. Vicente, Phys. Rev. E **59** 4490 (1999).
235. D.E. Postnov, S.K. Han, T.Y. Yim, O.V. Sosnovtseva, Phys. Rev. E **59** 3791 (1999).
236. A. Neiman, L. Schimansky-Geier, A. Cornell-Bell, F. Moss, Phys. Rev. Lett. **83**, 4896 (1999).
237. R. Graham, H. Haken, Z. Phys. **243** 289 (1971).
238. R. Graham, *Statistical Theory of Open Systems in Stationary Nonequilibruim Systems with Applications to Lasers and Nonlinear Optics*, Springer Tracts on Modern Physics, Vol. 66 (Springer, Berlin, Heidelberg 1973).
239. G.G. Izus, R.R. Deza, H.S. Wio, Phys. Rev. E **58** 93 (1998).
240. T. Alarcón, A. Pérez-Madrid, J.M. Rubí, Phys. Rev. E **57**, 4879 (1998).

241. B. Lindner, L. Schimansky-Geier, Phys. Rev. E **60**, 7270 (1999).
242. V.I. Melnikov, Phys. Rev. E **48**, 2481 (1993).
243. J. García-Ojalvo, L. Schimansky-Geier, Europhys. Letters **47**, 298 (1999).
244. J. García-Ojalvo, L. Schimansky-Geier, J. Stat. Phys. **101**, 473 (2000).
245. W.M. Kistler, R. Seitz, L. van Hemmen, Physica D **114**, 73 (1998).
246. F. Schweitzer, W. Ebeling, B. Tilch, Phys. Rev. Lett. **80**, 5044 (1998).
247. M. Smoluchowski, Physik. Z. **XIII**, 1069 (1912).
248. R. Feynman, R. Leighton, M. Sands, *The Feynman Lectures on Physics*, Vol. 46 (Addison-Wesley, New York 1963).
249. S. Flach, O. Yevtushenko, Y. Zolotaryuk, Phys. Rev. Lett. **84**, 2358 (2000).
250. M.O. Magnasco, Phys. Rev. Lett. **71**, 1477, 1993.
251. J. Maddox, Nature **365**, 203 (1993).
252. S. Leibler, Nature **370**, 412 (1994).
253. M.O. Magnasco, Phys. Rev. Lett. **72**, 2656 (1994).
254. C.R. Doering, W. Horsthemke, J. Riordan, Phys. Rev. Lett. **72**, 2984 (1994).
255. R. Bartussek, P. Hänggi, J. G. Kissner, Europhys. Lett. **28**, 459 (1994).
256. R.D. Astumian, M. Bier, Phys. Rev. Lett. **72**, 1766 (1994).
257. M.M. Millonas, M.I. Dykman, Phys. Lett. A **185**, 65 (1994).
258. J. Łuczka, R. Bartussek, P. Hänggi, Europhys. Lett. **31**, 431 (1995).
259. C.R. Doering, Nuovo Cimento D **17**, 685 (1995).
260. A. Mielke, Ann. Phys. **4**, 476 (1995).
261. P. Hänggi and R. Bartussek, "Brownian Rectifiers: How to Convert Brownian Motion into Directed Transport", In: *Nonlinear Physics of Complex Systems – Current Status and Future Trends*, ed. by J. Parisi, S. C. Müller, and W. Zimmermann (Springer, Berlin, Heidelberg 1996), p. 294.
262. T.C. Elston, C.R. Doering, J. Stat. Phys. **83**, 359 (1996).
263. M. Bier, R.D. Astumian, Phys. Rev. Lett. **76**, 4277 (1996).
264. F. Marchesoni, Phys. Rev. Lett. **77**, 2364 (1996).
265. M. Bier, Phys. Lett. A **211**, 12 (1996).
266. R. Bartussek, P. Reimann, P. Hänggi, Phys. Rev. Lett. **76**, 1166 (1996).
267. F. Jülicher, A. Ajdari, J. Prost, Rev. Mod. Phys. **69**, 1269 (1997).
268. R. D. Astumian, Science **276**, 917 (1997).
269. T. Hondou, Y. Sawada, Phys. Rev. Lett. **75**, 3269 (1995).
270. M.R. Evans, D.P. Foster, C. Godrèche, D. Mukamel, Phys. Rev. Lett. **74**, 208 (1995).
271. I. Derènyi, T. Viscek, Phys. Rev. Lett. **75**, 374 (1995).
272. T. Harms, R. Lipowsky, Phys. Rev. Lett. **79**, 2895 (1997).
273. I.M. Sokolov, A. Blumen, J. Phys. A **30**, 3021 (1997).
274. T.E. Dialynas, K. Lindenberg, G.P. Tsironis, Phys. Rev. E **56**, 3976 (1997).
275. C.R. Doering, Physica A **254**, 1 (1998).
276. G.P. Tsironis, K. Lindenberg, Adv. Struct. Biol. **5**, 271 (1998).
277. H. Qian, Phys. Rev. Lett. **81**, 3063 (1998).
278. V. Berdichevsky, M. Gitterman, Phys. Rev. E **60**, 7562 (1999).
279. Y. Aghababaie, G. I. Menon, and M. Plischke, Phys. Rev. E **59**, 2578 (1999).
280. M. Porto, M. Urbakh, and J. Klafter, Phys. Rev. Lett. **85**, 491 (2000).
281. P.S. Landa, P.V.E. McClintock, Phys. Rep. **323**, 4 (2000).
282. J. Prost, J.-F. Chauwin, L. Peliti, A. Ajdari, Phys. Rev. Lett. **72**, 2652 (1994).
283. W.D. Volkmuth, T. Duke, M.C. Wu, R.H. Austin, Phys. Rev. Lett. **72**, 2117 (1994).
284. T.A. Duke, R.H. Austin, Phys. Rev. Lett. **80**, 1552 (1998).
285. D. Ertas, Phys. Rev. Lett. **80**, 1548 (1998).
286. C. Kettner, P. Reimann, P. Hänggi, F. Müller, Phys. Rev. E **61**, 312 (2000).
287. J. Finer, R.M. Simmons, J.A. Spudich, Nature **368**, 113 (1994).

288. K. Svoboda, C.F. Schmidt, B.J. Schnapp, S.M. Block, Nature **365**, 721 (1993).
289. N. Cordova, B. Ermentrout, G. Oster, Proc. Natl. Acad. Sci. USA **89**, 339 (1992).
290. C. Peskin, G. Odell, G. Oster, Biophys. J. **65**, 316 (1993).
291. C.R. Doering, B. Ermentrout, G. Oster. Biophys. J. **69**, 2256 (1995).
292. J.F. Chauwin, G. Oster, B. Glick, Biophys. J. **74**, 1732 (1997).
293. A. Mogilner, G. Oster, Eur. J. Biophys. **28**, 235 (1998).
294. A. van Oudenaarden, S.G. Boxer, Science **285**, 1046 (1999).
295. A. van Oudenaarden, J.A. Theriot, Nat. Cell Biol. **1**, 493 (1999).
296. A. van Oudenaarden, J.A. Theriot, Biophys. J. **78**, 241A (2000).
297. R.D. Astumian, P.B. Chock, T.Y. Tsong, H.V. Westerhoff, Phys. Rev. A **39**, 6416 (1989).
298. D.S. Liu, R.D. Astumian, T.Y. Tsong, J. Biol. Chem. **265**, 7260 (1990).
299. T.D. Xie, P. Marszalek, Y. Chen, T.Y. Tsong, Biophys. J. **67**, 1247 (1994).
300. R.D. Astumian, M. Bier, Biophys. J. **70**, 637 (1996).
301. R.D. Astumian, J. Phys. Chem. **100**, 19075 (1996).
302. A. Ajdari, J. Prost, Proc. Natl. Acad. Sci USA **88**, 4468 (1991).
303. A. Ajdari, J. Prost, C. R. Acad. Sci. Paris **315**, 1635 (1992).
304. J.-F. Chauwin, A. Ajdari, J. Prost, Europhys. Lett. **27**, 421 (1994).
305. J. Rousselet, L. Salome, A. Ajdari, J. Prost, Nature **370**, 446 (1994).
306. L.P. Faucheux, L.S. Bourdieu, P.D. Kaplan, A.J. Libchaber, Phys. Rev. Lett. **74**, 1504 (1995).
307. L. Gorre, E. Ioannidis, P. Silberzan, Europhys. Lett. **33**, 267 (1996).
308. I. Zapata, B. Bartussek, F. Sols, P. Hänggi, Phys. Rev. Lett. **77**, 2292 (1996).
309. K. Seeger, W. Maurer, Solid State Commun. **27**, 603 (1978).
310. N.G. van Kampen, *Stochastic Processes in Physics and Chemistry* (North Holland, Amsterdam 1981).
311. J.M.R. Parrondo and P. Español, Am. J. Phys. **64**, 1125 (1996); K. Sekimoto, J. Phys. Soc. Jpn. **66**, 1234 (1997); J.M.R. Parrondo, J.M. Blanco, F.J. Cao, R. Brito, Europhys. Lett. **43**, 248 (1998); J.M.R. Parrondo. Phys. Rev. E. **57**, 7297 (1998); A. Parmeggiani, F. Jülicher, A. Ajdari, J. Prost, Phys. Rev. E **60**, 2127 (1999); I.M. Sokolov, Phys. Rev. E **63**, 021107 (2001).
312. J.A. Freund, L. Schimansky-Geier, Phys. Rev. E **60**, 1304 (1999).
313. I. Blekhman, *Synchronization of Dynamical Systems* (Nauka, Moscow 1971) (in Russian); *Synchronization in Science and Technology* (Nauka, Moscow 1981), English translation: (ASME Press, New York 1988).
314. M. Büttiker, Z. Phys. B **68**, 161 (1987).
315. P. Reimann, R. Bartussek, R. Häussler, P. Hänggi, Phys. Lett. A **215**, 26 (1996).
316. P. Hänggi, R. Bartussek, P. Talkner, J. Łuczka, Europhys. Lett. **35**, 315 (1996).
317. C. Jarzynski, O. Mazonka, Phys. Rev. E **59**, 6448 (1999).
318. P.S. Landa, A. Zaikin, L. Schimansky-Geier, Chaos Solitons Fractals, **12**, 1459 (2001).
319. R. F. Fox, I. R. Gatland, R. Roy, G. Vemuri, Phys. Rev. A **38**, 5938 (1988).
320. B. Lindner, Diploma thesis (Humboldt University, Berlin 1996).
321. R. Bartussek, Ratchets Driven by Colored Noise, In: *Stochastic Dynamics,* ed. by L. Schimansky-Geier, T. Pöschel, Lecture Notes on Physics, Issue 484 (Springer, Berlin, Heidelberg 1997), p. 69.
322. B. Lindner, L. Schimansky-Geier, P. Reimann, P. Hänggi, "Mass Separation by Ratchets", In: *Applied Nonlinear Dynamics and Stochastic Systems Near the Millenium,* ed. by J.B. Kadtke, A.R. Bulsara (AIP Press, Woodbury 1997), p. 309.
323. P. Jung, J. G. Kissner, P. Hänggi, Phys. Rev. Lett. **76**, 3436 (1996).

324. P. Reimann, M. Grifoni, P. Hänggi, Phys. Rev. Lett. **79**, 10 (1997).
325. F. Marchesoni, Phys. Lett. A **237**, 126 (1998).
326. P.S. Landa, Phys. Rev E **58**, 1325 (1998).
327. B. Lindner, L. Schimansky-Geier, P. Reimann, P. Hänggi, M. Nagaoka, Phys. Rev. E. **59**, 1417 (1999).
328. P. Hänggi, P. Jung, Adv. Chem. Phys. **89**, 239 (1995).
329. P. Jung, P. Hänggi, Phys.Rev. A **35**, 4464 (1987).
330. L. H'walisz, P. Jung, P. Hänggi, P. Talkner, L. Schimansky-Geier, Z. Phys. B **77**, 471 (1989); A. J. R. Madureira, P. Hänggi, V. Buonomano, and W. A. Rodrigues Jr., Phys. Rev. E **51**, 3849 (1995); R. Bartussek, A. J. R. Madureira, P. Hänggi, Phys. Rev. E **52**, 2149 (1995).
331. W. Wonneberger, H.-J. Breymayer, Z. Phys. B **56**, 241 (1984); H.-J. Breymayer, H. Risken, H. D. Vollmer, W. Wonneberger, Appl. Phys. B **28**, 335 (1982).
332. S.B.J. Eichcomb, A.J. McKane, Phys. Rev. E **51**, 2974 (1995).
333. R. Reimann, Phys. Rev. E **52**, 1579 (1995); K.M. Rattray, A.J. McKane, J. Phys. A **24**, 1215 (1991).
334. E. Pollak, P. Talkner, Phys. Rev. E **47**, 922 (1993).
335. D. Astumian, I. Derenyi. Phys. Rev. E **58**, 7781 (1998).
336. M. Bier, M. Kostur, I. Derenyi, R.D. Astumian, Phys. Rev. Lett. **83**, 903 (1999).
337. M. Kostur, L. Schimansky-Geier, Phys. Lett. A **265**, 337 (2000).
338. O.C. Zienkiewicz, R.L. Taylor, *The Finite Element Method*, Vols. I & II. (McGraw-Hill, New York 1989/91).
339. L. Schimansky-Geier, M. Kschicho, T. Fricke, Phys. Rev. Lett. **79**, 3335 (1997).
340. Hu Gang, A. Daffertshofer, H. Haken, Phys. Rev. Lett. **76**, 4874 (1996).
341. B. Derrida, J. Stat. Phys. **31**, 433 (1983).
342. I. Claes, C. Van den Broeck, J. Stat. Phys. **70**, 1215–1231 (1993).
343. L. Schimansky-Geier, S. Seefeld, V. Bucholtz, Ann. Phys. **9**, 705 (2000).
344. I. Derényi, R.D. Astumian, Phys. Rev. Lett. **80**, 4602 (1998).
345. L. Schimansky-Geier, M. Mieth, H. Rosé, H. Malchow, Phys. Lett. A **207**, 140 (1995).
346. R. Feistel, W. Ebeling, *Physics of Complex Systems* (Deutscher Verlag der Wissenschaften, Berlin 1989).
347. B. Lindner, M. Kostur, L. Schimansky-Geier, Fluct. Noise Lett. **1**, R25 (2001).
348. P. Reimann, C. Van den Broeck, H. Linke, P. Hänggi, J. M. Rubí, and A. Pérez-Madrid, Phys. Rev. Lett. **87**, 010602 (2001).
349. P. Reimann, R. Kawai, C. Van den Broeck, P. Hänggi, Europhys. Lett. **45**, 545 (1999).
350. C. Van den Broeck, P. Reimann, R. Kawai, P. Hänggi, "Coupled Brownian Motors", In: *Statistical Mechanics of Biocomplexity*, ed. by D. Reguera, J.M.G. Vilar, J.M. Rubí (Springer, Berlin, Heidelberg 1999), P 93.
351. J. Buceta, J. M. R. Parrondo, C. Van den Broeck, J. de la Rubia, Phys. Rev. E **61**, 6287 (2000).
352. C. Van den Broeck, I. Bena, P. Reimann, J. Lehmann, Ann. Phys. **9**, 713 (2000).

# Index

# Springer Series in Synergetics